Advanced Materials and Manufacturing Techniques for Biomedical Applications

Scrivener Publishing
100 Cummings Center, Suite 541J
Beverly, MA 01915-6106

Publishers at Scrivener
Martin Scrivener (martin@scrivenerpublishing.com)
Phillip Carmical (pcarmical@scrivenerpublishing.com)

Advanced Materials and Manufacturing Techniques for Biomedical Applications

Edited by

Arbind Prasad

Department of Science and Technology, Katihar Engineering College, Katihar, Bihar, India

Ashwani Kumar

Department of Mechanical Engineering, Technical Education Department (under Government of Uttar Pradesh), Kanpur, India

and

Manoj Gupta

Department of Mechanical Engineering, National University of Singapore, Singapore

Scrivener Publishing

WILEY

This edition first published 2024 by John Wiley & Sons, Inc., 111 River Street, Hoboken, NJ 07030, USA and Scrivener Publishing LLC, 100 Cummings Center, Suite 541J, Beverly, MA 01915, USA

For more information about Scrivener publications please visit www.scrivenerpublishing.com.

Wiley Global Headquarters
111 River Street, Hoboken, NJ 07030, USA

For details of our global editorial offices, customer services, and more information about Wiley products visit us at www.wiley.com.

Library of Congress Cataloging-in-Publication Data

ISBN 978-1-394-16619-0

Cover image: Pixabay.Com
Cover design Russell Richardson

Set in size of 11pt and Minion Pro by Manila Typesetting Company, Makati, Philippines

Printed in the USA

10 9 8 7 6 5 4 3 2 1

This book is dedicated to all engineers, researchers, health professionals and academicians

Contents

Preface

Advanced Materials and Manufacturing Techniques in Biomedical Applications covers the synthesis, processing, design, manufacturing, and characterization of advanced materials; self-healing, bioinspired, nature-resourced, nano biomaterials for biomedical applications; and manufacturing techniques such as rapid prototyping, additive manufacturing, etc. Biomaterials and the latest manufacturing techniques are the central themes of this book. The variety of chapters presented herein is fascinating and up-to-date, and this book will be invaluable to researchers in biomedical and engineering disciplines. The background information and the literature review provided in each chapter give an in-depth analysis of the present topic.

In origin, biomaterials can come from nature or be synthesized in the laboratory with a variety of approaches that use metals, polymers, ceramic, or composite materials. They are often used or adapted for various biomedical applications. Biomaterials are commonly used in scaffolds, orthopedic, wound healing, fracture fixation, surgical sutures, artificial organ developments, pins and screws to stabilize fractures, surgical mesh, breast implants, artificial ligaments and tendons, and drug delivery systems. The high quality of the material presented here will benefit those who seek to understand orthopedic biomaterials, processing, and other biomedical applications.

The sixteen chapters of this book thoroughly explore the latest information about biomaterials, such as self-healing, bioinspired, biomimetics, nature-resourced multifunctional materials, etc., and manufacturing techniques such as additive manufacturing and rapid prototyping for artificial organ developments. Each chapter provides an exhaustive literature review, solution, methodology, experimental setup, results validation, and future scope.

The book provides essential knowledge for the synthesis of biomedical products, development, nanomaterial properties, fabrication processes, and design techniques for different applications, as well as process design and optimization. Further, it encourages readers to discover and convert newly reported technologies into products and services for the future development of biomedical applications. This is an ideal book for upper-level undergraduate and graduate students, engineers, technologists, doctors, and researchers working in the area of biomedical engineering and manufacturing techniques. By lucidly presenting the latest information and research, this volume provides a foundational link to more specialized research work in biomedical engineering and applications.

The editors are appreciative of all contributors' collaboration and assistance. We also offer our sincere appreciation to Wiley and Scrivener Publishing for their support and guidance throughout the editorial process.

Editors
Dr. Arbind Prasad
Dr. Ashwani Kumar
Prof. (Dr.) Manoj Gupta

Acknowledgement

We express our gratitude to Wiley-Scrivener Publishing and the editorial team for their suggestions and support during completion of this book. We are grateful to all contributors and reviewers for their illuminating views on each book chapter presented in book *"Advanced Materials and Manufacturing Techniques for Biomedical Applications"*.

Section I

ADVANCED MATERIALS FOR BIOMEDICAL APPLICATIONS

1

Introduction to Next-Generation Materials for Biomedical Applications

Arbind Prasad[1]*, Sudipto Datta[2], Ashwani Kumar[3] and Manoj Gupta[4]

[1]Department of Mechanical Engineering, Katihar Engineering College (under DST, Govt. of Bihar), Katihar, Bihar, India
[2]Department of Materials Engineering, Indian Institute of Science, Bangalore, Karnataka, India
[3]Technical Education Department Uttar Pradesh, Kanpur, Uttar Pradesh, India
[4]Department of Mechanical Engineering, National University of Singapore, Lower Kent Ridge Road, Singapore

Abstract
Materials play a vital role in every field of applications. The demand for advance materials has always been in increasing mode in order to complete its intended applications. There are various types of materials which has now been utilized and researched by various researchers to give its best performance with low input. Self-healing materials, bioinspired materials, and smart nanomaterials shape memory materials, biomimetic materials, and composites. Now a days these materials are used for advanced healthcare such as cancer therapy, drug delivery, bioimaging, biosensors, antimicrobial applications, and tissue engineering. The resorbable composites are nowadays to use for bone fracture fixation in order to provide relief to the patients. Various materials and polymeric composite are processed for multifunctional tasks and also have multifunctional properties. This chapter deals with the various types of advance materials, processing techniques, market for the implants and materials, applications, latest status, challenges and future scope of these materials for the biomedical applications.

***Keywords*:** Bioabsorbable, biomedical, bioinspired, drug delivery, degradation

**Corresponding author*: arbind.iitg@gmail.com

Arbind Prasad, Ashwani Kumar and Manoj Gupta (eds.) Advanced Materials and Manufacturing Techniques for Biomedical Applications, (3–24) © 2024 Scrivener Publishing LLC

1.1 Introduction

The role of materials in almost all fields plays a significant role in order to perform its intended applications. In the biomedical applications the metallic biomaterials have lot of limitations such as leaching of metallic ions, corrosion, underlying bone damages, and stress shielding [1]. In order to eliminate the problems associated, researchers are working on various types of biomaterials which not only performs its intended function but also self-sustaining in nature. Clinical professional and researchers are looking for materials that provided (i) Sufficient mechanical properties (ii) corrosion less (iii) nontoxic (iv) biocompatible [2, 3]. In the second half of the twentieth century, materials scientists began to partner with physicians in order to develop novel biomaterials that were specifically designed for use within the human body. During this time, biomaterials were created that promoted specific responses by the surrounding tissues and also encourages the neo bone regeneration [4–6]. The examples of bioactive materials include calcium phosphate based inorganic materials such as Hydroxyapatite (HAp). The hydroxyapatite is having similar composition as of natural bone minerals. The HAp is brittle in nature thus alone this material is not sufficient to produce implants suitable for bone fracture fixations. Resorbable polymers are another class of second-generation biomaterials. The examples of such resorbable polymers include polylactic acid (PLA), Polycaprolactone (PCl), Poly D lactic acid, chitosan etc. The most prominent use of resorbable polymers is in scaffolds fabrication, implants productions, and drug-eluting. The main challenges associated with the scaffolds for recreating the artificial tissues that allow the nutrient transport and support cell proliferation. In recent years, several computer-aided additive and subtractive methods for processing biomaterials have been developed [7]. These techniques may be used to develop prostheses with patient-specific attributes, scaffolds for tissue engineering, as well as small-scale medical devices. The development and use of biomaterials are expected to dramatically increase over the coming decades which are mentioned in subsequent section. The industries based on biomaterials and bioimplants are based on the advanced composites processing, ceramics, metals and polymeric composites developments which has an enormous role to replace the damaged tissues, fracture fixations, encourages bone formation, self-healing and also consists of sufficient mechanical properties [8]. It is hoped that this special issue will stimulate interactions among the numerous stakeholders involved in the development of biomaterials, including physicians, surgeons, engineers and biologists, and will promote future research activities in this rapidly developing area. Figure 1.1 shows

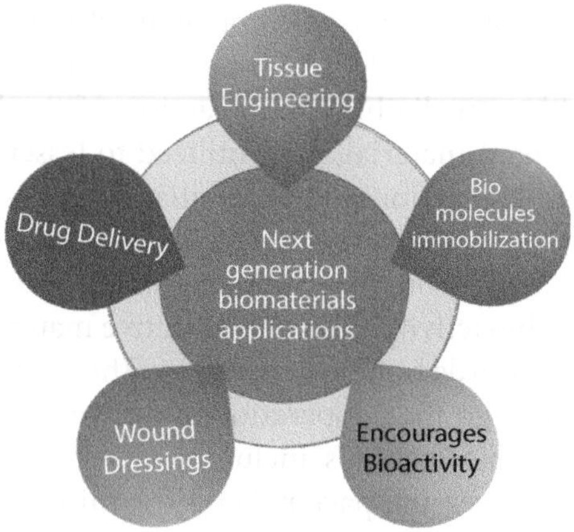

Figure 1.1 Next-generation biomaterials application in various areas in biomedical field.

the application areas, where next generation materials are establishing themselves in a sustainable manner.

In this chapter, various functional materials and their processing routes has been discussed. In the subsequent section, market and requirement of the next generation materials and their processed bioimplants has been analyzed and afterwards metal, polymeric biomaterials and their roles in these days, latest statue of these materials along with their application has been discussed. In the end of this chapter challenges associated with the processing and fabrication of these biomaterials and implants has been discussed in details.

1.2 Advanced Functional Materials

The advanced materials have their unique application based upon their intended application such as orthopedics fixations, dentistry, wound healing anti-inflammatory medicines, antimicrobial elements, and various other biomedical devices, implants and scaffolds. In initial days, researchers were using first generation materials such as Alumina, Zirconia and carbon etc. [9, 10]. The second-generation materials include mostly ceramics and phosphate-based materials such as calcium phosphates, calcium sulphate, iron oxides, calcium carbonates, hydroxyapatite, glasses, and its ceramics. The various protein molecules along with nanofibrils are generally used for Gels, films, fibers for enhancements of additional mechanical properties and biocompatibility. The evolution of biomaterials moves forwards from bioinertness to self-healing materials [11]. It was referenced that the first-generation materials were developed close to the year 1940, their objective was having properties of bioinertness. The bioinertness, the materials having reasonable actual properties to coordinate those of replaced tissue with a negligible toxic reaction of the host, so naturally idle, or almost dormant material, were utilized to lessen corrosion and to limit the insusceptible reaction and unfamiliar body response. The examples were gold fillings, wooden teeth, PMMA dental implants, steel, gold, ivory, bone plates and so on. The subsequent age was from 1970's their objective was about bioactivity [12]. The bioactive materials were observed more encouraging towards bone regeneration. The second era of biomaterials was created through the cooperation of doctor and specialists. These include most of the biomaterials including devices which has relevant application in heart valves and pacemakers. The third era was begun about the year 2000 their objective was to recover the original tissues. This age essentially presents the possibility of regenerative medication. Biomaterials are fit for stimulating one or a few elements of the neo tissues such as bone, muscle, ligament etc. to signaling, to induce the recovery of this tissue. For example: cell networks for 3D development and tissue regenerating. Tissue engineering includes the utilization of sub-atomic and cell science innovation to further apply the innovative materials to the patient for their worthy applications. The third-age materials incorporate bio glass-based materials and its composites, bioactive and biodegradable inorganic materials production, mesoporous materials, natural and inorganic composites. The necessity of the implants for the most part of the bodies relies on three things like mechanical properties, processing methods and compatibility. The cytocompatibility is the majority of the significant where inserts and all the obsession are taken consideration most extreme while applying for the capabilities [13–15]. Figure 1.2 shows the progress in materials and implants used for biomedical applications.

Cytocompatibility prompts the tissue response, changes in properties incorporates, mechanical, physical and chemical properties, degradation prompt nearby changes, and also systematic impacts. The mechanical properties of the mainly used biomaterials and implants incorporate generally elasticity, yield stress, ductility, toughness, time dependent deformation, creep ultimate strength, fatigue strength, hardness and wear resistance. The process for developing the implants and scaffolds. The manufacturing process held for fabrication of fixation, implants and scaffolds must be consistent to conform all the requirement laid by food and drug administration [16, 17]. The quality of raw materials, mechanical properties, compatibility, stability towards bearing load and ultimately cost will be the most important to take care. Thus, from Figure 1.2, it is clear that afterward year 2022, it is an era of smart materials which will not only self-healing but also shape

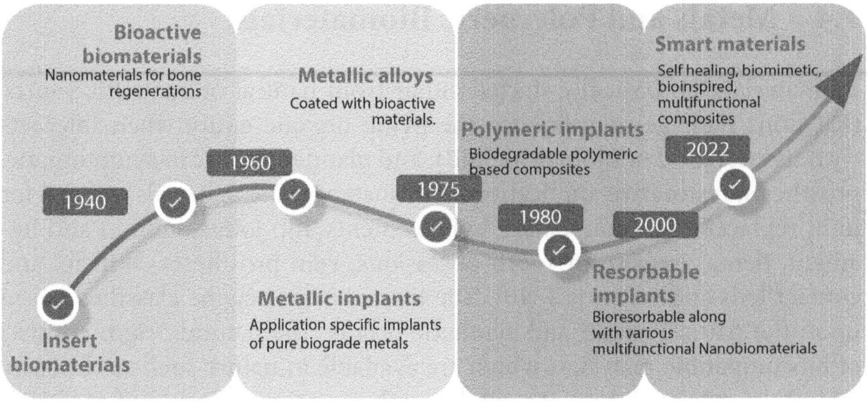

Figure 1.2 Progress in biomaterials in recent years in biomedical applications.

memory in nature as well. Various researchers are working on it in order to produce the advance biomaterials which would be multifunctional and also very much worthy for their targeted applications.

1.3 Market and Requirement of Next-Generation Materials

The patterns of bone cracks are expanding step by step every year. According to International osteoporosis foundation, In Asia, it is extended that more than around half of all osteoporotic hip break will happen in Asia continuously by the year 2050. Vertebral or spinal cracks are the most widely recognized fracture happening in 30–50% half of individuals beyond 50 years old and result in essentially expanded bleakness and mortality. Cracks happen when the bone is exposed to a power that is excessively solid for the unresolved issue. Assuming the bone is now debilitated, it doesn't take a lot of power to cause a break. According to the LANCET solid life span report, Globally, in 2019, there were 178 million new breaks. According to the fortune business experiences report, the worldwide muscular inserts market is expected to develop from USD 45.30 billion of every 2021 to USD 64.18 billion out of 2028 at a CAGR of 5.1% in the 2021–2028 period. Breaks in more established individuals might prompt diminished versatility, freedom, and capacity to complete every day capacities, moving into long haul care, constant torment, more terrible personal satisfaction, higher danger of death. For an unresolved issue appropriately, great blood supply after the injury, satisfactory adjustment, adequate new tissue arrangement.

1.4 Metals and Polymeric Biomaterials

Biomaterials are basically started either from natural or synthetic source that don't damage or influence the living organic entity when interacts with tissue, blood or biological fluids, and planned for use in symptomatic, prosthesis, restorative applications. For instance, the materials utilized for joint replacement, bone plates, bone cements, hip joints, tendon and ligament, dental embed for tooth obsessions, vein prostheses, skin fix and contact lens and so on [17–20]. The biomaterials may be classified based upon the natural origin and synthetic origin. The natural origin consists of biocompatible materials which are available in nature such as chitosan, hydroxyapatite, cellulose etc. while synthetic origin consists of materials which are widely processed in to the lab for their intended application such as PCL, PLA, PDLA and other based composites.

The addition of various natural and synthetic biomaterials enhances the biocompatibility, biodegradability and high potential to avoid resurgery [21, 22]. The general criterion for biomaterials selection is consists of mechanical and chemical properties such as strength, elastic modulus, fatigue strength, wear resistance and corrosion resistance. No undesirable biological effects-carcinogenic, toxic, allergenic or immunogenic. Possible to process, fabricate and sterilize with a good reproducibility, and cost effective [23–25]. Classification of synthetic biomaterials consists of metals, ceramic, polymers, semiconductor materials [26–29]. These materials have wide applicability in drug delivery devices, skin/cartilage, orthopedic screw fixations, dental implants, fracture fixations. Due to their excellent high strength, fracture toughness, relative ease of fabrication, good electrical conductivity made them use in various applications. The bio ceramics are specially designed and fabricated ceramics which can be used to repair and reconstruct the diseased, damaged or worn-out part of the body [30]. The resistance to microorganism, temperature, solvents, pH changes and high pressures is the advantage.

1.5 Bioabsorbable Biomaterials

A biomaterial is known as synthetic or natural material engineered to interrelate with natural structures thru modern medical usage [31]. It must be biocompatible significance that they make their purpose with a suitable host reaction [32]. Bio-polymers have an important perspective since flexibility in material science gives response to materials with excessive mechanical and chemical property variety [33–35]. Bioabsorbable polymers are of greatest attention since these types of materials are proficient to be packed up and defecated or adherence without abstraction or medical especially in surgical modification [36, 37]. From a fundamental material science viewpoint, the ability to moderate biomaterial to carry exclusive material characteristics is limitless until now needs important time and properties to comprehensive the investigation [38]. Since biomaterials are useful in the medical applications, several matters rise that can't be sufficiently recognized and deal with in earlier in vivo as well as *in vitro* investigations [39, 40]. The host response depends on the biomaterial properties of physical, chemical, and biological belongings. To enhance address the numerous concerns in biomaterial design and advanced development, biomaterial experts have essentially transformed their method of investigation [41, 42]. Particularly in the previous few time, there has been a modification in standards from researchers investigating individualistically on constricted investigation objectives to cooperative groups that simplify resolving better purposes. By uniting investigators with capability in biology, chemistry, medical practice, and engineering materials, biomaterials investigation has been capable of the improvement more quickly in the previous few years [43–45].

In biodegradable biomaterials, several significant properties must be well-thought-out. These materials are essential (a) not evoke a continued inciting response; [35, 38] (b) keep a degradation time corresponding with their utility; [41] (c) have suitable physical properties for their planned usage; [34, 38] (d) create non-toxic degradation produces which can be freely resorbed or defecated; and (e) comprise suitable permeability for premeditated use [46]. These belongings are importantly impacted several structures of polymeric degradable biomaterials as well as, but not restricted to: hydrophobicity, material science, surface charge, water adsorption, erosion, and degradation mechanism.

1.6 Processing of Bioabsorbable Polymeric Biomaterials

The processing of bioabsorbable polymeric biomaterials mostly held in injection molding machines. Some processing of polymers passes through shaping operations which consist of polymer pellet, powders, or resin which is transformed into preform or into final products using extrusion or blow molding process. Some of the techniques involve thermoforming or blow molding [47–52]. Figure 1.3 show the various techniques widely used for processing of next generation composite materials for biomedical applications. This sentence will be added here. The joining operation also happens for particular cases in which two or more parts of the polymer products are assembled

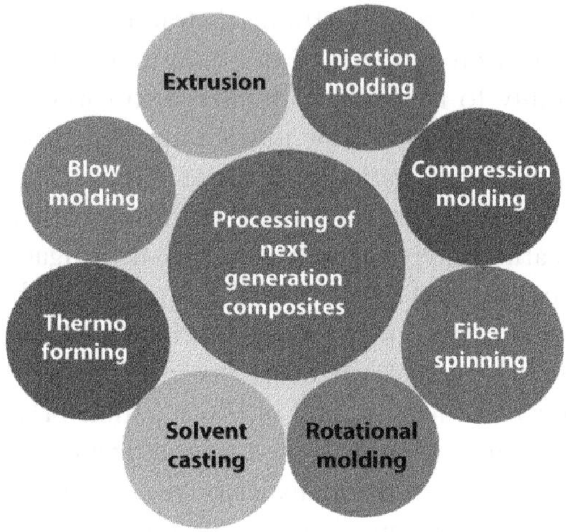

Figure 1.3 Processing of next-generation composites for biomedical applications.

physically or through bonding or welding operations. In Extrusion process, polymer is melted at a suitable temperature and pumped through a desired die in order to make the various shapes of the products and biomedical products. The extrusion process held in extruder comes in two mechanism i.e. single screw and twin-screw extruder. Generally, in polymer industries, single screw extruder is used whereas twin-screw extruder is widely used as mixing and compounding devices. Injection molding process is also one of the widely used commercially viable process for the production of various implants in a mass scale. In this process, the polymer melt is injected into a cavity in order to produce the products [53–58].

During injection molding process, various nanofillers along with base resorbable polymers is melt mixed in order to produce composites. Rotational molding is generally used in producing hollow objects. It can produce large parts having uniform thickness at cost effective approach. Blow molding is used to produce a hollow shape. In case of thermoforming, a plastic sheet is heated in order to reach a temperature that is slightly above the glass transition temperature for amorphous polymers and slightly below the melting temperature for semi crystalline polymers [59–64]. Generally, this process is widely used for amorphous materials because they have a wide temperature range. In the coating process, a liquid film is deposited on a substrate. it is generally used for providing an insulating coating over conducting wires. The solvent casting is used create the polymeric films and scaffolds. The polymer is first dissolved in a suitable solvent to have a viscous solution then after evaporation of the solvent, the polymeric film is obtained [65–69].

1.7 Application of Next-Generation Materials in Biomedical Applications

Biocompatibility is a precarious necessity for a biomaterial, which is a capability of a significant to purpose with a suitable host reaction in an exact use. Numerous physicochemical and biological features of a transplant material direct the host tissue's reaction to the material. For example, solubility material chemistry, hydrophobicity, degradation mechanism, molecular weight, hydrophobicity, surface energy, and lubricity, erosion, and profile and configuration of the implant can all affect biocompatibility. Prominently, a biodegradable object needs outstanding biocompatibility after a while as the biological, and physicochemical characteristics of a bioabsorbable biomaterial will fluctuate with instance and, therefore, the subsequent desolation produced can take variable stages of biocompatibility as related to the primary material [70–75].

As well as biocompatibility, numerous other significant belongings are deliberated when selecting a bioabsorbable biomaterial. Mainly, the stage of the degradation should correspond with the healing and regeneration progression to confirm the appropriate altering of the biological tissue. Additionally, the biocompatible biomaterial must sustain appropriate penetrability and processability for its planned use. As a final point, the mechanical characteristics of the biomaterial must be necessary to encourage regeneration throughout the daily tasks of patient, and any modification in biomechanical belongings because of degradation must reserve consistence with the regeneration or else healing procedure [76–78].

Certain the difficulty of the anatomical structure of human and the possibility of presentations that biomaterials (polymeric) are presently used for, no single polymeric arrangement can be measured the perfect biomaterial for all medicinal presentations. Therefore, current improvements in the synthesis of biodegradable biomaterial have been absorbed on the way to emerging and producing polymers with belongings tailored for precise biomedical uses. Furthermore, present advances integrating combinatorial and multifunctional methods in biomaterial design have enhanced the invention of original bioabsorbable biomaterials. Additional important issue in biomaterials investigation is the expansion of healing plans, three-dimensional (3D) scaffolds (porous) for modern biomedical and tissue engineering etc. Basically, they show a variation of organic and physicochemical characteristics and, consequently, can duplicate the belongings of dissimilar tissues, and these biomaterials are evaluated for usage as a) bone plates, large implants, and also bone screws; b) sutures and small implants; c) natural membranes for regeneration of tissue [79–82]. Furthermore, biodegradable materials can be applied to create nano or micro scale drug delivery devices for measured drug delivery in a diffusive or erosive way, or as an arrangement of together. As a result of the presence of an extensive variety of biomaterials, both synthetic and natural, fluctuating superiority of substance preparation, and a common shortage of relative investigates of dissimilar biomaterials for precise healthcare and tissue engineering applications, it is difficult to determine which biopolymer is the most perfect. Presently, in the area of biomedical engineering, impermanent, artificial scaffolds are being investigated and established for cells to keep to, segregate, and form fresh tissue. Though, as opposed to demanding to classify exact polymers for exact biomedical presentations, it appears that the upcoming biopolymer presentation is to use dissimilar groupings of polymers to grow hybrid biopolymers, which have an improved specificity summary for exact applications in drug delivery, tissue engineering, healthcare, etc. Certainly, several fresh revisions about polymers for biomedical application have addressed linking dissimilar biomaterials, through methods like grafting, blending, and chemical crosslinking responses and the consequences have been typically optimistic. Many challenges still are today through maximum biomaterials, for instance, the viability of mass production at a comparatively little cost in addition to defeating definite physiochemical boundaries of precise biomaterials; though, with the arrival of genetic and cloning engineering methods to definite together synthetic and natural biomaterials in a multiplicity of host arrangements, there endures a positive forthcoming for biomedical application [83–86].

As a result of the presence of an extensive variety of biomaterials, both synthetic and natural, fluctuating superiority of substance preparation, and a common shortage of relative investigates of dissimilar biomaterials for precise healthcare and tissue engineering applications, it is difficult to determine which biopolymer is the most perfect. Presently, in the area of biomedical engineering, impermanent, artificial scaffolds are being investigated and established for cells to keep to, segregate, and form fresh tissue. Though, as opposed to demanding to classify exact polymers for exact biomedical presentations, it appears that the upcoming biopolymer presentation is to use dissimilar groupings of polymers to grow hybrid biopolymers, which have an improved specificity summary for exact applications in drug delivery, tissue engineering, healthcare, etc. Certainly, several fresh revisions about polymers for biomedical application have addressed linking dissimilar biomaterials, through methods like grafting, blending, and chemical crosslinking responses and the consequences have been typically optimistic. Many challenges still are today through maximum biomaterials, for instance, the viability of mass production at a comparatively little cost in addition to defeating definite physiochemical boundaries of precise biomaterials; though, with the arrival of genetic and cloning engineering methods to definite together synthetic and natural biomaterials in a multiplicity of host arrangements, there endures a positive forthcoming for biomedical application [87–89].

1.8 Latest Status of Next Generation Materials in Biomedical Applications

1.8.1 Bioabsorbable Devices for Bone Tissue Engineering

Among all-natural tissues, bone basically acts as an extracellular matrix, which is basically made up of collagen and some mineral component like hydroxyapatite during the damage of the bone numerous surgeries are necessary for structural restoration and function of the damaged bone. The tissue engineering technique produces bone restoration. Among all materials, volumetric materials are required for sale growth in which blood vessels are formed. In the case of bio-resorbable polymer that can be enriched with hydroxyapatite and many growth factors can service bones morphogenic proteins. Those bone elements are highly effective in bone formation as well as stimulating bone reformation transplants basically consist of dissimilar types of isolated cells cultured on PLLA and PGA scaffold that can be explored as temporary substitutes of damage tissue portion. PLGA polymers act as biosorption materials and act as a newborn forming element at the damage site. In addition, PLGA can be completely biodegradable and can reserve the tissue in weeks or years subject to the polyester ratio present in the osteoblast sales can proliferate by a satisfactory pattern of sale addition and spreading over PLA, PGA, and PLGA scaffold. Furthermore, sales can grow over the polymer presence of phosphorus content in it. Osteoblast information is increased in the case of synthesis collagen. Similar results can be obtained when osteoblast cultures were performed over a PLGA scaffold. In this case mineralization of bones, and matrix can be observed. Interestingly bones meadow sale cultured on the porous PLGA scaffold are more prone in case of bone higher the porosity can eight higher bones growth rate, but the factors like Mechanical properties can be limited with an increase in porosity. In the case of in vitro and in Vivo results suggest that when the interconnections are greater than 300 micrometers, it can facilitate the vascularization of graft. Gugala *et al.* [32]

examined the absorption of protein and activity of osteoblast that cultured on non-porous and porous PDLA membranes for 3 weeks. In another study, the pattern of cell interaction with the substrate also depends on osteoblast culture and PLLA and PDLA [33]. Correct the problem differently with active ceramics has been used such as calcium phosphate with polymers to improve the mechanical strength of the scaffold. Ostioconductivity properties of the scaffold can be increased by composite matrix [33]. Porous PLLA and PLLA-hydroxyapatite composite matrix are seeded with osteoblasts cells. The sale work was found to penetrate deep into the PLA hydroxy appetite scaffold that was uniformly distributed over the scaffold. In sab composite scaffolds cell viability and cell proliferation are better during bone differentiation [35]. Another study can investigate the cells tend to adhere to hydroxyapatite and PCL hydroxyapatite composite and greater spreading in hydroxy appetite expose to the surface. The presence of hydroxy apatite can improve cellular by activity as well [90, 91].

1.9 Bioresorbable Devices for Skin Tissue Engineering

Numerous groups investigated the dissimilar methods to create dermal equivalent based on biological materials such as collagen, synthetic material, and epidermal layers [89]. But patient-specific s is soon slow growth of vascular structure. That consequence leads to a second surgical procedure to transplant the epidermal component for wound regeneration [90–95]. Keratinocytes and leukocytes are suitable substances for the migration of cells to provide an extracellular matrix. Fibroblast developed on collagen gel can produce extracellular matrix components such as fibronectin that can be resembled to reconstruct the connective tissue. But various culture conditions can change fibroblast behavior into the epithelial cells with the reduction in migration behavior of collagen matrix and that can produce the molecules such as collagen 4 and laminin. The skin graft is very flexible at the site of injury. It should not adhere to the deeper layer permitting free and functional joint movement [96–99]. Collagen base artificial dermatitis is the next successful target for skin grafting. This type of structure basically consists of chondroitin sulphate and collagen some non-inorganic compounds like chitosan will show slow degradation in an aqueous medium even presence of a lysosome. Furthermore, in addition of chitosan reduces the cell interactivity with collagen. fibroblast culture dance and porous three-dimensional PLLA membrane were found to add her on the polymer substrate to yield extracellular matrix molecules like collagen 4 and fibronectin [100–104].

1.10 Challenges and Perspectives

The challenges associated with coming next generation materials are the mechanical strength, surface compatibility, dug releases and control and cost effective. The processing is still in developing mode. In case of scaffold and temporary construct as an implant the blood supply must be continuous in all the tissue engineered constructs, which is now a challenge when the artificial organs the researchers are developing. Thus, researchers are using bioactive resorbable composites to enhance the vascularization of the soft tissues. Sometimes it was also observed that bacterial adhesion to biomaterials that causes biomaterials centered infection and leads to poor tissue integration which ultimately reduce the life period of the bioimplant devices [20–22]. However, the scientists are looking for the genetic basis for the development of the bioimplants and scaffolds which provides the foundation for molecular design of scaffolds for tissue engineering and preferably using minimally invasive surgery. In case of metallic implants, various limitation is still there so upgrade of the alloying along with composition of the metallic ions must be further investigated. Some of the common challenges seen in the next generation materials and devices are shown in Figure 1.4. The main aim of the researchers is now a day to find the best suitable

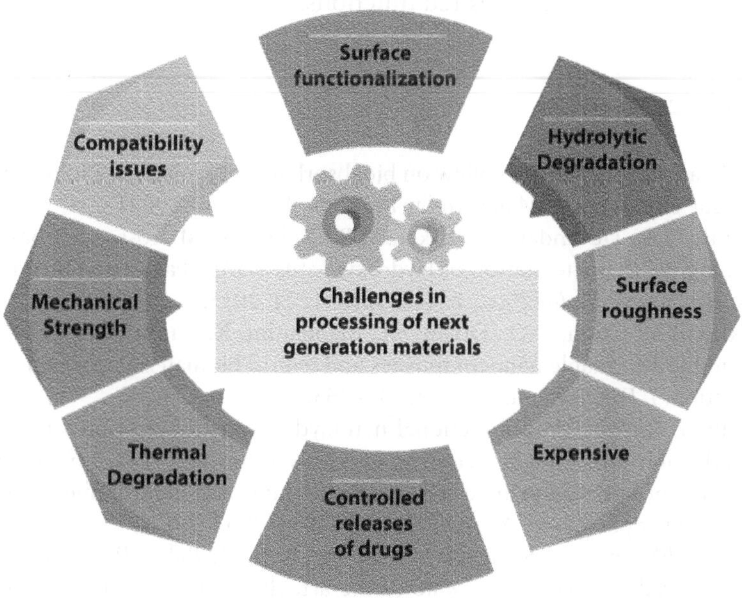

Figure 1.4 The challenges associated with processing of biocomposite for biomedical applications.

materials and bioimplant device which is not only approachable to all but also multifunctional in nature. In these regards, bioresorbable polymeric composites have established themselves as an alternative for their fruitful applications. The processing routes such as 3D printing, injection molding, additive manufacturing are extensively used these days to fulfill the huge demand of the biomedical implants and materials for their applications.

1.11 Conclusion

This chapter has summarized that for nearly two decades, some remarkable progress has been made in the processing routes and also in the production of biomaterials for various biomedical areas. There is huge demand for bio-implants which to fulfill is still a challenge in terms of cost effectiveness, mechanical strength, compatibility etc. Bioresorbable polymeric based composite and implants has established themselves in vast ways and still in rising phase. Various researchers are looking for environment friendly processing routes in order to produce these implants a cost-effective manner so that it would be approachable to all the needy patients. Various functional materials have a greater scope in the future in terms of shape memory, resorbable, self-healing biomimetic and also play a significant role in order to fulfill the desired functions.

References

1. Prasad, A., State of art review on bioabsorbable polymeric scaffolds for bone tissue engineering. *Mater. Today Proc.*, 44, 1391–1400, 2021.
2. Prasad, A., Devendar, B., Sankar, M.R., Robi, P.S., Micro-scratch based tribological characterization of Hydroxyapatite (HAp) Fabricated through fish scales. *Mater. Today Proc.*, 2, 4–5, 1216–1224, 2015.
3. Allo, B.A., Costa, D.O., Dixon, S.J., Mequanint, K., Rizkalla, A.S., Bioactive and biodegradable nanocomposites and hybrid biomaterials for bone regeneration. *J. Funct. Biomater.*, 3, 4, 432–463, 2012.
4. Gupta, A. *et al.*, Multifunctional nanohydroxyapatite-promoted toughened high-molecular-weight stereocomplex Poly(lactic acid)-based bionanocomposite for both 3D-printed orthopedic implants and high-temperature engineering applications. *ACS Omega*, 2, 7, 4039–4052, 2017.
5. RameshKumar, S., Shaiju, P., O'Connor, K.E., R. Babu, P., Bio-based and biodegradable polymers - state-of-the-art, challenges and emerging trends. *Curr. Opin. Green Sustain. Chem.*, 21, 75–81, 2020.

6. Prasad, A., Bhasney, S., Katiyar, V., Ravi Sankar, M., Biowastes processed hydroxyapatite filled poly(Lactic acid) bio-composite for open reduction internal fixation of small bones. *Mater. Today Proc.*, 4, 9, 10153–10157, 2017.
7. Prasad, A., Bioabsorbable polymeric materials for biofilms and other biomedical applications: Recent and future trends. *Mater. Today Proc.*, 44, 2447–2453, 2021.
8. Prasad, A., Sankar, M.R., Katiyar, V., State of art on solvent casting particulate leaching method for orthopedic scaffoldsfabrication. *Mater. Today Proc.*, 4, 2, 898–907, 2017.
9. Middleton, J.C. and Tipton, A.J., Synthetic biodegradable polymers as orthopedic devices. *Biomaterials*, 21, 23, 2335–2346, 2000.
10. Panda, N.N., Pramanik, K., Sukla, L.B., Extraction and characterization of biocompatible hydroxyapatite from fresh water fish scales for tissue engineering scaffold. *Bioprocess Biosyst. Eng.*, 37, 3, 433–40, 2014.
11. Ma, T., Wan, P., Cui, Y., Zhang, G., Li, J., Liu, J., Cytocompatibility of high nitrogen nickel-free stainless steel for orthopedic implants. *J. Mater. Sci. Technol.*, 28, 7, 647–653, 2012.
12. Arifin, A., Sulong, A.B., Muhamad, N., Syarif, J., Ramli, M., II, Material processing of hydroxyapatite and titanium alloy (HA/Ti) composite as implant materials using powder metallurgy: A review. *Mater. Des.*, 55, 165–175, 2014.
13. Stares, S. L. , Lourival, B., Fredel, M.C., Aragonês, A., Duek, E., Aragones, A., Self-reinforced bioresorbable polymer P (L/DL) LA 70:30 for the manufacture of craniofacial implant. *Polímeros (São Carlos. Impresso)*, 22, 378–383, 2012.
14. Besinis, A. *et al.*, Review of nanomaterials in dentistry: Interactions with the oral. *ACS Nano,* 9, 3, 2255–2289, 2015.
15. Kokubo, T., Kim, H.-M., Kawashita, M., Novel bioactive materials with different mechanical properties. *Biomaterials*, 24, 13, 2161–2175, 2003.
16. Satpathy, A. *et al.*, Bioactive Nano-Hydroxyapatite doped electrospun PVA-chitosan composite nanofibers for bone tissue engineering applications. *J. Indian Inst. Sci.*, 99, 3, 289–302, 2019.
17. Lei, K., Zhu, Q., Wang, X., Xiao, H., Zheng, Z., *In vitro* and *in vivo* characterization of a foam-like polyurethane bone adhesive for promoting bone tissue growth, *ACS Biomater. Sci. Eng.*, 5, 10, 5489–5497, 2019.
18. Parsons, A.J. *et al.*, Phosphate glass fibre composites for bone repair. *J. Bionic Eng.*, 6, 4, 318–323, 2009.
19. Lee, Y.K. *et al.*, Fabrication of macroporous scaffold using calcium phosphate glass for bone regeneration. *Key Eng. Mater.*, 254–256, 1079–1082, 2004.
20. Bhoi, S., Prasad, A., Kumar, A., Sarkar, R.B., Mahto, B., Meena, C.S., Pandey, C., Experimental study to evaluate the wear performance of UHMWPE and XLPE material for orthopedics application. *Bioengineering,* 9, 676, 2022, https://doi.org/10.3390/bioengineering9110676.

21. Bhoi, S., Kumar, A., Prasad, A., Meena, C.S., Sarkar, R.B., Mahto, B., Ghosh, A., Performance evaluation of different coating materials in delamination for micro-milling applications on high-speed steel substrate. *Micromachines*, *13*, 1277, 2022, https://doi.org/10.3390/mi13081277.
22. Sadeghzade, S., Emadi, R., Tavangarian, F., Naderi, M., Fabrication and evaluation of silica-based ceramic scaffolds for hard tissue engineering applications, Elsevier B.V, *Mater. Sci. Eng. C. Mater. Biol. Appl.*, 2017, 71, 431–438, 2016.
23. Sahoo, N.G., Pan, Y.Z., Li, L., Bin He, C., Nanocomposites for bone tissue regeneration. *Nanomedicine*, 8, 4, 639–653, 2013.
24. Jiang, T., Carbone, E.J., Lo, K.W.-H., Laurencin, C.T., Electrospinning of polymer nanofibers for tissue regeneration. *Prog. Polym. Sci.*, 46, 1–24, 2014.
25. Mulchandani, N., Prasad, A., Katiyar, V., Resorbable polymers in bone repair and regeneration, in: *Materials for Biomedical Engineering*, Elsevier Inc., Elsevier, Amsterdam, Netherlands, 87–125, 2019.
26. Kulper, S.A. *et al.*, A novel fracture mechanics model explaining the axial penetration of bone-like porous, compressible solids by various orthopaedic implant tips. *J. Mech. Behav. Biomed. Mater.*, 80, 128–136, 2018.
27. Macarini, L., Milillo, P., Mocci, A., Vinci, R., Ettorre, G.C., Poly-L-lactic acid-hydroxyapatite (PLLA-HA) bioabsorbable interference screws for tibial graft fixation in anterior cruciate ligament (ACL) reconstruction surgery: MR evaluation of osteointegration and degradation features. *Radiol. Med.*, 113, 8, 1185–97, 2008.
28. Sousa, R.A., Reis, R.L., Cunha, A.M., Bevis, M.J., Processing and properties of bone-analogue biodegradable and bioinert polymeric composites. *Compos. Sci. Technol.*, 63, 3–4, 389–402, 2003.
29. Friis, E.A., Decoster, T.A., Thomas, J.C., *Mechanical testing of fracture fixation devices*, Elsevier Ltd, Elsevier, Amsterdam, Netherlands, 2017, 131–141, 2017.
30. Agwuncha, S.C., Ray, S.S., Jayaramudu, J., Khoathane, C., Sadiku, R., Influence of boehmite nanoparticle loading on the mechanical, thermal, and rheological properties of biodegradable polylactide/poly(ε-Caprolactone) blends. *Macromol. Mater. Eng.*, 300, 1, 31–47, 2015.
31. Ulery, B.D., Nair, L.S., Laurencin, C.T., Biomedical applications of biodegradable polymers. *J. Polym. Sci. Part B Polym. Phys.*, *49*, 12, 832, 2011, https://doi.org/10.1002/polb.22259.
32. Vehof, J.W., Fisher, J.P., Dean, D., van der Waerden, J.P., Spauwen, P.H., Mikos, A.G., Jansen, J.A., Bone formation in transforming growth factor beta-1-coated porous poly(propylene fumarate) scaffolds. *J. Biomed. Mater. Res.*, *60*, 2, 241–251, 2002, https://doi.org/10.1002/jbm.10073.
33. Kenry, and Liu, B., Recent advances in biodegradable conducting polymers and their biomedical applications. *Biomacromolecules*, *19*, 6, 1783–1803, 2018, https://doi.org/10.1021/acs.biomac.8b00275.

34. Eldridge, J.H., Staas, J.K., Tice, T.R., Gilley, R.M., Biodegradable poly(DL-lactide-co-glycolide) microspheres. *Res. Immunol.*, *143*, 5, 557–581, 1992, https://doi.org/10.1016/0923-2494(92)80069-w.
35. Kumar, N., Ravikumar, M.N., Domb, A.J., Biodegradable block copolymers. *Adv. Drug Deliv. Rev.*, *53*, 1, 23–44, 2001, https://doi.org/10.1016/s0169-409x(01)00219-8.
36. Pişkin, E., Biodegradable polymers as biomaterials. *J. Biomater. Sci. Polym. Ed.*, *6*, 9, 775–795, 1995, https://doi.org/10.1163/156856295x00175.
37. Song, R., Murphy, M., Li, C., Ting, K., Soo, C., Zheng, Z., Current development of biodegradable polymeric materials for biomedical applications. *Drug Des. Devel. Ther.*, *12*, 3117–3145, 2018, https://doi.org/10.2147/DDDT.S165440.
38. Bat, E., Zhang, Z., Feijen, J., Grijpma, D.W., Poot, A.A., Biodegradable elastomers for biomedical applications and regenerative medicine. *Regen. Med.*, *9*, 3, 385–398, 2014, https://doi.org/10.2217/rme.14.4.
39. d'Orlyé, F., Trapiella-Alfonso, L., Lescot, C., Pinvidic, M., Doan, B.T., Varenne, A., Synthesis, characterization and evaluation of peptide nanostructures for biomedical applications. *Molecules (Basel, Switzerland)*, *26*, 15, 4587, 2021, https://doi.org/10.3390/molecules26154587.
40. Chmelař, J., Mrázek, J., Hermannová, M., Kubala, L., Ambrožová, G., Kocurková, A., Drmota, T., Nešporová, K., Grusová, L., Velebný, V., Biodegradable free-standing films from lauroyl derivatives of hyaluronan. *Carbohydr. Polym.*, *224*, 115162, 2019, https://doi.org/10.1016/j.carbpol.2019.115162.
41. Karadjian, M., Essers, C., Tsitlakidis, S., Reible, B., Moghaddam, A., Boccaccini, A.R., Westhauser, F., Biological properties of calcium phosphate bioactive glass composite bone substitutes: Current experimental evidence. *Int. J. Mol. Sci.*, *20*, 2, 305, 2019, https://doi.org/10.3390/ijms20020305.
42. Williams, C.K., Synthesis of functionalized biodegradable polyesters. *Chem. Soc. Rev.*, *36*, 10, 1573–1580, 2007, https://doi.org/10.1039/b614342n.
43. Vehof, J.W., Fisher, J.P., Dean, D., van der Waerden, J.P., Spauwen, P.H., Mikos, A.G., Jansen, J.A., Bone formation in transforming growth factor beta-1-coated porous poly(propylene fumarate) scaffolds. *J. Biomed. Mater. Res.*, *60*, 2, 241–251, 2002, https://doi.org/10.1002/jbm.10073.
44. Deng, M., Nair, L.S., Nukavarapu, S.P., Kumbar, S.G., Jiang, T., Weikel, A.L., Krogman, N.R., Allcock, H.R., Laurencin, C.T., *In situ* porous structures: A unique polymer erosion mechanism in biodegradable dipeptide-based polyphosphazene and polyester blends producing matrices for regenerative engineering. *Adv. Funct. Mater.*, *20*, 17, 2743–2957, 2010, https://doi.org/10.1002/adfm.201090073.
45. Xu, M., Banc, A., Woodruff, T.K., Shea, L.D., Secondary follicle growth and oocyte maturation by culture in alginate hydrogel following cryopreservation of the ovary or individual follicles. *Biotechnol. Bioeng.*, *103*, 2, 378–386, 2009, https://doi.org/10.1002/bit.22250.

46. Lloyd, A.W., Interfacial bioengineering to enhance surface biocompatibility. *Med. Device Technol.*, *13*, 1, 18–21, 2002.
47. Kohane, D.S. and Langer, R., Polymeric biomaterials in tissue engineering. *Pediatr. Res.*, *63*, 5, 487–491, 2008, https://doi.org/10.1203/01.pdr.0000305937.26105.e7.
48. Vert, M., Aliphatic polyesters: Great degradable polymers that cannot do everything. *Biomacromolecules*, *6*, 2, 538–546, 2005, https://doi.org/10.1021/bm0494702.
49. Gollwitzer, H., Ibrahim, K., Meyer, H., Mittelmeier, W., Busch, R., Stemberger, A., Antibacterial poly(D,L-lactic acid) coating of medical implants using a biodegradable drug delivery technology. *J. Antimicrob. Chemother.*, *51*, 3, 585–591, 2003, https://doi.org/10.1093/jac/dkg105.
50. Juncosa-Melvin, N., Matlin, K.S., Holdcraft, R.W., Nirmalanandhan, V.S., Butler, D.L., Mechanical stimulation increases collagen type I and collagen type III gene expression of stem cell-collagen sponge constructs for patellar tendon repair. *Tissue Eng.*, *13*, 6, 1219–1226, 2007, https://doi.org/10.1089/ten.2006.0339.
51. Khan, W., Muthupandian, S., Farah, S., Kumar, N., Domb, A.J., Biodegradable polymers derived from amino acids. *Macromol. Biosci.*, *11*, 12, 1625–1636, 2011, https://doi.org/10.1002/mabi.201100324.
52. Magnin, A., Pollet, E., Phalip, V., Avérous, L., Evaluation of biological degradation of polyurethanes. *Biotechnol. Adv.*, *39*, 107457, 2020, https://doi.org/10.1016/j.biotechadv.2019.107457.
53. Holmes, T.C., Novel peptide-based biomaterial scaffolds for tissue engineering. *Trends Biotechnol.*, *20*, 1, 16–21, 2002, https://doi.org/10.1016/s0167-7799(01)01840-6.
54. An, Y.H., Woolf, S.K., Friedman, R.J., Pre-clinical *in vivo* evaluation of orthopedic bioabsorbable devices. *Biomaterials*, 21, 24, 2635–2652, 2000.
55. Langer, R. and Vacanti, J.P., Tissue engineering. *Science*, 260, 920–926, 1993.
56. Reed, A.M. and Gilding, D.K., Biodegradable polymers for use in surgery – poly(glycolic)/poly(lactic acid) homo and copolymers 2: *In vitro* degradation. *Polymer*, 22, 342–346, 1981.
57. Ishaug, S.L., Yaszemski, M.J., Brizios, R., Mikos, A.G., Osteoblast function on synthetic biodegradable polymers. *J. Biomed. Mater. Res.*, 28, 1445–1453, 1994.
58. Ishaug-Riley, S.L., Crane, G.M., Gurlek, A., Miller, M.J., Yasko, A.W., Yaszemski, M.J., Mikos, A.G., Ectopic bone formation by marrow stromal osteoblast transplantation using poly(L-lactic-co-glycolic acid) foams implanted into the rat mesentery. *J. Biomed. Mater. Res.*, 36, 1–8, 1997.
59. Ishaug-Riley, S.L., Crane-Kruger, G.M., Yaszemski, M.J., Mikos, A.G., Three dimensional culture of rat calvarial osteoblasts in porous biodegradable polymers. *Biomaterials*, 19, 1405–1412, 1998.

60. Hutmacher, D.W., Schantz, J.T., Lam, C.X.F., Tan, K.C., Lim, T.C., State of the art and future directions of scaffold-based bone engineering from a biomaterials perspective. *J. Tissue Eng. Regen. Med.*, 1, 245–260, 2007.
61. Gugala, Z. and Gogolewski, S., Protein adsorption, attachment, growth and activity of primary rat osteoblasts on polylactide membranes with defined surface characteristics. *Biomaterials*, 25, 2341–2351, 2004, https://doi.org/10.1016/j.biomaterials.2003.09.013.
62. Simon Jr., C.G., Eidelmanb, N., Kennedya, S.B., Sehgala, A., Khatria, C.A., Washburna, N.R., Combinatorial screening of cell proliferation on poly(L-lactic acid)/poly(D,Llactic acid) blends. *Biomaterials*, 26, 6906–6915, 2005.
63. Ma, P.X., Zhang, R., Xiao, G., Franceschi, R., Engineering new bone tissue *in vitro* on highly porous poly(α-hydroxyl acids)/hydroxyapatite composite scaffolds. *Biomed. Mater. Res.*, 54, 284–293, 2001.
64. Rizzi, S.C., Heath, D.J., Coombes, A.G., Bock, N., Textor, M., Downes, S., Biodegradable polymer/hydroxyapatite composites: Surface analysis and initial attachment of human osteoblasts. *J. Biomed. Mater. Res.*, 55, 475–486, 2001.
65. El Ghalbzouri, A., Lamme, E.N., van Blitterswijk, C., Koopman, J., Ponec, M., The use of PEGT/PBT as a dermal scaffold for skin tissue engineering. *Biomaterials*, 25, 2987– 2996, 2004.
66. Dantzer, E. and Braye, F.M., Reconstructive surgery using an artificial dermis (Integra): Results with 39 grafts. *Br. J. Plast. Surg.*, 54, 659–664, 2001.
67. Mansbridge, J., Skin tissue engineering. *J. Biomater. Sci. Polym. Edn.*, 19, 955–968, 2008, https://doi.org/10.1163/156856208784909417.
68. Santos Jr., A.R., Dolder, H., Wada, M.L.F., Effects of fetal calf serum and dexamethasone in the differentiation of fibroblastic cells cultured on collagen I gel. *J. Submicrosc. Cytol. Pathol.*, 35, 35–42, 2003. PMID: 12762650.
69. Haas, V.R., Santos Jr., A.R., Wada, M.L.F., Behaviour of fibroblastic cells cultured in collagen I using the sandwich technique. *Cytobios*, 106, 255–267, 2001. PMID: 11545452.
70. Stern, R., McPherson, M., Longaker, M.T., Histologic study of artificial skin used in the treatment of full-thickness thermal injury. *J. Burn. Care Rehabil.*, 11, 7–13, 1990.
71. Yannas, I.V. and Burke, J.F., Design of an artificial skin 1: Basic design principles. *J. Biomed. Mater. Res. (Appl. Biomat.)*, 14, 65–81, 1980.
72. Yannas, I.V., Burke, J.F., Gordon, P.L., Huang, C., Rubenstein, R.H., Design of an artificial skin 2: Control of chemical composition. *J. Biomed. Mater. Res. (Appl. Biomat.)*, 14, 107–131, 1980.
73. Burke, J.F., Yannas, I.V., Quinby, W.C., Jr, Bondoe, C.C., Jung, W.K., Successful use of physiologicaly acceptable artificial skin in the treatment of extensive burn injury. *Ann. Surg.*, 194, 413–428, 1981.

74. Ma, L., Gao, C., Mao, Z., Zhou, J., Shen, J., Hu, X., Han, C., Collagen/chitosan porous scaffold with improved biostability of skin tissue engineering. *Biomaterials*, 24, 4833–4841, 2003.
75. Santos Jr., A.R., Barbanti, S.H., Duek, E.A.R., Dolder, H., Wada, R.S., Wada, M.L.F., Growth and differentiation of Vero cells on poly(L-lactic acid) membranes of different pore diameters. *Artif. Organs*, 25, 7–13, 2001.
76. Ma, J., Wang, H., He, B., Chen, J., A prelimilary *in vitro* study on the fabrication and tissue engineering applications of a novel chitosan bilayer material as a scaffold of human neofetal dermal fibroblasts. *Biomaterials*, 22, 331–336, 2001.
77. Gangwar, A.K.S., Rao, P.S., Kumar, A., Bio-mechanical design and analysis of femur bone. *Mater. Today: Proc.*, 44, Part 1, 2179–2187, 2021, https://doi.org/10.1016/j.matpr.2020.12.282.
78. Gangwar, A.K.S., Rao, P.S., Kumar, A., Patil, P.P., Design and analysis of femur bone: BioMechanical aspects. *J. Crit. Rev.*, 6, 4, 133–139, 2019.
79. Kumar, A., Mamgain, D.P., Jaiswal, H., Patil, P., Modal analysis of hand arm vibration (Humerus Bone) for biodynamic response using varying boundary conditions based on FEA. *Adv. Intell. Syst. Comput.*, 308, 169–176, 2015, https://doi.org/10.1007/978-81-322-2012-1_18.
80. Kumar, A., Behmad, S.I., Patil, P., Vibration characterization and static analysis of cortical bone fracture based on finite element analysis. *Eng. Autom. Probl.*, 3, 115–119, 2014. No 3-2014, UDC- 621.
81. Kumar, A., Jaiswal, H., Garg, T., Patil, P., Free vibration modes analysis of femur bone fracture using varying boundary conditions based on FEA. *Proc. Mater. Sci.*, 6, 1593–1599, 2014, https://doi.org/10.1016/j.mspro.2014.07.142.
82. Kumar, A., Gori, Y., Rana, S., Sharma, N.K., Yadav, B., FEA of humerus bone fracture and healing, in: *Advanced Materials for Biomechanical Applications*, 1st ed, CRC Press, Florida, United States, 2022, https://doi.org/10.1201/9781003286806.
83. Kumar, A., Datta, A., Kumar, A., Recent advancements and future trends in next-generation materials for biomedical applications, in: *Advanced Materials for Biomedical Applications*, vol. Chapter 1, A. Kumar, Y. Gori, A. Kumar, C.S. Meena, N. Dutt (Eds.), pp. 1–19, CRC Press, Boca Raton, FL, USA, 2022, https://doi.org/10.1201/9781003344810-1.
84. Prasad, A., Chakraborty, G., Kumar, A., Bio-based environmentally benign polymeric resorbable materials for orthopedic fixation applications, in: *Advanced Materials for Biomedical Applications*, vol. Chapter 15, A. Kumar, Y. Gori, A. Kumar, C.S. Meena, N. Dutt (Eds.), pp. 251–266, CRC Press, Boca Raton, FL, USA, 2022, https://doi.org/10.1201/9781003344810-15.
85. Datta, A., Kumar, A., Kumar, A., Kumar, A., Singh, V.P., Advanced materials in biological implants and surgical tools, in: *Advanced Materials for Biomedical Applications*, vol. Chapter 2, A. Kumar, Y. Gori, A. Kumar, C.S. Meena, N. Dutt (Eds.), pp. 21–43, CRC Press, Boca Raton, FL, USA, 2022, https://doi.org/10.1201/9781003344810-2.

86. Kumar, A., Gangwar, A.K.S., Kumar, A., Meena, C.S., Singh, V.P., Dutt, N., Prasad, A., Gori, Y., Biomedical study of femur bone fracture and healing, in: *Advanced Materials for Biomedical Applications*, vol. Chapter 14, A. Kumar, Y. Gori, A. Kumar, C.S. Meena, N. Dutt (Eds.), pp. 235–250, CRC Press, Boca Raton, FL, USA, 2022, https://doi.org/ 10.1201/9781003344810-14.
87. Patil, P.P., Gori, Y., Kumar, A., Tyagi, M., Experimental analysis of tribological properties of polyisobutylene thickened oil in lubricated contacts. *Tribol. Int.*, 159, 106983, 2021, https://doi.org/10.1016/j.triboint.2021.106983.
88. Kumar, A., Rana, S., Gori, Y., Sharma, N.K., Thermal contact conductance prediction using FEM based computational techniques, in: *Advanced Computational Methods in Mechanical and Materials Engineering*, pp. 183–220, CRC Press, Boca Raton, FL, USA, 2021, https://doi.org/10.1201/9781003202233-13.
89. Kumar, R., Kumar, A., Kant, L., Prasad, A., Bhoi, S., Meena, C.S., Singh, V.P., Ghosh, A., Experimental and RSM-based process-parameters optimisation for turning operation of EN36B Steel. *Materials*, 16, 339, 2023, https://doi.org/10.3390/ma16010339.
90. Prasad, A., Chakraborty, G., Kumar, A., Gajrani, K.K., Introduction to biodegradable polymers, in: *Biodegradable Composites for Packaging Applications*, vol. Chapter 1, A. Prasad, A. Kumar, K.K. Gajrani (Eds.), pp. 1–12, CRC Press, Boca Raton, FL, USA, 2022, https://doi.org/10.1201/9781003227908-.
91. Chakraborty, G., Prasad, A., Kumar, A., Processing of biodegradable composites, in: *Biodegradable Composites for Packaging Applications*, vol. Chapter 3, A. Prasad, A. Kumar, K.K. Gajrani (Eds.), pp. 33–48, CRC Press, Boca Raton, FL, USA, 2022, https://doi.org/ 10.1201/9781003227908-3.
92. Kumar, A., Rana, S., Gori, Y., Sharma, N.K., Thermal contact conductance prediction using FEM based computational techniques, in: *Advanced Computational Methods in Mechanical and Materials Engineering*, vol. Chapter11, A. Kumar, Y. Gori, N. Dutt, Y.K. Singla, A. Maurya (Eds.), pp. 183–220, CRC Press, Boca Raton, FL, USA, 2022.
93. Sharma, V.K., Kumar, V., Joshi, R.S., Kumar, A., Effect of REOs on tribological behavior of aluminum hybrid composites using ANN, in: *Additive Manufacturing in Industry 4.0: Methods, Techniques, Modeling and Nano Aspects*, vol. Chapter 9, V.K. Sharma, A. Kumar, M. Gupta, V. Kumar, D.K. Sharma, S.K. Sharma (Eds.), pp. 153–168, CRC Press, Boca Raton, FL, USA, 2022.
94. Bansal, A., Bhardwaj, H.K., Sharma, V.K., Kumar, A., Static and dynamic behavior analysis of Al-6063 alloy using modified Hopkinson bar, in: *Additive Manufacturing in Industry 4.0: Methods, Techniques, Modeling and Nano Aspects*, vol. Chapter 6, V.K. Sharma, A. Kumar, M. Gupta, V. Kumar, D.K. Sharma, S.K. Sharma (Eds.), pp. 107–124, CRC Press, Boca Raton, FL, USA, 2022.
95. Singh, S., Kumar, A., Behura, S.K., Verma, K., Challenges and opportunities in nanomanufacturing, in: *Nanomanufacturing and Nanomaterials Design:*

Principles and Applications, vol. Chapter 02, S. Singh, S.K. Behura, A. Kumar, K. Verma (Eds.), pp. 17–30, CRC Press, Boca Raton, FL, USA, 2022.

96. Srivastava, S., Verma, D., Thusoo, S., Kumar, A., Singh, V.P., Kumar, R., Nanomanufacturing for energy conversion and storage devices, in: *Nanomanufacturing and Nanomaterials Design: Principles and Applications*, vol. Chapter 10, S. Singh, S.K. Behura, A. Kumar, K. Verma (Eds.), pp. 165–173, CRC Press, Boca Raton, FL, USA, 2022.
97. Singh, V.P., Dwivedi, A., Karn, A., Kumar, A., Singh, S., Srivastava, S., Srivastava, K., Nanomanufacturing and design of high-performance piezoelectric nanogenerator for energy harvesting, in: *Nanomanufacturing and Nanomaterials Design: Principles and Applications*, vol. Chapter 15, S. Singh, S.K. Behura, A. Kumar, K. Verma (Eds.), pp. 241–242, CRC Press, Boca Raton, FL, USA, 2022.
98. Gori, Y., Verma, R.P., Kumar, A., Patil, P., FEA based fatigue crack growth analysis. *Mater. Today Proc.*, 46, 10575–10581, 2021, https://doi.org/10.1016/j.matpr.2021.01.319.
99. Patil, P.P. and Kumar, A., Design and FEA simulation of omega type coriolis mass flow sensor. *Int. J. Control Theory Appl.*, 9, 40, 383–387, 2017.
100. Patil, P., Sharma, S.C., Paliwal, V., Kumar, A., ANN modeling of Cu type omega vibration based mass flow sensor. *Proc. Technol.*, 14, 260–265, 2014, https://doi.org/10.1016/j.protcy.2014.08.034.
101. Patil, P., Sharma, S.C., Jaiswal, H., Kumar, A., Modeling influence of tube material on vibration based EMMFS using ANFIS. *Proc. Mater. Sci.*, 6, 1097–1103, 2014, https://doi.org/10.1016/j.mspro.2014.07.181.
102. Paul, A.K., Prasad, A., Kumar, A., Review on artificial neural network and its application in the field of engineering. *J. Mech. Eng.: PRAKASH*, 1, 1, 53–61, 2022, https://doi.org/10.56697/JMEP.2022.1107.
103. Rana, S., Kumar, A., Gori, Y., Patil, P., Design and analysis of thermal contact conductance. *J. Crit. Rev.*, 6, 5, 363–370, 2019.
104. Rana, S. and Kumar, A., FEA based design and thermal contact conductance analysis of steel and Al rough surfaces. *Int. J. Appl. Eng. Res.*, 13, 16, 12715–12724, 2019.

2

Advanced Materials for Surgical Tools and Biomedical Implants

Sudipto Datta[1*] **and Ranjit Barua**[2]

[1]*Department of Materials Engineering, Indian Institute of Science, Bangalore, Karnataka, India*
[2]*Centre for Healthcare Science and Technology, Indian Institute of Engineering Science and Technology, Howrah, West Bengal, India*

Abstract

With the advancement of technologies, many tools like assistive, surgical, and medical are developed for enlightening human life. The surgical and medical tools have enormously grown in the previous few years. By 2023 about an 8.6% (CAGR) growth rate is expected in medical tools. For improving the quality of patient, medical and surgical tools development technologies are integrated from the areas of bioengineering, biotechnology, pharmaceutics, and chemistry. Accidents, aging, sports injuries, and various other injuries lead to the requirement of IVD, ophthalmic, cardiologic, dental, and orthopedic implants as well as diagnostic instruments. The material to be implanted should be biocompatible and its toxicity profile, immunological response, and its capability to incorporate with the surrounding tissues without causing any systemic effects and undesirable local on the patient should be checked. In this book chapter biomaterials, implants like dental implants, cardiovascular implants, ophthalmic implants, and orthopedic implants are briefly discussed. The book chapter also gives some light on various materials used to develop the bio-implants like polymers, ceramics, metals, composites, and nanomaterials.

***Keywords*:** Metals, polymers, ceramics, composite, nanomaterials

**Corresponding author*: sudiptodatta1990@gmail.com

Arbind Prasad, Ashwani Kumar and Manoj Gupta (eds.) Advanced Materials and Manufacturing Techniques for Biomedical Applications, (25–36) © 2024 Scrivener Publishing LLC

2.1 Introduction

For a longer duration, biomaterials are being used in medical areas, although noticeable progress was done in the 1940s in the biomaterials area. Further, a significant improvement was noticed in implant devices and therapeutic medical technologies in the last 25 years [1]. From past to present there are many changes in the medical field, metallic implants are being changed to regular tissues and their byproducts, also valves of mechanical types are changed to valves of prosthetic types which are developed from harvested porcine valves and bovine tissue. The problem with these tissues is that their life is only 5–7 years and that they fail after their implantation. Various researchers have shown that mechanical and chemical changes of these types of valves can rectify their output outcome; also these types of implants are used in patients of elder age. Moreover, the expertise of healthcare is transformed with the concept of bioengineering over the last 25 years for the technology of peacemaker in the management of arrhythmia [4]. Also with the development of software algorithms and sensor technology, the pacemakers are improved that automatically respond to changing the level of the physical activities of patients so that it can change the rate of simulation rate depending upon the patient's condition. The first commercially available cardiac resynchronization therapy (CRT) devices were seen in 2001 [2]. These devices are the patient's life savior for chronic heart failure. The advancement of these implant technologies gave treatment options for neurological abnormalities which is a life-threatening diseases. Medicines are not allowed to enter the central nervous system (CNS) by the blood-brain barrier (BBB), for this reason, programmable implanted pumps are being used for transferring accurate doses of drugs similar to morphine to reduce extensive pain. For the relief of cancer pain and treatment of non-malignant cancer, these implantable pumps are commonly used [3]. Parkinson's degenerative disease treatment using the deep brain stimulation method is one of the most amazing applications in biomedical engineering over the past decade. This technology uses a thin electrode insertion to the brain's targeted location and the other electrode part is connected to the high-frequency stimulator [4]. Because of the simulations, the symptoms like slowness, rigidity, and uncontrolled tremor disappeared, which ultimately affects the patient's life quality. The use of metal spinal cages for the treatment of degenerative disk disease using bone strategies of bone in-growth is one of the current advancements. In this method, bone morphogenetic protein (rBMP) collagen plugs are placed in cages of metal which promotes the growth of bone for spinal disk repair [5]. Polymers both synthetic and natural have good healthcare applications. For making contact lenses, polymethylmethacrylate (PMMA) is commonly used. For soft tissue reconstruction and vascular system engineering, polytetrafluoroethylene (PTFE) biocompatible polymers are being used. Various other polymer types which are used commonly in medical applications are polyurethanes, polycarbonates, and polypropylene. The use of bioactive molecules and biomaterials for drug delivery appeared in the late 1070s and it is still in use and research is still going on in this area. The main basic drug delivery models are of four types they are site-specific, pulmonary, transdermal, and oral. The pulmonary delivery system uses nasal spray aerosol-based systems. Transdermal delivery is used for skin drug delivery by diffusion process but one limitation is that it is only restricted for only smaller molecules like nicotine. In many pharmaceutical applications, the most common mode of drug delivery is oral delivery. Implants of polymers for drug delivery are mainly focused and it is expected that further development will come in the coming years which will improve healthcare. Already products which work on this concept are available in the market like prostate cancer treatment by Lupron Depot and fertility regulation by Norplant. More progress in this area is expected in the upcoming years which will significantly improve consumer and market applications. From a healthcare point of view, biomaterials can be differentiated into three categories: a) Derived from nature (plant and animal-derived); b) Artificial (metals, polymers, ceramics, and composites); and c) Hybrid or Semi-synthetic materials. All three biomaterial types are currently being used in healthcare for many years but subsequent progress has improved their usefulness in clinics and healthcare. For applications where load bearing is needed commonly used biomaterials are metals. A few examples of these are screws and wires in plates of fracture fixation as well as artificial joints. For hip replacement, the femoral components which are normally used are mainly fabricated from titanium alloys, Co-Ni-Mo alloys, and Co-Cr-Mo. Polymeric implants are used in liver parts, tracheal tubes, facial prostheses, and heart components. Ultrahigh molecular weight polyethylene (UHMWPE) applications are in the joints of the shoulder, knee, and hips. In Table 2.1, all the biopolymers that are used in clinics and healthcare are listed in detail [6]. For filling applications and dental implants, ceramics are commonly used. Because of the low fracture toughness in ceramics, they are not being used in load-bearing materials applications. Because of their high strength and low density, composite materials are commonly used in prosthetic limbs. Some composite materials like polymethyl methacrylate-glass filler and bisphenol A-glycidyl-quartz/silica are commonly used in dental restorations [7]. Polymers that are naturally extracted like hyaluronic acid, alginate, gelatin,

Table 2.1 Normally applied polymers for healthcare applications.

No.	Name of the material	Applications
1	KF/PMA, PET/PHEMA, GF/PU, KF/PE	Tendons/ligaments
2	PET/collagen, PET/PU	Abdominal wall prosthesis
3	PET/PU, CF/PTFE, PTFE/PU	Cartilage replacement
4	Cells/PTFE, PET/collagen, Cells/PET	Vascular grafts
5	Polymethylmethacrylate	Bone Cement
6	Bioglass/PS, Bioglass/PU, CF/PS	Spine plates, cages, rods, discs, screws

and collagen are generally used in healthcare areas for 3D scaffold fabrication which supports cell proliferation and cell growth. This 3D structured scaffold resembles the natural host tissue so these materials are extremely used in various tissue engineering and regenerative medicine applications [8]. These materials have low mechanical strength so these are not used in applications where load-bearing applications are needed. To improve their mechanical properties these materials are chemically modified. For example chains of collagen modified with hydroxyl-lysine and lysine [9], and PEGylated fibrinogen (PF) [10], etc. Biomaterials of hybrid types have also shown considerable development for healthcare region applications.

2.2 Application of Bioengineering to Healthcare

Using the concept of regenerative medicine and tissue engineering, the area of healthcare is fast developing, and for regeneration, the replacement paradigm shift is being seen. For repairing the damaged tissues tissue engineering works in two steps a) with cells only (autologous cells, stem cells, etc.), b) with 3D construct having and not having cells that resemble natural tissue. Tissue engineering and regenerative medicine assure remarkable medical care for millions of patients per year at low medical budgets. Conventional methods of healthcare like organ transplants can be handled by tissue engineering and/or newer advanced technologies. Tissue-engineered and regenerated organs can avoid the difficulties connected with the transplantation of organs which can be a blessing for future healthcare complex problems. This chapter covers only a few areas. For cardiac repair, myocardial tissue engineering, which is a recent tissue engineering and cell therapy are being used. In various research, it has been attempted to redevelop the unwell heart with a bioengineered tissue patch pre-seeded with implantation of epicardial BM-derived mesenchymal stem cells or bone marrow cells [11]. For this particular application, both synthetic as well as natural polymers can be used. PGA, collagen, PLGA, and fibrin can be used [12]. For rat models, the cardiac patch sandwich is developed by a combination of bovine pericardia which is an animal-extracted tissue with the cell sheet that is engineered for regeneration of the ischemic heart. In Figure 2.1 simple step is shown for the cardiac patch fabrication. For the repairing of infracted myocardium, injectable biomaterials are being used in situ cardiac tissue engineering [13]. This method has huge benefits similar to the non-invasive and also encourages refurbishment by rectifying benefactor cell neoangiogenesis and neoangiogenesis [14, 15]. Other sources of the cell can be from adipose imitative stem cells which impulsively distinguish into useful cardiomyocytes. Another source of cells is cardiac stem cells (CSC) which are also being used for repairing cardiac damage. Cardiac stem cells are extracted and isolated by many researchers from murine heart or human heart biopsies using techniques of explants culture known as cardio-spheres (CSs). These cardio-spheres (CSs) have possible to differentiate for forming myocardium bands with changeable organization degree *in vivo*. This technology of myocardial bio-engineering through stem cells of cardiac is in the development stage but the tissue engineering method by a scaffold-free technique using CSCs was presented [16]. In rat models, the cell sheets have been noticed to re-establish the infracted myocardium functions. Besides these techniques, other forthcoming techniques for repairing ischemic hearts, nanotechnology, and gene therapy hold good capabilities for the future. From the clinical point of view techniques with minimal invasion will always be favored.

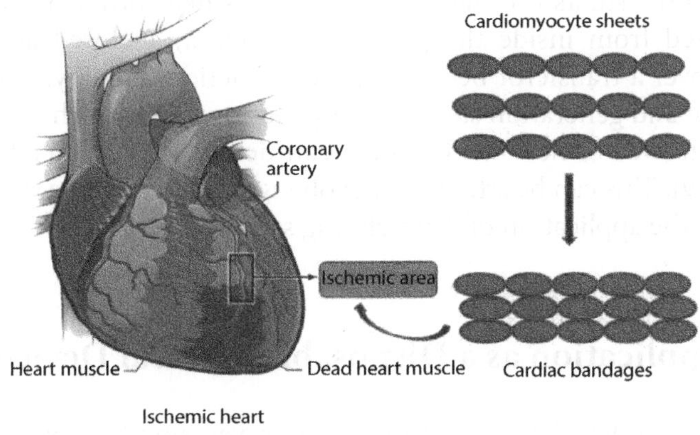

Figure 2.1 Graphic illustration of cardiac bandages for ischemic heart [11].

2.3 Application in Musculoskeletal and Orthopedic Medicines

Ceramics (Al_2O_3, $Ca_3(PO_4)_2$, etc.), Metals (stainless steel, Ti alloys, etc.), polymers (Si), and combinations of material are extensively being applied for orthopedic uses [17]. However, with the introduction of biomaterials (3rd generation) for this area, considerable development has been witnessed in this part. This novel type of material has the capability of an osteoinductive effect, which can be familiarized into the supports by approaches like external adjustment, a combination of different growth influences like VEGF, BMP, and TGFβ, and planting stem cells of bone marrow. Current investigation work showed that BMP-2 and VEGF co-loaded constructs improved vascularization along with the construction of a new bone. In recent times gene therapy is reconnoitered to moderate the properties osteoinductive of transcriptional and growth factors. Another influence that is significant in bone material (soft tissue) and is existence modified is the resorption of the scaffolds. Continued transfer of molecules from a permeable scaffold is a developing conception and can have inclusive inferences for upcoming medical science. Arrangement of the biomolecules could be strained these can improve tissue incorporation and fresh bone construction [18]. One of the cutting-edge methods for the treatment of atherosclerotic wounds is the establishment of suitable stents. On the other hand, metal stents have restricted medical usage as numerous patients practice. Recently drug-eluting stents have arisen as possible biomaterials that can transport the drugs through a measured way for decreasing cell proliferation by succeeding establishment of a metal stent.

Figure 2.2 shows the required drugs or the drugs that can also be combined into the stents through several methods like entrenched as well as unrestricted from inside the polymeric material, enclosed and unconstrained over a transferor because of the restrictions employing the 1st as well as the 2nd generation stents of drug-eluting types which are prepared into the growth of new stents which can decrease the difficulties of the patient [19]. This can be attained by evolving a biocompatible stent. Figure 2.3 shows the application of drug-eluting stents.

2.4 Application as a Disposable Medical Device

Globally as modern health technology improves, simultaneously an enormous load is pressurizing on the surroundings. Huge amounts of clinical

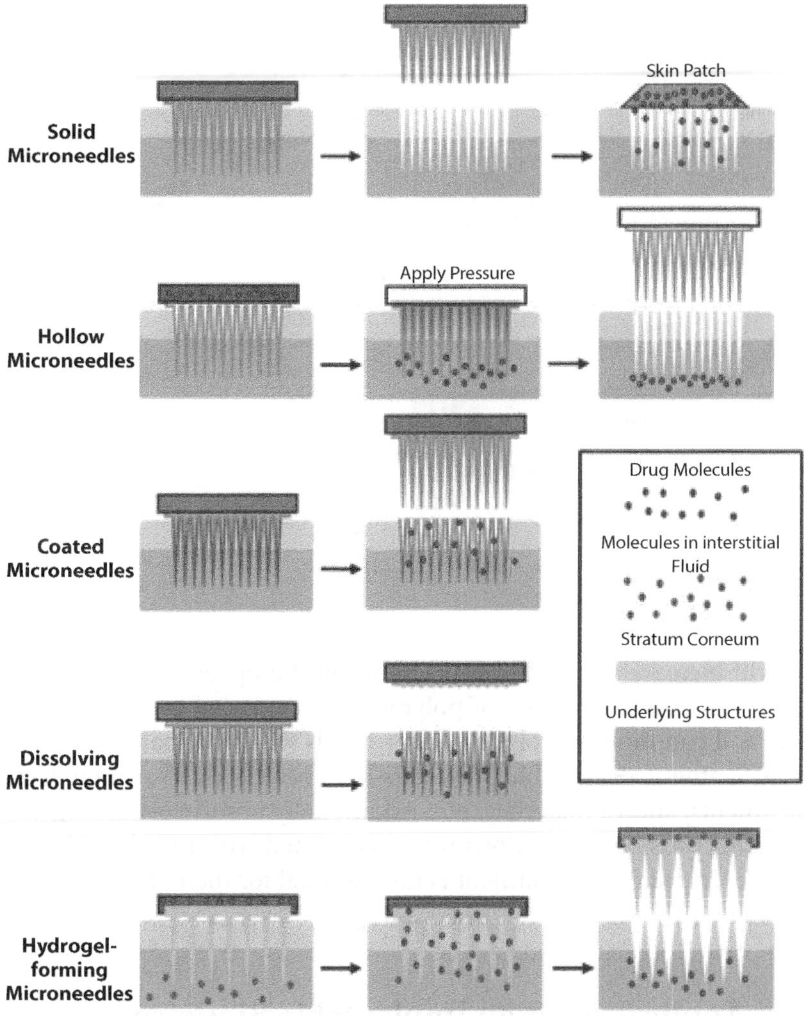

Figure 2.2 Schematic illustration of different types of microneedles use in drug delivery systems [14].

equipment that are made up of non-degradable biomaterials pretense a severe ecological and also profitable matter. The part of biomedical engineering and biomaterials, cutting-edge technology practically modifies the material's biodegradability to reduce the harmfulness of the material. Biodegradable materials like polyglycolic acid (PGA), and PLA, have appeared as favorable materials for usage as not reusable medical devices and are sustainable [20]. These biodegradable polymers are existence extensively applied for the manufacture of throwable medical instruments

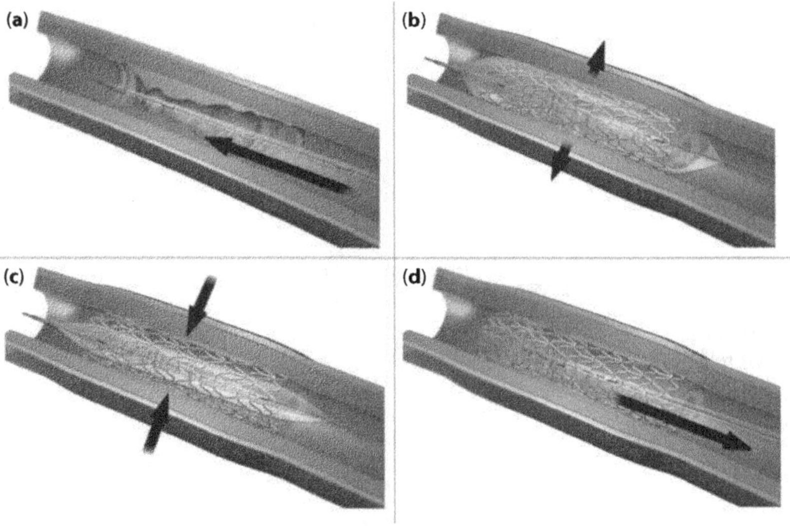

Figure 2.3 Application of drug-eluting stents [21].

that will increase the healthcare facilities in the upcoming years and will be harmless. As these types of polymers get degraded by hydrolytic produces and common hydrolysis of ester bonds which are harmless to living tissue. Synthetic degradable polyesters have been applied as bone fixation instruments, suture materials, etc. Additionally absorbable arrangements have numerous compensations when associated with metal implants, for example, 2nd surgical treatment is not essential for their elimination from the patient [21].

2.5 Application as an Implantable Biosensor

Biosensors work on the idea of bio MEMS which are developing as a probable non-invasive method to identify and study the biomarkers. These sensors are medically applied aimed at the initial recognition of the diseases of cardiovascular, diabetes, etc. [22, 23]. An extensively examined fluorescence sensor is developed on Con A, which is a plant-sugar-binding protein [24]. On the other hand, other types of GCMS are diborononic or boronic acid-related sensors which are fixed below the skin, and the fluorescence could be detected by the skin. Biosensors for the detection of cholesterol are used for the detection of total and free cholesterol in the body, for the initial identification of cardiovascular diseases. However, the technological improvements in microelectronics are liable to accelerate the improvement of further advanced biosensors. This expertise has enormous applicability for the initial recognition of cancer which can raise the continued existence of a patient.

2.6 Conclusions

Present healthcare technology and diagnostics method have numerous restrictions that are they are costly and has restricted exactness or there is no approach to cure certain illnesses (like cancer); thus, a huge demand to advance the present healthcare abilities. The investigation is being implemented to recover the remaining approaches and for the progress of new methods. Biomedical engineering is one of the greatest feasible possibilities which can recover the current healthcare consequence. It uses tissue engineering and biomedical ideas for the restoration of injured tissue. The main objective at this time is to restore and heal the organ or tissue rather than eliminate it. Biomedical engineering has also revealed good development in diagnostics and innovative approaches are currently advanced that can create the recognition informal and precise. With the present development in biomaterials, it is expected upcoming healthcare which will be obtainable at a reasonable value and with healthier facilities.

References

1. Rahimtoola, S.H., Choice of prosthetic heart valve for adult patients. *J. Am. Coll. Cardiol.*, *41*, 6, 893–904, 2003, https://doi.org/10.1016/S0735-1097(02)02965-0.
2. Young, J.B., Abraham, W.T., Smith, A.L., Leon, A.R., Lieberman, R., Wilkoff, B., Canby, R.C., Schroeder, J.S., Liem, L.B., Hall, S., Wheelan, K., Multicenter InSync ICD Randomized Clinical Evaluation (MIRACLE ICD) Trial Investigators, Combined cardiac resynchronization and implantable cardioversion defibrillation in advanced chronic heart failure: The MIRACLE ICD Trial. *JAMA*, *289*, 20, 2685–2694, 2003.
3. Rawicki, B., Treatment of cerebral origin spasticity with continuous intrathecal baclofen delivered via an implantable pump: Long-term follow-up review of 18 patients. *J. Neurosurg.*, *91*, 5, 733–736, 1999.
4. Benabid, A.L., Koudsie, A., Benazzouz, A., Piallat, B., Krack, P., Limousin-Dowsey, P., Lebas, J.F., Pollak, P., Deep brain stimulation for Parkinson's disease. *Adv. Neurol.*, *86*, 405–412, 2008.

5. Boden, S.D., Kang, J., Sandhu, H., Heller, J.G., Use of recombinant human bone morphogenetic protein-2 to achieve posterolateral lumbar spine fusion in humans: A prospective, randomized clinical pilot trial: 2002 Volvo Award in clinical studies. *Spine*, *27*, 23, 2662–2673, 2002.
6. Williams, D.F., Titanium for medical applications, in: *Titanium in Medicine*, pp. 13–24, Springer, Berlin, Heidelberg, 2001.
7. Rice, J.J., Martino, M.M., De Laporte, L., Tortelli, F., Briquez, P.S., Hubbell, J.A., Engineering the regenerative microenvironment with biomaterials. *Adv. Healthc. Mater.*, *2*, 1, 57–71, 2001, 2013.
8. Sosnik, A. and Sefton, M.V., Semi-synthetic collagen/poloxamine matrices for tissue engineering. *Biomaterials*, *26*, 35, 7425–7435, 2005.
9. Ben-David, D., Srouji, S., Shapira-Schweitzer, K., Kossover, O., Ivanir, E., Kuhn, G., Müller, R., Seliktar, D., Livne, E., Low dose BMP-2 treatment for bone repair using a PEGylated fibrinogen hydrogel matrix. *Biomaterials*, *34*, 12, 2902–2910, 2013.
10. Chen, C.H., Wei, H.J., Lin, W.W., Chiu, I., Hwang, S.M., Wang, C.C., Lee, W.Y., Chang, Y., Sung, H.W., Porous tissue grafts sandwiched with multilayered mesenchymal stromal cell sheets induce tissue regeneration for cardiac repair. *Cardiovasc. Res.*, *80*, 1, 88–95, 2008.
11. Martinez, E.C. and Kofidis, T., Adult stem cells for cardiac tissue engineering. *J. Mol. Cell. Cardiol.*, *50*, 2, 312–319, 2011, https://doi.org/10.1016/j.yjmcc.2010.08.009.
12. Leor, J., Amsalem, Y., Cohen, S., Cells, scaffolds, and molecules for myocardial tissue engineering. *Pharmacol. Ther.*, *105*, 2, 151–163, 2005.
13. Planat-Bénard, V., Menard, C., André, M., Puceat, M., Perez, A., Garcia-Verdugo, J.M., Pénicaud, L., Casteilla, L., Spontaneous cardiomyocyte differentiation from adipose tissue stroma cells. *Circ. Res.*, *94*, 2, 223–229, 2004.
14. Turner, J.G., White, L.R., Estrela, P., Leese, H.S., Hydrogel-forming microneedles: Current advancements and future trends. *Macromol. Biosci.*, *21*, 2, 2000307, 2021. https://doi.org/10.1002/mabi.202000307.
15. Bartosh, T.J., Wang, Z., Rosales, A.A., Dimitrijevich, S.D., Roque, R.S., 3D-model of adult cardiac stem cells promotes cardiac differentiation and resistance to oxidative stress. *J. Cell. Biochem.*, *105*, 2, 612–623, 2008.
16. Barrere, F., Mahmood, T.A., De Groot, K., Van Blitterswijk, C.A., Advanced biomaterials for skeletal tissue regeneration: Instructive and smart functions. *Mater. Sci. Eng.: R: Rep.*, *59*, 1-6, 38–71, 2008, https://doi.org/10.1016/j.mser.2007.12.001.
17. Bose, S., Roy, M., Bandyopadhyay, A., Recent advances in bone tissue engineering scaffolds. *Trends Biotechnol.*, *30*, 10, 546–554, 2012.
18. Fattori, R. and Piva, T., Drug-eluting stents in vascular intervention. *Lancet (London, England)*, *361*, 9353, 247–249, 2003.
19. Tian, H., Tang, Z., Zhuang, X., Chen, X., Jing, X., Biodegradable synthetic polymers: Preparation, functionalization and biomedical application. *Prog. Polym. Sci.*, *37*, 2, 237–280, 2012.

20. Lasprilla, A.J., Martinez, G.A., Lunelli, B.H., Jardini, A.L., Filho, R.M., Poly-lactic acid synthesis for application in biomedical devices - A review. *Biotechnol. Adv.*, *30*, 1, 321–328, 2012, https://doi.org/10.1016/j.biotechadv.2011.06.019.
21. Burt, H.M. and Hunter, W.L., Drug-eluting stents: A multidisciplinary success story. *Adv. Drug Deliv. Rev.*, *58*, 3, 350–357, 2006.
22. Ren, F., Pearton, S.J., Kang, B.S., Chu, B.H., AlGaN/GaN high electron mobility transistor based sensors for bio-applications, in: *Biosensors for Health, Environment and Biosecurity*, pp. 15–68, IntechOpen, UK, 2011, 2006.
23. Ballerstadt, R., Evans, C., Gowda, A., McNichols, R., Fiber-coupled fluorescence affinity sensor for 3-day *in vivo* glucose sensing. *J. Diabetes Sci. Technol.*, *1*, 3, 384–393, 2007.
24. Arya, S.K., Datta, M., Malhotra, B.D., Recent advances in cholesterol biosensor. *Biosens. Bioelectron.*, *23*, 7, 1083–1100, 2008.

3

Insights into Multifunctional Smart Hydrogels in Wound Healing Applications

Sriparna De[1]*, Dipankar Das[1], Arbind Prasad[2], Ashwani Kumar[3] and Dipankar Chattopadhyay[4]

[1]*Department of Allied Health Sciences, Brainware University, Kolkata, India*
[2]*Mechanical Engineering Department, Katihar Engineering College (Department of Science & Technology, Govt. of Bihar), Katihar, Bihar, India*
[3]*Technical Education Department Uttar Pradesh, Kanpur, Uttar Pradesh, India*
[4]*Department of Polymer Science & Technology, University of Calcutta, Kolkata, West Bengal, India*

Abstract

Hydrogels are convenient and employed in biomedicine. The ability of hydrogel dressings to reduce inflammation has significantly increased in recent years, addressing a wide range of clinical concerns raised in ongoing efforts to expedite wound healing. With chronic wounds like diabetes and severe burn wounds, where harmful endogenous or exogenous factors can interfere with inflammatory response and impede the healing process, wound healing is a cascaded and incredibly difficult process. Dressings for wounds are required to start the healing process. Engineered wound dressings may greatly reduce healing time, the risk of recurrent infections, the illness, and cost associated with such infections, and other factors. Due to its benefits, hydrogels (with various architectural, physico-chemical, and biological properties) have been extensively investigated as platforms for wound dressing. Herein the chapter portrays the recent significant advancement on fabrication and architectural framework of hydrogels for wound healing applications. Furthermore, we emphasize the significance of conductive nanomaterial and their relevance to wound dressing as well as monitoring. In order to address a few issues, such as bacterial growth suppression, enhanced re-epithelialization, vascularization, and generally accelerated the active process of wound healing, hydrogel design for wound healing applications pledges to offer cutting edge technological explication. Also addressed is how hydrogel wound dressings are used

**Corresponding author*: sriparna.de2@gmail.com

Arbind Prasad, Ashwani Kumar and Manoj Gupta (eds.) Advanced Materials and Manufacturing Techniques for Biomedical Applications, (37–60) © 2024 Scrivener Publishing LLC

to treat different kinds of wounds, including incisional and excisional wounds. Moreover, chronic wounds are treated, with an emphasis on infected, burnt, and diabetic wounds. Based on ongoing improvement, the emerging trend of intelligentization for hydrogel wound dressings in wound healing applications was also investigated.

Keywords: Self-healing, anti-inflammatory, hydrogel, chronic wound, antibacterial, biomedicine

3.1 Introduction

Skin renders the largest organ in the body, providing the protective barrier from external elements. Due to characteristics features the skin has become one of the most sensitive tissues of direct interaction with the outside environment [1–3]. A wound is defined as an injury or disintegration of skin tissue caused by a variety of factors such as heat damage, trauma, and chronic ulcerations caused by diabetes mellitus, pressure, and venous stasis. Scar tissue accumulation and the characteristic absence of skin appendages like hair, apocrine glands, and sebaceous glands are three indicators that this process is occurring. [3] Hemostasis, inflammation, proliferation, and remodeling are four cumulative that work together to repair skin wounds [4]. The management of wound care may change depending on a number of factors, including extreme inflammation, burns, or an event that results in large-area skin tissue loss, infections, diabetes, and others. In addition to acting as a temporary barrier against infections in the periphery, wound dressings can facilitate skin cell remodeling, the following influx and synchronization of host tissues, and considerable impact on wound healing [4, 5]. Clot development is the pivotal tissue response phase to a wound to restrict bleeding from the wound site. Released cytokines oversight blood flow to the wound bed and lymphocytes and macrophages are promoted to invade infection [6, 7]. The series of inflammatory response is associated with angiogenesis, fibroblast proliferation followed by collagen deposition. Granulation leads to the development of higher cell content and blood capillaries wherein remodeling includes the replacement of scar tissues for reconstructing the epidermal structure. The ideal wound dressing should have the following properties: good tissue compatibility, adequate mechanical strength, sufficient moisture retention capacity, and a suitable microstructural environment to encourage cell adhesion, proliferation, and differentiation. Many wound dressings, including semipermeable membranes, semipermeable foams, hydrocolloids, and hydrogels, have been produced to support wound healing [8, 9] (Figure 3.1). Due to its excellent

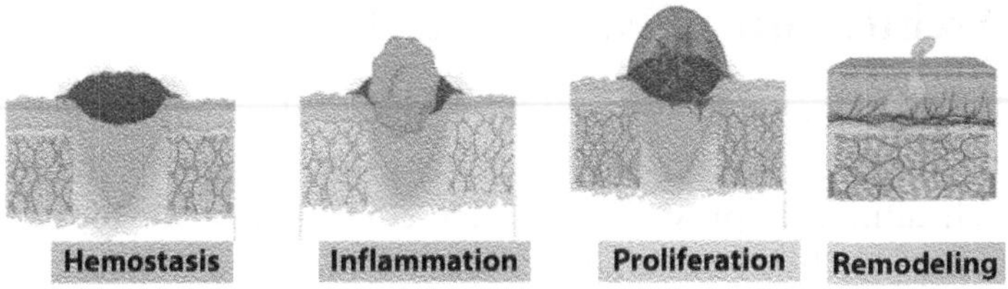

Figure 3.1 Schematic diagram of four consecutive steps of wound healing.

hydrophilicity, biocompatibility, and three-dimensional (3D) porous structure similar to extracellular matrix (ECM), hydrogels have emerged as one of the most cutthroat candidates for wound dressings. This has also aroused significant attention among the research community [5, 8, 9].

The term 'hydrogel' was first coined in the year of 1894 followed by the reported synthesis of poly(2-hydroxyethylmethacrylate) hydrogel by DuPont scientists in the year of 1936 [10, 11]. In the year of 1980, hydrogel research trend shifted from simple synthesis to the formulation of stimuli responsive hydrogels with respect to pH and temperature. The careful tweaking of hydrogel production resulted in the development of pH/thermo-responsive hydrogels that depend on physiological inputs. In rational view water-rich structures formed by the crosslinking of hydrophilic monomers are known as hydrogels [12, 13]. The hydrophilic three-dimensional (3D) network topologies imitate the extracellular matrix of biological soft tissues and swells in water fluids via chemical and physical cross-linkers. Depending on their binding modes hydrogels can be embodied as physical and chemical hydrogels promoting covalently cross-linked network stability. Synthetic hydrogels can be categorized into five categories according to their hydrophilic chemical composition poly (hydroxyalkyl methacrylates), poly(acrylamide), poly(N-vinyl pyrrolidone), poly(acrylic acid), and poly(vinyl alcohol whereas many naturally derived hydrophilic polymers such as chitosan, gelatin, hyaluronic acid, alginate, dextran, peptides and polysaccharides have been admirable candidates for hydrogel miniaturization [14]. By amalgamating these multifunctional hydrogels that mimic native tissue ECM favoring suitable microenvironments for accelerating wound healing applications. As one of the quickly evolving techniques, nano-reinforced hydrogels have been studied in the development of multifunctional hydrogels with fascinating characteristics for wound healing applications. Present chapter highlights the rational designing, architecture and functional attributes of wound dressing materials and their versatile archetype and prospect towards chronic wound care management [1–4, 14].

3.2 Architecture of Fabricated Hydrogels

The fascinating assembly and architectonics of the hydrogel designs for wound healing depends on cell attachment, spreading and cell proliferation which ultimately provides favorable environment for bio-mimicked native tissue [15, 16]. For enhanced cell filtration and nutrient transportation hydrogels should possess porous morphology to mimic ECM compounds. Wound dressing material also has the property of water retention. Although for efficient cell attachment and proliferation cascade, cells maintain the humid environment henceforth dressing patch should absorb the excessive exudates that could otherwise facilitate the bacterial growth impede the healing process [17].

Biocompatibility is another pivotal and influential factor for designing wound dressing material. Utilization of Polysaccharides is effectively considered as having superior biocompatibility. The use of toxic initiators for cross-linking agent in biopolymeric tissue scaffolds, hydrogels, and composites should be avoided to minimize the deleterious immunological, toxic, or foreign body response [17]. In case of chronic wound, visceral bleedings and deep fissured wound biodegradability is bottom-line approach wherein hydrogels lose their functionality and degrade successively during cell regeneration after homeostasis. Also, cleavable bonds and groups ultimately assist the process wherein hydrolysis leads to ultimate degradation of hydrogel breakdown. In question of mechanical strength, double network (DN) hydrogels comprised of brittle and ductile interpenetrating networks promote superior flexibility and stretchability around the extremities like ankle, knee, wrist joints etc. [18]. Simultaneously, hydrogel adhesion properties have foremost important characteristic property for high adhesiveness, toughness, and cell affinity. Enriched catechol chemistry-based hydrogels advocate promising application for skin trauma and suturing. Gou and co-workers fabricated antibacterial injectable hydrogel consisting of quaternate chitosan and Pluronic F127 illustrated improved mechanical properties and efficient hemostatic performance endorsing suitable and stable addressability to the underlying skin tissues [18].

Furthermore, self-healing hydrogels possess the ability to work in sol-gel state. After injection the sol state relocates to gel state via *in-situ* cross-linking. While the hydrogel is in its sol state, therapeutic drugs or cells can be incorporated within it for delivery to the body for controlled release or wound healing. Additionally Self-healing properties, particularly for hydrogels with weak elasticity and injectable form could extend the durability of hydrogels by automatically repairing deformation or damage. The major concerns regarding the chronic wound healing involves hypoxia, impaired angiogenesis, ROS damage still lies in a dormant phase. Using Pluronic F127, oxidative hyaluronic acid, and poly-L-lysine, Mao and colleagues created a multifunctional hydrogel that could carry exosomes of adipose-derived mesenchymal stem cells to help complete skin regeneration and improve chronic wound healing [19]. Exosome-loaded hydrogel could be a promising candidate and assist in the subsequent healing process step-wise including moderate wound closure rates, rapid angiogenesis, smooth re-epithelization followed by collagen deposition within the wound framework.

3.3 Bactericidal Effect on Wound Repair

3.3.1 Historical Perspective

A wound is a rupture in the skin's epithelial integrity and may be accompanied by a disruption of the structure and functionality of the normal tissue beneath the skin. The dynamic process of regaining the anatomic function of live tissue is known as wound healing which consists of hemostasis, inflammation, proliferation, and remodeling. Robson *et al.* reported that wounds undergo healing processes in an organized and timely manner and maintain anatomic and functional integrity, they are termed acute and are categorized as chronic when they either do not develop in a timely and ordered manner or do not achieve sustained anatomical and functional integrity [20]. Jones and his coworkers suggested that wound healing is a complicated chain of activities, and an imbalance can result in a failure to heal, with substantial implications for patients and healthcare systems [21].

3.3.2 Microbial Influence on Wound Healing

Wounded tissue provides a chance for microflora to attach, colonize, penetrate, and infect nearby healthy tissue. Bacteria are found in the surface and deep tissues of all wounds and bacterial products like endotoxin, and metalloproteinase may influence all wound healing stages (from inflammation to remodeling) [21]. According to the report provided by Robson *et al.* Hippocrates noticed that inflammation modifies wound healing and reported about the infection of aseptic surgical wounds as repair mechanisms were arrested [1, 4, 11]. In the study of clinical surgical practice, infection is the end product of microorganism entry, growth, metabolic activity, and subsequent pathophysiologic effects in the patient's tissues [22]. Fan *et al.* stated that skin acts as a natural barrier that protects internal organs from the external environment but loses its protecting mechanism upon damage. As a result, microorganisms can easily invade the protecting barrier and begin to establish colonies, resulting in serious wound infection and preventing the healing process [22]. Most importantly the minor presence of organisms in a wound is less important than the level of bacterial growth.

3.3.3 Wound Tissue Healing Strategies: Case Study

Recently, Mohammad Hadi *et al.* reported that the bacterial proliferation and colony formation effectively break the process of wound healing with 10^5 population of bacteria per gram of tissue [23, 24]. Based on the aforementioned findings of Robson's team numerous clinical and experimental studies have shown that bacterial growth is predominantly influenced wound infection wherein substantial accumulation of 100,000 organisms per gram of tissue [20]. Current study focus on the study of hydrogels with chitosan showed MIC value of higher than 188 microgram per ml against *S. aureus* [23]. In tissue, various researchers have developed cost effective bioabsorbable polymeric implants for orthopedic and biomedical applications [25–28].

3.3.4 Degradation of Wound Healing Factors

In case of wound healing, there is a perfect balance between the immunological system of the host and the pathologic effects of bacteria. If microorganisms are present in a large amount or the microorganism contains virulence factor for example beta-hemolytic *Streptococcus sp.*, etc., it can hamper the wound healing process. Bacterial growth has been shown to impair wound contraction in both acute and chronic wounds due to the release of bacterial enzymes and metalloproteinase that may degrade fibrin as well as wound growth factors. Fibronectin, one of the important fibrin matrix is needed for fibroblast migration as well as phagocytic activity of macrophages. The studies of Lindsey [29] focused on vitronectin as a wound-healing protein is responsible for regulating the fibrinolytic balance of new endothelial cells during angiogenesis. The presence of bacteria and wound infection easily imbalance this process. Phagocytic activity of the wound healing process is inhibited by the action of bacterial proteases. It has been found that leucocyte activity followed by chemotaxis is suppressed by the *Pseudomonas aeruginosa* alkaline proteases. Similarly, *S. aureus* cysteine protease staphopain B (SspB) is responsible for reducing the chemotactic activity of neutrophils. Cytokine activity is also inhibited by the action of bacteria present in the wound and suppresses the wound healing process. Some kinds of proteases that are synthesized by *P. aeruginosa* (Protease IV and Protease LasB) are able to degrade IgA and IgG, as a result, bacteria can easily invade the wound region. Healing of chronic wounds is directly related to the high bacterial load within wound tissue [29].

If chronic wounds contain necrotic tissue, it can act as a growth medium for bacteria. As reported pro-inflammatory mediator levels are elevated for a prolonged period due to the presence of bacterial endotoxin and destroying the glycosaminoglycan matrix needed for wound healing and also stimulating tissue necrosis [26]. Bacteria as a replicating organism within a wound stimulate host injury either directly (bacterial pathogenicity) or indirectly (to eliminate invasive microorganisms by the host immunological system) [27]. stated that in the case of direct mechanism, endotoxin secreted from the cells of bacteria elevates the level of proinflammatory cytokines like interleukin 1β and TNFα and it is responsible for the imbalance in levels of matrix metalloproteinase and their inhibitors [27]. The bacterial population is present within the chronic or acute wound, secreted matrix of extracellular polysaccharide (EPS) to enclose themselves, it is termed biofilms. The report of Jones *et al.*, 2004 stated that an alternate growth pattern, as well as metabolism, is seen in biofilms due to a change in phenotype and as a result, bacteria are less sensitive to innate immunity and therapeutic drugs which ultimately restricts the wound healing process [21]. The study by Wall *et al.* 2002 stated that bacterial supernatants and proteases not only inhibit re-epithelization but are also responsible for the degradation of dermal proteins and polysaccharides [28]. As per the study of Lindsay *et al.* 2017 due to the production of intracellular and extracellular microbial protease, host protein, growth factor and receptors are degraded which in turn stimulates microbial dissemination into underlying soft tissue and slows the process of wound healing [29].

3.3.5 pH and Wound Healing: Impact of Bacteria

The wound healing process is also affected by the change in pH which in turn increases the toxicity of bacterial by-products. A study report by Shi *et al.* 2011 suggested that the activity of proteases such as collagenase of *Clostridium sp.* reached a peak level at a pH value of around 8.5 [1, 4, 11, 30]. The alkalinity of the skin wound region supports the growth of bacterial infection. According to the Steven *et al.* urease, an enzyme secreted by *Klebsiella sp.*, *P. aeruginosa* is important to reduce the deleterious effect of the acidic environment of the wound as the enzyme urease release ammonia from urea. This phenomenon helps in microbial proliferation [31].

Another study report by Mohammad Hadi *et al.* 2023 showed that antimicrobial polymer Poly-ethylene-imine (PEI) contain amine units and is able to protonate at physiological pH. The positive charge of PEI interacts with negatively charged bacterial cell membrane causing cell membrane damage or cell polarization [23]).

3.4 New Frontiers of Hydrogels in Wound Dressing Applications

3.4.1 Hemostatic Hydrogel as Wound Dressing

Hemostasis is the preliminary stage of wound healing. As a result, hydrogel dressings with hemostatic properties are beneficial for accelerating wound healing. According to studies, the hemostatic function of hydrogels is maintained by both the physical sealing and the stimulation of coagulation cascades brought about by the absorption of wound extract. A wound bed can be impenetrable to produce hemostasis. The adhesive hydrogels can also be permanently attached to the wound site, reducing the danger of infection brought on by the wound's interaction with the environment. Among biopolysaccharides dextran, CS50, Sodium Alginate (SA) have been recorded as adhesive hydrogels. CS with a specific positive charge can improve adherence to biological tissues. Additionally amino-functionalization leads to tune the bioadhesibility after grafting onto CS backbone [36].

At healthy pH levels, 36,125-Polylysine (EPL) is likewise a cationic polymer with outstanding biological adherence to a variety of biological surfaces. The production of *in situ* hydrogels by photo-cross-linking has also been reported to promote covalent bonding that can contribute to some degree of tissue adherence. Mercapto-tissue responses have also been observed to improve tissue adherence. Moreover, adding certain long-chain alkyl groups to macromolecules' primary chains can result in hydrophobic interactions with the alkyl groups in subcutaneous adipose tissue, producing an adhesion effect. Encouraged by super adhesion of marine mussels in the humid environment researchers have proposed the *in situ* biomimetic dopamine-modified EPL-PEG-based hydrogel (PPD hydrogel) by using horseradish peroxidase (HRP) crosslinking. In another study hyaluronan (HA) hydrogel patches grafted by catechol/pyrogallol efficiently utilized for wound healing applications. Several dopamine based hydrogels exhibit good adhesion properties, injectable, conductive, antibacterial, hemostatic, and antioxidant, illustrating considerable potential to improve the healing of infected skin wounds/lesions [37–41].

Simultaneously other bioinspired adhesive hydrogels includes Andrias davidianus skin secretion where higher value of shear adhesion was observed (26.66 ± 8.22 kPa) in contrast to acrylate based synthetic glue and fibrin glue [42, 43]. Chitosan is well recognized for tuned wound healing applications. Several modification and functionalization promotes activation of platelets to yield coagulation factors favoring hemorrhage hemostasis and wound healing. CS/gelatin/PVA composite hydrogel exhibit the potential addition of positively charged components to hydrogels can enhance their electrostatic contact with platelets, blood cells, and plasma fibronectin, cause platelet instigation and blood cell clumping and ultimately have the impact of expedite coagulation [44, 45]. Glycidyl methacrylate-based multifunctional cryogel with improved CNT-quaternization CS. These cryogels have been shown to have a high blood absorption ability, quick blood trigger shape recovery, and strong mechanical capacity. Also, in the lethal compressed hemorrhage model of rabbit liver and mouse liver injury, the cryogel demonstrated outstanding hemostatic efficacy. Furthermore porous cryogels consist of oxidative cross-linking dopamine with chitosan, quaternary ammonium chitosan with gelatin confer a superb hemostatic effect favoring the facilitation of tissue repair [46–48].

3.4.2 Anti-Oxidant and Anti-Inflammatory Hydrogel Wound Dressing

The second phase of wound healing, inflammation, highlights on eradicating debris and destroying microorganisms. Controlling inflammation is a pivotal component of wound dressings since adequate control of inflammation is vital for wound repair. High levels of oxidative stress brought on by excessive inflammation can result in an increase in reactive oxygen species (ROS), which include superoxide anions, hydroxyl radicals, nitric oxide and hydrogen peroxide [49–52]. ROS can set off chain reactions that can cause DNA and protein oxidation as well as lipid peroxidation, which can kill cells. Free radicals can no longer harm the body because of antioxidants' ability to catch and neutralize them. Hence, hydrogels having antioxidant capabilities might undoubtedly aid in the healing process. Natural antioxidant made of polyphenols is more stable and tolerant of long-term storage. Versatile natural polyphenols includes tea polyphenols,

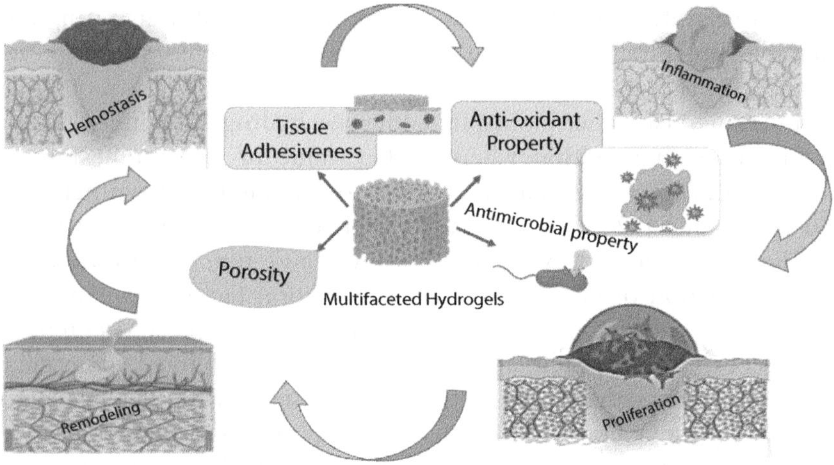

Figure 3.2 Multifaceted application prospects of hydrogels on wound healing.

resveratrol, anthocyanins and some flavonoids164 have been incorporated within hydrogels promoting the development of antioxidative hydrogels aiding wound healing [53–56] (Figure 3.2).

Recently, the simple lyophilization and photo cross-link method (SL/GelMA) implemented to fabricate a novel form of miniaturized scaffold based on tomatoes (*Solanum lycopersicon*) and gelatin methacryloyl (GelMA). By utilizing various antioxidants found in tomatoes, including carotenoids, flavonoids, phenolic acids, vitamin E, and vitamin C, SL/GelMA is able to control the level of ROS, reduce oxidative stress factor in the wound bed, accelerate cell passage and angiogenesis, aid in collagen synthesis, and stimulate the wound enclosure process [57]. Specifically, a microchannel structure based on quaternized chitosan (QCS) and carboxylated cellulose nanofibers (CCNF) produced by utilizing tannic acid and copper complex to create the microchannels in the hemostatic sponge (CQTC) (crosslinking agent). The developed 3D porous structure of the CQTC possess superior absorption capacity along with significant hemostasis capacity, both confirmed by consecutive test protocol e.g. liquid absorption test and hemostasis experiment. Furthermore, CQTC enhanced wound healing and displayed significant antibacterial properties. As a clinical wound hemostatic material, this permeable and microchannel hemostatic sponge has numerous application possibilities. [58] Oshin Jacintha D'souza and co-workers developed naturally derived PVA/Chitosan/*Basella alba* stem extract (BAE) based bio-composite film prominent antibacterial activity and cell adhesbility could be employed as a promising biomaterial in wound care management [59].

3.4.3 Antibacterial Hydrogel Wound Healing

Hydrogel loaded with antibiotic is very effective measure to reduce bacterial load in wound area followed by it able to reduce resistance power of bacteria against antibiotics, if hydrogel administrate topically. Nanoparticle based hydrogels offer equal strength to fight against bacteria present in wound [32]. Silver nanoparticle in hydrogel with graphene having Ag to graphene ratio 5:1 (Ag5P1) possesses highest bacterial killing property against both *E. coli* and *S. aureus*. Zinc contains nanoparticle ZnONP-based hydrogel accelerate wound healing as Zn ion improve migration of keratinocyte toward the wound site. Tetracycline as an antibiotic also can be introduced into ZnONP based hydrogel to increase antibacterial property of the hydrogel [1, 4, 11, 23]. Different inorganic nanoparticles which are incorporated with hydrogel like TiO_2, CeO_2, CdsSe, FeO, ZNS, Cu show great wound healing efficiency. Antibiotic loaded hydrogel is very popular nowadays as hydrophilic nature of hydrogel provides proper environment to store antibiotic and controlled release of small molecules of antibiotics namely gentamicin, ciprofloxacin, vancomycin, amoxicillin [33]. Hydrogel not necessarily always involve the incorporation of antiseptics or any antimicrobial substance, but depends on its own intrinsic antimicrobial properties like cationic quaternary ammonium compound or chitosan which react with anionic phospholipid membrane of bacterial cell and results in lysis of bacterial cell [34, 35]. Furthermore, owing to great antioxidant and reactive oxygen species scavenging activity of cerium oxide, our team has successfully fabricated folate conjugated ceo_2 and investigated its potential efficacy as a novel spray formulation towards wound healing applications. The good synergism between folic acid and nanoceria could be highly promising for further tailoring of pharmacokinetic process evaluation in various biomedical applications [36].

3.4.4 Self-Healing Hydrogel Wound Dressing

Furthermore *in situ* synthesis, adhesion, and additional characteristics that promote the rapid hemostasis wherein hydrogels serve a significant tactic to prevent external bacterial infection of the wound. Regardless of the amount of tissue activity or external force applied, typical hydrogels are highly susceptible towards mechanical flexibility including fracture and impact strength. In addition to weakening or even losing their own properties, hydrogel breakage will lead to the infiltration of alien microbes and lead to wound infection. So, it's crucial to maintain the hydrogel dressings' structural integrity while a wound is healing. As a result, the idea of self-healing hydrogels is put forward as an Avant-grade material and heal physico-structural damage on its own. The majority of self-healing hydrogels are made via dynamic and reversible chemical linkages to create a network of cross-linking hydrogels. Consequently, in this section, we will provide a summary of how self-healing hydrogels are used to repair damaged tissues. Physical self-healing hydrogels renovate fabric networks by taking into account dynamic noncovalent interactions (such as hydrophobic interactions, host-guest interactions, hydrogen bonds, crystallization, polymer-nanocomposite interconnecting networking and various intermolecular interactions) between molecules, oligomers, or polymer residual chain. Gelatin methacrylate/tannic acid and curcumin encapsulated gelatin modified bacterial cellulose self-healable patch was developed for transdermal drug delivery system [60, 61]. Also based on host–guest interactions various self-healing hydrogels such as cyclodextrins (CD) and NIPAM, amantadine, or silk fibroin were typically examined and reported [62–64].

However, in the case of chemical self-healing hydrogels, various remodeling networks, including water-soluble carboxymethyl chitosan (CMC) and the aldehyde of cellulose nanocrystal (DACNC), functionalized HA (HAAD) and F127 triblock copolymers (BAF127), quaternized CS (QCS), and benzaldehyde-terminated PF127 was identified exhibiting good mechanical strength effectively utilized for self-healing applications [65, 66].

Nowadays self-healing hydrogels based on dopamine have just come to light. Because to the covalent/noncovalent bond cohabitation properties of dopamine during self-cross-linking, dopamine-based cross-linking hydrogels exhibit both structural toughness and self-healing properties. Dopamine-grafted HA and PDA-coated GO were used to particularly synthesis dopamine cross-linked hydrogels that highlighted the hydrogels' intrinsic ability to self-heal by virtue of covalent and noncovalent interactions. The majority of the self-healing properties still come from the dynamic Schiff base, and there are still very few macromolecular frameworks that can be used to develop Schiff base based hydrogels, notwithstanding the fact that the variety of self-healable hydrogels for wound dressings has amplified pointedly in current centuries. Consequently, further strategies must be strategically applied in the future for self-healing hydrogel dressings. Also, the current self-healing hydrogels need to be strengthened in future by enhancing sufficient mechanical and structural integrity suitable for wound care [67].

3.4.5 Conductive Hydrogel Wound Dressing for Wound Monitoring

Skin battery works concomitantly via switching positive and negative charges in the presence of defects or wounds. The endogenic electrokinetic at the wound site is also hypothetically function as a steering signal that instigate the cells migration into the wound site during healing process. Research have manifested that coupling an electrode on the wound bed with external current to mimic the endogenous dynamic current at the wound site inspires the migration of inflammatory mediators and expedite up the recovery of wound [68, 69]. Several research group reported injectable conductive hydrogel dressing material wherein poly(ethylene glycol) copolymer conjugated with modified CS based-g-polytetraaniline, HA-graft-aniline tetramer and dopamine encapsulated gelatin and CNT used as effective dressing for repairing of microbial wound defects. Simultaneously polypyrrole, polythiophene derivatives prompted versatile applicability towards wound closure, angiogenesis and tissue regeneration [70, 71].

Recently, electroactive polysaccharides-based hydrogel based on carboxymethylated chitosan and sodium alginate was reported as a potential candidate in promoting fibroblasts proliferation followed by angiogenesis and collagen deposition under electrical stimulation [72–75]. Similarly, sodium hyaluronate based self-powered system was investigated as portable and flexible wound healing candidate for biomedical applications [76–78]. Moreover, the adjuvant and polyelectrolyte sodium hyaluronate would accelerate wound healing, promote cell growth, and improve the electrochemical characteristic functions of supercapacitors. Furthermore, to monitor the subsequent altered parameters near wound compact sensor system is now prime demand and strategy needs to be implemented accordingly. 3D printing technology was efficiently utilized wherein pH sensor array triggered alginate/mesoporous resin beads cater real time data about wounds, such as bacterial contamination as well as antibiotic expulsion [79–83]. Current research trends follow on -demand drug delivery in wound management system. Based on the concept pH and temperature sensor was highly compatible for diagnosis of wound chronicity and its inflammation index. In this context Zhu *et al.* advanced the development of a multifunctional zwitterionic hydrogel for the purpose of monitoring the health of wounds in diabetic patients. This hydrogel has the ability to monitor glucose and pH levels. The wound dressing provides adequate moisture to the wound site that aids in the treatment of diabetic lesions. Therefore this versatile wound dressing illustrate potential prominence and showcased to manage diabetes clinically and cure persistent wounds [85–90].

3.4.6 Chronic Wound Dressing

Diabetes, the most prevalent metabolic disorder significantly conceived health and economic dilemma. The impaired wound healing of diabetic patients resulted the generation of tortuous pathophysiology including the alteration of immune, vascular, biochemical system components. This considerably leads to amputations in chronic wound, making interventions to accelerate diabetic wound healing more critical. Diabetic wound healing possesses typical characteristic features such as extensive inflammation, hyperglycemia, hypoxia, weak vascularization followed by poor cellular penetration and tissue granulation. Current research will portray the strategic approach encountered during the wound healing process for prolonged diabetic patients. Thermoresponsive hydrogel comprised of triblock copolymers based on PEG, PLGA effectively used diabetic wound healing. Importantly various cytokines and autologous platelet-derived growth factors (GF) impringed hydrogels can be used in adorning angiogenesis and endothelial progenitor cell homing [91–97]. Tiny extracellular vesicles, or exosomes, are cell-based vesicles carrying mRNA molecules along with protein and lipid biomolecules successively used to heal diabetic lesions. Furthermore, glucose responsive hydrogel system has been developed by Zhao and his coworkers wherein incorporation of fibroblast, therapeutic drug delivery and cellular therapy is synergistically evaluated in detail [98–100]. Wound vascularization and hypoxia are also two key driven factors for slow healing of diabetic lesions. Increased levels of glucose hinder the normal blood flow followed by the impairment of angiogenesis ultimately constrict the healing process. Versatile usage of multifunction collagen molecules assists in tissue maturation whereas incorporation of silicon ion directly supervises the upregulation of vascular endothelial growth factor (VEGF) facilitating vascular regeneration in diabetic lesions. Furthermore, hyaluronic acid (HA) oligosaccharides residues have widely explored in various conceptual contexts to inspire endothelial cell proliferation scheme and relocation phenomenon thereby induce angiogenesis and employed in conjunction with VEGF to tune blood vessel reorganization [101–108].

3.5 Conclusion and Future Perspectives

Hydrogels have emerged as the most competitive options in recent decades, with their versatile usage in wound dressing. Consistent emphasis and fine-tuning of clinical prerequisite, the functions of hydrogels have widely explored for plethora of clinical applications. Despite their potential importance, multifunctional hydrogels continue to explore challenges in wound healing applications. Creating complex multifunctional hydrogels with wound monitoring tools such as pathogenic infection detectors and smart *on-demand* drug release systems might be a viable option in the future. This strategy necessitates real-time monitoring of the wound environment in a closed-loop fashion. DNA based methodological avenues can be employed to fabricate biomacromolecules backbone in conjunction with DNA encoding proteins as wound dressing patch. Also current research should focus on the amelioration of cell-laden hydrogels for scarless tissue renewal. Moreover, electrotherapy and real-time wound monitoring supported by conductive biomaterials will be significantly scrutinized in the coming decades thanks to the development of nanogenerators and bioelectronics. We anticipate that conductive biomaterials will endure considerable innovation for wound healing in perspective of these accomplishments and the burgeoning growth of new technological advancement. Epitomizing the above findings, the conceptual research appeals the aptitude of hydrogels dressing material towards clinical translational and marketization to nourish a better future for wound care management [109–117].

References

1. Dąbrowska, A.K., Spano, F., Derler, S., Adlhart, C., Spencer, N.D., Rossi, R.M., The relationship between skin function, barrier properties, and body-dependent factors. *Skin Res. Technol.*, 24, 165–174, 2018.
2. Gurtner, G.C., Werner, S., Barrandon, Y., Longaker, M.T., Wound repair and regeneration. *Nature*, 453, 314–321, 2008.
3. Beanes, S.R., Dang, C., Soo, C., Ting, K., Skin repair and scar formation: The central role of TGF-β. *Expert Rev. Mol. Med.*, 5, 1–22, 2003.
4. Rosique, R.G., Rosique, M.J., Farina Junior, J.A., Curbing inflammation in skin wound healing: A review. *Int. J. Inflam.*, 2015, 316235, 2015.
5. Koehler, J., Brandl, F.P., Goepferich, A.M., Hydrogel wound dressings for bioactive treatment of acute and chronic wounds. *Eur. Polym. J.*, 100, 1–11, 2018.
6. Gupta, A., Kowalczuk, M., Heaselgrave, W., Britland, S.T., Martin, C., Radecka, I., The production and application of hydrogels for wound management: A review. *Eur. Polym. J.*, 111, 134–151, 2019.
7. Xiang, J., Shen, L., Hong, Y., Status and future scope of hydrogels in wound healing: Synthesis, materials and evaluation. *Eur. Polym. J.*, 130, 109609, 2020.
8. Hu, H. and Xu, F.-J., Rational design and latest advances of polysaccharide-based hydrogels for wound healing. *Biomater. Sci.*, 8, 2084–2101, 2020.

9. Kamoun, E.A., Kenawy, E.-R.S., Chen, X.A., Review on polymeric hydrogel membranes for wound dressing applications: PVA-based hydrogel dressings. *J. Adv. Res.*, 8, 217–233, 2017.
10. Van Bemmelen, J.M. and Anorg, Z., Der hydrogel und das kristallinische Hydrat des Kupfer-oxydes. *Allg. Chem.*, 5, 466–483, 1894.
11. du Pont de Nemours, E., I, Methacrylate resins. *Ind. Eng. Chem.*, 28, 1160–1163, 1936.
12. Chen-Chow, P.C. and Frank, S.G., *In vitro* release of lidocaine from pluronic F-127 gels. *Int. J. Pharm.*, 8, 89–99, 1981.
13. Liang-chang, D., Qi, Y., Hoffman, A.S., Controlled release of amylase from a thermal and pH-sensitive, macroporous hydrogel. *J. Control. Release*, 19, 171–177, 1992.
14. Das, S.K., Parandhaman, T., Dey, M.D., Biomolecule-assisted synthesis of biomimetic nanocomposite hydrogel for hemostatic and wound healing applications. *Green Chem.*, 23, 629–669, 2021.
15. Zhang, Y.S. and Khademhosseini, A., Advances in engineering hydrogels. *Science*, 356, 6337, 5 May 2017.
16. Guimarães, C.F., Ahmed, R., Marques, A.P., Reis, R.L., Demirci, U., Engineering hydrogel-based biomedical photonics: Design, fabrication, and applications. 33, 23, 2006582, June 10, 2021.
17. Pereira, R.F. and Bártolo, P.J., 3D bioprinting of photocrosslinkable hydrogel constructs. 132, 48, Special Issue: Manufacturing of advanced biodegradable polymeric components, *J. Appl. Polym. Sci.*, December 20, 2015.
18. Asadi, N., Pazoki-Toroudi, H., Del Bakhshayesh, A.R., Akbarzadeh, A., Davaran, S., Annabi, N., Multifunctional hydrogels for wound healing: Special focus on biomacromolecular based hydrogels. *Int. J. Biol. Macromol.*, 170, 728–750, 2021.
19. Mao, C., Xiang, Y., Liu, X., Cui, Z., Yang, X., Yeung, K.W.K., Pan, H., Wang, X., Chu, P.K., Wu, S., Photo-inspired antibacterial activity and wound healing acceleration by hydrogel embedded with Ag/Ag@AgCl/ZnO nanostructures. *ACS Nano*, 11, 9010–9021, 2017.
20. Robson, M.C., Wound infection, a failure of wound healing caused by an imbalance of bacteria, wound healing. *Surg. Clin. North Am.*, 77, 3, 637–650, 1997.
21. Jones, S.G., Edwards, R., Thomas, D.W., Inflammation and wound healing: The role of bacteria in immune-regulation of wound healing. *Int. J. Low. Extrem. Wound*, 3, 4, 201–208, 2004.
22. Fan, Z., Liu, B., Wang, J., Zhang, S., Lin, Q., Gong, P., Ma, L., Yang, S., A novel wound dressing based on Ag/graphene polymer hydrogel: Effectively kill bacteria and accelerate wound healing. *Adv. Funct. Mater. J.*, 24, 25, 3933–3943, 2014.
23. Norahan, M. H., Pedroza-Gonzalez, S. C., Sanchez-Salazar, M.G., és Alvarez, M. M.,Trujillo de Santiago, G., Structural and biological engineering of 3D hydrogels for wound healing, *Bioact. Mater.*, 24, 2023, 197–235, 2023.

24. Krizek, T.K., Robson, M.C., Kho, E., Bacterial growth and skin graft survival. *Surg. Forum.*, 18, 518–519, 1967.
25. Lindsay, S., Oates, A., Bourdillon, K., The detrimental impact of extracellular bacterial proteases on wound healing. *Int. Wound J.*, 14, 6, 1237–1247, 2017.
26. Stadelmann, W.K., Digenis, A.G., Tobin, G.R., Impediments to wound healing. *Excerpta Med.*, 176, Suppl 2A, 39S- 47S, 1998.
27. Stephens, P., Wall, I.B., Wilson, M.J. *et al.*, Anaerobic cocci populating the deep tissues of chronic wounds impair cellular wound healing responses *in vitro*. *Br. J. Dermatol.*, 148, 456–66, 2003.
28. Wall, I.B., Davies, C.E., Hill, K.E. *et al.*, Potential role of anaerobic cocci in impaired human wound healing. *Wound Repair Regen.*, 10, 346–53, 2002.
29. Lindsay, S., Oates, A., Bourdillon, K., The detrimental impact of extracellular bacterial proteases on wound healing. *Int. Wound J.*, 14, 6, 1742–4801, 2017.
30. Shi, L., Ramsay, S., Ermis, R., Carson, D., pH in the bacteria contaminated wound and its impact on Clostridium histolyticum collagenase activity implications for the use of collagenase wound debridement agents. *J. Wound Ostomy Continence Nurs.*, 38, 514–21, 2011.
31. Percival, S.L., McCarty, S., Hunt, J.A., Woods, E.J., The effects of pH on wound healing, biofilms, and antimicrobial efficacy. *Wound Repair Regen.*, 22, 174–186, 2014.
32. Liu, W.-S., Liu, Y., Gao, J., Zheng, H., Lu, Z.-M., Li, M., Biomembrane-based nanostructure- and microstructure-loaded hydrogels for promoting chronic wound healing. *Int. J. Nanomed.*, 18, 385–411, 2023.
33. Kapusta, O., Jarosz, A., Stadnik, K., Giannakoudakis, D.A., Barczyński, B., Barczak, M., Antimicrobial natural hydrogels in biomedicine: Properties, applications, and challenges—A concise review. *Int. J. Mol. Sci.*, 24, 2191, 2023.
34. Fan, F., Saha, S., Hanjaya-Putra, D., Biomimetic hydrogels to promote wound healing. *Front. Bioeng. Biotechnol.*, 3, 718377, 2021.
35. Committee on Control of Surgical Infections of the Committee on Pre and Postoperative Care of the American College of Surgeons, *Manual on control of infection in surgical patients*, JB Lippincott, Philadelphia, 1976.
36. Sarkar, K., Dutta, K., Chatterjee, A., Sarkar, J., Das, D., Prasad, A., Chattopadhyay, D., Acharya, K., Das, M., Verma, S.K., De, S., Nanotherapeutic potential of antibacterial folic acid-functionalized nanoceria for wound-healing applications. *Nanomedicine*, 18, 2, 109–123, 2023, https://doi.org/10.2217/nnm-2022-0233.
37. Hoque, J., Prakash, R.G., Paramanandham, K., Shome, B.R., Haldar, J., Biocompatible injectable hydrogel with potent wound healing and antibacterial properties. *Mol. Pharm.*, 14, 1218–1230, 2017.
38. Pan, H., Fan, D., Duan, Z., Zhu, C., Fu, R., Li, X., Non-stick hemostasis hydrogels as dressings with bacterial barrier activity for cutaneous wound healing. *Mater. Sci. Eng. C*, 105, 110118, 2019.

39. Wang, R., Li, J.Z., Chen, W., Xu, T.T., Yun, S.F., Xu, Z., Xu, Z.Q., Sato, T., Chi, B., Xu, H.A., Biomimetic mussel-inspired ε-poly-l-lysine hydrogel with robust tissue-anchor and anti-capacity. *Adv. Funct. Mater.*, 27, 1604894, 2017.
40. Konieczynska, M.D., Villa-Camacho, J.C., Ghobril, C., PerezViloria, M., Tevis, K.M., Blessing, W.A., Nazarian, A., Rodriguez, E.K., Grinstaff, M.W., On-demand dissolution of a dendritic hydrogel based dressing for second-degree burn wounds through thiolthioester exchange reaction. *Angew. Chem., Int. Ed.*, 55, 9984–9987, 2016.
41. Wang, R., Li, J.Z., Chen, W., Xu, T.T., Yun, S.F., Xu, Z., Xu, Z.Q., Sato, T., Chi, B., Xu, H.A., Biomimetic mussel-inspired ε-poly-l-lysine hydrogel with robust tissue-anchor and anti-infection capacity. *Adv. Funct. Mater.*, 27, 1604894, 2017.
42. Deng, J., Tang, Y.Y., Zhang, Q., Wang, C., Liao, M., Ji, P., Song, J.L., Luo, G.X., Chen, L., Ran, X.H., Wei, Z.M., Zheng, L.W., Dang, R.Y., Liu, X., Zhang, H.M., Zhang, Y.S., Zhang, X.M., Tan, H.A., Bioinspired medical adhesive derived from skin secretion of andrias davidianus for wound healing. *Adv. Funct. Mater.*, 29, 1809110, 2019.
43. Cui, C.Y., Wu, T.L., Gao, F., Fan, C.C., Xu, Z.Y., Wang, H.B., Liu, B., Liu, W.G., An autolytic high strength instant adhesive hydrogel for emergency self-rescue. *Adv. Funct. Mater.*, 28, 1804925, 2018.
44. Wang, C., Niu, H., Ma, X., Hong, H., Yuan, Y., Liu, C., Bioinspired, injectable, quaternized hydroxyethyl cellulose composite hydrogel coordinated by mesocellular silica foam for rapid, noncompressible hemostasis and wound healing. *ACS Appl. Mater. Interfaces*, 11, 34595–34608, 2019.
45. Venault, A., Bai, Y.-W., Dizon, G.V., Chou, H.-Y.E., Chiang, H.-C., Lo, C.-T., Zheng, J., Aimar, P., Chang, Y., Healing kinetics of diabetic wounds controlled with charge-biased hydrogel dressings. *J. Mater. Chem. B*, 7, 7184–7194, 2019.
46. Zhao, X., Guo, B.L., Wu, H., Liang, Y.P., Ma, P.X., Injectable antibacterial conductive nanocomposite cryogels with rapid shape recovery for noncompressible hemorrhage and wound healing. *Nat. Commun.*, 9, 2784, 2018.
47. Li, M., Zhang, Z., Liang, Y., He, J., Guo, B., Multifunctional tissue-adhesive cryogel wound dressing for rapid nonpressing surface hemorrhage and wound repair. *ACS Appl. Mater. Interfaces*, 12, 35856–35872, 2020.
48. Huang, Y., Zhao, X., Zhang, Z., Liang, Y., Yin, Z., Chen, B., Bai, L., Han, Y., Guo, B., Degradable gelatin-based ipn cryogel hemostat for rapidly stopping deep noncompressible hemorrhage and simultaneously improving wound healing. *Chem. Mater.*, 32, 6595–6610, 2020.
49. Jaiswal, M., Gupta, A., Agrawal, A.K., Jassal, M., Dinda, A.K., Koul, V., Bi-layer composite dressing of gelatin nanofibrous mat and poly vinyl alcohol hydrogel for drug delivery and wound healing application: *In vitro* and *in vivo* studies. *J. Biomed. Nanotechnol.*, 9, 1495–1508, 2013.
50. Wang, P., Huang, S., Hu, Z., Yang, W., Lan, Y., Zhu, J., Hancharou, A., Guo, R., Tang, B., *In situ* formed anti-inflammatory hydrogel loading plasmid DNA

encoding VEGF for burn wound healing. *Acta Biomater.*, 100, 191–201, 2019.

51. Priprem, A., Damrongrungruang, T., Limsitthichaikoon, S., Khampaenjiraroch, B., Nukulkit, C., Thapphasaraphong, S., Limphirat, W.J.A.P., Topical niosome gel containing an anthocyanin complex: A potential oral wound healing in rats. *AAPS PharmSciTech*, 19, 1681–1692, 2018.
52. Soares, R.D.F., Campos, M.G.N., Ribeiro, G.P., Salles, B.C.C., Cardoso, N.S., Ribeiro, J.R., Souza, R.M., Leme, K.C., Soares, C.B., de Oliveira, C.M., Elston, L.B., Fonseca, C.C.P., Ferreira, E.B., Rodrigues, M.R., Duarte, S.M.S., Paula, F.B.A., Development of a chitosan hydrogel containing flavonoids extracted from Passiflora edulis leaves and the evaluation of its antioxidant and wound healing properties for the treatment of skin lesions in diabetic mice. *J. Biomed. Mater. Res., Part A*, 108, 654–662, 2020.
53. Pereira, R., Carvalho, A., Vaz, D.C., Gil, M.H., Mendes, A., Bartolo, P., Development of novel alginate based hydrogel films for wound healing applications. *Int. J. Biol. Macromol.*, 52, 221–230, 2013.
54. Oryan, A., Alemzadeh, E., Mohammadi, A.A., Moshiri, A., Healing potential of injectable aloe vera hydrogel loaded by adipose derived stem cell in skin tissue-engineering in a rat burn wound model. *Cell Tissue Res.*, 377, 215–227, 2019.
55. Huang, Y., Shi, F., Wang, L., Yang, Y., Khan, B.M., Cheong, K.L., Liu, Y., Preparation and evaluation of Bletilla striata polysaccharide/carboxymethyl chitosan/Carbomer 940 hydrogel for wound healing. *Int. J. Biol. Macromol.*, 132, 729–737, 2019.
56. Chen, X., Peng, L.-H., Shan, Y.-H., Li, N., Wei, W., Yu, L., Li, Q.-M., Liang, W.-Q., Gao, J.-Q., Astragaloside IV-loaded nanoparticle enriched hydrogel induces wound healing and anti-scar activity through topical delivery. *Int. J. Pharm.*, 447, 171–181, 2013.
57. Fang, Y., Li, G., Huang, C., Huang, K., Zhao, Y., Nie, T., Wu, J., Tomato based gelatin methacryloyl hydrogel as an effective natural and low-cost scaffold for accelerative wound healing. *Int. J. Biol. Macromol.*, 229, 123–135, 28 February 2023.
58. Chen, J., Zhao, L., Ling, J., Yang, L.-Y., Ouyang, X-k., A quaternized chitosan and carboxylated cellulose nanofiber-based sponge with a microchannel structure for rapid hemostasis and wound healing. *Int. J. Biol. Macromolecules*, 233, 123631, 1 April 2023.
59. D'souza, O.J., Gasti, T., Hiremani, V.D., Pinto, J.P., Contractor, S.S., Shettar, A.K., Olivia, D., Arakera, S., Masti, S.P., Chougale, R.B., Basella alba stem extract integrated poly (vinyl alcohol)/chitosan composite films: A promising bio-material for wound healing. *Int. J. Biol. Macromol.*, 225, 673–686, 15 January 2023.
60. Liu, B., Wang, Y., Miao, Y., Zhang, X., Fan, Z., Singh, G., Zhang, X., Xu, K., Li, B., Hu, Z., Xing, M., Hydrogen bonds autonomously powered gelatin methacrylate hydrogels with super-elasticity, selfheal and underwater

self-adhesion for sutureless skin and stomach surgery and E-skin. *Biomaterials*, 171, 83–96, 2018.
61. Khamrai, M., Banerjee, S.L., Paul, S., Samanta, S., Kundu, P.P., Curcumin entrapped gelatin/ionically modified bacterial cellulose based self-healable hydrogel film: An eco-friendly sustainable synthesis method of wound healing patch. *Int. J. Biol. Macromol.* 122, 940–953, 2019.
62. Deng, Z.X., Guo, Y., Zhao, X., Ma, P.X., Guo, B.L., Multifunctional stimuli-responsive hydrogels with self-healing, high conductivity, and rapid recovery through host-guest interactions. *Chem. Mater.*, 30, 1729–1742, 2018.
63. Zhang, B., He, J., Shi, M., Liang, Y., Guo, B., Injectable self healing supramolecular hydrogels with conductivity and photothermal antibacterial activity to enhance complete skin regeneration. *Chem. Eng. J.*, 400, 125994, 2020.
64. Yu, R., Yang, Y., He, J., Li, M., Guo, B., Novel supramolecular self-healing silk fibroin-based hydrogel via host–Guest interaction as wound dressing to enhance wound healing. *Chem. Eng. J.*, 417, 128278, 2021.
65. Huang, W., Wang, Y., Huang, Z., Wang, X., Chen, L., Zhang, Y., Zhang, L., On-demand dissolvable self-healing hydrogel based on carboxymethyl chitosan and cellulose nanocrystal for deep partial thickness burn wound healing. *ACS Appl. Mater. Interfaces*, 10, 41076–41088, 2018.
66. Li, Z., Zhou, F., Li, Z., Lin, S., Chen, L., Liu, L., Chen, Y., Hydrogel cross-linked with dynamic covalent bonding and micellization for promoting burn wound healing. *ACS Appl. Mater. Interfaces*, 10, 25194–25202, 2018.
67. Liang, Y.P., Zhao, X., Hu, T.L., Chen, B.J., Yin, Z.H., Ma, P.X., Guo, B.L., Adhesive hemostatic conducting injectable composite hydrogels with sustained drug release and photothermal antibacterial activity to promote full-thickness skin regeneration during wound healing. *Small*, 15, 1900046, 2019.
68. Zhao, M., Song, B., Pu, J., Wada, T., Reid, B., Tai, G., Wang, F., Guo, A., Walczysko, P., Gu, Y., Sasaki, T., Suzuki, A., Forrester, J.V., Bourne, H.R., Devreotes, P.N., McCaig, C.D., Penninger, J.M., Electrical signals control wound healing through phosphatidylinositol-3-oh kinase-γ and PTEN. *Nature*, 442, 457–460, 2006.
69. Korupalli, C., Li, H., Nguyen, N., Mi, F.L., Chang, Y., Lin, Y.J., Sung, H.W., Conductive materials for healing wounds: Their incorporation in wound dressings, characterization, and perspectives. *Adv. Healthc. Mater.*, 10, 2001384, 2021.
70. Liang, Y., Zhao, X., Hu, T., Han, Y., Guo, B., Mussel-inspired, antibacterial, conductive, antioxidant, injectable composite hydrogel wound dressing to promote the regeneration of infected skin. *J. Colloid Interface Sci.*, 556, 514–528, 2019.
71. Qu, J., Zhao, X., Liang, Y.P., Xu, Y.M., Ma, P.X., Guo, B.L., Degradable conductive injectable hydrogels as novel antibacterial, antioxidant wound dressings for wound healing. *Chem. Eng. J.*, 362, 548–560, 2019.
72. Zhao, L., Feng, Z., Lyu, Y., Yang, J., Lin, L., Bai, H., Li, Y., Feng, Y., Chen, Y., Electroactive injectable hydrogel based on oxidized sodium alginate and

carboxymethyl chitosan for wound healing. *Int. J. Biol. Macromol.*, 230, 123231, 1 March 2023.
73. Wang, X., Zhang, W., Zhou, Q., Ran, F., Integrating supercapacitor with sodium hyaluronate based hydrogel as a novel all-in-one wound dressing: Self-powered electronic stimulation. *Chem. Eng. J.*, 452, Part 3, 139491, 15 January 2023.
74. Wang, X., Zhang, W., Zhou, Q., Ran, F., Integrating supercapacitor with sodium hyaluronate based hydrogel as a novel all-in-one wound dressing: Self-powered electronic stimulation. *Chem. Eng. J.*, 452, Part 3, 139491, 15 January 2023.
75. Mirani, B., Pagan, E., Currie, B., Siddiqui, M.A., Hosseinzadeh, R., Mostafalu, P., Zhang, Y.S., Ghahary, A., Akbari, M., An advanced multifunctional hydrogel-based dressing for wound monitoring and drug delivery. *Adv. Healthc. Mater.*, 6, 1700718, 2017.
76. Zhu, Y., Zhang, J., Song, J., Yang, J., Du, Z., Zhao, W., Guo, H., Wen, C., Li, Q., Sui, X., Zhang, L.A., Multifunctional pro-healing zwitterionic hydrogel for simultaneous optical monitoring of pH and glucose in diabetic wound treatment. *Adv. Funct. Mater.*, 30, 1905493, 2020.
77. Zhao, L., Niu, L., Liang, H., Tan, H., Liu, C., Zhu, F., pH and glucose dual-responsive injectable hydrogels with insulin and fibroblasts as bioactive dressings for diabetic wound healing. *ACS Appl. Mater. Interfaces*, 9, 37563–37574, 2017.
78. Zhu, Y., Hoshi, R., Chen, S., Yi, J., Duan, C., Galiano, R.D., Zhang, H.F., Ameer, G.A., Sustained release of stromal cell derived factor-1 from an antioxidant thermoresponsive hydrogel enhances dermal wound healing in diabetes. *J. Control. Release*, 238, 114–122, 2016.
79. Zhao, L., Niu, L., Liang, H., Tan, H., Liu, C., Zhu, F., pH and glucose dual-responsive injectable hydrogels with insulin and fibroblasts as bioactive dressings for diabetic wound healing. *ACS Appl. Mater. Interfaces*, 9, 37563–37574, 2017.
80. Wang, T., Zheng, Y., Shi, Y., Zhao, L., pH-responsive calcium alginate hydrogel laden with protamine nanoparticles and hyaluronan oligosaccharide promotes diabetic wound healing by enhancing angiogenesis and antibacterial activity. *Drug Deliv. Transl. Res.*, 9, 227–239, 2019.
81. Bhoi, S., Prasad, A., Kumar, A., Sarkar, R.B., Mahto, B., Meena, C.S., Pandey, C., Experimental study to evaluate the wear performance of UHMWPE and XLPE material for orthopedics application. *Bioengineering*, 9, 676, 2022, https://doi.org/10.3390/bioengineering9110676.
82. Bhoi, S., Kumar, A., Prasad, A., Meena, C.S., Sarkar, R.B., Mahto, B., Ghosh, A., Performance evaluation of different coating materials in delamination for micro-milling applications on high-speed steel substrate. *Micromachines*, 13, 1277, 2022, https://doi.org/10.3390/mi13081277.
83. Gupta, A. *et al.*, Multifunctional nanohydroxyapatite-promoted toughened high-molecular-weight stereocomplex Poly(lactic acid)-based

bionanocomposite for both 3D-printed orthopedic implants and high-temperature engineering applications. *ACS Omega*, 2, 7, 4039–52, 2017.
84. Prasad, A., Bioabsorbable polymeric materials for biofilms and other biomedical applications: Recent and future trends. *Mater. Today: Proc.*, 44, 2447–53, 2021a, https://doi.org/10.1016/j.matpr.2020.12.489.
85. Prasad, A., State of art review on bioabsorbable polymeric scaffolds for bone tissue engineering. *Mater. Today: Proc.*, 44, 1391–1400, 2021b, https://doi.org/10.1016/j.matpr.2020.11.622.
86. Prasad, A., Bhasney, S., Katiyar, V., Ravi Sankar, M., Biowastes processed hydroxyapatite filled Poly (lactic acid) bio-composite for open reduction internal fixation of small bones. *Mater. Today: Proc.*, 4, 9, 10153–57, 2017. http://linkinghub.elsevier.com/retrieve/pii/S2214785317311410.
87. Prasad, A., Devendar, B., Ravi Sankar, M., Robi, P.S., Micro-scratch based tribological characterization of Hydroxyapatite (HAp) fabricated through fish scales. *Mater. Today: Proc.*, 2, 4–5, 1216–24, 2015. http://linkinghub.elsevier.com/retrieve/pii/S2214785315002795.
88. Prasad, A., Ravi Sankar, M., Katiyar, V., State of art on solvent casting particulate leaching method for orthopedic scaffolds fabrication. *Mater. Today: Proc.*, 4, 2, 898–907, 2017. http://dx.doi.org/10.1016/j.matpr.2017.01.101.
89. Prasad, A., Biomaterial-based nanofibers scaffolds in tissue engineering application, in: *Functional Biomaterials*, pp. 245–264, Springer, Singapore, 2022.
90. Gangwar, A.K.S., Rao, P.S., Kumar, A., Bio-mechanical design and analysis of femur bone. *Mater. Today: Proc.*, 44, Part 1, 2179–2187, 2021. https://doi.org/10.1016/j.matpr.2020.12.282.
91. Gangwar, A.K.S., Rao, P.S., Kumar, A., Patil, P.P., Design and analysis of femur bone: BioMechanical aspects. *J. Crit. Rev.*, 6, 4, 133–139, 2019.
92. Kumar, A., Mamgain, D.P., Jaiswal, H., Patil, P., Modal analysis of hand arm vibration (Humerus Bone) for biodynamic response using varying boundary conditions based on FEA. *Adv. Intell. Syst. Comput.*, 308, 169–176, 2015, https://doi.org/10.1007/978-81-322-2012-1_18.
93. Kumar, A., Behmad, S.I., Patil, P., Vibration characterization and static analysis of cortical bone fracture based on finite element analysis. *Eng. Automation Problems*, 115–119, 2014. No 3-2014, UDC- 621.
94. Kumar, A., Jaiswal, H., Garg, T., Patil, P., Free vibration modes analysis of femur bone fracture using varying boundary conditions based on FEA. *Proc. Mater. Sci.*, 6, 1593–1599, 2014, https://doi.org/10.1016/j.mspro.2014.07.142.
95. Kumar, A., Gori, Y., Rana, S., Sharma, N.K., Yadav, B., FEA of humerus bone fracture and healing, in: *Advanced Materials for Biomechanical Applications*, 1st ed, CRC Press, 2022, https://doi.org/10.1201/9781003286806.
96. Kumar, A., Datta, A., Kumar, A., Recent advancements and future trends in next-generation materials for biomedical applications, in: *Advanced Materials for Biomedical Applications*, vol. Chapter 1, A. Kumar, Y. Gori,

A. Kumar, C.S. Meena, N. Dutt (Eds.), pp. 1–19, CRC Press, Boca Raton, FL, USA, 2022, https://doi.org/10.1201/9781003344810-1.

97. Prasad, A., Chakraborty, G., Kumar, A., Bio-based environmentally benign polymeric resorbable materials for orthopedic fixation applications, in: *Advanced Materials for Biomedical Applications*, vol. Chapter 15, A. Kumar, Y. Gori, A. Kumar, C.S. Meena, N. Dutt (Eds.), pp. 251–266, CRC Press, Boca Raton, FL, USA, 2022, https://doi.org/10.1201/9781003344810-15.
98. Datta, A., Kumar, A., Kumar, A., Kumar, A., Singh, V.P., Advanced materials in biological implants and surgical tools, in: *Advanced Materials for Biomedical Applications*, vol. Chapter 2, A. Kumar, Y. Gori, A. Kumar, C.S. Meena, N. Dutt (Eds.), pp. 21–43, CRC Press, Boca Raton, FL, USA, 2022, https://doi.org/10.1201/9781003344810-2.
99. Kumar, A., Gangwar, A.K.S., Kumar, A., Meena, C.S., Singh, V.P., Dutt, N., Prasad, A., Gori, Y., Biomedical study of femur bone fracture and healing, in: *Advanced Materials for Biomedical Applications*, vol. Chapter 14, A. Kumar, Y. Gori, A. Kumar, C.S. Meena, N. Dutt (Eds.), pp. 235–250, CRC Press, Boca Raton, FL, USA, 2022, https://doi.org/ 10.1201/9781003344810-14.
100. Patil, P.P., Gori, Y., Kumar, A., Tyagi, M., Experimental analysis of tribological properties of polyisobutylene thickened oil in lubricated contacts. *Tribol. Int.*, 159, 106983, 2021, https://doi.org/10.1016/j.triboint.2021.106983.
101. Kumar, A., Rana, S., Gori, Y., Sharma, N.K., Thermal contact conductance prediction using FEM based computational techniques, in: *Advanced Computational Methods in Mechanical and Materials Engineering*, pp. 183–220, CRC Press, Boca Raton, FL, USA, 2021, https://doi.org/ 10.1201/9781003202233-13.
102. Kumar, R., Kumar, A., Kant, L., Prasad, A., Bhoi, S., Meena, C.S., Singh, V.P., Ghosh, A., Experimental and RSM-based process-parameters optimisation for turning operation of EN36B steel. *Materials*, 16, 339, 2023, https://doi.org/ 10.3390/ma16010339.
103. Prasad, A., Chakraborty, G., Kumar, A., Gajrani, K.K., Introduction to biodegradable polymers, in: *Biodegradable Composites for Packaging Applications*, vol. Chapter 1, A. Prasad, A. Kumar, K.K. Gajrani (Eds.), pp. 1–12, CRC Press, Boca Raton, FL, USA, 2022, https://doi.org/10.1201/9781003227908-1.
104. Chakraborty, G., Prasad, A., Kumar, A., Processing of biodegradable composites, in: *Biodegradable Composites for Packaging Applications*, vol. Chapter 3, A. Prasad, A. Kumar, K.K. Gajrani (Eds.), pp. 33–48, CRC Press, Boca Raton, FL, USA, 2022, https://doi.org/ 10.1201/9781003227908-3.
105. Kumar, A., Rana, S., Gori, Y., Sharma, N.K., Thermal Contact conductance prediction using FEM based computational techniques, in: *Advanced Computational Methods in Mechanical and Materials Engineering*, vol. Chapter 11, A. Kumar, Y. Gori, N. Dutt, Y.K. Singla, A. Maurya (Eds.), pp. 183–220, CRC Press, Boca Raton, FL, USA, 2022, doi: 10.1201/9781003202233-13.
106. Sharma, V.K., Kumar, V., Joshi, R.S., Kumar, A., Effect of REOs on tribological behavior of aluminum hybrid composites using ANN, in: *Additive*

Manufacturing in Industry 4.0: Methods, Techniques, Modeling and Nano Aspects, vol. Chapter 9, V.K. Sharma, A. Kumar, M. Gupta, V. Kumar, D.K. Sharma, S.K. Sharma (Eds.), pp. 153–168, CRC Press, Boca Raton, FL, USA, 2022.

107. Bansal, A., Bhardwaj, H.K., Sharma, V.K., Kumar, A., Static and dynamic behavior analysis of Al-6063 alloy using modified Hopkinson bar, in: *Additive Manufacturing in Industry 4.0: Methods, Techniques, Modeling and Nano Aspects*, vol. Chapter 6, V.K. Sharma, A. Kumar, M. Gupta, V. Kumar, D.K. Sharma, S.K. Sharma (Eds.), pp. 107–124, CRC Press, Boca Raton, FL, USA, 2022.
108. Singh, S., Kumar, A., Behura, S.K., Verma, K., Challenges and opportunities in nanomanufacturing, in: *Nanomanufacturing and Nanomaterials Design: Principles and Applications*, vol. Chapter 02, S. Singh, S.K. Behura, A. Kumar, K. Verma (Eds.), pp. 17–30, CRC Press, Boca Raton, FL, USA, 2022.
109. Srivastava, S., Verma, D., Thusoo, S., Kumar, A., Singh, V.P., Kumar, R., Nanomanufacturing for energy conversion and storage devices, in: *Nanomanufacturing and Nanomaterials Design: Principles and Applications*, vol. Chapter 10, S. Singh, S.K. Behura, A. Kumar, K. Verma (Eds.), pp. 165–173, CRC Press, Boca Raton, FL, USA, 2022.
110. Singh, V.P., Dwivedi, A., Karn, A., Kumar, A., Singh, S., Srivastava, S., Srivastava, K., Nanomanufacturing and design of high-performance piezoelectric nanogenerator for energy harvesting, in: *Nanomanufacturing and Nanomaterials Design: Principles and Applications*, vol. Chapter 15, S. Singh, S.K. Behura, A. Kumar, K. Verma (Eds.), pp. 241–242, CRC Press, Boca Raton, FL, USA, 2022.
111. Gori, Y., Verma, R.P., Kumar, A., Patil, P., FEA based fatigue crack growth analysis. *Mater. Today Proc.*, 46, 20, 10575–10581, 2021, https://doi.org/10.1016/j.matpr.2021.01.319.
112. Patil, P.P. and Kumar, A., Design and FEA simulation of omega type coriolis mass flow sensor. *Int. J. Control. Theory Appl.*, 9, 40, 383–387, 2017.
113. Patil, P., Sharma, S.C., Paliwal, V., Kumar, A., ANN modeling of Cu type omega vibration based mass flow sensor. *Proc. Technol.*, 14, 260–265, 2014, https://doi.org/10.1016/j.protcy.2014.08.034.
114. Patil, P., Sharma, S.C., Jaiswal, H., Kumar, A., Modeling influence of tube material on vibration based EMMFS using ANFIS. *Proc. Mater. Sci.*, 6, 1097–1103, 2014, https://doi.org/10.1016/j.mspro.2014.07.181.
115. Paul, A.K., Prasad, A., Kumar, A., Review on artificial neural network and its application in the field of engineering. *J. Mech. Eng.: PRAKASH*, 1, 1, 53–61, 2022, https://doi.org/10.56697/JMEP.2022.1107.
116. Rana, S., Kumar, A., Gori, Y., Patil, P., Design and analysis of thermal contact conductance. *J. Crit. Rev.*, 6, 5, 363–370, 2019.
117. Rana, S. and Kumar, A., FEA based design and thermal contact conductance analysis of steel and Al rough surfaces. *Int. J. Appl. Eng. Res.*, 13, 16, 12715–12724, 2019.

4

Natural Resource-Based Nanobiomaterials: A Sustainable Material for Biomedical Applications

Monika Singh[1]*, Murchana Changmai[2], Tabli Ghosh[3] and Anugraha Karwa[2]

[1]*Department of Forestry and Wood Technology, Linnaeus University, Växjö, Sweden*
[2]*Department of Chemical Engineering, Birla Institute of Technology and Science Pilani, Dubai Campus, Dubai, United Arab Emirates*
[3]*Department of Food Engineering and Technology, Tezpur University, Guwahati, India*

Abstract

With the onset of depletion of petroleum-based polymer, bio-based materials have recently been in high demand due to their extensive range of applications in the areas of biomedical engineering, pharmaceuticals, and others due to their inherent plastic characteristics such as biocompatibility, biodegradability, abundance, and most are non-toxic in nature. Applications of biopolymers are limited in biomedical applications because of its poor solubility in water, batch to batch variation and poor melt processabilities, etc. However, these limitations can be overcome taking into consideration the combine following approaches: (i) either modification of biopolymer or introducing other biopolymer/nano additive, and (ii) applying nanotechnology and nanoscience. The improvement in functionalities, physicochemical and biological properties with enhanced performances of nanostructured biomaterials, over their macrostructure counterparts are their unique features as nature in nanoscience and nanotechnology. Their benefits expand their potential biomedical applications in the diverse areas such as tissue engineering, drug delivery, wound healing, biosensors, and others such as single or in combination with other biopolymers or additives.

In this chapter, extraction of natural resource-based nanobiomaterials consisting of cellulose, chitosan, starch, and lignin and silk are elaborated. Several techniques have been used to produce nano particles from renewable resources.

**Corresponding author*: 21.monika.singh@gmail.com

Arbind Prasad, Ashwani Kumar and Manoj Gupta (eds.) Advanced Materials and Manufacturing Techniques for Biomedical Applications, (61–102)

Mostly, hydrophilic structures present a variety of surface functional groups to combine with bioactive compounds, target organs of the human body, and drugs. As we know that, properties of these biopolymers are unlike, the disparities would yield in many possibilities for their modifications. Also, variation in selection of source, processing methods, and chemical modification using harsh chemical to green effect the size, shape, morphology, and other properties of nanobiocomposites, and finally their end use in various applications. Nanobiocomposite mainly exists as a mixture of two or more components referred to as a bio-matrix wherein the more concentrated phase forms the continuous phase and the lesser concentration phase forms the reinforcement phase. In general, this combination refines the mechanical, thermal, and other properties of sustainable biomaterial. This chapter addressed the potential of nature resource biomaterial and their nanocomposites with emphasis on their applications in the biomedical and health sectors.

***Keywords*:** Biomaterial, biomedical, biodegradable, sustainable

4.1 Introduction

Recently, demand for biomass has risen to fabricate renewable and high-performance bio-based materials and for minimizing the excessive usage of non-renewable and conventional plastic. Biopolymer derived from renewable resources, due to their inexpensiveness, availability, biodegradability, biocompatibility, and non-toxicity, are very promising candidates for high value applications like biomedical and pharmaceutical areas [1]. In addition to these characteristics, advanced biopolymeric materials could be fabricated by altering the polymer components, which would generate more goal-oriented applications. These biopolymers comprise polysaccharide like cellulose, starch, chitosan, proteins like silk, and lignin etc. Also, structure of biopolymers resembles the structure of extracellular matrix (ECM) macromolecules which permit them to show biocompatibility and perform effectively within the host [1, 2]. Due to their excellent features, biopolymers have attracted much attention in diverse applications such as tissue engineering, wound healing, and drug delivery etc. [2]. Recently, the benefits of biopolymers are less emission of carbon di oxide, municipal solid waste, and dependency on conventional plastics.

Nanoscience innovations in almost every area of the science and nanotechnologies further make human life more comfortable. With the progression in the field of nanoscience and nanotechnology and the urge of new and advanced nanostructured material, the conventional research practices and many application domains are getting restructured [3]. After chemical modification, size of biopolymer is decreased from micrometer to nanometers, various factors such as surface energy (surface to volume ratio), biological effectiveness, and adsorption capacity change. Besides physical and chemical properties, optical, thermodynamic, and magnetic properties can also alter to tune with the target-oriented applications [4]. Nano biopolymers have a size of nanoscale ranging1 to 100 nm and can be modified further. In this chapter, the most common verities of such nanostructured biopolymers include cellulose, starch, chitosan, lignin, and silk are discussed.

Nanobiocomposites form a fascinating interdisciplinary area that merge nanotechnology, biology, and material science [5, 6] and provide plausible remedies to many diseases. Nanobiocomposites used in medical sectors have added most advanced features, by incorporating nanostructures into biopolymer, which offers enhanced properties such as hemocompatibility, nontoxicity, mechanical properties and their application is beneficial both economic and environmental point of view. Furthermore, biopolymer combining different kinds of nanostructures such as nanomaterials, nanofillers, metal/metal oxide, and mixed biopolymers are appropriate candidates for utilizing in biological applications [4]. Nowadays, scientist and researchers are more interested in green nanobiocomposite. Green nanobiocomposite are a class of composite in which their materials, especially materials used in both (matrix and reinforcing agent) extracted from the renewable resources and enhance the existing characteristics of them [7]. It is well known fact that the optimum composition of polymeric matrix and reinforcing agent may produce nanobiocomposite possessing the finest characteristics of each component. In this regard, special focused have been given to the application of nano biopolymer and their composite in tissue engineering, drug delivery, wound healing, pharmaceutics, biosensors, antibacterial materials, and others [1].

In tissue engineering, one of the feature requirements of nano biopolymer and their biocomposites are to be an ideal scaffold with low cytotoxicity, ability to transport nutrients within the cell matrix, and imitate collagen rich extracellular matrix (ECM). The cell cultural, proliferation, and the growth can be improved by imitating the biochemical and physical nature of tissues by reconstructing the natural ECM applying scaffolds showing similar surface topology [1, 8]. Finally, scaffold must be naturally decomposed into the biological fluids with time, hence minimizing surgical intervention for its removal. The search for the expansion of controlled drug delivery system for nursing human disease could emanate the side effects of conventional drug of usage, like poor effectiveness, low bioavailability, issue of target-oriented delivery, and toxicity [9]. The emerging of nanomedicine proposes advantages of time targeted specific drug delivery, controlled release, high payload, and precise treatment of disease. Therefore,

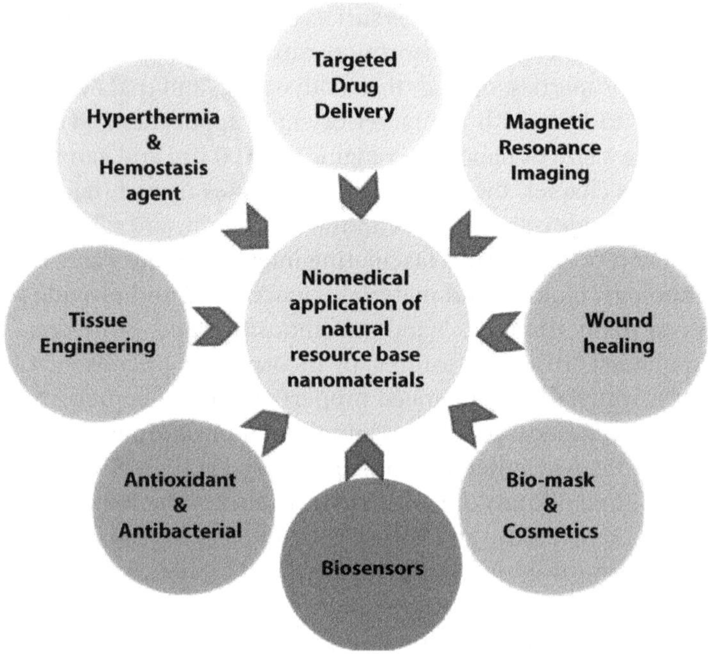

Figure 4.1 Biomedical application of natural resource-based bionanomaterials.

use of nanobiomaterials and their nanobiocomposites in drug delivery prime aims at controlled target delivery of drug, desired release profiles, maximizing transport of therapeutic agent to a specific site, and preventing side effects. For drug delivery, nano biopolymers and its composites are an excellent choice because it shows improvement in properties such as bioavailability, efficiency, ease of functionality, drug stability, improve targeting ability, and extend drug's influence in the target organ. These nano biopolymers can inhibit any enzymatic degradation before release a therapeutic agent in the targeted organ.

The advancement of nanotechnology has resulted in the development of nano biopolymer and their nanocomposite for applying in skin repairing. Once a trauma is suffered, the infected skin should be immediately covered with a bio-based dressing, which should be able to maintain a moist environment along with nutrient management, easily sterilized, permitting gaseous exchange, preventing swelling, and growth control of the fibroblast [10]. It has been found that nano biopolymers such as polysaccharide and proteins are the most widely used in wound dressing. Because the large surface energy of nano biopolymer, they show better activity. In addition, it presents enhancement in mechanical strength, biodegradability, biocompatibility, easy applicability, and cost effective. Beside these properties, soft and flexible texture, elasticity, chemical and physical stability, antibacterial properties, and long shelf life are also required for wound healing applications [11]. These materials can be utilized in various biomedical sectors where they are aimed to function up to a specific time duration and design to eliminate by the body (mostly by hydrolysis or mediated by metabolic method) after their use. In this chapter, special emphasis has been given to the extraction of nano biopolymer and the applications of nano-biopolymers and their nanobiocomposite in tissue engineering, drug delivery, wound healing, biosensor, antioxidant, and others biomedical applications, as shown in Figure 4.1.

4.2 Natural Resource-Based Biopolymer

The non-sustainable nature and adverse environmental effect of the non-renewable fossil resources have made the demand for bio-renewable natural resources to surge helping to move towards a sustainable planet. Biopolymers are the organic substances easily available in nature resource. According to their origin, there are three main categories of biopolymer such as (i) natural biopolymer (extracted by the cells of living organism), (ii) microbial biopolymer (produced by microorganism) and (iii) synthetic biopolymer. Based on monomeric units and their molecular structure, they can also classify (i) polysaccharides (linear or branched carbohydrates chains), (ii) proteins (composed of amino acid) and (iii) polynucleotide (long chain of nucleic acid). Natural biopolymer composed of monomeric units that are covalently linked to produce high molecular weight molecules. Among these, polysaccharides and proteins are the most used natural biopolymers. Polysaccharides, mainly achieved from plants, animal, microbes, and algae. They can be further divided based on charges: anionic (i.e., alginate and xanthum gum), cationic (i.e., chitosan) and non-ionic (i.e., cellulose and lignin etc.). On the other hand, proteins composed of one or many long chains of amino acids in their composition. Owing to the availability of different types of functional groups such as carboxyl, amino and hydroxyl groups, natural biopolymers can be easily modified to overcome their limitations, such as poor solubility in common solvent and batch to batch variation.

Natural biopolymers, exhibit good biocompatibility and bioactivity, and easily obtained from a wide range of renewable resources, such as wood, grasses, plant, tunicates, stalks, and crustacean etc. Among these plants and woods are recognized as richest bio-source consisting of cellulose, hemicellulose, and lignin [12, 13]. In this chapter, some used natural biopolymers are cellulose, lignin, starch, chitosan, and silk, which are discussed here.

4.2.1 Cellulose

Cellulose, a biopolymer, is the most prevalent renewable biomass on earth. In general, cellulose gives a structural role to plant and bacteria. The chemical structure of cellulose is consisting of a linear chain of homopolysaccharide glucose monomers linked by β (1→4) glycosidic bonds [14]. There is formation of intra or inter-molecular hydrogen bonds by the hydroxyl groups on each monomer units, with the adjacent glucose units present in the same or nearby chains [12]. Nanocellulose, a ubiquitous material extracted from native cellulose, has gained much attention owing to its significant physicochemical properties such as high surface area, high aspect ratio, good mechanical strength, easy to surface modification, and excellent biological characteristics like biodegradability, biocompatibility, and low cytotoxicity. Disintegration of the large cellulose units give rise to nanocellulose by rupturing the molecular hydrogen bonding (inter and intra molecular) present between the chain segment in cellulose. Nanocellulose can be isolated from broad range of sources, mainly including plants, animals, and bacteria. Among plants, wood, bamboo, banana leaf, rice husk, flax, hemp, sisal leaves, ramie, wheat straw, sugarcane bagasse, potato tuber, algae, and coconut husk have been widely investigated as a potential source for the extraction of nanocellulose [12]. Marine animals (turnicates) are also rich in cellulose. It has recently been produced from ginger, coffee seeds, lemon seeds, tea stalk, durian rind waste, and *Phragmites australis*, etc. Different approaches such as chemical, mechanical, and biological process have been used to synthesize nanocellulose. It can also be extracted from microorganism (bacteria, fungi, and algae). Cellulose nanocrystals (CNCs), Cellulose nanofibers (CNFs) and bacterial nanocellulose (BNC) are the three main groups into which nanocellulose can be categorized based on the source and structure [12]. In addition to help deliver drugs, proteins and plasmids, tissue engineering, wound healing, biosensors, and cosmetics are some other applications of nanocellulose. Recent studies prove that all variants of nanocellulose have wonderful opportunity not only in the drug delivery system, but also in various parts of pharmaceutical industry and in biomedicine in the coming future [15, 16].

4.2.2 Lignin

Lignin is an integral part of lignocellulosic biomass, and the second most abundant natural biomacromolecules next to cellulose in the biosphere. The major presence of lignin is in the lamella of plants, and it is responsible for incorporating physical structure to plants in general. It is characterized by a highly branched amorphous structure [17]. The molecular structure of lignin is extremely complex and irregular in nature, composed of randomly cross-linked phenylpropanoid groups (coniferyl, para-cumaryl, and sinapyl alcohol). On basis of these monolignol units, the building block guaiacyl, p-hydroxyphenyl, and syringyl are generated. These building blocks are further interconnected by a numerous bond, mainly ether bonds such as aryl- or phenyl ether biphenyl, pinoresinol or diphenyl ethane and carbon-carbon bond [1, 18]. The molecular structure of lignin is typically dependent on the botanical origin and their molecular structure diversity. It has large number of active sites such as phenolic and aliphatic hydroxyl groups, methoxy groups, phenyl groups, and carbonyl groups. Due to the availability of numerous functional groups (phenolic and aliphatic groups), lignin has the inherent capacity to easily undergo modifications as desired. These functionalize sites can further chemically modify by graft copolymerization, oxidation, sulfonation, or hydroxy methylation reactions etc. Apart from providing physical structure to plants, it can be used in various application because of its aromatic nature, different type of reactive functional group, high thermal stability, antimicrobial, antioxidation, antifungal properties, biodegradability, low cytotoxicity, and low cost. It is also more stable when exposed to chemical or biological attacks when compared to materials like cellulose and hemicellulose [19]. Other unique properties include viscoelasticity, fire-retardant, ability to absorb UV light and a promising material for renewable energy. Due to such varied properties of lignin, huge demands can be witnessed in the field of bio-based products and bio-based applications.

Lignin can be extracted using various methods such as Orgaosolv, Sulfite, Kraft, Soda, and enzymatic hydrolysis methods. Among these approach, Soda, Kraft, and Sulfite process are widely accepted at industrial scale to extract high yield of lignin [18]. Studies confirmed that the Organsolv might be favorable process to produce lignin of narrow molecular weight distribution and good solubility. Main outcome of lignin extraction is to break down the lignin into low molecular weight species. The key parameters that play a vital role during extraction process include temperature, time, pressure, pH, solvent capacity to solubilize lignin molecules etc. There also seems to be substantial problems associated with the downstream processing and valorization of lignin which limits its utilization in fabrication of highly value add products [20]. To overcome these challenges, remarkable efforts has been done in extraction, modification, and functionalization of lignin for the preparation of scaffold and drug encapsulations. However, it is relatively less explored for potential biomedical application owing to its complex nature, and molecular structural diversity.

4.2.3 Starch

Starch is a promising material when it comes to biodegradable polymers, with the most common sources of starch being potato, peas, tapioca, corn, maize, rice, and wheat to name a few [21]. Starch is mostly present as a granular structure in its native form which is mostly an extract obtained from plants. Starch is a polysaccharide consisting of repeating unit of amylose and amylopectin. Amylose has a linear structure formed by long chains of α (1→4) linked D-glucose unit. Amylopectin is highly branched and high molecular weight polymer, and has the same backbone structure of amylose, contains many α (1→6) linked glucose unit [22]. In native starch, there exists both crystalline (array of double-helical amylopectin side chain) and amorphous regions (branching regions of amylopectin) alternately arranged in the lamellae. Hence, native starch is semicrystalline in nature. Numerous processes have been dedicated specifically for the extraction of starch like acid, alkali or enzymatic process and refining of starch such as sieving, wet grinding, drying, etc. [23]. The inherent properties of starch such as excellent biocompatibility, biodegradability, renewability, easily available, non-toxicity, and low price make it suitable candidate as medical polymeric material. The native starch extracted from distinctive plants has some drawbacks such as poor thermal decomposition, thermal resistance, shear resistance, and high tendency towards retrogradation which limits its industrial application. To overcome these challenges, starch can be further modified to enhance its functional properties. Generally, starch modification is achieved through the following routes: (i) physical, (ii) chemical, and (iii) enzymatic methods [24]. Chemical modification usually includes acid hydrolysis, etherification, esterification, crosslinking/grafting, resulting in profoundly modified physicochemical properties of starch, whereas physical treatment involving heat-moisture, pregelatinization and dry heat treatment [25, 26]. Also, enzymatic hydrolysis can be used to modify starch. Starch properties such as gelatinization, swelling, solubility, enzymatic digestibility, rheological and mechanical play a major role while selecting starch source for biomedical application such as bone replacement and fixation. The starch characteristics can be significantly altered by chemical functionalization, which enhance its applicability in biomedical sectors [26]. The functionalized starch contains more reactive sites to transfer biologically active material, which can be metabolized in the human body.

4.2.4 Chitosan

Chitosan is composed of β-(1→4)-linked D-glucosamine and N-Acetyl-D-glucosamine [27, 28]. It belongs to the class of linear polysaccharide extracted from various organisms including crustaceans, mollusks, insects, algae, fungi, and bacteria [29]. The crustaceans-based sources of chitosan are crabs, shrimps, prawns, lobsters, crayfish, etc. [29, 30]. The chitosan from targeted natural sources can be extracted *via* chemical, biological, and other methods [29]. The chemical extraction of chitosan includes: (i) demineralization, (ii) deproteinization, (iii) discoloration, and (iv) deacetylation. Prior to these treatments, the targeted raw materials for chitosan sources are processed, washed, dried, and homogenized. The chemical methods to extract chitin are demineralization, deproteinization, and decolorization. The dried raw materials are demineralized using hydrochloric acids, nitric acid, sulfuric acid, and acetic acid [29]. In the demineralization process, the dependent factors are acid concentration, and treatment time, which depends on the chitin sources. After demineralization, the treated shell wastes are deproteinized using sodium hydroxide. Further, the waste after demineralization is bleached using acetone or ethanol or sodium hypochlorite solution or chloroform or others. The deacetylation of chitin provides chitosan [31] and is carried out using sodium hydroxide or potassium hydroxide. The biological method of chitosan production includes (i) the use of organic acid producing bacteria for demineralization, (ii) protease producing bacteria for deproteinization, and (iii) deacetylation of chitin is done using chitin deacetylase [27]. It is noteworthy to mention that chitosan has many active amino and hydroxyl groups that can undergo different chemical reactions. Chemical modification method such hydroxylation, acylation, alkylation, carboxylation, and esterification are well known approach for producing modified chitosan. In addition, pendant groups were introduced into the chitosan backbone after the chemical reaction, which further demolish its crystal structure and therefore improving the solubility of the modified chitosan [32]. In this way, chitosan is extracted from various sources, which is having high functionality and can be modified for versatile application such as agriculture, biomedical, packaging, cosmetics, and, industries etc. These chitosan derivatives with enhanced physicochemical (antimicrobial, antioxidant, biodegradable, and nontoxic) and biological characteristics (mucoadhesive, and hemocompatibility) are well suited for act as carriers in the pharmaceuticals, and biomedical engineering [33].

4.2.5 Silk

Silk is a type of natural protein fiber, produced from silkworms, lacewing, wasp, honeybee, spiders, etc. [34, 35]. The two main proteins of silk are sericin (20 to 30%) and fibroin (70-80%) [35, 36]. Among them, fibroin is approved by FDA for fabricating some biomedical devices [37]. The sericin protein in silk consists of amorphous amino acids and is hydrophilic in nature [38]. Silk fibroin is fabricated by removing sericin protein (silk gum) from raw silk by a process known as degumming [35, 39]. The process of degumming aims the hydrolytic or enzymatic method, where cleavage of the peptide bonds in sericin occurs [40]. The hydrolysis methods involve the use of neutral, acidic, or alkaline conditions. The extracted silk fibroin after removing sericin protein has the properties of excellent mechanical property, slow biodegradability, biocompatibility, etc. Among the various type of silks, silkworm silk (*Bombyx mori* silk) is the most used in biomedical field [41]. As we know that, silk is most widely used to make sutures because of its excellent mechanical strength and biodegradability. Because of its unique fundamental properties and its versatility, silk is an appropriate candidate for bone regeneration, skin-based materials, wound dressing, drug delivery, ocular, nerve regeneration, vascular tissue engineering, and cartilage applications [42].

4.3 Extraction of Nature Resource-Based Nanomaterials

In this section, extraction of nature resource-based nanomaterials such as cellulose, lignin, starch, chitosan, and silk has been discussed in brief.

4.3.1 Extraction of Cellulose-Based Nanostructures

Depending on the extraction method and morphological structures, they are categorized into three forms: CNC, CNF, and BNC, as mentioned in section 2.1. In case of CNC and CNF, first pretreatment is required to remove the lignin and hemicellulose before hydrolysis and second, to deconstruct the cellulosic hierarchical structure. CNC can be extracted by chemically induced destructing process, such as strong acid hydrolysis, hydrolysis assisted by ultrasound, enzymatic hydrolysis, and high-pressure homogenization, removing the amorphous segments of the fibrils while preserving its crystalline segments intact [12, 43]. Among these techniques, acid hydrolysis is the mainly used strategy to hydrolyze amorphous part which is performed under controlled conditions. CNC has rod-like or needle like crystalline structured; each rod can be considered as a rigid cellulosic crystal with no defects. They are obtained by subjecting different types of acids such as sulphuric acid, nitric acid, phosphoric acid, and hydrochloric acid etc. They are mostly produced using concentrated sulphuric acid at commercial scale. It is well known fact that the anionic sulfate moieties attached with surface of nanocrystals (-SO_3-), whose negative charge promote the dispersion of CNC in water [12]. Physiochemical and morphological characteristics can be influenced by some important parameters such as type of acid, acid concentration, acid to solid ratio, temperature, processing time during hydrolysis [44]. It is required to optimize these parameters to obtain desired set of properties and high yield of CNC.

However, CNF is typically extracted from cellulose strands, using several strategies such as physical, physiochemical (acid/alkali hydrolysis, TEMPO mediated oxidation), high shear mechanical medications (machining, crushing and homogenization), or by a combination of these methods [43]. Among them, mechanical disintegration process is widely accepted which includes high pressure homogenization before or after enzymatic hydrolysis or chemical treatment, which produce a gel like substance composed of an entangled network of CNF. Several mechanical shearing actions can successfully delaminate individual microfibrils from cellulosic fibers [45]. Temp-mediated oxidation, developed by Isogai, is the most popular method to extract CNF from wood pulp [46]. CNF possess long thread-like bundles of cellulose stabilized by hydrogen bond between the hydroxyl groups and the oxygen atoms of adjacent molecules. Their morphology and physicochemical properties based on source of native cellulose, processing conditions such as time, and temperature etc.

BNC is mainly extracted by rod shape bacteria (*Acetobacter xylinum*), in a pure form of cellulose. It can also be extracted using other microorganism including *Acanthamoeba*, *Acetobacter*, *Achromobacter*, Agrobacterium, *Alcaligenes*, *Pseudomonas*, *Rhizobium*, *Sarcina*, and algal species, *Chaetomorpha*, *Cladophora*, *Rhizoclonium*, and *Microdiction*. It is an effective approach to produce good quality nanocellulose with high amount of yield [12]. BNC extracted from Styela claves has been examined for absorbable hemostats, tissue scaffolds, stitching fiber, and hemodialysis. During extraction, microfibrils are produced and further aggregates ribbons. These ribbons (nanofiber) subsequently entangled to generate stable network structure. One of the most specific characteristics of BNC is its purity. They possess high aspect ratio, unique mechanical properties, biodegradability, and biocompatibility. In addition, BNC has weblike structure which resembles, to some extent, the characteristics of extracellular matrix. Therefore, BNC is suited for a wide range of biomedical application such as tissue engineering, wound healing, cosmetics, and drug delivery.

Nanocellulose derived from bacteria can easily be differentiated from the nanocellulose derived from plants, as their purity level is higher. In other words, BNC is extracted as pure cellulose. CNC has high crystallinity index, whereas CNF has both crystalline and amorphous domains. On the contrary, the molecular structures for both, plant-derived and bacteria-derived nanocellulose remain similar [45]. However, researchers are more focused on optimizing the existing techniques for extraction of different form of nanocellulose and developing alternative approach, i.e., green techniques, which can aid the production process and favor the nanocellulose with advanced properties.

4.3.2 Extraction of Lignin-Based Nanostructures

Nano scaled lignin thus can be utilized to develop new properties by the increased surface area to volume ratio where most of the new interactions would be governed by the surface properties [47]. Recently, lignin nanoparticle (LNP) can be synthesized from various resources using a combination of physical or chemical treatments. Different methods such as acid precipitation, polymerization, self-assembly, solvent exchange, microbial and enzyme mediated, antisolvent precipitation, alkaline precipitation, ultrasonication, interfacial crosslinking and emulsion, thermal stabilization and freeze drying, homogenization has been uncovered to obtain LNP [48]. The method of extraction and the source of lignin would in turn profoundly influence on the final properties of the extracted lignin, which influence the performance and application of the nanomaterial. Viability for surface modifications due to the presence of large number of functional groups (phenolic, thiol, and aliphatic hydroxyl) and nanoscale tunability of lignin makes it a potential candidate for preparation of high value-added applications of nanolignin. However, the utilization of lignin and LNP has been very limited till recent years due to problems of isolation and purification of lignin, and lack of well-defined molecular structure [49]. Generally, hazardous chemical solvents are utilized for the extraction of LNP, therefore it is beneficial to replace them with green solvents, i.e., water and room temperature ionic liquid. It is necessary to develop environment friendly LNP and further optimized properties for targeted applications.

4.3.3 Extraction of Starch-Based Nanostructures

The extraction of nanostarch mainly focusses on obtaining either nanostarch crystals or starch nano particles. Different methods such as acid hydrolysis, enzymatic hydrolysis, gamma radiation and physical methods have been developed to extract nanostarch [50, 51]. Among these, acid hydrolysis of native starch is the most widely used and adopted. The acid hydrolysis of native starch proposes two steps, (i) fast and (ii) slow hydrolysis. Amorphous segments of starch are attacked in fast hydrolysis, whereas the crystalline regions are eroded during the slow hydrolysis. The starch nano crystals are achieved by removing the amorphous regions from native starch by acid hydrolysis using strong acid like sulfuric acid. A method developed by Concha *et al.* [52] says that if native starch is hydrolyzed with sulfuric acid at 40 °C for long period of time (3-10 days) to achieve nanostarch. Longer hydrolysis resulted in smaller sized starch nanocrystals due to partial degradation of crystalline areas of the lamellae along with the complete degradation of the amorphous regions. Maize starch was found to be an appropriate candidate for producing starch nano crystals. Crystallinity is inversely proportional to the amylose content. Waxy maize (1% amylose) gives the best crystalline nanostarch followed by normal maize (15% amylose) or high amylose maize (70% amylose). Using modified methods, such as combination of hydrolysis with ultrasonication reduced the preparation time of starch nano crystals to a great extent [53].

Nanostarch has received high attention of late due to its enhanced surface properties when compared to the native starch granules. These enhanced surface properties were due to the reduction in size or the increase in surface energy of the nanostarch particles. Nanostarch may be available in the form of nano crystals or nano particles. Nano crystals of starch are obtained mostly by treating the starch granules with acid or amylase for an extended time. The nano particles of starch on the other hand are prepared by disruption of micron sized starch called as top-down process or by interaction between starch chains called as bottom-up process. Due to the difference in the preparation process, the starch nano crystals have a higher degree of crystallinity imparted to them in comparison to the starch nano particles [53].

4.3.4 Extraction of Chitosan-Based Nanostructures

As mentioned in earlier section, chitosan in its structure has chemical functional groups which are modified to fabricate nanochitosan. Chitosan based nanostructures are fabricated using various synthesis routes such as ionotropic gelation, microemulsion method, reverse micellar method, polyelectrolyte complexes, emulsion cross linking, etc. [29]. The nanostructures of chitosan are available as nanoparticles, nanofibers, nanocrystals, etc. Nanochitosan has the properties of non-toxicity, biocompatibility, biodegradability, antimicrobial, antioxidant property, etc., which make it a potential candidate in biomedical application [28]. It has unique properties due to presence of primary amines along the backbone in its chemical structure. It is mentioned that chitosan is decomposed in human body in form of nontoxic amino sugars that are simply absorbed by biological medium in easy way. In ionotropic gelation method of nanochitosan formation, chitosan, a polycation and sodium tripolyphosphate, a polyanion cross linked with one another [28, 54]. The method involves the electrostatic interactions between positive and negative charged ions. The factors affecting the nanochitosan fabrication are pH, mixing methods, proportions of polycation and polyanion, etc. On the other hand, the microemulsion method of chitosan nanoparticle formation includes the covalent cross linking of chitosan and functional cross-linking agent [55]. The process involves covalent cross-linking method, where, the cross-linking agents such as polyethylene glycol, glutaraldehyde, epichlorohydrin, dicarboxylic, acid, etc. On the otherhand, the polyelectrolyte method of chitosan nanoparticle formulation occurs through electrostatic interactions between anion and cation, where the mechanism involve is chitosan with positively charged groups form polyelectrolyte complexes with polyanion. Besides, the other methods such as complex coacervation method, acid hydrolysis, emulsion cross linking method, etc. are used to fabricate chitosan nanoparticles. The chitosan nanoparticle and their nanobiocomposites have wide usability in drug delivery, antibacterial agents, vaccine delivery, wound healing, tissue engineering etc.

4.3.5 Extraction of Silk Nanostructures

The silk-based nanostructures are generally fabricated using acid hydrolysis method [35]. Additionally, the silk nanofibers are also developed from natural silk fibers using sulfuric acid hydrolysis along with low intensity ultrasonication [56]. In this process, the parameters affecting the silk nanofibers production are amount of acid, treatment time, temperature, etc. Silk nano-disc can also be fabricated using acid hydrolysis of silk fibroin with crystallinity of 95% [35]. Additionally, the silk fibroin and its nanostructures are extensively used in biomedical application for available noteworthy properties such as mechanical property, thermal stability, surface property, etc. The various forms of silk fibroin are used in biomedical application such as micro/nano spheres, nanoparticles, films, fibers, sponge, hydrogel, etc. The desirable properties of silk fibroin are water solubility, biodegradability, biocompatibility, mechanical strength, etc., which make it a promising candidate in biomedical application.

4.4 Biomedical Applications of Nature Resource-Based Nanomaterials and Their Nanobiocomposites

4.4.1 Nanocellulose in Biomedical Application

Cellulose-based bionanocomposites have been developed for drug delivery, wound healing, tissue engineering, and other application etc. as presented in Table 4.1. The cellulose films showcase two major functions,

Table 4.1 Cellulose in biomedical applications.

SI. no.	Materials	Applications	Ref
1.	CNC functionalized PNIPAAm hydrogel	Deliver the drug for wound dressing	[68]
2.	CNC, Chitosan nanoparticles	Controlled release of drug	[69]
3.	CNC, Alginate	Oral drug delivery	[70]
4.	Alginate, Fe_3O_4 modified CNC (hydrogel)	Drug delivery	[71]
5.	Branched polyethyleneimine, Tempo oxidized CNF, Citric acid as a crosslinking agent (aerogels)	Controlled release of targeted drug	[72]
6.	CNF, Chitosan	Drug carrier for transdermal delivery of ketorolac tromethamine	[73]
7.	Curcumin, Cyclodextrin, CNC	Delivery of curcumin drug in human prostate cancer cell line	[74]
8.	PCL, CNC	Release of tetracycline drug	[75]
9.	CNC, chitosan	Delivery of curcumin drug for stomach	[76]
10.	CNF, Fe_3O_4-Ag_2O	Release of etoposide, and methotrexate drug for skin cancer treatment	[77]
11.	CNF, Chitosan	Bone tissue regeneration	[78]
12.	Nanocellulose, polyvinyl pyrrolidone, honey	Wound dressing	[79]
13.	OCNF, chitosan	Absorbable hemostasis	[80]

(Continued)

Table 4.1 Cellulose in biomedical applications. (*Continued*)

SI. no.	Materials	Applications	Ref
14.	Tempoxidized CNFPEG, ZnO	Hemostasis and antibacterial agent	[81]
15.	CNC, alginate, gelatin (scaffold)	Bone tissue engineering	[82]
16.	Nanoparticle of MgO incorporating in PCL, gelatine nanocellulose	Dental tissue engineering	[83]
17.	BNC, polypropylene	Repair hernia surgery	[84]
18.	BNC extracted in fermented tea medium with herbal plants like arjuna, neem, giloy, and curry leaves	Wound dressing, scaffolding, and herbal biomask	[62]
19.	Silk fibroin, TEMPO-CNC	Hemostatic applications	[85]
20.	BC, nano HAP, Fe3O4	Bone tissue engineering	[60]
21	PEG incorporated CNC, PLA	Bone tissue engineering	[86]

PCL: Polycaprolactone, PEG: polyethylene glycol, Fe3O4: iron oxide, MgO: Magnesium oxide, ZnO: Zinc oxide, OCNF: oxidized cellulose nanofiber, Ag2O: silver oxide, PLA: poly (lactic acid).

which are wound healing and absorption of exudates released from anti-inflammatory doses. The advancement in loading and unloading of medications which do not adhere to cellulose nanocrystals are done by surface alteration. Measurable drug release can be assisted by a polymer formed by methyl cellulose and polyvinyl alcohol via cross-linking with C_3H_5ClO compound [57]. Additionally, for measurable release of anticancer drugs, nanocellulose can be coated with cetrimonium bromide. Moreover, to address the arterial embolization, bio-decomposable cellulose microspheres were synthesized by researchers which carry doxorubicin (a chemotherapy drug). These show a medication discharge within eight hours post which it reached a level and concluded in 24h in the rabbit model used [57].

Nowadays, grafts made of cellulose are substituting the use of synthetic grafts. In a study, it was seen that till a month, there were no side effects seen in rats, sheep and pigs when their carotid arteries were replaced with cellulose-based vessels [57]. The mechanical strength for the cardiovascular tissues and the nanocellulose and polyvinyl alcohol-based composites are observed to be very similar. Formation of a layer of arterial intima and fibroblast growth were seen after 12 months in rats because of using nanocellulose grafts in blood vessel, but they are very beneficial in replacing the porcine arteries [58]. A successful trial of creating hollow tubes made up of a composite of chitosan polymer and nanocellulose having comparable mechanical strength with human arteries was carried out by researchers. It is concluded that grafts made up of this composite can be used in coronary artery [59].

Lot of studies are being done for producing cellulose containing nanofibrous meshes which find application for wound dressing. A nanocellulose-nano silver composite was produced by Diez *et al.* [170] where a nanocellulose film was dipped in solution of silver nanoclusters protected by poly (methacrylic acid) (PMMA) [60]. This surface functionalization of nanofibrillated cellulose by metal nanoclusters give a good antibacterial property. The bacterial nanocellulose (BNC)-grafted poly (acrylic acid) (AA)-graphene oxide (BNC/P(AA)/GO) composite hydrogel synthesized by Chen *et al.* (2019) using electron beam irradiation is appropriate as wound dressing material as it is biocompatible and show rise in cell attachment and proliferation [61]. Assessment of various parameters show improvement in swelling capacity, water retention ratio and bio-adhesion while the mechanical properties are maintained.

A lot of studies have been done to create films with totally natural antibacterial nanocellulose, by modifying bacterial nanocellulose (BNC) and producing in fermented tea medium containing bio-extracts which are obtained from herbal plants. The synthesized BNC were impregnated into the bio-extracts of Arjuna (*Terminalia arjuna*), Giloy (*Tinospora cordifolia*), Ashwagandha (*Withaniasomnifera*), Neem (*Azadirachta indica*) and Curry leaves (*Murrayakoenigii*) to acquire the antibacterial property. A wonderful antibacterial action was shown by these modified BNC, against Escherichia coli (*E. coli*) and *Aerococcus viridians*. With the inhibition zone of 27.08 mm for E. coli and 26.34 mm for *A. viridans*, the highest antibacterial activity is shown by *T. arjuna* modified BNC. Conversely, with inhibition zone of 14 mm and 14.2 mm for E. coli and *A. viridans* respectively, the lowest antibacterial activity was shown by *M. koenigii* modified BNC. The BNC film's high hydrophilic nature is reflected from their water retention, moisture content and porosity values and help prove their suitability for various biomedical applications such as antibacterial wound dressing and herbal biomask [62].

NCC-fibrin nanocomposites compose of homogeneously dispersed oxidized NCC (ONCC) present in fibrin matrix to develop blood vessel mimic. Carbonyl groups on NCC can be increased by periodate oxidation of NCC which assists its cross-linking with fibrin. ONCC brings in the strength and fibrin provide elasticity to the composite. Atomic Force Microscopy (AFM) can be used to determine the highest strength and elongation of nanocomposites. The results from influencing the degree of oxidation of NCC and NCC-to-fibrin ratio shows a variation in strength and elongation of nanocomposites. Thus, they can be customized to cater various mechanical properties of native blood vessels [63].

Upon surgery, the wounds are delivered with human adipose mesenchymal stem cells (hASC) which help in reducing the inflammatory reactions and other complications caused by postoperative infections and supports in wound healing. However, injecting the stem cells many times reduces the efficacy of the treatment because often the cell retention rate is low and there are high chances for deposition of cells in other organs. Thus, development of carries to which the cells have high affinity is essential to improve cell delivery. Moreover, two major concerns in using hASC during surgery are that it can bring immune reactions or can transmit pathogen-caused infections as the hASC are trusting on animal origin components. But the plant-extracted native cellulose nanofibers (NFC) allow the formation of glutaraldehyde cross-linked threads (NFC-X) having good mechanical strength when exposed to wet cell culture or even surgery situations. The hASC attached on the NFC-X retain the undifferentiated state and the bioactivity. Thus, it is predictable that these NFC-X provides a useful generic platform as a surgical bionanomaterial [64].

The biocompatibility of the polyvinyl-alcohol (PVA)/CNC composite hydrogel was studied by Lenses Tummala *et al.* [171] in the form of contact lens containing HCE-2 (human corneal epithelial) cells. Seeding HCE-2 cells with and without simulated tear fluid (STF) was used to evaluate cell response to hydrogel. To determine protein adsorption onto the hydrogel's surface is done using protein elution and quantification once the composites are nurtured with STF. Due to the interactions between negatively charged CNCs and positively charged lysozyme shows affinity to protein adsorption hence PVA-CNC hydrogels are apt for use as an ophthalmic biomaterial [65].

Moreover, to immobilize the biomolecules, nanocellulose are an apt material and very beneficial in biosensors. Single-step bio-template technique in water suspension provides a gold BNC composite having excellent biocompatibility with conductivity. Preserving the biological function and capturing Horseradish peroxidase (HRP) enzyme can be done easily by its thin nanofibers. Carboxylating layers and activating them by coupling with two proteins provide an activated nanocellulose layer to help in trapping antihuman immunoglobulin G antibody through physical adsorption. The selectivity and adhesion of these layers are enhanced by conjugating peptide with certain affinity to human antibody to the grafted polymer and by employing some amino acid sequences respectively [57].

Bacterial nanocellulose (BNC) has extensive application in cosmetic preparation. It has anti-wrinkle, stabilizing and thickening properties which is helpful to make sunscreen sprays. It can also be used to fabricate hairs, eyebrows, eye lashes and nails due to its good coating properties. Various studies suggest that BNC can be used as a brightening agent, base foundation liquid to remove makeup, face mask sheets and scrub [66]. Ophthalmic contact lenses can be manufactured using nanocellulose. The polyvinyl alcohol/CNF based microporous hydrogel containing more than 90% of water exhibited high transparency in nature, UV-blocking, and elasticity like collagen [67].

4.4.2 Nanolignin in Biomedical Application

Lignin-based bionanocomposites have been developed for wound healing, drug delivery, tissue engineering, and hyperthermia application etc. as shown in Table 4.2. Traditional materials like ceramics, metals and synthetic polymers used for manufacturing of medical implants have a disadvantage of inflammation and immunogenic. Biopolymers like lignin can be a suitable alternative to such traditional materials due to their biocompatibility and biodegradability. A large number of ear implants are manufactured using lignin via 3D printing. Lignin has also been successfully used in the replacement of traditional metallic stents and catheters. The

Table 4.2 Lignin in biomedical applications.

SI. no.	Materials	Application	Ref
1.	Polyacrylamide, LNP (hydrogel)	Tissue engineering or regeneration, artificial muscles	[96]
2.	Nanoparticles of CuO, CNF using alkali lignin	Wound dressing	[17]
3.	Carboxylated LNP with block copolymer of PEG, Polyhistidne	Targeted delivery of anticancer drug	[88]
4.	Alkali LNP, Resveratrol drug with Fe_3O_4 nanoparticle	Stable Drug carrier	[97]

(*Continued*)

Table 4.2 Lignin in biomedical applications. (*Continued*)

SI. no.	Materials	Application	Ref
5.	Fe or Fe_3O_4 incorporated LNP	Good magnetic properties, Drug release	[98]
6.	Lignosulfonate, chitosan nanoparticle	Excellent antimicrobial activity	[99]
7.	Lignin-mediated Ag nanoparticle in carrageenan	Wound dressing	[100]
8.	Fluorescent Lignin hollow nanoparticle	Release of anticancer drug	[101]
9.	Pure LNP, Fe-LNP, and Fe_3O_4-LNP	Drug delivery, Fe_3O_4-LNP for cancer therapy, magnetic resonance imaging, magnetic targeting	[102]
10.	PVA, Ag loaded lignin nanoparticle, CNC	Wound dressing	[103]
11.	Curcumin loaded LNP	Wound healing	[104]
12.	Lignin-based PU foam, Ag nanoparticle	Wound dressing	[105]
13.	Lignin-capped Ag nanoparticle, polyurethane	Wound healing	[106]
14.	Lignin, chitosan nanoparticle	Delivery of anticancer drug	[107]
15.	Lignin, thermoresponsive nanogel	Antioxidant agent for wound healing	[95]
16.	PLA, LNF	For protecting cells from oxidative stress	[108]
17.	Magnetic GO functionalized lignin	For hyperthermia in anticancer therapy	[109]

CuO: copper oxide, PEG: polyethylene glycol, Fe: iron, PVA: Polyvinyl alcohol, PU: polyurethane foam, LNF: lignin nanofiber, GO: graphene oxide.

highly advantageous antimicrobial properties of lignin also make it a suitable candidate for coatings on implants [87].

Figueiredo *et al.* successfully functionalized the carboxylated LNP with block copolymer of poly (ethylene glycol),poly (histidine), and cell-penetrating peptide. The developed nanoparticles showed good stability in physiological media, good size distribution, and low toxicity in the tested cell line [88]. It could be used for responsive and targeted delivery of anticancer drug. A composite of lignin/chitosan nanoparticle were prepared by blending the two in a microreactor causing an electrostatic assembly between carboxyl group of lignin and amino group of chitosan. The prepared composite has a size of 180 nm and could be efficiently used as a nano drug carrier. Anticancer drug utilizing Curcumin and Docetaxal assembled on the composite in a microreactor, could very easily release drugs in an acidic environment of the tumor of a cancerous cell [89]. Ligno-phenols characterized by high phenol content and high stability have been found to be advantageous in obesity control. The lignophenols were able to reduce the secretion of oleate induced apolipoprotein-β and reduce the cholesterol in HepG2 (Liver Hepatocellular Carcinoma) Cell Line [90].

Lignin of late has also gained interests among researchers to be used as a replacement or regeneration component for tissues or a particular organ. Lignins in the form of nano fibers mimic the native extracellular matrix. Electrospun nanofibers are also widely recommended in tissue engineering applications due to their fibrous and porous arrangements. Amount of lignin on the spun fibers helped in controlling physical properties such as modulus of elasticity, diameter etc. [91]. Nanocomposite of lignin-based materials are also developed for tissue engineering due to their unique properties, biodegradable, and biocompatible nature. A hydrogel of lignin nanoparticle with polyacrylamide has been developed for use in tissue engineering or regeneration and artificial muscles [92].

A composite wound dressing material was prepared using carrageenan matrix, lignin, silver nanoparticles and cations such as calcium chloride, magnesium chloride etc. by the one pot synthesis method. Crosslinking was enhanced due to the presence of cations resulting in good flexibility, swelling ratio and release rate of silver nanoparticles to a wound was observed when evaluated on rats. The wound was found to be substantially recovered when the epidermal and mature glands were analyzed in 2 weeks of the wound dressing application [93]. Another composite of LNP with Curcumin were also used for wound dressing application. The antimicrobial property of Curcumin is well known but the lack of permeability through skin made it difficult to harness the benefits of Curcumin. Hence, a composite was prepared which showed potent antimicrobial activity against gram positive bacteria *Staphylococcus aureus in vitro.* In a period of 12 days the wound achieved almost full contraction exhibiting excellent dermal wound closure [94].

As we know that the presence of aromatic rings and hydroxy groups helps in the incorporation of antioxidant properties to lignin. Nanolignins were prepared using supercritical anti-solvent precipitation method. The obtained lignin was used at a relatively higher concentration than bulk lignin which resulted in high antioxidant properties and high free radical scavenging activity. As a result, this lignin could be well used in pharmaceutical industries and food packaging industries where the antioxidant and free radical scavenging plays a vital role [92]. Lignin extracted from rice husk were incorporated into thermoresponsive nanogels made of polyethylene glycol, polypropylene glycol, and polydimethylsiloxaneto form a nanogel with excellent antioxidant properties reducing active oxygen levels in human hepatic cells, LO2. Rheological properties remained unaltered due to the addition of lignin; however, the critical micelle concentration of the polymer underwent a slight increase [95]. This study confirms that lignin loaded nanogel could act as antioxidant agent and help to heal the wound. A novel nanocomposite of lignin with magnetic graphene oxide exhibits its potential application for hyperthermia in cancer treatment.

4.4.3 Nanostarch in Biomedical Application

Starch-based bionanocomposites have been developed for wound healing, drug delivery, tissue engineering, bone regeneration, and hemostatic application etc. as displayed in Table 4.3. One of the most widely used applications of starch would be for starch-based bone cements and scaffolds. Starch being biodegradable was sufficient to provide the necessary structural support and thereafter would eventually disappear aiding in the growth of new bone during the healing period. Scaffolds are also widely designed using starch as a temporary base material which would eventually biodegrade and help in new tissue formation [110]. Shakir *et al.* (2015) reported the fabrication of nanobiocomposite from nanohydroxyapatite, chitosan and starch using co-precipitation method for making scaffolds [111]. They found that lower swelling rate of ternary nanobiocomposite exhibits a good mechanical strength supportive of bone tissue growth. This study also showed improvement in antimicrobial properties.

Table 4.3 Starch in biomedical applications.

SI. no.	Materials	Applications	Ref
1.	Chitosan, poly (vinyl pyrrolidone), nanostarch-SA	Antibacterial patch for wound healing	[117]
2.	PCL, nanostarch	PH responsive drug delivery	[118]
3.	Iron oxide mediated starch nanoparticle	Anticancer drug delivery	[119]
4.	Starch, MWCNT	Tissue scaffold or bone regeneration	[120]
5.	PVDF, Starch nano fiber	Tissue engineering	[121]
6.	Nano HAP, chitosan, starch	Bone tissue engineering	[122]
7.	Porous starch, CNF	Tissue engineering	[123]
8.	Starch, nano GO	Bone tissue engineering	[124]
9.	PVA, starch coated silver nanoparticles	Antibacterial scaffold, Wound dressing	[125]
10.	Ag embedded starch nanofiber, polyvinyl alcohol	Soft tissue engineering	[126]
11.	Carrageenan coated starch, CNF	Hemostatic application	[116]
12.	Oxidized starch, CuO	Drug delivery	[127]

MWCNT: multiwalled carbon nano tube, PVDF: Poly (vinylidene fluoride), HAP: hydroxyapatite, GO: graphene oxide, PVA: Polyvinyl alcohol, Ag: Silver, CuO: Copper Oxide, SA: stearic acid.

Bare starch has been suggested to be used as bone cement, but the mechanical strength available is somewhat limited. To overcome this shortfall a composite of starch-hydroxyapatite composite with property resemblance to mineral bone has been prepared. Due to properties such as biocompatibility and osteoconductivity, the hydroxyapatite (HA) promoted cell regeneration without intermediate connective tissues. The only disadvantage is that hydroxyapatite is quite brittle and hence it gets difficult to shape the composite when compared to the flexible bare starch. the addition of starch not only helps in providing flexibility to hydroxyapatite through interaction of hydroxy groups of starch and Ca^{2+} ions of HA but also imparts porosity mimicking the actual mineral bone [110].

Starch based polymers are also widely acceptable in the field of drug delivery systems. The advantage being excellent deliverability process without the requirement for removal of the system by a later surgery due to the inherent biodegradable property of starch. However, due to restraints in the field of mechanical properties, the starch has not been fully utilized in biomedical field. The way forward is the preparation of composite starch which can to a great extent overcome the shortfalls of native starch [111]. Starch functionalized graphene nanosheets are also another variety of starch composites which have been explored for drug delivery of hydroxy-camptothecin (HCPT). The anticancer drug (HPT) is loaded on the composite which exhibited a very high drug loading capacity. The composite starch/graphene nanosheets-HCPT were encapsulated in the cytoplasm by SW-620 cancer cells. The starch aided in the in vitro release of HCPT and had an added therapeutic efficacy [112]. Curcumin has been found to have very good cytotoxicity towards cancer cells, however, it is hydrophobic in nature and degrades rapidly. Thus, a composite of starch loaded with Curcumin were synthesized. It was suggested that curcumin encapsulation in a starch matrix helped substantially the cyto toxic effects on MG 63 cells and thus can be suggested as a potential drug delivery system for osteosarcoma [113]. The disadvantages associated with available wound dressings such as low water vapor transmission, inadequate adsorption, disposability and recycle restraints led researchers to prepare a nanocomposite starch-based dressing which could overcome the said disadvantages. Starch/polyvinyl alcohol/zinc oxide particles were prepared with an excellent antibacterial effect. The dressing also inhibited the growth and penetration of *Escherichia coli, Staphylococcus aureus* etc. with good physical, chemical and mechanical properties [114]. Similar dressings named as antimicrobial membranes were synthesized using starch, polyvinyl alcohol, and chitosan with embedded nano zinc oxide. Studies suggested the said membrane dressing to be more effective during the early stages of healing wherein the membrane also acts as a topical application too inhibit bacterial activity while healing [115]. Tavakoli *et al.* (2017) [116] successfully developed a nanobiocomposite of carrageenan coated starch with CNF for hemostatic application. Their experiment result shows superior mechanical properties, good blood clotting ability, and adjustable degradation rate.

4.4.4 Nanochitosan in Biomedical Application

The biomedical applications of nanochitosan are wound healing, drug delivery, anticancer application, tissue engineering, tissue regeneration, antimicrobial agent, and antitumor activity, etc. as represented in Table 4.4. Chitosan is considered as a unique polysaccharide for drug delivery as it having primary amino groups provide the cationic characters providing easy chemical modifications [128, 129]. The various form of chitosan is used in drug delivery application such as nanocomposites, films, nanofibers, hydrogel, microspheres, nanoparticles, films, etc. [130]. The nanocomposite based on chitosan and montmorillonite delivers mucoadhesive properties, good biocompatibility, which is potential in drug delivery system [131]. Additionally, chitosan nanoparticles are widely used in mucosal drug delivery, parenteral drug delivery, in non-viral gene delivery, vaccine delivery, etc. [132]. It can also be used in hemodialysis and renal failure. The nanocomposite film based on chitosan and organic rectorite can be used for drug-delivery system for drug-controlled release carrier [133]. The release kinetics of chitosan complexed materials are important parameter for cancer application. In this context, the proper release and drug delivery systems for targeted and non-targeted system are beneficial to use as medicine in cancer treatment. In this context, the nano-complex based on chitosan and polyoxometalate when used to treat cancer cells generates intracellular reactive oxygen species (ROS) *via* ROS-mediated stress causing death of cancer cells [134]. Moreover, onion extract encapsulated on nanochitosan is reported to be a protective agent against antitumor,

Table 4.4 Chitosan in biomedical application.

Sl. no.	Materials	Biomedical application	Ref
1.	Nanochitosan dots, Bio-cellulose (Hydrogel)	Wound healing treatment	[136]
2.	Nanochitosan, Fish Collagen, Henna extract	Accelerating wound healing Control of skin pathogens	[144]
3.	Curcumin, Chitosan nanoparticle, Collagen, Alginate, Scaffolds	Diabetic wound healing	[138]
4.	Chitosan nanoparticle, Vaccarin	Wound healing application Biomedically relevant system	[137]
5.	Nanochitosan, Onion extract	Anticancer agent	[135]

(*Continued*)

Table 4.4 Chitosan in biomedical application. (*Continued*)

Sl. no.	Materials	Biomedical application	Ref
6.	Chitosan, Organic rectorite, (Film)	Drug delivery	[133]
7.	Chitosan, Montmorillonite	Drug delivery	[131]
8.	Chitosan, Polyvinyl alcohol, Cloisite 30B, Glutaraldehyde as a crosslinking agent	Controlled release of anticancer drug curcumin	[145]
9.	Chitosan, Polyoxometalate, (Nanocomplex)	Anticancer application	[134]
10.	Nanochitosan, Silk fibroin, Hyaluronic acid, (3D porous scaffold)	Tissue engineering	[143]
11.	Nanochitosan, Polypyrrole, Alginate, (Scaffold)	Neural Tissue engineering	[142]
12.	Chitosan, Gelatin, Nano hydroxyapatite (Porous scaffold)	Tissue engineering	[141]
13.	Hydroxyapatite, Chitosan phosphate	Bone tissue engineering	[139]
14.	Chitosan, Pectin, CNC (Hydrogel)	Scaffold for cartilage tissue regeneration	[146]
15.	Chitosan, BNC (Hydrogel)	Antibacterial wound dressing	[147]
16.	Chitosan, Nano alginate (Membrane)	Anticancer drug	[148]
17.	Chitosan nanoparticle, BNC	Biomedical application	[149]
18.	Poly lactide co-glycolide, chitosan nanoparticle	Brain targeted intranasal delivery	[150]
19.	Self-assembled lecithin, chitosan nanoparticles	Oral insulin delivery	[151]
20.	Chitosan, CNF	Neural regeneration	[152]

and cancer [135]. The cytotoxicity of onion extract can be improved using nanochitosan encapsulation. However, chitosan-based hydrogel is used in wound healing and drug delivery system for unique properties such as moist environment, biocompatible, faster wound healing, protect from infection, etc. The mechanism for chitosan-based hydrogel provides scaffold for cell growth, protein for healing, forms barrier against infection, encourages natural blood clotting, etc., which is beneficial property to promote wound healing [136]. The nanochitosan dots and bacterial cellulose-based nanocomposites hydrogel can be used in wound healing treatment. As bacterial infections are a problem in wound healing, thus, the use of gamma irradiation produced nanochitosan dots encapsulated in bacterial cellulose can be a potential in would healing treatment providing bactericidal activity. Additionally, Hou *et al.* has developed vaccarin and chitosan nanoparticle-based hydrogel which also promotes wound healing application [137]. The nanohybrid scaffold developed by curcumin loaded chitosan nanoparticles impregnated into collagen-alginate scaffolds are also targeted for wound healing, where the combination of curcumin, chitosan, and collagen are better for wound healing capability [138]. Additionally, chitosan in combination with other materials are used in tissue engineering application such as (i) Nanocomposite of hydroxyapatite and chitosan phosphate in bone tissue engineering, [139] (ii) Chitosan, montmorillonite, hydroxyapatite in bone tissue engineering, [140] (iii) Chitosan, gelatin, nano based hydroxyapatite in tissue engineering, [141] (iv) Nanochitosan, polypyrrole, alginate in tissue engineering, [142] (v) Nanochitosan, silk fibroin, hyaluronic acid in tissue engineering etc. [143]. Based on this, chitosan and derivatives having numerous beneficial properties such as antioxidant property, antimicrobial property, antiviral property, anticancer property, has make it a promising candidate for biomedical application. Chitosan derivatives are also used in developing antiviral coating.

4.4.5 Nanosilk in Biomedical Application

It is well known that silk-based smart materials are used for versatile areas of biomedical such as drug delivery, tissue engineering, wound dressing, sutures, and cancer therapy, etc. as shown in Table 4.5. The electrospinning of polyblend of poly (methyl methacrylate) with silk fibroin, and polyethylenimine can be used in denture materials having antibacterial properties and improved mechanical properties [153]. Additionally, the micro and nanosphere of silk fibroin has the properties of biodegradability, biocompatibility, excellent mechanical property, tailored drug release activity [154, 155]. Electrospun functionalized silk nanofibers with tailored properties are a promising agent to be used as drug carriers [156]. The silk/polyvinyl alcohol blend films-based microsphere and nanosphere are used in drug delivery and other biomedical application [154]. Recently, the bionanocomposites scaffold based on silk nanodisc and poly (lactic acid) was fabricated using melt mixing via electrospinning method and it is used for

Table 4.5 Silk in biomedical application.

Sl. no.	Materials	Biomedical application	Ref
1.	Functionalized silk fibroin nanofibers	Drug carriers	[156]
2.	Silk, gold nanoparticles	Tissue engineering	[161]
3.	Silk, HAP nano particles (Scaffold)	Bone tissue engineering	[163]
4.	Magnetic nanoparticle, Silk, Chitosan	To help cells and tissue	[164]
5.	Silk nano disc, PLA based scaffolds	Cancer therapy	[157]
6.	Silk, Silver nanoparticles	Surgical sutures for wound healing	[165]
7.	Nanosilk fibroin, anti-inflammatory micro-RNA	Diabetic wound healing	[158]

cancer therapy [157]. Niemiec *et al.* [158] reported that developed nanosilk enhance the strength of diabetic skin and help to improve the wound healing effectively. Silk hydrogel and sponges containing nano titanium dioxide, graphene oxide has also been developed for bone tissue engineering [159, 160]. Silk/gold nanoparticles hydrogel have also been produced for photothermal treatments of infections [161]. Hilton *et al.* [162] confirmed that nanosilk enhanced the bio-mechanical characteristics of diabetic skin in human. In this way, silk in combination with other biomaterials are used for various biomedical application.

4.5 Other Applications

In the biomedical field, 3D printing is a modern technique which offers manufacturing of any customized product with complex 3D structures within a short time. This technique has been successfully used in tissue engineering, reconstructive surgery, and implants where it shows a high demand for complex 3D construct. In 3D bioprinting, main component is bio-ink, which is biocompatible with living cells, biodegradable, printable, and mechanical strong after printing. CNC has optimum rheological behavior, low density, high strength, and young modulus which make it an ideal choice for 3D bioprinting. Lan *et al.* in his study reported the use of CNF-alginate hydrogel as a bio-ink for printing human meniscus [166].

The enhanced biocompatibility and enabled cellular proliferation from BNC/PCL bio-nanocomposite were expected for the delivery of growth drugs to help in the application of skin tissue. A green ink printed bionanocomposite composed of TEMPO oxidized BNC, laponite nano clay, sodium alginate (SA) was developed for potential tissue engineering which represent mechanical strength as well as long-lasting performance in the delivery of protein [167]. 3D printed CNF reinforced xylan bionanocomposite has been developed as hydrogel and present promising use in tissue engineering [168]. Researchers showed that the 3D printed bio-nanocomposite hydrogel composed of CNF and gelatin methacrylamide (GMA) exhibit effectively printing of the human nose [169]. After crosslinking with GMA and generating the covalent bond using an APS/TEMED initiator, printed structure of the human nose developed a diffuse structure and resultant bionanocomposite displayed a highly elastic and mechanical stable network. NFC bio-link employed in the regeneration of auricular cartilage issue.

References

1. Biswas, M.C., Jony, B., Nandy, P.K., Chowdhury, R.A., Halder, S., Kumar, D., Ramakrishna, S., Hassan, M., Ahsan, M.A., Hoque, M.E., Recent advancement of biopolymers and their potential biomedical applications. *J. Polym. Environ.*, 30, 51, 2022.
2. Gobi, R., Ravichandiran, P., Babu, R.S., Yoo, D.J., Biopolymer and synthetic polymer-based nanocomposites in wound dressing applications: A review. *Polymers*, 13, 112, 2021.
3. Kanaoujiya, R., Saroj, S.K., Srivastave, S., Chaudhary, M.K., Renewable polysaccharide and biomedical application of nanomaterials. *J. Nanomater.*, 25, 1, 2022.
4. Chausali, N., Saxena, J., Prasad, R., Recent trends in nanotechnology application of biobased packaging. *J. Agric. Res.*, 7, 100257, 2022.
5. Porter, A.L. and Youtie, J., How interdisciplinary is nanotechnology? *J. Nanopart. Res.*, 11, 1023, 2009.
6. Rahim, M., Haris, M.R.H.M., Saqib, N.U., An overview of polymeric nanobiocomposites as targeted and controlled release devices. *Biophys. Rev.*, 12, 1223, 2020.
7. Adeosun, S.O., Lalwal, G., I, Balogun, S.A., Akpan, E., I, Review of green polymer nanocomposites. *J. Miner. Mater. Character. Eng.*, 11, 385, 2012.
8. Yanya, E.B., Amirul, A.A., Abdul, K.H.P.S., Olaiya, N.G., Iqbal, M.O., Jummaat, F., Atty, S.A.K., Adnan, A.S., Insights into the role of biopolymer aerogel scaffolds in tissue engineering and regenerative medicine. *Polymers*, 13, 1612, 2021.

9. Adepu, S. and Ramakrishna, S., Controlled drug delivery system: Current status and future directions. *Molecules*, 26, 5905, 2021.
10. Dhivya, S., Padma, V.V., Santhin, E., Wound dressing-a review. *Biomedicine*, 5, 22, 2015.
11. Pascual, A.M.D. and Vicente, A.L.D., Wound healing bionanocomposites based on castor oil polymeric films reinforced with chitosan modified ZnO nanoparticles. *Biomacromolecules*, 16, 2631, 2015.
12. Patil, T.V., Patel, D.K., Dutta, S.D., Ganguly, K., Santra, T.S., Lim, K.T., Nanocellulose a versatile platform: From the delivery of active molecules to tissue engineering applications. *Bioact. Mater.*, 9, 566, 2022.
13. Barhoum, A., Jeevanandam, J., Ratogi, A., Samyn, P., Boluk, Y., Dufresne, A., Danquah, M.K., Bechelany, M., Plant celluloses, hemicelluloses, lignins, and volatile oils for the synthesis of nanoparticles and nanostructured materials. *Nanoscale*, 12, 22845, 2020.
14. Shen, R., Xue, S., Xu, Y., Liu, Q., Feng, Z., Ren, H., Zhai, H., Kong, F., Research progress and development demand of nanocellulose reinforced polymer composites. *Polymers*, 12, 2113, 2020.
15. Pradeep, H.K., Patel, D.H., Onkarappa, H.S., Pratiksha, C.C., Prasanna, G.D., Role of nanocellulose in industrial and pharmaceutical sectors-A review. *Int. J. Biol. Macromol.*, 207, 1038, 2020.
16. Liang, L., Huang, C., Ragauskas, A.J., Nanocellulose-based materials for biomedical applications. *JSM Chem.*, 5, 1048, 2017.
17. Haider, Md. K., Ullah, A., Sarwar, M.N., Saito, Y., Sun, L., Park, S., Kim, I.S., Lignin mediated *in situ* synthesis of CuO nanoparticles on cellulose nanofibers: A potential wound dressing material. *Int. J. Biol. Macromol.*, 173, 315, 2017.
18. Tang, Q., Qian, Y., Yang, D., Qiu, X., Qin, Y., Zhou, M., Lignin based nanoparticles: A review on their preparation and applications. *Polymers*, 12, 2471, 2020.
19. Kumar, R., Butreddy, A., Kommineni, N., Reddy, P.G., Bunekar, N., Sarkar, C., Dutt, S., Mishra, V.K., Aadil, K.R., Mishra, Y.K., Oupicky, D., Kaushik, A., Lignin: Drug/gene delivery and tissue engineering applications. *Int. J. Nanomed.*, 16, 2419, 2021.
20. Zhang, Z., Terrasson, V., Guenin, E., Lignin nanoparticles and their nanocomposites. *Nanomaterials*, 11, 1336, 2021.
21. Waterschool, J., Gomand, S.V., Fierens, E., Delcour, J.A., Production, structure, physicochemical and functional properties of maize, cassava, wheat, potato, and rice starches. *Starch*, 67, 14, 2014.
22. Yu, X., Chen, L., Jin, Z., Jiao, A., Research progress of starch based biodegradable materials: A review. *J. Mater. Sci.*, 56, 11187, 2021.
23. Dufresne, A., Thomas, S., Pothan, L.A. (Eds.), *Biopolymer Nanocomposites: Processing, Properties, and Applications*, First Edition, John Wiley & Sons, Inc, Hoboken, NJ, USA, 2013.

24. Fan, Y. and Picchioni, F., Modification of starch: A review on the application of "green" solvents and controlled functionalization. *Carbohydr. Polym.*, 241, 116350, 2020.
25. Haq, F., Yu, H., Wang, L., Teng, L., Haroon, M., Khan, R.U., Mehmood, S., Amin, B.U., Ullah, R.S., Khan, A., Nazir, A., Advances in chemical modifications of starches and their applications. *Carbohydr. Res.*, 476, 12, 2019.
26. Garcia, M.A.V.T., Garcia, C.F., Faraco, A.A.G., Pharmaceutical and biomedical applications of native and modified starch: A review. *Starch*, 7, 72, 2020.
27. Ghosh, T. and Katiyar, V., Chitosan-based edible coating: A customise practice for food protection, in: *Advances in Sustainable Polymers*, p. 167, Springer, Singapore, 2019.
28. Ghosh, T. and Katiyar, V., Nanochitosan functionalized hydrophobic starch/guar gum biocomposite for edible coating application with improved optical, thermal, mechanical, and surface property. *Int. J. Biol. Macromol.*, 211, 116, 2022.
29. Ghosh, T., Mondal, K., Katiyar, V., Chitosan based nanostructured materials in edible food packaging, in: *Nanotechnology in Edible Food Packaging*, p. 101, Springer, Singapore, 2021.
30. Duan, C., Meng, X., Meng, J., Khan, M., I, Dai, L., Khan, A., Ni, Y., Chitosan as a preservative for fruits and vegetables: A review on chemistry and antimicrobial properties. *J. Bioresour. Bioprod.*, 4, 11, 2019.
31. Monika, Pal, A.K., Bhasney, S.M., Bhagabati, P., Katiyar, V., Effect of dicumyl peroxide on a poly (lactic acid) (PLA)/poly (butylene succinate) (PBS)/ functionalized chitosan based nanobiocomposite for packaging: A reactive extrusion study. *ACS Omega.*, 3, 13298, 2018.
32. Ardean, C., Davidescu, C.M., Neme, N.S., Negrea, A., Ciopec, M., Duteanu, N., Negrea, P., Seiman, D.D., Musta, V., Factors influencing the antibacterial activity of chitosan and chitosan modified by functionalization. *Int. J. Mol. Sci.*, 22, 7449, 2021.
33. Gupta, A., Pal, A.K., Patwa, R., Dhar, P., Katiyar, V., Green composites with excellent barrier properties, in: *Advanced Green Composites*, p. 321, John Wiley & Sons, Inc, Hoboken, NJ, USA, 2018.
34. Patil, P.P., Reagan, M.R., Bohara, R.A., Silk fibroin and silk-based biomaterial derivatives for ideal wound dressings. *Int. J. Biol. Macromol.*, 164, 4613, 2020.
35. Ghosh, T., Mondal, K., Giri, B.S., Katiyar, V., Silk nanodisc based edible chitosan nanocomposite coating for fresh produces: A candidate with superior thermal, hydrophobic, optical, mechanical and food properties. *Food Chem.*, *360*, 130048, 2021.
36. Punyamoonwongsa, P., Klayya, S., Sajomsang, W., Kunyanee, C., Aueviriyavit, S., Silk sericin semi-interpenetrating network hydrogels based on PEG-diacrylate for wound healing treatment. *Int. J. Polym. Sci.*, 2, 1, 2019.
37. CaO, Y., Liu, F., Chen., Y., Yu, T., Lou, D., Guo, Y., Li, P., Wang, Z., Ran, H., Drug release from core shell PVA/silk fibroin nanoparticles fabricated by one step electrospraying. *Sci. Rep.*, 7, 11913, 2017.

38. Kundu, S.C., Dash, B.C., Dash, R., Kaplan, D.L., Natural protective glue protein, sericin bioengineered by silkworms: Potential for biomedical and biotechnological applications. *Prog. Polym. Sci.*, 33, 998, 2008.
39. Borkotoky, S.S., Ghosh, T., Patwa, R., Katiyar, V., Silk nanocrystal reinforced poly (lactic acid) based microcellular foam: Impact on porous structure, crystallinity, thermomechanical and surface property. *Mater. Today Commun.*, *27*, 102258, 2021.
40. Gulrajani, M.L., Degumming of silk. *Rev. Prog. Coloration*, 22, 79, 1992.
41. Kundu, B., Rajkhowa, R., Kundu, S.C., Wang, X., Silk fibroin biomaterials for tissue regenrations. *Adv. Drug Deliv. Rev.*, 65, 457, 2013.
42. Zielinskal, S.G. and Sionkowska, A., How to improve physicochemical properties of silk fibroin materials for biomedical applications?-blending and crosslinking of silk fibroin- a review. *Materials*, 14, 1510, 2021.
43. Zinge, C. and Kandasubramanian, B., Nanocellulose based biodegradable polymers. *Eur. Polym. J.*, 133, 109758, 2020.
44. Monika, Dhar, P., Katiyar, V., Thermal degradation kinetics of polylactic acid/acid fabricated cellulose nanocrystal based bionanocomposites. *Int. J. Biol. Macromol.*, 104, 827, 2017.
45. Lin, N. and Dufresne, A., Nanocellulose in biomedicine: Current status and future prospect. *Eur. Polym. J.*, 59, 302, 2014.
46. Isogai, A., Saito, T., Fukuzumi, H., Tempo-oxidized cellulose nanofibers. *Nanoscale*, 3, 71, 2011.
47. Hussin, M.H., Appaturi, J.N., Poh, N.E., Latif, N.H.F., Brossel, N., Devin, I.Z., Firda, H.V., Syamani, A., Fatriasari, W., Solihat, N.N., Karimah, A., Herilswanto, A., Sekeri, A.H., Mohamadlbrahim, M.N., A recent advancement on preparation, characterization and application of nanolignin. *Int. J. Biol. Macromol.*, 200, 303, 2022.
48. Chauhan, P.S., Lignin nanoparticles: Ecofriendly and versatile tool for new era. *Bioresour. Technol.*, 9, 100374, 2020.
49. Beisl, S., Miltner, A., Friedl, A., Lignin from micro to nanosize: Production methods. *Int. J. Biol. Mol. Sci.*, 2017, 18, 1244.
50. Lamanna, M., Morales, N.J., Garcia, N.L., Goyanes, S., Development and characterization of starch nanoparticles by gamma radiation: Potential application as starch matrix filler. *Carbohydr. Polym.*, 97, 90, 2013.
51. Campelo, P.H., Ana, A.S., Pedrosa, M.T., Clerici, S., Starch nanoparticles: Production methods, structure, and properties for food applications. *Curr. Opin. Food Sci.*, 33, 136, 2020.
52. Concha, B.B.S., Acevedo, E.A., Santiago, M.C.N., Perez, L.A.B., Garcia, H.S., Ramirez, J.A., Acid hydrolysis of waxy starches with different granule size for nanocrystal production. *J. Cereal Sci.*, 79, 2018, 193.
53. Lu, H. and Tian, Y., Nanostarch: Preparation, modification, and application in pickering emulsions. *J. Agric. Food Chem.*, 69, 6929, 2021.
54. Koukaras, E.N., Papadimitriou, S.A., Bikiaris, D.N., Froudakis, G.E., Insight on the formation of chitosan nanoparticles through ionotropic gelation with tripolyphosphate. *Mol. Pharm.*, 9, 2856, 2012.

55. Martins, A.F., de Oliveira, D.M., Pereira, A.G., Rubira, A.F., Muniz, E.C., Chitosan/TPP microparticles obtained by microemulsion method applied in controlled release of heparin. *Int. J. Biol. Macromol.*, 51, 1127, 2012.
56. Hu, Y., Yu, J., Liu, L., Fan, Y., Preparation of natural amphoteric silk nanofibers by acid hydrolysis. *J. Mater. Chem. B.*, 7, 1450, 2019.
57. Nehra, P. and Chauhan, R.P., Ecofriendly nanocellulose and its biomedical applications: Current status and future prospect. *J. Biomater. Sci. Polym. Ed.*, 32, 112, 2021.
58. Schumann, D.A., Wippermann, J., Klemm, D.O., Kramer, F., Koth, D., Kosmeh, H., Wahlers, T., Gelani, S.S., Artificial vascular implants from bacterial cellulose: Preliminary results of small arterial substitutes. *Cellulose*, 16, 877, 2009.
59. Azevedo, E.P., Retarekar, R., Raghavan, M.L., Kumar, V., Mechanical properties of cellulose: Chitosan blends for potential use as a coronary artery bypass graft. *J. Biomater. Sci. Polym. Ed.*, 24, 239, 2013.
60. Joseph, B., Sagrika, V.K., Sabu, C., Kalarikkal, N., Thomas, S., Cellulose nanocomposites: Fabrication and biomedical applications. *J. Bioresour. Bioprod.*, 5, 223, 2020.
61. Chen, X.Y., Low, H.R., Loi, X.Y., Merel, L., Amin, M., Iqbal, M.C., Fabrication and evaluation of bacterial nanocellulose/poly(acrylic acid)/graphene oxide composite hydrogel: Characterization and biocompatibility studies for wound dressing. *J. Biomed. Mater. Res. B. Appl. Biomater.*, 107, 2140, 2019.
62. Sharma, C. and Bhardwaj, N.K., Fabrication of natural origin antibacterial nanocellulose films using bio-extract for potential use in biomedical industry. *Int. J. Bio. Macromol.*, 145, 914, 2020.
63. Brown, E.E., Hu, D., Lail, N.A., Zhang, X., Potential of nanocrystalline cellulose-fibrin nanocomposites for artificial vascular graft applications. *Biomacromolecules*, 14, 1063, 2013.
64. Mertaniemi, H., Lucea, C.E., Garcia, A.S., Gandia, C., Makitie, A., Partanen, P., Ikkala, O., Yliperttula, M., Human stem cell decorated nanocellulose thread for biomedical applications. *Biomaterials*, 82, 208, 2016.
65. Joseph, B., Sagrika, V.K., Sabu, C., Kalarikkal, N., Thomas, S., Cellulose nanocomposites: Fabrication and biomedical applications. *J. Bioresour. Bioprod.*, 5, 223, 2020.
66. Pradeep, H.K., Patel, D.H., Onkarappa, H.S., Pratiksha, C.C., Prasanna, G.D., Role of nanocellulose in industrial and pharmaceuticals – A review. *Int. J. Biol. Macromol.*, 207, 2022, 1038.
67. Tummala, G.K., Joffre, T., Rojas, R., Persson, C., Mihranyan, A., Strain-induced stiffening of nanocellulose-reinforced poly (vinyl alcohol) hydrogels mimicking collagenous soft tissues. *Soft Matter*, 13, 3936, 2017.
68. Zubik, K., Singhsa, P., Wang, Y., Manuspiya, H., Narain, R., Thermoresponsive poly (N-isopropylacrylamide)-cellulose nanocrystal hybrid hydrogels for wound dressing. *Polymers*, 9, 2017, 119.
69. Abo-Elseoud, W.S., Hassan, M.L., Sabba, M.W., Basha, M., Hassan, E.A., Fadel, S.M., Chitosan nanoparticles/cellulose nanocrystals nanocomposites

as a carrier system for the controlled release of repaglinide. *Int. J. Biol. Macromol.*, 111, 604, 2018.
70. Thomas, D., Latha, M.S., Thomas, K.K., Synthesis and *in vitro* evaluation of alginate-cellulose nanocrystal hybrid nanoparticles for the controlled oral delivery of rifampicin. *J. Drug Deliv. Sci. Technol.*, 46, 392, 2018.
71. Supramaniam, J., Adnan, R., MohdKaus, N.H., Bushra, R., Magnetic nanocellulose alginate hydrogel beads as potential drug delivery system. *Int. J. Biol. Macromol.*, 118, 640, 2018.
72. Fiorati, A., Turco, G., Travan, A., Caneva, E., Pastori, N., Cametti, M., Punta, C., Melone, L., Mechanical and drug release properties of sponge from cross-linked cellulose nanofiber. *ChemPlusChem*, 82, 848, 2017.
73. Sarkar, G., Orasugh, J.T., Saha, N.R., Roy, I., Bhattacharyya, A., Chattopadhyay, A.K., Rana, D., Chattopadhyay, D., Cellulose nanofibrils/chitosan based transdermal drug delivery vehicle for controlled release of ketorolac tromethamine. *New J. Chem.*, 41, 15312, 2017.
74. Ntoutoume, G.M.A.N., Granet, R., Mbakidi, J.P., Bregier, F., Leger, D.Y., Dugas, C.F., Lequart, V., Joly, N., Liagre, B., Chaleix, V., Sol, V., Development of curcumin cyclodextrin/cellulose nanocrystal complexes: New anticancer drug delivery systems. *Bioorg. Med. Chem. Lett.*, 26, 941–945, 2016.
75. Hivechi, A., Bahrami, S.H., Siegel, R.A., Drug release and biodegradability of electrospun cellulose nanocrystal reinforced polycaprolactone. *Mater. Sci. Eng. C.*, 94, 929, 2019.
76. Patel, D.K., Dutta, S.D., Shin, W.C., Ganguly, K., Lim, K.T., Fabrication and characterization of 3D printable nanocellulose based hydrogels for tissue engineering. *RSC Adv.*, 11, 7466, 2021.
77. Fakhri, A., Tahami, S., Nejad, P.A., Preparation of Fe_3O_4-Ag_2O quantum dots decorated cellulose nanofibers as a carrier of anticancer drugs for skin cancer. *J. Photochem. Photobiol. B Biol.*, 175, 83, 2017.
78. Maharjan, B., Park, J., Kaliannagou, V.K., Awasthi, G.P., Joshi, M.K., Park, C.H., Kim, C.S., Regenerated cellulose nanofiber reinforced chitosan hydrogel scaffolds for bone tissue engineering. *Carbohydr. Polym.*, 251, 117023, 2021.
79. Taher, M.A., Zahan, K.A., Rajaie, M.A., Ring, L.C., Hamin, N.S.M.N., Nee, T.W., Yenn, T.W., Nanocellulose as drug delivery system for honey as antimicrobial wound dressing. *Mater. Today: Proc.*, 31, 14, 2020.
80. Sukul, M., Ventura, R.D., Bae, S.H., Choi, H.J., Lee, S.Y., Lee, B.T., Plant derived oxidized nanofibrillar cellulose chitosan composite as an absorbable hemostat. *Mater. Lett.*, 197, 150, 2017.
81. Shefa, A.A., Taz, M., Hossain, M., Kim, Y.S., Lee, S.Y., Lee, B.T., Investigation of efficiency of a novel, zinc oxide loaded TEMPO-oxidized cellulose nanofiber-based hemostat for topical bleeding. *Int. J. Biol. Macromol.*, 126, 786, 2019.
82. Yan, H., Chen, X., Feng, M., Shi, Z., Layer-by-layer assembly of 3D alginate-chitosan-gelatin composite scaffold incorporating bacterial cellulose nanocrystals for bone tissue engineering. *Mater. Lett.*, 209, 492, 2017.

83. Peng, W., Ren, S., Zhang, Y., Fan, R., Zhou, Y., Li, L., Xu, X., Xu, Y., MgO nanoparticles incorporated PCL/gelatin derived coaxial electrospinning nanocellulose membranes for periodontal tissue regeneration. *Front. Bioeng. Biotechnol.*, 9, 1, 2021.
84. Sales, I.A., Sanchez, S.R., Traeger, K., Weis, C., Laomaine, A., Turon, P., Roig, A., *In vivo* soft tissue reinforcement with bacterial nanocellulose. *Biomater. Sci.*, 9, 3040, 2021.
85. Shefa, A.A., Taz, M., Lee, S.Y., Lee, B.T., Enhancement of hemostatic property α plant derived oxidized nanocellulose silk fibroin-based scaffolds by thrombin loading. *Carbohydr. Polym.*, 208, 168, 2019.
86. Zhang, C., Salick, M.R., Cordie, T.M., Ellingham, T., Dan, Y., Turng, L.S., Incorporation of poly (ethylene glycol) grafted cellulose nanocrystals in poly (lactic acid) electrospun nanocomposite fibers as potential scaffolds for bone tissue engineering. *Mater. Sci. Eng. C*, 49, 463, 2015.
87. Mohan, J., Stewart, S.A., Espinosa, E., Rosal, A., Rodriguez, A., Larraneta, E., Donnelly, R.F., Robles, J.D., Cellulose nanofibers, and other biopolymers for biomedical applications. A review. *Appl. Sci.*, 10, 65, 2020.
88. Figueiredo, P., Lintinen, K., Hirvonen, J.T., Kostiainen, M.A., Santos, H.A., Properties and chemical modifications of lignin: Towards. Lignin based nanomaterials for biomedical applications. *Prog. Mater. Sci.*, 93, 233, 2018.
89. Chai, Y., Wang, Y., Li, B., Qi, W., Su, R., He, Z., Microfluidic synthesis of lignin/chitosan nanoparticles for the PH-responsive delivery of anticancer drugs. *Langmuir*, 23, 7219, 2021.
90. Vinardell, M.P. and Mitjans, M., Lignin and their derivatives with beneficial effects on human health. *Int. J. Mol. Sci.*, 18, 1219, 2017.
91. Kumar R., Butreddy A., Kommineni N., Reddy P.G., Bunekar N., Sarkar C., Dutt S., Mishra Y.K., Oupicky D., Kaushik K., Lignin: Drug/gene delivery and tissue engineering applications. *Dove Med. Press*, 2021, 2419, 2021.
92. Figueiredo, P., Lintinen, K., Hirvonen, J.T., Kostiainen, M.A., Santos, H.A., Properties and chemical modifications of lignin: Towards lignin based nanomaterials for biomedical applications. *Prog. Mater. Sci.*, 93, 233, 2018.
93. Jaiswal, L., Shankar, S., Rhim, J.W., Hahm, D.H., Lignin mediated green synthesis of AgNPs in carrageenan matrix for wound dressing applications. *Int. J. Biol. Macromol.*, 159, 859, 2020.
94. Algahtani, M.S., Algahtani, A., Kazi, M., Ahmad, M.Z., Alahmari, A., Alsenaidy, M.A., Syed, R., Wound healing potential of curcurmin loaded lignin nanoparticles. *J. Drug Deliv. Sci. Technol.*, 60, 2020, 102020.
95. Xu, J., Xu, J.J., Lin, Q., Jiang, L., Zhang, D., Li, Z., Ma, B., Zhang, C., Li, L., Kai, D., Yu, H.D., Loh, X.J., Lignin incorporated nanogel serving as an antioxidant biomaterial for wound healing. *ACS Appl. Biomater. Sci.*, 4, 3, 2021.
96. Chen, Y., Zheng, K., Niu, L., Zhang, Y., Liu, Y., Wang, C., Chu, F., Highly mechanical properties nanocomposite hydrogels with biorenewable lignin nanoparticles. *Int. J. Biol. Macromol.*, 128, 414, 2019.

97. Dai, L., Liu, R., Hu, L.Q., Zou, Z.F., Si, C.L., Lignin nanoparticles as a novel green carrier for the efficient delivery of resveratrol. *ACS Sustain. Chem. Eng.*, 5, 8241, 2017.
98. Figueiredo, P., Lintinen, K., Hirvonen, J.T., Kostiainen, A., Santos, H.A., Properties and chemical modifications of lignin: Towards lignin-based nanomaterials for biomedical applications. *Prog. Mater. Sci.*, 93, 233, 2018.
99. Pandey, R.P., Rasool, K., Rasheed, P.A., Gomez, T., Pasha, M., Mansour, S.A., Lee, O.S., Mahmoud, K.A., One step synthesis of an antimicrobial framework based on covalently crosslinked chitosan/Lignosulfate nanosphere. *Green Chem.*, 22, 678, 2020.
100. Jaiswal, L., Shankar, S., Rhim, J.W., Hahm, D.H., Legnin mediated green synthesis of AgNP in carrageenan matrix for wound dressing applications. *Int. J. Biol. Macromol.*, 159, 859, 2020.
101. Zhou, Y., Han., Y., Li, G., Xiong, F., Chu, F., Lignin based fluorescence hollow nanoparticles: Their preparation, characterization, and encapsulation properties of doxorubicin. *Int. J. Biol. Macromol.*, 165, 2136, 2020.
102. Figueiredo, P., Lintinen, K., Kiriazis, A., Hynninen, V., Liu, Z., Ramos, T.B., Rahikkala, A., Correia, A., Kohout, T., Sarmento, B., Kauhaluoma, J.Y., Correia, A., Kohout, T., Sarmento, B., Kauhaluoma, J.Y., Hirvonen, J., Ikkala, O., Kostiainen, M.A., Santos, H.A., *In vitro* evaluation of biodegradable lignin-based nanoparticles for drug delivery and enhanced antiproliferation effect in cancer cells. *Biomaterials*, 121, 97, 2017.
103. Yang, W., Xu, F., Ma, X., Guo, J., Li, C., Shen, S., Puglia, D., Chen, J., Xu, P., Kenny, J., Ma, P., Highly toughened PVA/nanocellulose hydrogels with antioxidative and antibacterial properties triggered by lignin-Ag nanoparticles. *Mater. Sci. Eng. C*, 129, 112385, 2021.
104. Alqahtani, M.S., Alqahtani, A., Kazi, M., Ahmad, M.Z., Alahmari, A., Alsenaidy, M.A., Syed, R., Wound healing potential of curcumin loaded lignin nanoparticles. *J. Drug Deliv. Sci. Technol.*, 60, 102020, 2020.
105. Li, S., Zhang, Y., Ma, X., Qiu, S., Chen, J., Lu, G., Jia, Z., Zhu, J., Yang, Q., Chen, J., Wei, Y., Antimicrobial lignin-based polyurethane/Ag composite foams for improving healing. *Biomacromolecules*, 23, 1622, 2022.
106. Morena, A.G., Stefanov, I., Ivanova, K., Rafael, S.P., Soto, M.S., Tzanov, T., Antibacterial polyurethane foams with incorporated lignin capped silver nanoparticles for chronic wound treatment. *Ind. Eng. Chem. Res.*, 59, 4504, 2020.
107. Chai, Y., Wang, Y., Li, B., Qi, W., Su, R., He, Z., Microfluidic synthesis of lignin/chitosan nanoparticles for the pH responsive delivery of anticancer drug. *Langmuir*, 37, 7219, 2021.
108. Kai, D., Ren, W., Tian, L., Chee, P.L., Liu, Y., Ramakrishna, S., Loh, X.J., Engineering poly (lactide)-lignin nanofibers with antioxidant activity for biomedical application. *ACS Sustain. Chem. Eng.*, 4, 5268, 2016.
109. Keihan, R.E., Asgharnasi, S., Aliabadi, H.A.M., Tahmasebi, B., Radinekiyan, F., Maleki, A., Bahreinizad, H., Mahdavi, M., Alavijeh, M.S., Saber, R., Mendez, S.L., Shalan, A.E., Magnetic graphene oxide-lignin nanobiocomposite:

A novel, ecofriendly and stable nanostructure suitable for hyperthermia in cancer therapy. *RSC Adv.*, 12, 3593, 2022.

110. Zakaria, N.H., Muhammad, N., Abdullah, M.M.A.B., Potential of starch nanocomposite for biomedical applications. *IOP Conference Series: Materials Science and Engineering*, 209, 1, 2017.
111. Shakir, M., Jolly, R., Khan, M.S., Iram, N., Khan, H.M., Nano-hydroxyapatite/chitosan-starch nanocomposite as a novel bone construct: Synthesis and *in vitro* studies. *Int. J. Biol. Macromol.*, 80, 282, 2015.
112. Liu, K., Wang, Y., Li, H., Duan, Y., A facile one pot synthesis of starch functionalized graphene as nanocarrier for pH sensitive and starch mediated drug delivery. *Colloids Surf. B Biointerfaces*, 128, 86, 2015.
113. Baniani, D.D., Zahedifar, P., Bagheri, R., Solouk, A., Curcurmin loaded starch micro/nano particles for biomedical application: The effects of preaparation parameters on release profile. *Starch*, 71, 1800305, 2019.
114. Delavari, M.M. and Stiharu, I., Preparing and characterizing novel biodegradable starch/PVA based films with nanosized zinc oxide particles for wound dressing applications. *Appl. Sci.*, 12, 4001, 2022.
115. Baghaie, S., Khorasani, M.T., Zarrabi, A., Moshtaghian, J., Wound healing properties of PVA/Starch/chitosan hydrogel membranes with nano zinc oxide as antibacterial wound dressing material. *J. Biomater. Sci.*, 28, 2220, 2017.
116. Tavakoli, S., Kharaziha, M., Nemati, S., Kalateh, A., Nanocomposite hydrogel based on carrageenan coated starch/cellulose nanofibers as a hemorrhage control material. *Carbohydr. Polym.*, 251, 117013, 2017.
117. Ponnguzhali, R., Basha, K., Kumari, S., Nanostarch reinforced with chitosan/poly (vinyl pyrrolidone) blend for *in vitro* wound healing application. *Polym. Plast. Technol. Eng.*, 57, 1400, 2017.
118. Reshmi, C.R., Sheeja, P.S., Sujith, A., Poly (ε-caprolactone/nanostarch composite nanofibrous wound dressing with antibacterial property and pH stimulus drug release. *Cellulose*, 29, 427, 2022.
119. Likhitkar, S. and Bajpai, A.K., Magnetically controlled release of cisplatin from superparamagnetic starch nanoparticles. *Carbohydr. Polym.*, 87, 300, 2012.
120. Fama, L.M., Pettarin, V., Goyanes, S.N., Bernal, C.R., Starch/multi walled carbon nanotubes composites with improved mechanical properties. *Carbohydr. Polym.*, 83, 1226, 2011.
121. Amini, M., Haddadi, S.A., Ghaderi, S., Ramazani, A., Ansarizadeh, M.H., Preparation and characterization of PVDF/Starch nanocomposite nanofiber using electrospinning method. *Mater. Today: Proc.*, 5, 15613, 2018.
122. Shakir, M., Jolly, R., Khan, M.S., Ha, I., Khan, M., Nano-hydroxyapatite-chitosan nanocomposite as a novel bone construct: Synthesis and *in vitro* studies. *Int. J. Biol. Macromol.*, 80, 282, 2015.

123. Nasrabadi, B.N., Mehrasa, M., Rafienia, M., Bonakdar, S., Behzad, T., Gavanji, S., Porous starch/cellulose nanofibers composite prepared by salt leaching technique for tissue engineering. *Carbohydr. Polym.*, 108, 232, 2014.
124. Wu, D., Samanta, A., Srivastava, R.K., Hakkarainen, M., Starch derived nanographene oxide paves the way for electrospinnable and bioactive starch scaffolds for bone tissue engineering. *Biomacromolecules*, 18, 1582, 2017.
125. Akturk, A., Taygun, M.E., Guler, F.K., Goller, G., Kuckbayrak, S., Fabrication of antibacterial polyvinylalcohol nanocomposite mats with soluble starch coated silver nanoparticles. *Colloids Surf A*, 562, 255, 2019.
126. Wadke, P., Chhabra, R., Jain, R., Dandekar, P., Silver embedded starch based nanofibrous mats for soft tissue engineering. *Surf. Interfaces*, 8, 137, 2017.
127. Gholamali, I., Hosseini, S.N., Alipour, E., Yadollahi, M., Preparation and characterization of oxidized starch/CuO nanocomposite hydrogels applicable in drug delivery system. *Starch*, 71, 1800118, 2019.
128. Bernkop-Schnürch, A. and Dünnhaupt, S., Chitosan-based drug delivery systems. *Eur. J. Pharm. Biopharm.*, 81, 463, 2012.
129. Hu, L., Sun, Y., Wu, Y., Advances in chitosan-based drug delivery vehicles. *Nanoscale*, 5, 3103, 2013.
130. Ali, A. and Ahmed, S., A review on chitosan and its nanocomposites in drug delivery. *Int. J. Biol. Macromol.*, *109*, 273, 2018.
131. Salcedo, I., Aguzzi, C., Sandri, G., Bonferoni, M.C., Mori, M., Cerezo, P., Caramella, C., *In vitro* biocompatibility and mucoadhesion of montmorillonite chitosan nanocomposite: A new drug delivery. *Appl. Clay Sci.*, *55*, 131, 2012.
132. Nagpal, K., Singh, S.K., Mishra, D.N., Chitosan nanoparticles: A promising system in novel drug delivery. *Chem. Pharm. Bull.*, *58*, 1423, 2010.
133. Wang, X., Du, Y., Luo, J., Lin, B., Kennedy, J.F., Chitosan/organic rectorite nanocomposite films: Structure, characteristic and drug delivery behavior. *Carbohydr. Polym.*, 69, 41, 2007.
134. Menon, D., Thomas, R.T., Narayanan, S., Maya, S., Jayakumar, R., Hussain, F., Nair, S.V., A novel chitosan/polyoxometalate nano-complex for anti-cancer applications. *Carbohydr. Polym.*, 84, 887, 2011.
135. Alzandi, A.A., Naguib, D.M., Abas, A.S.M., Onion extract encapsulated on nanochitosan: A promising anticancer agent. *J. Gastrointest. Cancer*, 53, 211, 2022.
136. Zmejkoski, D.Z., Marković, Z.M., Budimir, M.D., Zdravković, N.M., Trišić, D.D., Bugárová, N., Marković, B.M.T., Photoactive and antioxidant nanochitosan dots/biocellulosehydrogels for wound healing treatment. *Mater. Sci. Eng. C*, 122, 111925, 2021.
137. Hou, B., Qi, M., Sun, J., Ai, M., Ma, X., Cai, W., Qiu, L., Preparation, characterization and wound healing effect of vaccarin-chitosan nanoparticles. *Int. J. Biol. Macromol.*, 165, 3169, 2020.
138. Karri, V.V.S.R., Kuppusamy, G., Talluri, S.V., Mannemala, S.S., Kollipara, R., Wadhwani, A.D., Malayandi, R., Curcumin loaded chitosan nanoparticles

impregnated into collagen-alginate scaffolds for diabetic wound healing. *Int. J. Biol. Macromol.*, 93, 1519, 2016.

139. Pramanik, N., Mishra, D., Banerjee, I., Maiti, T.K., Bhargava, P., Pramanik, P., Chemical synthesis, characterization, and biocompatibility study of hydroxyapatite/chitosan phosphate nanocomposite for bone tissue engineering applications. *Int. J. Biomater.*, 1687, 512417, 2009.
140. Katti, K.S., Katti, D.R., Dash, R., Synthesis and characterization of a novel chitosan/montmorillonite/hydroxy-apatite nanocomposite for bone tissue engineering. *Biomed. Mater.*, 3, 034122, 2008.
141. Thariga, S., Subashini, R., Pavithra, S., Meenachi, P., Kumar, P., Balashanmugam, P., Senthil Kumar, P., *In vitro* evaluation of biodegradable nHAP/Chitosan/Gelatin based scaffold for tissue engineering application. *IET Nanobiotechnol.*, 13, 301, 2019.
142. Manzari-Tavakoli, A., Tarasi, R., Sedghi, R., Moghimi, A., Niknejad, H., Fabrication of nanochitosan incorporated polypyrrole/alginate conducting scaffold for neural tissue engineering. *Sci. Rep.*, 10, 1, 2020.
143. Gokila, S., Gomathi, T., Vijayalakshmi, K., Sukumaran, A., Sudha, P.N., Development of 3D scaffolds using nanochitosan/silk-fibroin/hyaluronic acid biomaterials for tissue engineering applications. *Int. J. Biol. Macromol.*, 120, 876, 2018.
144. Tayel, A.A., Ghanem, R.A., Al-Saggaf, M.S., Elebeedy, D., Abd El Maksoud, A.I., Application of fish collagen-nanochitosan-henna extract composites for the control of skin pathogens and accelerating wound healing. *Int. J. Polym. Sci.*, 1, 2021.
145. Parida, U.K., Nayak, A.K., Binhani, B.K., Nayak, P.L., Synthesis and characterization of chitosan-polyvinyl alcohol blended with cloisite 30B for controlled release of the anticancer drug curcumin. *J. Biomater. Nanobiotechnol.*, 2, 414, 2011.
146. Ghorbani, M., Roshangar, L., Rad, J.S., Development of reinforced chitosan/pectin scaffold by using the cellulose nanocrystals as nanofillers: An injectable hydrogel for tissue engineering. *Eur. Polym. J.*, 130, 109697, 2020.
147. Zhang, P., Chen, L., Zhang, Q., Hong, F.F., Using *in situ* dynamic cultures to rapidly biofabricate fabric reinforced composite of chitosan/bacterial nanocellulose for antibacterial wound dressing. *Front. Microbiol.*, 7, 1, 2016.
148. Maan, G.K. and Bajpai, A.K., Investigation of *in vitro* release of cisplatin from electrostatically crosslinked chitosan alginate nanoparticle. *Synth. React. Inorg. Metal Org. Nano Met. Chem.*, 46, 1532, 2016.
149. Wei, Z., Feng, P.P., Hong, F., Cao, Z., Ji, Y., A novel approach for efficient fabrication of chitosan nanoparticles embedded bacterial nanocellulose conduits. *Carbohydr. Polym.*, 2021, 1, 264.
150. Kaur, S., Manhas, P., Swami, A., Bhandari, R., Sharma, K.K., Jain, R., Kumar, R., Pandey, S.K., Kuhad, A., Sharma, R.K., Wangoob, N., Bioengineered

PLGA-chitosan nanoparticles for brain targeted intranasal delivery of antiepileptic TRH analogues. *Chem. Eng. J.*, 346, 630, 2018.
151. Liu, L., Zhou, C., Xia, X., Liu, Y., Self-assembled lecithin/chitosan nanoparticles for oral insulin delivery: Preparation and functional evaluation. *Int. J. Nanomed.*, 11, 761, 2016.
152. Cheng, K.C., Huang, C.F., Wei, Y., Hsu, S.H., New chitosan-cellulose nano fiber self-healing hydrogels to correlates self-healing properties of hydrogels with neural regeneration effects. *NPG Asia Mater.*, 11, 1, 2019.
153. Karatepe, U.Y. and Ozdemir, T., Improving mechanical and antibacterial properties of PMMA via polyblend electrospinning with silk fibroin and polyethyleneimine towards dental applications. *Bioact. Mater.*, 5, 510, 2020.
154. Wang, X., Yucel, T., Lu, Q., Hu, X., Kaplan, D.L., Silk nanospheres and microspheres from silk/pva blend films for drug delivery. *Biomaterials*, 31, 1025, 2010.
155. Wenk, E., Merkle, H.P., Meinel, L., Silk fibroin as a vehicle for drug delivery applications. *J. Control. Release.*, 150, 128, 2011.
156. Farokhi, M., Mottaghitalab, F., Reis, R.L., Ramakrishna, S., Kundu, S.C., Functionalized silk fibroin nanofibers as drug carriers: Advantages and challenges. *J. Control. Release.*, 321, 324, 2020.
157. Patwa, R., Soundararajan, N., Mulchandani, N., Bhasney, S.M., Shah, M., Kumar, A., Katiyar, V., Silk nanodisc: A natural material for cancer therapy. *Biopolymers*, 109, 1, 2018.
158. Niemiec, S.M., Louiselle, A.E., Hilton, S.A., Dewberry, L.C., Zhang, L., Azeltine, M., Xu, J., Singh, S., Sakthivel, T.S., Seal, S., Liechty, K.W., Zgheib, C., Nanosilk increases the strength of diabetic skin and delivers CNP-miR146a to improve wound healing. *Front. Immunol.*, 11, 1, 2020.
159. Kim, J.H., Sheikh, F.A., Ju, H.W., Park, H.J., Moon, B.M., Lee, O.J., Park, C.H., 3D silk fibroin scaffold incorporating titanium dioxide nanoparticle for tissue engineering. *Int. J. Biol. Macromol.*, 68, 158, 2014.
160. Sun, J., Li, L., Xing, F., Yang, Y., Gong, M., Liu, G., Wu, S., Luo, R., Duan, X., Liu, M., Zou, M., Xiang, Z., Graphene oxide modified silk fibroin/nanohydroxyapatite scaffold loaded with urine derived stem cells for immunomodulation and bone regeneration. *Stem Cell Res. Ther.*, 12, 1, 2021.
161. Marin, C.B., Fitzpatrick, V., Kaplan, D.L., Landoulsi, J., Guenin, E., Egles, C., Silk polymers and nanoparticles: A powerful combination for the design of versatile biomaterials. *Front. Chem.*, 8, 1, 2020.
162. Hilton, S.A., Zgheib, C., Hodges, M.M., Dewberry, L.C., Seal, S., Liechty, K.W., Nanosilk improves the biomechanical properties of human diabetic skin. *J. Am. Coll. Surg.*, 227, 108, 2018.
163. Kim, M.H., Kim, B.S., Lee, J., Cho, D., Kwon, O.H., Park, W.H., Silk fibroin/hydroxyapatite composite hydrogel induced by gamma-ray irradiation for bone tissue engineering. *Biomater. Res.*, 21, 1, 2017.
164. Aliramaji, S., Zamanian, A., Mozafari, M., Super paramagnetic responsive silk fibroin/chitosan/magnetite scaffolds with tunable pore structure for

bone tissue engineering applications. *Mater. Sci. Eng. C Mater. Bio. Appl.*, 70, 736, 2017.
165. Baygar, T., Characterization of silk sutures coated with propolis and biogenic silver nanoparticles: An ecofriendly solution with wound healing potential against surgical site infections. *Turk J. Med. Sci.*, 50, 258, 2020.
166. Lan, X., Ma, Z., Szojka, A.R.A., Kunze, M., Sierra, A.M., Vyhlidal, J., Boluk, Y., Adesida, A.B., Tempo-oxidized cellulose nanofiber-alginate hydrogel as a bioink for human meniscus. *Front. Bioeng. Biotechnol.*, 9, 1, 2021.
167. Wei, J., Wang, B., Li, Z., Wu, Z., Zhang, M., Sheng, N., Liang, Q., Wang, H., Chen, S., A 3D printable TEMPO-oxidized bacterial cellulose/alginate hydrogel with enhanced stability via nanoclay incorporation. *Carbohydr. Polym.*, 238, 1, 2020.
168. Markstedt, K., Escalante, A., Toriz, G., Gatenholm, P., Biomimetic inks based on cellulose nanofibrils and crosslinkablexylans for 3D printing. *ACS Appl. Mater. Interfaces*, 9, 40878, 2017.
169. Shin, S., Park, S., Park, M., Jeong, E., Na, K., Youn, H.J., Hyun, J., Cellulose nanofibers for the enhancement of printability of low viscosity gelatin derivatives. *Bioresources*, 2941, 12, 2017.
170. Diez, I., Eronen, P., Sterberg, M., Linder, M.B., Ikkala, O., Ras, R.H.A., Functionalization of nanofibrillated cellulose with silver nanoclusters: flourescene and antimicrobial activity. *Macromol. Sci.*, 11, 1185, 2011.
171. Tummala., G. K., Lopes, V. R., Mihranyan, A. M., Ferraz, N., Biocompatibility of nanocellulose reinforced PVA hydrogel with human corneal epithelial cells for ophthalmic applications. *J. Funct. Biomater.*, 10, 35, 2019.

5

Biodegradable Magnesium Composites for Orthopedic Applications

Anshu Dubey[1*], Satish Jaiswal[2], S. Vincent[3†] and Vignesh Kumaravel[1]

[1]*International Centre for Research on Innovative Biobased Materials (ICRI-BioM)—International Research Agenda, Lodz University of Technology Żeromskiego, Lodz, Poland*

[2]*Biomaterials and Multiscale Mechanics Laboratory, Department of Metallurgical and Materials Engineering, Indian Institute of Technology Roorkee, Uttarakhand, India*

[3]*Department of Mechanical Engineering, BITS Pilani Dubai Campus, Dubai, United Arab Emirates*

Abstract

The new generation of biodegradable magnesium (Mg) implants has drawn substantial attention because of their similar physical properties (such as elastic modulus and relative density) to human bone. Due to this, Mg tends to overcome the stress shielding effect and secondary surgery in contrast to conventional metallic implants (stainless steel, titanium, and titanium alloys). However, the degradation rate of Mg is higher (standard electrode potential of -2.73 V) in the physiological environment, which ultimately produces a large amount of hydrogen gas. Hence, it exponentially induces a decrease in mechanical strength and failure of the implants with impaired bone healing. Several efforts have been taken, such as alloying, surface modifications, and nanocomposites, to control the onset degradation to address these concerns. Among them, Mg nanocomposites have gained much interest in controlling the biocorrosion of Mg. The appropriate shape, size, and reinforcement amount significantly alter nanocomposites' functional properties. Various bioceramics such as metal oxides (MgO and ZnO), tricalcium phosphate (TCP), and hydroxyapatite (HA) have been used as bioceramics reinforcements in the Mg matrix for developing bioresorbable Mg-based composites. The synergy

**Corresponding author*: anshu.dubey@p.lodz.pl

†*Corresponding author*: vincent@dubai.bits-pilani.ac.in

Arbind Prasad, Ashwani Kumar and Manoj Gupta (eds.) Advanced Materials and Manufacturing Techniques for Biomedical Applications, (103–134) © 2024 Scrivener Publishing LLC

of load transfer from matrix to hard reinforcement and mismatch of elastic modulus between matrix and reinforcement can maintain the physical and mechanical properties in the biocorrosion environment and mechanical integrity. Further, apart from the reinforcements for proper composite design, the fabrication route is also an essential criterion as it directly influences the functional properties of the composites. *In vitro* and *in vivo* characteristics of Mg-based nanocomposites are highly based on the kind and quantity of reinforcements. Additionally, some of the techniques (additive manufacturing, friction stir processing, and spark plasma sintering) facilitate rapid fabrication, grain refinement, and the high performance of metal matrix composites. To illustrate the crucial updating of research and development on Mg-based nanocomposites for biomedical applications, attempts are made in this chapter.

***Keywords*:** Magnesium, biodegradable, composite, orthopedic

5.1 Introduction

Bone fracture due to accidents, diseases, and sports injuries increase tremendously, rendering bone implants during the major clinical procedure. However, bone tissue is capable of regeneration in case of minor bone defects or trivial injuries, whereas, for severe damage, implants are required to assist bone regeneration. Patients have surgery to maintain their daily routine and quality of life. Over the last two decades, the orthopedic implant market has been in high demand. The global biomaterials market had a net worth of US$94.1 billion in 2012 and is forecasted to reach $595 billion by 2024 billion in 2017 [1]. The exponential increase in the demand for biomaterials has also led to the development of bone implants. The implant should be designed morphologically to mimic and support bone and osteogenesis in nature. For an ideal implant, the parameters involved are biocompatibility, better mechanical behavior, and biodegradability [2]. In the medical field, modern techniques, advanced materials design, and architecture of medical devices have been established in the last two decades. Of late, several materials are being explored, such as polymers, ceramics, composites, and metals, for orthopedic applications.

In this chapter, the primary focus is on the research investigating using magnesium-based nanocomposite material systems developed for orthopedic accessories. The discussion is systematically directed at realistic design strategies and crucial methods of implant fabrication. The process of recreating the bone tissue microenvironment is also highlighted.

The effect of various Mg-based materials systems and their functional properties regarding the orthopedic application are comprehensively summarized. Different fabrication strategies are mentioned in detail to give a clear picture of the correlation between the developed material's structural and functional characteristics for bone restoration. An effort is also made to investigate and represent the mechanism of specific properties of developed materials systems. Finally, a summary of the current trends in clinical translations and future perspectives of this expanding field is also included.

5.1.1 Biomaterials for Bone Implants

Polymers, ceramics and metallic materials are widely used as potential candidates for bone fracture management. However, metallic implants are widely preferred for load-bearing orthopedic applications due to their strength, fracture toughness, and fatigue resistance. Commercially available cobalt-chromium alloys, stainless steel (316L) and titanium and its alloys are non-degradable materials widely used for permanent internal fracture management. These materials are permanent implants due to their inert nature, as the implant stays at the fracture site even after healing the bone fracture. There are limitations with permanent implants, such as toxic metallic ions leaching out, leading to necrosis in local tissue. The major disadvantage of permanent implants is the higher elastic modulus compared to natural bone. The stress shielding phenomenon is caused by the elastic modulus mismatch between native bone and permanent implants, which has an impact on how the bone remodels after a bone fracture. Furthermore, removing the implant after healing a bone fracture is another painful operation for the patient. Thus, the alternative solution is biodegradable material-based implants which would degrade with the healing of bone tissue. Therefore, the degradable implants avoid the requirement of a second surgery as it degrades with bone tissue regeneration. Degradable polymer materials limit the choice of researchers since their mechanical properties are lower than metallic materials.

Of late, magnesium (Mg) based biodegradable orthopedic materials have gained attention due to the similarity of elastic modulus to bone (Figure 5.1). Magnesium makes up a large portion of the bone in the human body and is the fourth most common cation in the human body. A healthy adult weighing 70 kg needs 24–30 g of magnesium for daily activities, and a daily dose of 310–420 mg is advised for sustaining health [3]. Without causing harm, the surplus Mg ions can be carried via the circulatory system

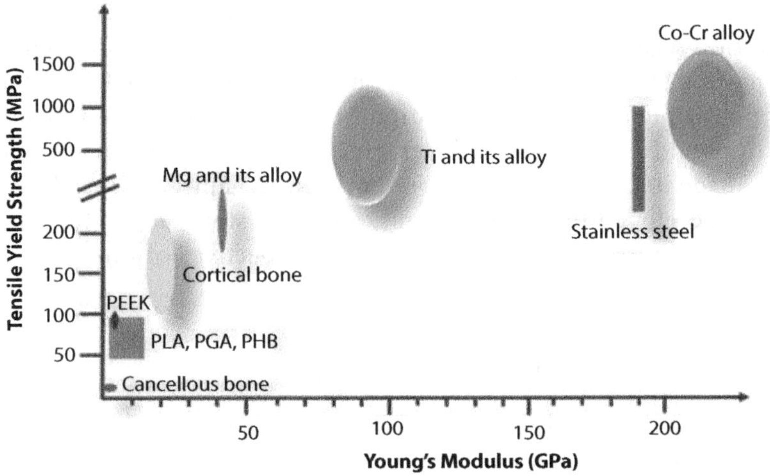

Figure 5.1 Mechanical features of commercially available implants, natural bone and Mg alloys. Reproduced with permission from ref [5]. Copyright 2020 John Wiley & Sons.

and eliminated by urine [4]. Additionally, the degradation product of Mg implants in the physiological environment does not produce any toxic elements and can be excreted through the urinary tract. Hence, Mg-based degradable orthopedic implants can be a game changer for fracture-fixing accessories. Although Mg-based implants have many benefits, the major drawback of Mg is its high corrosion rate in the physiological environment. Several approaches are adopted to improve the corrosion behavior of Mg, such as alloying, surface modifications, and fabrication of composites. Interestingly, the fabrication of composite materials offers to tune the functional properties of composites by changing the shape, size or composition of reinforcement materials in the matrix. A thorough understanding of production processes and reinforcement ingredients is necessary to fabricate tailored biodegradable Mg-based composites.

5.1.2 Magnesium: A Smart Material

Mg-based orthopedic implants have been used for clinical purposes for over a century. Austrian physician Erwin Payr introduced the Mg plates and screws implants in 1900. The evidence of early-century use of Mg for bone fixation in orthopedic applications is mentioned in Figure 5.2. Additionally, Mg was applied as an intramedullary stabilizer for pseudarthrosis and irreparable bone fractures. Lambotte performed surgery using Mg plates and 6 steel screws on a patient suffering from pseudarthrosis in 1906; due to its

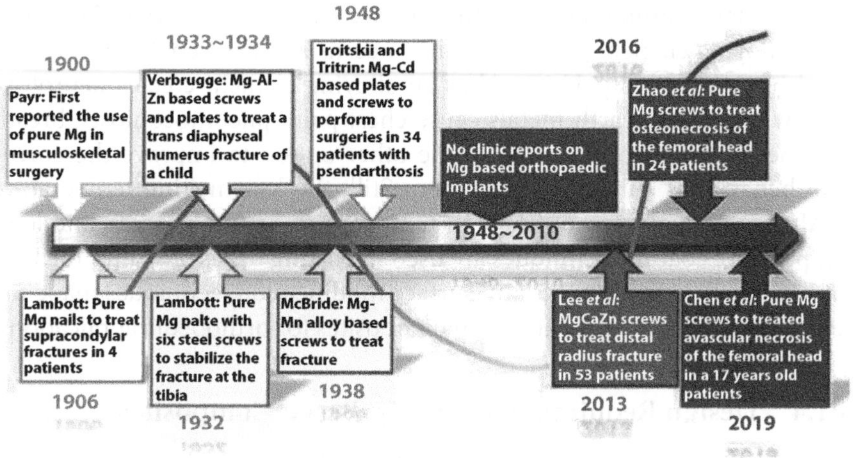

Figure 5.2 The case studies on the use of magnesium for fixing bones since the 20th century [9]. Reproduced with permission from ref [9]. Copyright 2020 Elsevier.

highly corrosive nature, the Mg plate degraded rapidly [6]. A revision was performed due to the Mg plate's high corrosion rate, which led to fracture instability. The researcher also used a magnesium nail to treat supracondylar fractures in four kids. The joints were healed without any further complications. The study concluded that the degradation rate is one year for Mg. In 1938 Mg plates and screws were used in clinical trials as a degradable fixation device [7]. Mcbride noticed that the plates made of Mg were highly susceptible to degradation than screws, and many other operations were performed to fix the devices. Spindle-shaped Mg sheets were used to treat human fractures in 1940 [8]. According to reports, one patient's recovery went smoothly, while the other's resulted in an excessive gas cavity formation. The plates and screws made of Mg-Cd alloy were implanted in thirty-four patients with pseudarthrosis [6]. The results showed complete degradation of implants and healing of bone tissue. It was postulated that degradation of Mg raising the pH of the surrounding environment, but Mg implantation neutralizes the acidic environment and stimulates the fracture callus.

5.1.3 Materials and Methods

As was already established, magnesium corrodes quite quickly, with a common electrode potential of -2.73 V in a physiological environment. Additives were used to improve the degradation behavior, alloying, surface modification of Mg.. Metals for instance aluminium (Al), zinc (Zn), zirconium (Zr), calcium (Ca), Cerium (Ce), yttrium (Y), and manganese (Mn)

[10–14] have commonly used to enhance the corrosion rate. However, these alloying elements raised health concerns and even could not provide good biocompatibility. Using Al as an alloying element creates neurotoxicity, whereas rare earth metals cause chest pain, pneumoconiosis, breath shortness, skin lesions, etc. [15]. The other option to improve the corrosion behavior is surface modification. Still, the low melting temperature of Mg prohibits the use of the high-temperature coating process, and low-temperature coating techniques are used in coating flakes off. Thus, the limitations of alloying elements and surface modification techniques made the researchers fabricate the Mg-based composites for orthopedic applications.

5.1.4 Design Requirements for Mg-Based Composites

High-performance magnesium-based biodegradable nanocomposites require innovative production methods, careful formulation, and in-depth design. When designing nanocomposites, it is crucial to take into account the demands of the final application, choose the appropriate reinforcement, and choose the most effective fabrication method. The designer must focus on the materials' compatibility, adhesion, and degree of reinforcement dispersion in the matrix while selecting the reinforcement materials for the Mg matrix. Prior to starting on the design formulation of the nanocomposites, it is essential to have a thorough understanding of these interrelated aspects. The design strategy for creating appropriate nanocomposites is shown in Figure 5.3. The important variables that have a direct impact on the design of nanocomposites are the choice of reinforcing materials, the degree of reinforcement distribution and interfacial interaction, the architecture's hierarchy, and the method of processing.

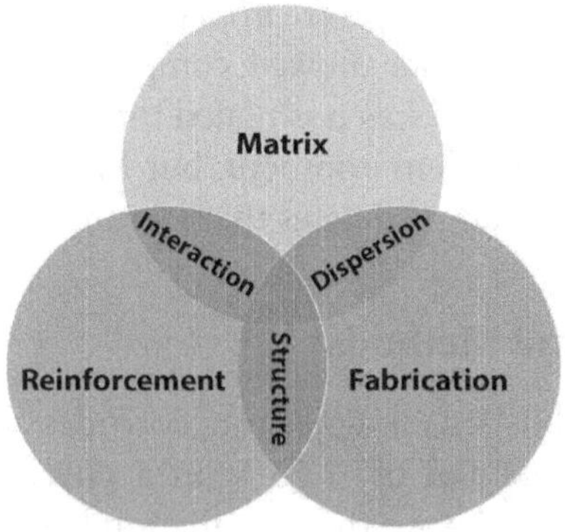

Figure 5.3 The design strategy for creating high-performance magnesium-based nanocomposites.

The following reaction is a result of the degrading behavior of implants made of magnesium:

$$Mg + 2H_2O \rightarrow Mg(OH)_2 + H_2 \tag{5.1}$$

Around the implant locations, the uncontrolled reaction causes an excessive hydrogen gas emission, forming pockets of gas and reducing the biocompatibility of the implant. Thus, it is essential to control the biocorrosion and mechanical behavior to implement implants successfully [16–19]. Various compositions, shapes and sizes of reinforcement can tailor the corrosion and mechanical properties of magnesium-based composites [6, 20]. The following criteria must be met for reinforcements used in magnesium-based composites:

1. There should be biodegradable or bioabsorbable reinforcement.
2. It should be osteogenic and biocompatible to prevent bone fracture.
3. Reinforcement should also enhance the corrosion behavior and mechanical properties

5.1.5 Types of Reinforcements

In order to create structural Mg-based composites, a variety of reinforcement types are employed. When selecting a suitable reinforcement, it should be non-toxic. Recently, various ceramics have been developed into Mg-based composites for orthopedic applications as reinforcements. Ceramics used as reinforcement for the biomaterials are classified as bio-inert, bioactive, and bioresorbable. Researchers have chosen resorbable ceramics as the Mg-based degradable composites to enhance the corrosion, mechanical behavior, and biocompatibility. Figure 5.4 shows the schematic on the design requirement for Mg-based composite to the orthopedic application utilizing bioactive glasses, hydroxyapatite (HAP), tricalcium phosphate (TCP), and other materials as reinforcement in the magnesium matrix.

Hydroxyapatite (HA)

The natural bioceramic HA($Ca_5(PO_4)_3(OH)$) which is inorganic component of human bone with and calcium phosphate ratio 1.67. HA also known as

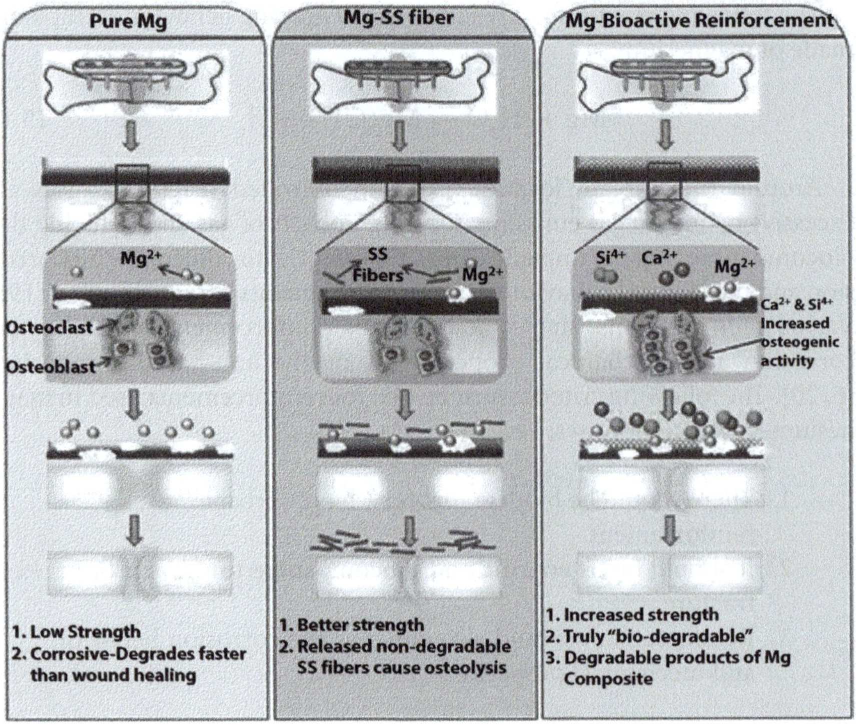

Figure 5.4 The distinction between Mg/degradable bioactive composite and Mg/non-degradable composite for orthopedic purposes is shown schematically. Reproduced with permission from ref [21] Copyright 2020 John Wiley & Sons.

for slower degradation rate in physiological environment [22]. Further, HA is also known for its bioactivity and biocompatibility [23]. The application of free standing HA is limited because of brittle characteristics for load bearing application. For orthopedic applications, it works best as a reinforcement in Mg-based metal matrix composites. The mechanical characteristics and corrosion behavior of the Mg-based composites are improved by the presence of hydroxyapatite in the Mg matrix.

β-Tricalcium Phosphate (β-TCP)

In 1920 β-Tricalcium Phosphate was introduced by Albee and Morrison. β-Tricalcium Phosphate has gained much attention in the orthopedic industries for bone fracture [24]. Compared to HA, -TCP has a lower calcium phosphate ratio (1.5); that is why it has a higher rate of degradation compared to HA [25, 26]. Studies have proven that β-TCP is bioactive, bioresorbable, and osteoconductive properties.

Bioactive Glass (BG)

L. L. Hench introduced the use of bioactive glass as a degradable bioceramic in early 1970 [27]. Calcium oxide, sodium oxide, phosphorus pentoxide(P_2O_5), and silicon dioxide make up bioactive glass [27]. Many bioactive glass modifications have been performed to improve the degradation and bond-forming ability. Bioactive glass can form enduring connections with bone and promote bone regrowth at implant sites [28]. *In vivo* bioglass bioactivity involves the leaching out of calcium and silicon ions with the formation of a hydroxycarbonate apatite layer [29]. The apatite layer is similar to bone apatite, which promotes osseointegration [30]. Bioactive glass has shown enormous potential to prove its candidacy as a material of reinforcement in biodegradable magnesium-based composites.

Calcium Polyphosphate (CPP)

Calcium polyphosphate shows high biocompatibility due to the higher ratio of Ca/P (i.e., 0.5). Condensed calcium phosphate, or CPP, is made up of PO_4 tetrahedral units [31]. It has drawn the attention of researchers as a controlled biodegradable bioceramic in a physiological environment with similar characteristics to the bone [32]. The CPP degradation products can be easily metabolized. It has been found during *in vivo*, the degradation rate can be controlled with proper particle size selection. Hence, CPP presents another reinforcement material for magnesium-based biocomposite.

Fluoroapatite (FA)

The chemical formula of fluorapatite is $Ca_5(PO_4)_3F$ which is hard crystalline phosphate. It is predominantly found in the tooth enamel [33]. It has been reported that FA shows comparatively slow biodegradation, good adhesion of cells, and enhanced alkaline phosphatase activity (ALP) compared to hydroxyapatite during *in vitro* studies [34]. Moreover, fluorapatite promotes bone-like apatite formation, which means adding fluoride crystallization and mineralization of calcium phosphate. It can be then remarked that fluorapatite also has attractive reinforcement for the Mg-based composites.

Zinc Oxide (ZnO)

ZnO nanoparticles are widely used in biomedical applications [35]. Many studies have shown that ZnO has osteointegration and better antimicrobial properties [36]. It is an extensively used as food additive even listed by FDA [37]. Using ZnO as reinforcement material can enhance the mechanical and corrosion properties [38].

Bredigite

The chemical formula of bredigite bioceramic is $Ca_7MgSi_4O_{16}$ with a group $CaO–SiO_2–MgO$ [39]. It has better biocompatibility and promotes cell integration [40]. It also tends to form an apatite layer on the implant surface [41]. Further, it has mechanical properties almost like the cortical bone, which is why it can be applied in load-bearing applications. Thus, bredigite can be used as a reinforcement material for the magnesium matrix.

Magnesium Oxide (MgO)

A key component of bioactive glass, MgO is a bioactive ceramic with superior mechanical and thermal properties [42]. Additionally, it exhibits antibacterial characteristics, and the *in vivo* breakdown products resemble Mg [43]. Thus, it also improves the mechanical and biological properties. The extent of interfacial interaction between reinforcement (MgO) and matrix (Mg) tunes the functional properties.

Titanium Dioxide (TiO_2)

TiO_2 is a bioceramic that is both chemically and physiologically stable. It exists in rutile, anatase, and brookite, three distinct crystallographic forms [44]. Recently, titanium dioxide has been utilized as reinforcement in metal and polymer matrix to improve the mechanical and corrosion behavior [45]. Titanium Dioxide coatings can enhance the bioactivity and corrosion resistance of alloys based on magnesium metal. The unique combination of 25% rutile and 75% anatase phase is used in biomedical applications known as Aeroxide TiO_2 P25.

Graphene Nanoplatelets (GNPs)

Carbonaceous materials, for example, graphene, nanodiamonds, and porous carbon are attention in the field of biomaterials due to their unique physical, chemical, and biocompatibility [46]. In the realm of biomaterials, the unique feature of graphene has made it distinguish itself among the carbon allotropes. It is also best known for chemical stability [47]. Covalent bonding and van der Waals forces between the atoms make it stronger in sheet form. A single graphene atomic layer can inhibit corrosion and act as a barrier to gas molecules. Additionally, GNPs have better biocompatibility and good antimicrobial properties. Thus, GNPs can be added to the Mg matrix as reinforcements to enhance the material's resistance to corrosion as well as its mechanical and biological behavior.

5.2 Materials and Methods

The main aim is to obtain the final product for the proper homogenization of reinforcement particles in the metal matrix along with appropriate densification. The adequate tuning of mechanical, structural, and degradation can be obtained by the distribution of reinforcement. The following are the sub-sections that give clear images of the efficacy of various fabrication techniques for Mg-based composites.

5.2.1 Powder Processing Route

It is common practice to produce magnesium-based composites using the powder processing technique. The homogeneous dispersion of reinforcement materials in the composite matrix at comparatively lower processing temperatures to casting and melting is the advantage of the powder processing route. The mixing of powder, and fabrication of green compact, followed by sintering, are the main steps in the powder processing method. The preparation of the green compacts is generally at room temperatures, while sintering is at controlled atmospheres such as vacuum or argon. Typically, the conventional route is followed for sintering, but they have advanced powder method such as spark plasma (SPS) and hot-pressing sintering (HP) in which pressure and sintering can be performed simultaneously.

Many studies have used HA as reinforcement in the Mg matrix to enhance the corrosion and mechanical behavior of Mg-based composites fabricated through the powder processing route. Jaiswal *et al.* [48], Mg-3Zn/xHA (x= 0, 2, 5, and 10 wt%) composites were fabricated using the traditional powder processing method. When compared to a Mg composite without HA, it was shown that the HA content with 5 wt% HA increased the ultimate compressive strength by 23%. Further, improvement in hardness and corrosion behavior (40%) was also observed with the same incorporation of HA. The same group performed a similar study through spark plasma sintering and observed that the incorporation of HA increased up to 15% into the Mg matrix [49]. The improvement in corrosion behavior was 60% approximately. Both studies found that the agglomeration problem starts after a certain amount of HA. Other studies also noticed a similar pattern [50, 51]. Figure 5.5 shows the microstructure of HA reinforced in Mg matrix along with agglomeration in 10 wt % HA incorporation in magnesium. In a study, spark plasma sintering was used to fabricate β- -TCP-based Mg composites [52]. There was no reaction found between Mg and β-TCP. Tests involving prolonged immersion and electrochemistry revealed that the β-TCP encouraged the formation of a calcium phosphate on composites surface. However, from β-TCP Ca diffused into Mg matric forms the galvanic couple, increasing the corrosion rate. Further, a biocompatibility study showed that 10 vol% of reinforcement β-TCP in magnesium has a higher number of cell adhesion. Another study used wt % of β-TCP (0, 5, 10, and 15 wt %) with ZK61 alloy using spark plasma sintering in which uniform distribution of β-TCP in magnesium matrix was obtained with improved corrosion, mechanical, and biocompatibility behaviors [53]. The maximum percentage of β-TCP could be added to 5 wt %; after that, content accumulation of β-TCP can be seen. The benefits of SPS led to better interfacial bonding between matrix and reinforcements, increasing the mechanical properties. Biocompatibility tests showed no toxic behavior on L-929 cells. The use of bioactive glass in Mg matrix was also fabricated through spark plasma sintering with the various composition 0, 5, 10 and 15 wt % bioactive glass (SiO_2 (57.4%), CaO (21.6%), Na_2O (10.3%), P_2O_5 (3.4%) and TiO_2 (1.3%)) [54]. The hypothesis made that the pressure and temperature lead to the flow of bioactive glass in the form of viscous liquid and eventually coat the boundary of Mg particles. The agglomeration problem was noted with Mg/15 wt % BG composite. The composition with 10 wt % of BG in Mg showed significantly good properties and improved corrosion behavior. Another reinforcement, fluorapatite (FA), was fabricated through the blending-pressing-sintering method through biomedical applications with the varying composition of 10, 20, and 30 wt % FA in AZ91 alloy [55]. Since FA is a harder phase, the compressive yield strength improved for 20 wt% FA in AZ91 alloy by restricting the movement of the dislocations. Even though the compressive yield strength increased, the ductility decreased significantly. The corrosion resistance increased with the content of FA in the AZ91 matrix and enhanced the precipitation of the apatite layer on surfaces. 20 wt% composition of FA in the AZ91 matrix showed significant mechanical and corrosion behavior improvement.

In another study, a novel composite with ZnO and Mg matrix was prepared by *in situ* reaction followed by the hot press sintering technique [56]. The intermetallic blends with Mg-Zn and MgO forms. The increase in tensile strength with 2.2 times of Mg composites after adding ZnO is because of the grain refinement. In addition, the intermetallic (Mg-Zn) and magnesium oxide also contributed to the strengthening effect. In comparison to pure Mg, the ductility dropped from 7.43 to 6%. It is caused by the matrix's inclusion of hard phase reinforcement material such intermetallic Mg-Zn and MgO. Compared to pure Mg, the corrosion resistance was improved by 4 times.

Using a pressure-assisted sintering technique, Mg was blended with 0, 20, and 40% of the volume [57]. Bredigite is employed as a reinforcing phase because silicon and magnesium-containing ceramics can slow down the corrosive process of magnesium and improve biological behavior. The microstructure obtained was isotropic, and the Mg powder was spherical and surrounded by a homogeneous distribution of small bredigite particles at 40 vol % of bredigite. Compared to pure Mg and natural bone, bredigite content with 20 vol% in magnesium increases the compressive strength. The enhancement in mechanical properties is attributed because of the uniform dispersion of hard reinforcement in the Mg matrix, which inhibits dislocation movement. The addition beyond 20 vol% decreases the compressive strength because of the agglomeration of the reinforcement phase, which shows the weak bonding between the matrix and reinforcement. The degradation experiments showed that the Mg with 20 vol% exhibits better biocorrosion resistance in comparison to pure magnesium. The limitation of bredigite particles, which diverts the pathway in the Mg matrix and changes the OH layer into a layer consisting of Mg-, Ca-, P-, and Si-containing compounds, may be the cause of the improved degrading behavior.

Graphene oxide (GO) is another reinforcement with Mg matrix to overcome the corrosion behavior of Mg [58]. However, the capacity to offer adequate protection is constrained by the poor interfacial interaction between GO and Mg. With the help of the selective laser melting (SLM) technique, reduced graphene oxide (RGO) in the AZ61 matrix was produced. The interface property showed the strong bonds between RGO and MgO. Grain refinement was obtained when the content of RGO and MgO increased from 0 wt % to 4 wt % in the Mg matrix. Further, the mechanical properties such as hardness and compressive strength increased due to grain refinement and good interfacial bonding (90 ± 2 to 115 ± 2 HV and 30% higher than the base matrix). The agglomeration of reinforcement led to a decrease in compressive strength after a certain amount.

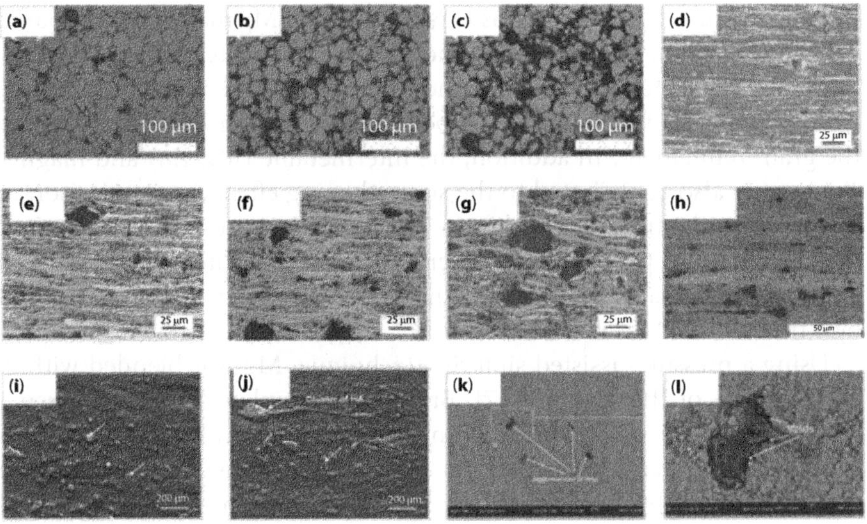

Figure 5.5 Microstructures obtained by microwave sintering of (a) Mg/5HAP, (b) Mg/10HAP, and (c) Mg/15HAP through powder processing, followed by extrusion, (d) Mg/0HAP, (e) Mg/5HAP, (f) Mg/10HAP, and (g) Mg/15HAP are shown in SEM and optical pictures (h) Mg/20HAP composite longitudinal direction microstructure. (i) Mg/10HAP and (j) Mg/30HAP (k) agglomeration of HA shown by arrows (l) Mg with 12.5 wt% HA and 10 wt% MgO composites (Black inclusions in the AZ91D matrix show agglomerated hydroxyapatite particles via hot press and hot extrusion). Reproduced with permission from ref [59]. Copyright 2020 John Wiley & Sons.

5.2.2 Casting Route

It is the conventional fabrication method that is used for the fabrication of composites and alloys. The mixture consisting of reinforcement in the liquid metal is stirred continuously. Stirring is performed in induction and mechanically. The liquid melt is poured into a preheated mold with the desired shape, followed by solidification. The cast product is the solidified component which will be ejected to complete the casting process. A large volume casting technique is a convenient way to produce in bulk. Mg with HA was prepared through casting and extrusion process with various content of HA (0, 5, 10, and 15 wt. % HAP) [22]. With the increase in the addition of HA the refined microstructure was obtained. After the addition of 15 wt.% HA the mechanical strength decreased by 27%, while the compressive strength increased as the HA content increased [22].

The agglomeration of HA particles was observed containing higher amounts HA particles. Nanohydroxyapatite was used as reinforced in Mg composites using the casting and extrusion process [60]. The corrosion behavior and cytocompatibility of the HA reinforced in Mg composites are found to improve as compared to pure Mg-Zr-Zn alloy (Figure 5.6a). The nanohydroxyapatite powder in the Mg-Zr-Zn alloy facilitate a calcium phosphate layer caused by adsorption of Ca^{2+} and PO^{4-}, which reduces the severe pitting corrosion. The biocompatibility test showed that the cell adhesion and proliferation on the surface of composite were comparatively higher than Mg alloy. The squeeze casting technique was used to fabricate the porous AZ91/HAP composites. A significant improvement was observed for AZ91/HAP composite in the compressive strength as compared to AZ91 alloy [23]. It has been reported that after the 470 h of soaking in SBF, ~60% of total mass was remained as shown in immersion study Figure 5.6 (b).

The 1 wt% β-TCP reinforcement was used in Mg–2Zn–0.5Ca in biodegradable composite fabricated through high shear solidification and equal channel angular extrusion (ECAP) [61]. The combined effect of equal angular extrusion and high shear solidification results in grain refinement and breaks the agglomerated β-TCP agglomerated into finer particles with

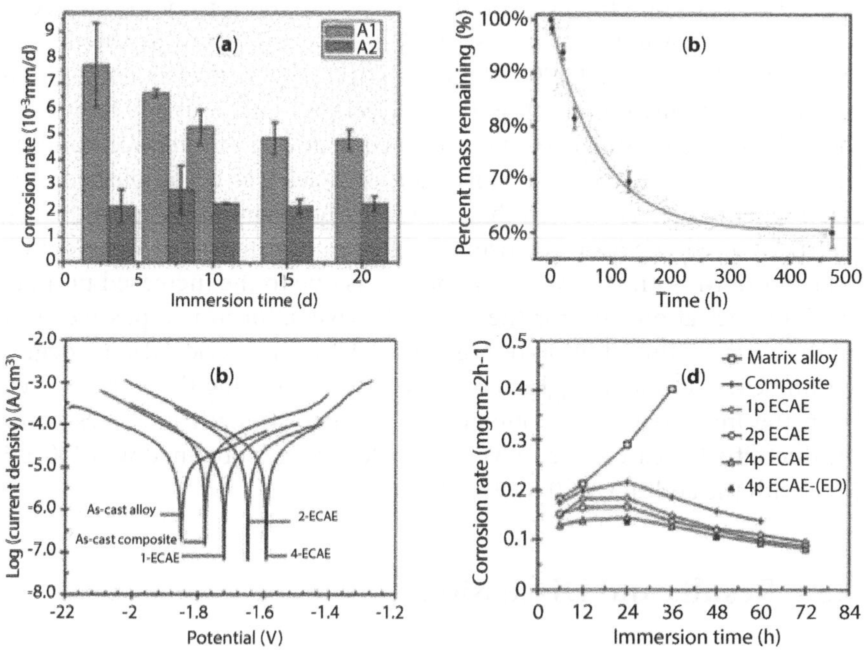

Figure 5.6 (a) Degradation rate of Mg–Zn–Zr/n-HAP composite (b) mass gain of AZ91/HAP composite after soaking in simulated body fluid (c) electrochemical and (d) degradation behavior of high shear solidified and ECAE Mg-2Zn-0.5Ca/1β-TCP composite after soaking in simulated body fluid. Reproduced with permission from ref [59]. Copyright 2020 John Wiley & Sons.

uniform dispersion in Mg matrix. Increased corrosion resistance was found which was attributed due to refined grain structure and formation of the deposited layer of β-TCP. Figure 5.6 (c) and (d) show that the uniform distribution of β-TCP increased the corrosion in comparison to the Mg matrix.

Very few studies are available on the Mg-BG composites through the casting route. The ZK30 alloy was reinforced with BG with various content (5, 10, and 20 wt %) by a semisolid high-pressure casting [62]. A uniform distribution of BG particles were observed in ZK30 alloy matrix during the microstructure analysis. Some porosity was also reported in composite microstructure. The ultimate compressive strength deteriorated with the higher content of BG. The apatite layer precipitation was observed on the composite surfaces during the immersion testing in Earle's balanced salts E-MEM.

In another study, MgO with various compositions was incorporated in Mg-3Zn-0.2Ca fabricated through a novel high shear technique followed by extrusion [63]. With an increase in MgO particles, a measurable reduction in mean grain size was seen in the cast composites. It was discovered that the MgO nanoparticles act as heterogeneous nucleation nucleants. The host extrusion enhanced grain refinement, and for composites containing 0.5 wt% MgO, the smallest grain size (1.5 ± 0.2 μm)was attained. In the extruded Mg3Zn0.2Ca/0.5MgO composite, the issue of agglomeration by the reinforcement phase MgO was discovered. Without any fissures in the link between the matrix and reinforcements, good interface bonding between the Mg matrix and MgO nanoparticles was accomplished. The electrochemical tests showed that 0.2 wt.% MgO nanoparticles in the Mg matrix showed superior corrosion behavior. The improved degradation behavior of Mg–Zn–Ca/MgO composites is due to the increased number of MgO nucleation sites for the $Mg(OH)_2$ layers, forming a passive layer on the surface. The addition of MgO after 0.5 wt% deteriorates the degradation behavior due to agglomeration of MgO, reducing the passive layer. With the increased MgO content, the ultimate strength increases while the ductility decreases. The maximum yield strength obtained was 300 ± 3 MPa with the addition of 0.5 wt% of MgO.

5.3 Results and Discussion

Physiochemical Characterization of Biodegradable Nanocomposites

An ideal combination of properties is required to engineer multi-functional materials for bone tissue regeneration. The researchers focus on developing a biodegradable composite with excellent biocompatibility, suitable porosity, significant mechanical properties, and controlled biodegradability *in vivo* for potential application in bone tissue regeneration. Moreover, the properties mentioned above should be in a predefined range for their best performance in orthopedics. Thus, in this section, efforts have been made to elaborate on the effect of different reinforcement and fabricating techniques on the mechanical, degradation, and biocompatibility behavior of biodegradable nanocomposites.

5.3.1 Biodegradation Study

According to conventional wisdom, an implant's capacity for *in vivo* bone bonding and calcification are correlated. For a sequence of chemical processes to start at the exposed surface in the physiological environment and generate an apatite layer after implantation, the implants must be bioactive. The establishment of a link with the native bone requires this activity. Additionally, the breakdown behavior of composites in the case of biodegradable nanocomposite implants should correspond with the mending of native bone tissues. The most effective method for confirming composites' potential to regenerate bone tissue is to examine how they degrade in a physiological context. Consequently, it is crucial that the biodegradable nanocomposites go through the degrading test to make sure that implant materials are safe. Simulated bodily fluids (SBF) and phosphate buffer saline are the two most often utilized degradation testing media (PBS). Proteins (fetal bovine serum) have also been used in SBF in recent investigations on Mg-based nanocomposites to simulate the genuine surrounding bodily fluids [64]. The weight loss of biodegradable composites is the main metric used to assess the breakdown behavior of these materials. On the behavior of the nanocomposites' biodegradation, there are numerous publications accessible [64, 65]. Due to the physiological environment's negative standard potential of -2.4 V, pure Mg degrades quickly. The corrosion behavior of magnesium includes the dissolving of anodic metal and the decomposition of cathodic water, which causes the development of hydrogen gas and alkalization of the corrosion medium [66–68]. When submerged in water, the magnesium hydroxide layer forms on the surface of the magnesium matrix. Chlorine in the immersion media causes the conversion of magnesium hydroxide into magnesium chloride, which encourages the breakdown of the Mg surface and causes severe pitting biocorrosion on the magnesium implant. The overall degrading behaviors are as follows:

$$Mg \rightarrow Mg^{+2} + 2e^{-} \tag{5.2}$$

$$2H_2O + 2e^- \rightarrow 2OH^- + H_2(g) \quad (5.3)$$

$$Mg(s) + 2H_2O \rightarrow Mg(OH)_2(s) + H_2(g) \quad (5.4)$$

$$Mg + 2Cl^- \rightarrow MgCl_2 \quad (5.5)$$

$$Mg(OH)_2(s) + 2Cl^-(aq) \rightarrow MgCl_2 + H_2(g) \quad (5.6)$$

The above equations can be summarized in Figure 5.7.

Due to homoepitaxy, HA as a reinforcement provides a thick layer of apatite deposition on the composite surfaces [70]. One of the intriguing research conducted by the authors' team evaluated the biodegradation behavior of Mg-3Zn/HA nanocomposite in two distinct immersion mediums (simulated bodily fluid (SBF) and FBS) for 3, 7, and 14 days. [65]. According to reports, adding nanoHA to the magnesium matrix significantly improved the corrosion resistance. The number of immersion days affects weight gain in terms of the creation of the apatite layer during immersion experiments. After a given number of days, the rate of apatite layer development stabilizes after initially increasing with the number of days. The degradation behavior of Mg-3Zn/HA nanocomposite made using the SPS approach is also reported, as is the impact of nanoHA shape. [71]. In this study, the scientists reinforced two distinct nanoHA forms (spherical and cylindrical) in a matrix of magnesium and zinc. In comparison to a matrix (Mg-3Zn) and round nanoHA (15 wt.%), the addition

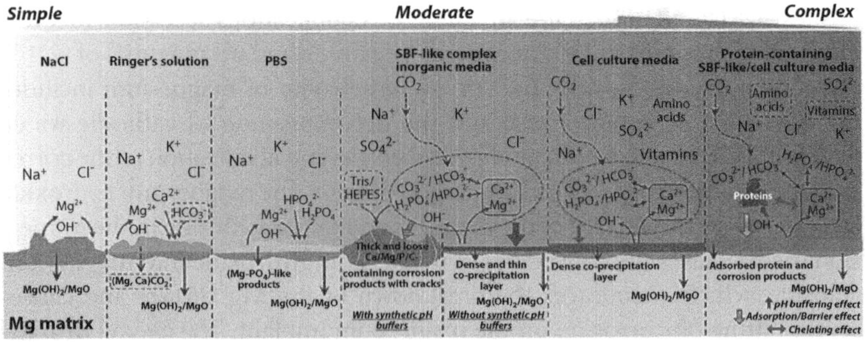

Figure 5.7 The chemical reactions to reveal the corrosion behavior of Mg in the physiological environment [69] Reproduced with permission from ref [69]. Copyright 2015 Elsevier.

of 15 wt.% cylindrical nanoHA in the Mg matrix increased the corrosion resistance by 63 and 27%. The ability of both forms of HA (round and spherical) to precipitate apatite was roughly comparable. Figure 5.8 displays the mass gain, rate of degradation, and SEM images of the apatite layer on nanocomposites. Better degradation resistance was made possible by the integration of nanoHA and increased grain boundary areas brought about by the refining of the grains in the nanocomposites.

In another report, micro galvanic corrosion existed in Mg/β-TCP composite produced through SPS [52]. Indeed the β-TCP promotes the Ca-P layer, but then after a certain amount of β-TCP, Ca starts diffusing into the

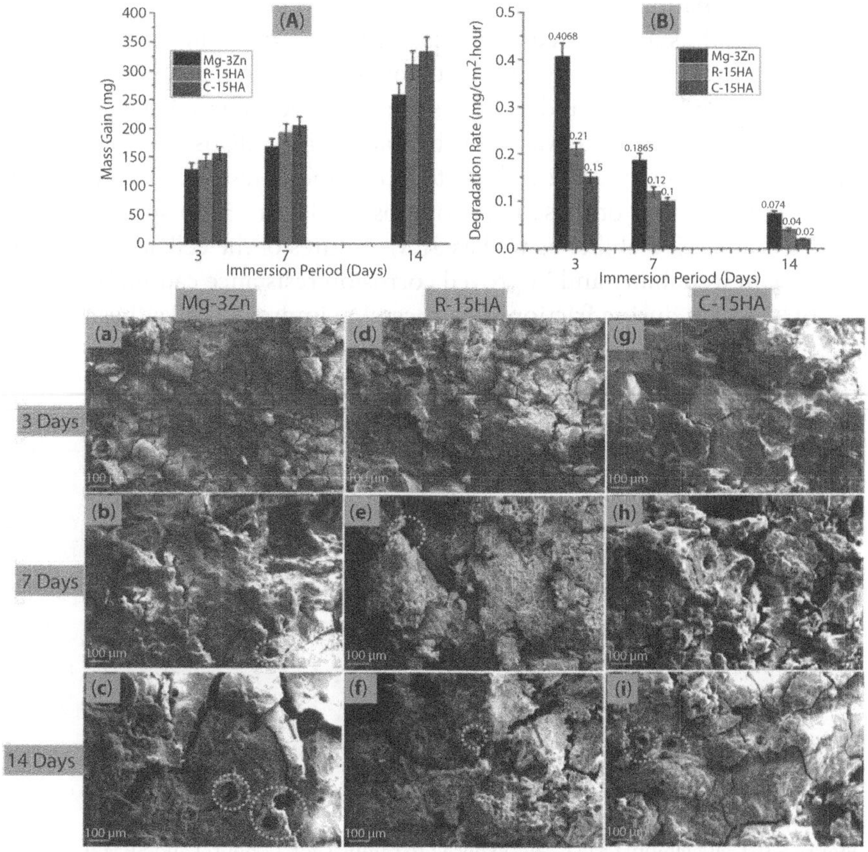

Figure 5.8 The mass gain behavior (A), the degradation resistance (B), and SEM image of apatite layers (a–i) formed on the biodegradable nanocomposites up to 14 days of immersion in modified simulated body fluids (m-SBF) [65] Reproduced with permission from ref [65]. Copyright 2020 Elsevier.

Mg matrix during fabrication, hence microgalvinc corrosion. Although a concentration of βTCP was 20 wt% in the Mg matrix, it showed higher cell adhesion, which resembles the effect of β-TCP on bone marrow-derived mesenchymal stem cells BMSC cell lines.

5.3.2 Biocompatibility

When designing orthopedic implants, biocompatibility is a crucial factor to take into account. The intended biomaterial shouldn't trigger an immune reaction. Additionally, the existence of toxicity in biodegradable nanocomposites has the potential to render the entire biological system inert. The created implants are subjected to a number of *in vitro* tests to evaluate their biocompatibility behavior. Cell viability, MTT assay, DAPI staining, and live/dead assay are frequently used in these procedures. To evaluate the cytocompatibility of the implants for bone tissue regeneration in this context, MG-63, hFOB, and hBMSCs are intensively investigated cell lines. Mg-nHA nanocomposite was made by Ratnasunil *et al.* using a friction stir manufacturing method [72]. The cytocompatibility test was used to evaluate the effectiveness of these nanocomposites. Figure 5.9 displays the outcomes of the cell viability as well as SEM pictures of the cell adhesion. The increase in wettability and improved corrosion resistance caused by adding nanoHA to Mg during friction stir processing improved cell adhesion.

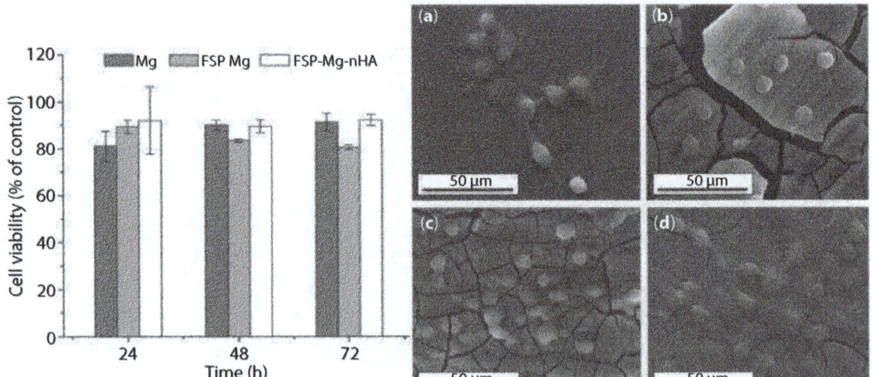

Figure 5.9 Cell viability test (L cells exposed to samples for 72 hours) and SEM image of cell adhesion on (a) control, (b) Mg, (c), friction stirred Mg (d) friction stirred Mg-nanoHA composite [72] Reproduced with permission from ref [72]. Copyright 2020 Elsevier.

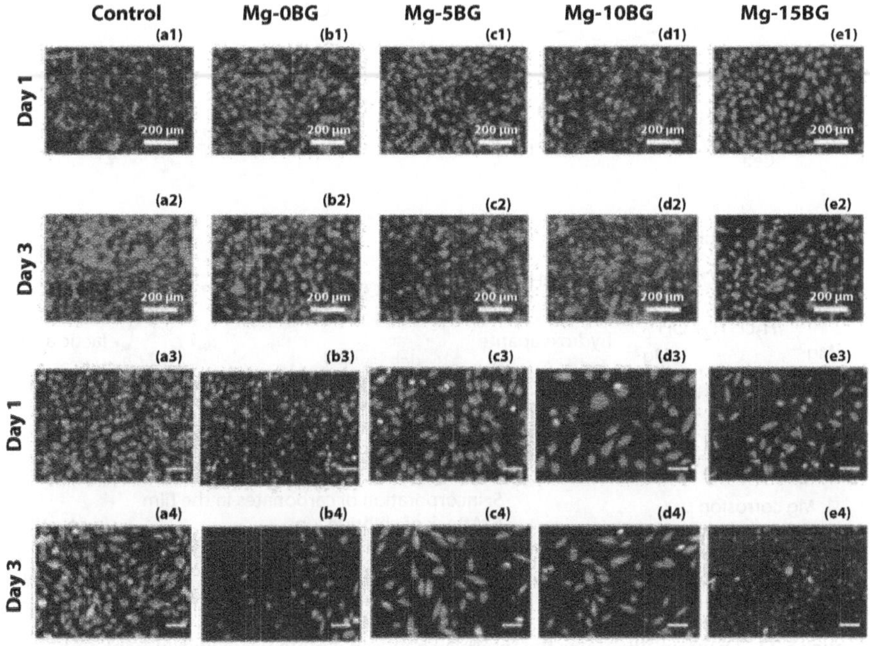

Figure 5.10 Fluorescent images of Mg/BG composite live-dead assay on MG-63 cells through hot pressing extracted after 1 (a1–e1) and 3 days (a2–e2). Green color denotes live cells whereas red color dead cells. Reproduced with permission from ref [21]. Copyright 2017 John Wiley & Sons.

Xiong *et al.* [51] reported the alkaline phosphatase activity (ALP) activity of Mg-HAP composites and showed better cell adhesion and enhanced biocompatible behavior of the composites. Dutta *et al.* [68] also showed the bioactive of magnesium- BG by Alamar and dead-alive assay and reported that Mg with 10 wt% BG composites showed better cytocompatibility. Further, another study showed that in the MTT assay of 24, 48, and 72 h, the BG increased the relative growth rate of L929 cells in Figure 5.10 due to the enhanced corrosion resistance [73].

The processes in the schematic shown in Figure 5.11 can be noted as follows:

- As discussed above the corrosion of magnesium release hydrogen gas, Mg^{2+} ions, and alkalization.
- The hydroxide ions produced during the cathodic reaction are detrimental to cells

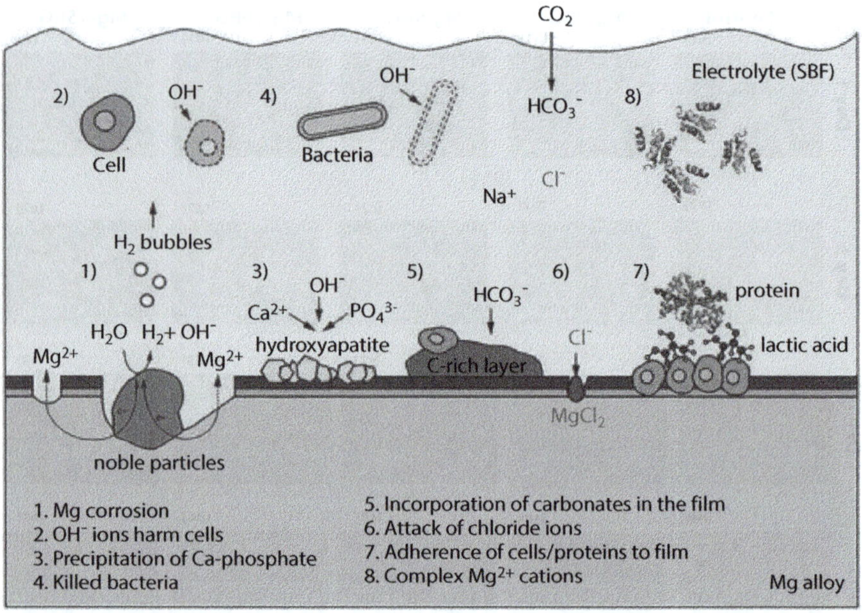

Figure 5.11 Diagram illustrates the interplay between a biological environment and corroding magnesium alloy [74] Reproduced with permission from ref [74]. Copyright 2017 Elsevier.

- The calcium phosphates formed on the surface are due to an increase in the level of pH
- The increase in the alkaline behavior kills the bacteria
- Carbonates are introduced surface of the layers incorporated
- Magnesium chloride is formed on the attack of chlorine
- Immediate action is protein adsorption followed by cells attachment on the Mg surface. lactic acid is produced by the cells
- Proteins might complex the Mg^{2+} cations.

5.3.3 *In Vivo* Assessment of the Nanocomposites for Tissue Compatibility

The ability of nanocomposites to control cellular behavior has spurred a flood of study in the orthopedic sector. There is undoubtedly some sort of response between the inorganic or organic components in the biological milieu when bone tissues come into touch with biomaterials in living body fluids. According to the most recent studies on the nano-bio interface, healthy cells' cell membranes can easily be breached by nanoscale materials, which can then enter the vasculature and be distributed throughout numerous tissues [75].

However, the most important factor in assessing whether nanocomposites are suitable for use in orthopedic applications is their biocompatibility. European Society for Biomaterials defined biocompatibility as "*the ability of a material to perform with an appropriate host response in a specific application*" [75]. The latest definition of biocompatibility "*is the ability of a biomaterial to perform its desired function concerning medical therapy, without eliciting any undesirable local or systemic effect in the recipient or beneficiary of that therapy, but generating the most appropriate beneficial cellular tissue response in that specific situation, and optimizing its clinically relevant performance*" [76]. Biocompatibility is thought to be determined either *in vivo* through implantation or injection of the test material, or *in vitro* utilizing appropriate cells that are important in regeneration processes. These two standardized tests are used to evaluate a material's biological behavior. According to the standard practice, three crucial variables should be taken into account: the chemical analysis of the hazardous item, biosafety information, and historical usage of the material.

The International Organization for Standardization (ISO) technical committee was established in the early 1990s with the goal of developing a set of standards for biological evaluation of medical devices on a globally harmonized basis. In accordance with ISO 10993-1 (2018), the degree of toxicity of medical devices is determined by certain characteristics and the extent of exposure. Other negative characteristics include hypersensitivity, irritability, chronic severe toxicity, genotoxicity, carcinogenicity, reproductive toxicity, and developmental toxicity [77]. The mechanical characteristics, wettability, and surface chemistry of biomaterials play a role in how cells react to nanocomposites. In addition to physical characteristics, the biological properties of the material play a crucial role in encouraging the adhesion and differentiation of non-forming cells in order to achieve bone regeneration. Hydrophilicity, surface energy, surface area, the material's conjugates, and functional groups also have an impact on cellular functions and, consequently, the biocompatibility of *in vivo* tissue. Based on the characteristics of the nanocomposite surface and how it interacts with cells, the materials can be standardized for bone tissue engineering (BTE). Genome sequencing can be used to conduct additional characterization using gradient energy technology [78]. Eventually, an evaluation of *in vivo* tissue compatibility is reached as a result of the complexity of the cellular response to nanocomposites. By changing a single parameter, new nanomaterials can be created with highly controlled properties. However, it is very difficult to concurrently analyze the destiny of the cells after nanocomposite stimulation. If the rate of degradation of the components of the nanocomposite varies, there are restrictions on its efficacy. Due to this, the experimental system is incoherent, and the analysis of the findings must take this into serious consideration. Various sorts of biocomposites that were created to direct osteoblast response are highlighted in some papers.

5.4 Conclusion and Future Outlook

Small-fracture bone repair is still a challenging problem that requires significant funding for orthopedic implant research and development. Temporary biodegradable implants are much more effective at healing and stabilizing broken bones than permanent non-degradable ones. Researchers studying biomaterials have become interested in temporary orthopedic implants because of their ability to degrade. Biodegradable composite materials that are used for bone healing enable a crawling bridge to fix shattered bone and offer a favorable environment for cells and growth factors to aid in the procedure. The development of a biodegradable implant that has received clinical approval must first undergo extensive, long-term biocompatibility testing. Corrosion rate must be adjusted to coincide with bone regeneration rate. Furthermore, the implants must possess important mechanical qualities. A temporary implant that has received clinical approval must be biodegradable, osteoconductive, low immunogenic, and antimicrobial. Numerous novel implant materials have been researched for potential use in orthopedic disciplines as a result of technological advancements. However, the Mg-based biodegradable implants continue to offer important benefits for bone regenerations (removing the issue of re-surgery and diminishing the stress shielding effect). The exceptional qualities of natural adhesion ligands that improve cell adhesion, outstanding mechanical strength, corrosion resistance, and cellular interactions have also been demonstrated by Mg-based biodegradables. The development of a smart biodegradable implant is still based on these ordinary biodegradable implants, despite the fact that they also feature important necessary qualities. The most recent development in fracture-fixing accessories is implants constructed of nanocomposites, which take advantage of the advantages of various biodegradable materials. In addition to acting intelligently in response to various signals in the microenvironment during bone regeneration, biodegradable nanocomposite implants should also have a biomimetic design that extends beyond the traditional implants themselves. Innovative implants that were created employing cutting-edge fabrication methods and clever biodegradable nanocomposite materials would be an example of current development in the field of bone tissue fixing implants. Researchers in the fields of biomaterials, chemistry, materials science, and engineering are working together to create intelligent, biodegradable nanocomposite implants for orthopedic use.

References

1. Zheng, Y.F., Gu, X.N., Witte, F., Biodegradable metals. *Mater. Sci. Eng. R Rep.*, 77, 1–34, 2014, https://doi.org/10.1016/j.mser.2014.01.001.
2. Abed, E. and Moreau, R., Importance of melastatin-like transient receptor potential 7 and magnesium in the stimulation of osteoblast proliferation and migration by platelet-derived growth factor. *Am. J. Physiol. - Cell Physiol.*, 297, C360–C368, 2009, https://doi.org/10.1152/ajpcell.00614.2008.
3. Saris, N.E.L., Mervaala, E., Karppanen, H., Khawaja, J.A., Lewenstam, A., Magnesium: An update on physiological, clinical and analytical aspects. *Clin. Chim. Acta*, 294, 1–26, 2000, https://doi.org/10.1016/S0009-8981(99)00258-2.
4. Yamamoto, A. and Hiromoto, S., Effect of inorganic salts, amino acids and proteins on the degradation of pure magnesium *in vitro*. *Mater. Sci. Eng. C*, 29, 1559–1568, 2009, https://doi.org/10.1016/j.msec.2008.12.015.
5. Wang, J.L., Xu, J.K., Hopkins, C., Chow, D.H.K., Qin, L., Biodegradable magnesium-based implants in orthopedics—A general review and perspectives. *Adv. Sci.*, 7, 1902443, 2020, https://doi.org/10.1002/advs.201902443.
6. Witte, F., The history of biodegradable magnesium implants: A review. *Acta Biomater.*, 6, 1680–1692, 1938, 2010, https://doi.org/10.1016/j.actbio.2010.02.028.
7. McBride, E.D., Magnesium screw and nail transfixion in fractures. *South Med. J.*, 31, 508 514, https://doi.org/10.1097/00007611-193805000-00010.
8. Maier, O., Über die Verwendbarkeit von Leichtmetallen in der Chirurgie (metallisches Magnesium als Reizmittel zur Knochenneubildung). *Dtsch Zeitschrift Für Chir*, 253, 552–556, 1940, https://doi.org/10.1007/BF02794837.
9. Yang, Y., He, C., Dianyu, E., Yang, W., Qi, F., Xie, D. *et al.*, Mg bone implant: Features, developments and perspectives. *Mater. Des.*, 185, 108259, 2020, https://doi.org/10.1016/j.matdes.2019.108259.
10. Saha, P., Roy, M., Datta, M.K., Lee, B., Kumta, P.N., Effects of grain refinement on the biocorrosion and *in vitro* bioactivity of magnesium. *Mater. Sci. Eng. C*, 57, 294–303, 2015, https://doi.org/10.1016/j.msec.2015.07.033.
11. Brar, H.S., Wong, J., Manuel, M.V., Investigation of the mechanical and degradation properties of Mg-Sr and Mg-Zn-Sr alloys for use as potential biodegradable implant materials. *J. Mech. Behav. BioMed. Mater.*, 7, 87–95, 2012, https://doi.org/10.1016/j.jmbbm.2011.07.018.

12. Zhang, S., Zhang, X., Zhao, C., Li, J., Song, Y., Xie, C. *et al.*, Research on an Mg-Zn alloy as a degradable biomaterial. *Acta Biomater.*, 6, 626–640, 2010, https://doi.org/10.1016/j.actbio.2009.06.028.
13. Zhang, E., He, W., Du, H., Yang, K., Microstructure, mechanical properties and corrosion properties of Mg-Zn-Y alloys with low Zn content. *Mater Sci. Eng. A*, 488, 102–111, 2008, https://doi.org/10.1016/j.msea.2007.10.056.
14. Zhang, X., Yuan, G., Mao, L., Niu, J., Fu, P., Ding, W., Effects of extrusion and heat treatment on the mechanical properties and biocorrosion behaviors of a Mg-Nd-Zn-Zr alloy. *J. Mech. Behav. BioMed. Mater.*, 7, 77–86, 2012, https://doi.org/10.1016/j.jmbbm.2011.05.026.
15. Yuen, C.K. and Ip, W.Y., Theoretical risk assessment of magnesium alloys as degradable biomedical implants. *Acta Biomater.*, 6, 1808–1812, 2010, https://doi.org/10.1016/j.actbio.2009.11.036.
16. Seong, J.W. and Kim, W.J., Development of biodegradable Mg-Ca alloy sheets with enhanced strength and corrosion properties through the refinement and uniform dispersion of the Mg2Ca phase by high-ratio differential speed rolling. *Acta Biomater.*, 11, 531–542, 2015, https://doi.org/10.1016/j.actbio.2014.09.029.
17. Ding, Y., Li, Y., Lin, J., Wen, C., Effects of zirconium and strontium on the biocorrosion of Mg-Zr-Sr alloys for biodegradable implant applications. *J. Mater. Chem. B*, 3, 3714–3729, 2015, https://doi.org/10.1039/c5tb00433k.
18. Ding, Y., Lin, J., Wen, C., Zhang, D., Li, Y., Mechanical properties, corrosion, and biocompatibility of Mg-Zr-Sr-Dy alloys for biodegradable implant applications. *J. BioMed. Mater. Res. - Part B Appl. Biomater.*, 106, 2425–2434. 2018, https://doi.org/10.1002/jbm.b.34051.
19. Li, Y., Wen, C., Mushahary, D., Sravanthi, R., Harishankar, N., Pande, G. *et al.*, Mg-Zr-Sr alloys as biodegradable implant materials. *Acta Biomater.*, 8, 3177–3188, 2012, https://doi.org/10.1016/j.actbio.2012.04.028.
20. Allen, M.J., Myer, B.J., Millett, P.J., Rushton, N., The effects of particulate cobalt, chromium and cobalt-chromium alloy on human osteoblast-like cells *in vitro*. *J. Bone Joint Surg. Br.*, 79-B, 475–482, 1997, https://doi.org/10.1302/0301-620x.79b3.0790475.
21. Dutta, S., Devi, K.B., Gupta, S., Kundu, B., Balla, V.K., Roy, M., Mechanical and *in vitro* degradation behavior of magnesium-bioactive glass composites prepared by SPS for biomedical applications. *J. BioMed. Mater. Res. - Part B Appl. Biomater.*, 107B, 352–365, 2019, https://doi.org/10.1002/jbm.b.34127.
22. Khanra, A.K., Jung, H.C., Hong, K.S., Shin, K.S., Comparative property study on extruded Mg-HAP and ZM61-HAP composites. *Mater. Sci. Eng. A*, 527, 6283–6288, 2010, https://doi.org/10.1016/j.msea.2010.06.031.
23. Chen, B., Yin, K.Y., Lu, T.F., Sun, B.Y., Dong, Q., Zheng, J.X. *et al.*, AZ91 magnesium Alloy/Porous hydroxyapatite composite for potential application in bone repair. *J. Mater. Sci. Technol.*, 32, 858–864, 2016, https://doi.org/10.1016/j.jmst.2016.06.010.

24. Yu, X., Tang, X., Gohil, S.V., Laurencin, C.T., Biomaterials for bone regenerative engineering. *Adv. Healthc. Mater.*, 4, 1268–1285, 2015, https://doi.org/10.1002/adhm.201400760.
25. Kasten, P., Beyen, I., Niemeyer, P., Luginbühl, R., Bohner, M., Richter, W., Porosity and pore size of β-tricalcium phosphate scaffold can influence protein production and osteogenic differentiation of human mesenchymal stem cells: An *in vitro* and *in vivo* study. *Acta Biomater.*, 4, 1904–1915, 2008, https://doi.org/10.1016/j.actbio.2008.05.017.
26. Gao, P., Zhang, H., Liu, Y., Fan, B., Li, X., Xiao, X. *et al.*, Beta-tricalcium phosphate granules improve osteogenesis *in vitro* and establish innovative osteo-regenerators for bone tissue engineering *in vivo*. *Sci. Rep.*, 6, 23367, 2016, https://doi.org/10.1038/srep23367.
27. Jones, J.R., Review of bioactive glass: From hench to hybrids. *Acta Biomater.*, 9, 4457–4486, 2013, https://doi.org/10.1016/j.actbio.2012.08.023.
28. El-Rashidy, A.A., Roether, J.A., Harhaus, L., Kneser, U., Boccaccini, A.R., Regenerating bone with bioactive glass scaffolds: A review of *in vivo* studies in bone defect models. *Acta Biomater.*, 62, 1–28, 2017, https://doi.org/10.1016/j.actbio.2017.08.030.
29. Hench, L.L., Opening paper 2015- some comments on bioglass: Four eras of discovery and development. *BioMed. Glas.*, 1, 1, 2015, https://doi.org/10.1515/bglass-2015-0001.
30. Turdean-Ionescu, C., Stevensson, B., Grins, J., Izquierdo-Barba, I., García, A., Arcos, D. *et al.*, Composition-dependent *in vitro* apatite formation at mesoporous bioactive glass-surfaces quantified by solid-state NMR and powder XRD. *RSC Adv.*, 5, 86061–86071, 2015, https://doi.org/10.1039/c5ra13410b.
31. Qiu, K., Wan, C.X., Zhao, C.S., Chen, X., Tang, C.W., Chen, Y.W., Fabrication and characterization of porous calcium polyphosphate scaffolds. *J. Mater. Sci.*, 41, 2429–2434, 2006, https://doi.org/10.1007/s10853-006-5182-2.
32. Grynpas, M.D., Pilliar, R.M., Kandel, R.A., Renlund, R., Filiaggi, M., Dumitriu, M., Porous calcium polyphosphate scaffolds for bone substitute applications *in vivo* studies. *Biomaterials*, 23, 2063–2070, 2002, https://doi.org/10.1016/S0142-9612(01)00336-2.
33. Niu, Y., Cao, L., Wei, J., Ma, Y., Song, S., Weng, W. *et al.*, Development of a bioactive composite of nano fluorapatite and poly(butylene succinate) for bone tissue regeneration. *J. Mater. Chem. B*, 2, 1174–1181, 2014, https://doi.org/10.1039/c3tb21371d.
34. Fathi, M.H. and Mohammadi Zahrani, E., Mechanical alloying synthesis and bioactivity evaluation of nanocrystalline fluoridated hydroxyapatite. *J. Cryst. Growth*, 311, 1392–1403, 2009, https://doi.org/10.1016/j.jcrysgro.2008.11.100.
35. Zhang, Y., Nayak, T., Hong, H., Cai, W., Biomedical applications of zinc oxide nanomaterials. *Curr. Mol. Med.*, 13, 1633–1645, 2013, https://doi.org/10.2174/1566524013666131111130058.

36. Hao, L., Hu, Y., Zhang, Y., Wei, W., Hou, X., Guo, Y. *et al.*, Enhancing the mechanical performance of poly(ether ketone)/zinc oxide nanocomposites to provide promising biomaterials for trauma and orthopedic implants. *RSC Adv.*, 8, 27304–27317, 2018, https://doi.org/10.1039/c8ra01736k.
37. Espitia, P.J.P., Soares N de, F.F., Coimbra JS dos, R., de Andrade, N.J., Cruz, R.S., Medeiros, E.A.A., Zinc oxide nanoparticles: Synthesis, antimicrobial activity and food packaging applications. *Food Bioprocess Technol.*, 5, 1447–1464, 2012, https://doi.org/10.1007/s11947-012-0797-6.
38. Khatami, M., Varma, R.S., Zafarnia, N., Yaghoobi, H., Sarani, M., Kumar, V.G., Applications of green synthesized Ag, ZnO and Ag/ZnO nanoparticles for making clinical antimicrobial wound-healing bandages. *Sustain Chem. Pharm.*, 10, 9–15, 2018, https://doi.org/10.1016/j.scp.2018.08.001.
39. Wu, C. and Chang, J., Synthesis and *in vitro* bioactivity of bredigite powders. *J. Biomater. Appl.*, 21, 251–263, 2007, https://doi.org/10.1177/0885328206062360.
40. Yi, D., Wu, C., Ma, B., Ji, H., Zheng, X., Chang, J., Bioactive bredigite coating with improved bonding strength, rapid apatite mineralization and excellent cytocompatibility. *J. Biomater. Appl.*, 28, 1343–1353, 2014, https://doi.org/10.1177/0885328213508165.
41. Wu, C., Chang, J., Wang, J., Ni, S., Zhai, W., Preparation and characteristics of a calcium magnesium silicate (bredigite) bioactive ceramic. *Biomaterials*, 26, 2925–2931, 2005, https://doi.org/10.1016/j.biomaterials.2004.09.019.
42. Goh, C.S., Gupta, M., Wei, J., Lee, L.C., Characterization of high performance Mg/MgO nanocomposites. *J. Compos. Mater.*, 41, 2325–2335, 2007, https://doi.org/10.1177/0021998307075445.
43. Lei, T., Ouyang, C., Tang, W., Li, L.F., Zhou, L.S., Enhanced corrosion protection of MgO coatings on magnesium alloy deposited by an anodic electrodeposition process. *Corros. Sci.*, 52, 3504–3508, 2010, https://doi.org/10.1016/j.corsci.2010.06.028.
44. Paulauskas, I.E., Modeshia, D.R., Ali, T.T., El-Mossalamy, E.H., Obaid, A.Y., Basahel, S.N. *et al.*, Photocatalytic activity of doped and undoped titanium dioxide nanoparticles synthesised by flame spray pyrolysis. *Platin. Met. Rev.*, 57, 32–43, 2013, https://doi.org/10.1595/147106713X659109.
45. Pc, E., Radhakrishnan, G., Emarose, S., Investigation into physical, microstructural and mechanical behaviour of titanium dioxide nanoparticulate reinforced magnesium composite. *Mater. Technol.*, 36, 575–584, 2021, https://doi.org/10.1080/10667857.2020.1782050.
46. Munir, K., Wen, C., Li, Y., Graphene nanoplatelets-reinforced magnesium metal matrix nanocomposites with superior mechanical and corrosion performance for biomedical applications. *J. Magnes. Alloy*, 8, 269–290, 2020, https://doi.org/10.1016/j.jma.2019.12.002.
47. Kirkland, N.T., Schiller, T., Medhekar, N., Birbilis, N., Exploring graphene as a corrosion protection barrier. *Corros. Sci.*, 56, 1–4, 2012, https://doi.org/10.1016/j.corsci.2011.12.003.

48. Jaiswal, S., Kumar, R.M., Gupta, P., Kumaraswamy, M., Roy, P., Lahiri, D., Mechanical, corrosion and biocompatibility behaviour of Mg-3Zn-HA biodegradable composites for orthopaedic fixture accessories. *J. Mech. Behav. BioMed. Mater.*, 2018, https://doi.org/10.1016/j.jmbbm.2017.11.030.
49. Dubey, A., Jaiswal, S., Haldar, S., Roy, P., Lahiri, D., Mg-3Zn/HA biodegradable composites synthesized via spark plasma sintering for temporary orthopedic implants. *J. Mater. Eng. Perform.*, 78, 442–454, 2019, https://doi.org/10.1007/s11665-019-04315-5.
50. del Campo, R., Savoini, B., Muñoz, A., Monge, M.A., Garcés, G., Mechanical properties and corrosion behavior of Mg-HAP composites. *J. Mech. Behav. BioMed. Mater.*, 39, 238–246, 2014, https://doi.org/10.1016/j.jmbbm.2014.07.014.
51. Xiong, G., Nie, Y., Ji, D., Li, J., Li, C., Li, W. *et al.*, Characterization of biomedical hydroxyapatite/magnesium composites prepared by powder metallurgy assisted with microwave sintering. *Curr. Appl. Phys.*, 16, 830–836, 2016, https://doi.org/10.1016/j.cap.2016.05.004.
52. Narita, K., Tian, Q., Johnson, I., Zhang, C., Kobayashi, E., Liu, H., Degradation behaviors and cytocompatibility of Mg/β-tricalcium phosphate composites produced by spark plasma sintering. *J. BioMed. Mater Res. - Part B Appl. Biomater.*, 107B, 2238–2253, 2019, https://doi.org/10.1002/jbm.b.34316.
53. Cui, Z., Zhang, Y., Cheng, Y., Gong, D., Wang, W., Microstructure, mechanical, corrosion properties and cytotoxicity of betacalcium polyphosphate reinforced ZK61 magnesium alloy composite by spark plasma sintering. *Mater. Sci. Eng. C*, 2019, https://doi.org/10.1016/j.msec.2019.02.050.
54. Huan, Z., Leeflang, S., Zhou, J., Zhai, W., Chang, J., Duszczyk, J., *In vitro* degradation behavior and bioactivity of magnesium-Bioglass® composites for orthopedic applications. *J. BioMed. Mater. Res. - Part B Appl. Biomater.*, 99, 1035–1047, 2012, https://doi.org/10.1002/jbm.b.31968.
55. Razavi, M., Fathi, M.H., Meratian, M., Microstructure, mechanical properties and bio-corrosion evaluation of biodegradable AZ91-FA nanocomposites for biomedical applications. *Mater. Sci. Eng. A*, 527, 6938–6944, 2010, https://doi.org/10.1016/j.msea.2010.07.063.
56. Lei, T., Tang, W., Cai, S.H., Feng, F.F., Li, N.F., On the corrosion behaviour of newly developed biodegradable Mg-based metal matrix composites produced by *in situ* reaction. *Corros. Sci.*, 54, 270–277, 2012, https://doi.org/10.1016/j.corsci.2011.09.027.
57. Dezfuli, S.N., Huan, Z., Mol, A., Leeflang, S., Chang, J., Zhou, J., Advanced bredigite-containing magnesium-matrix composites for biodegradable bone implant applications. *Mater. Sci. Eng. C*, 79, 647–660, 2017, https://doi.org/10.1016/j.msec.2017.05.021.
58. Shuai, C., Wang, B., Bin, S., Peng, S., Gao, C., Interfacial strengthening by reduced graphene oxide coated with MgO in biodegradable Mg composites. *Mater. Des.*, 191, 108612, 2020, https://doi.org/10.1016/j.matdes.2020.108612.

59. Dutta, S., Gupta, S., Roy, M., Recent developments in magnesium metal-matrix composites for biomedical applications: A review. *ACS Biomater. Sci. Eng.*, 6, 4748–4773, 2020, https://doi.org/10.1021/acsbiomaterials.0c00678.
60. Ye, X., Chen, M., Yang, M., Wei, J., Liu, D., *In vitro* corrosion resistance and cytocompatibility of nano-hydroxyapatite reinforced Mg-Zn-Zr composites. *J. Mater. Sci. Mater. Med.*, 21, 1321–1328, 2010, https://doi.org/10.1007/s10856-009-3954-3.
61. Huang, Y., Liu, D., Anguilano, L., You, C., Chen, M., Fabrication and characterization of a biodegradable Mg-2Zn-0.5Ca/1β-TCP composite. *Mater. Sci. Eng. C*, 54, 120–132, 2015, https://doi.org/10.1016/j.msec.2015.05.035.
62. Huan, Z., Zhou, J., Duszczyk, J., Magnesium-based composites with improved *in vitro* surface biocompatibility. *J. Mater. Sci. Mater. Med.*, 21, 3163–3169, 2010, https://doi.org/10.1007/s10856-010-4165-7.
63. Lin, G., Liu, D., Chen, M., You, C., Li, Z., Wang, Y. *et al.*, Preparation and characterization of biodegradable Mg-Zn-Ca/MgO nanocomposites for biomedical applications. *Mater. Charact.*, 144, 120–130, 2018, https://doi.org/10.1016/j.matchar.2018.06.028.
64. Dubey, A., Jaiswal, S., Lahiri, D., Assessment of biomechanical stability and formulation of a statistical model on magnesium based composite in two different milieus. *J. Mech. Behav. BioMed. Mater.*, 111, 103980, 2020, https://doi.org/10.1016/j.jmbbm.2020.103980.
65. Jaiswal, S., Dubey, A., Haldar, S., Roy, P., Lahiri, D., Differential *in vitro* degradation and protein adhesion behaviour of spark plasma sintering fabricated magnesium-based temporary orthopaedic implant in serum and simulated body fluid. *BioMed. Mater.*, 15, 015006, 2020, https://doi.org/10.1088/1748-605X/ab4f8b.
66. Witte, F., Hort, N., Vogt, C., Cohen, S., Kainer, K.U., Willumeit, R. *et al.*, Degradable biomaterials based on magnesium corrosion. *Curr. Opin. Solid State Mater. Sci.*, 12, 63–72, 2008, https://doi.org/10.1016/j.cossms.2009.04.001.
67. Staiger, M.P., Pietak, A.M., Huadmai, J., Dias, G., Magnesium and its alloys as orthopedic biomaterials: A review. *Biomaterials*, 27, 1728–34, 2006, https://doi.org/10.1016/j.biomaterials.2005.10.003.
68. Dutta, S., Bavya Devi, K., Roy, M., Processing and degradation behavior of porous magnesium scaffold for biomedical applications. *Adv. Powder Technol.*, 28, 3204–3212, 2017, https://doi.org/10.1016/j.apt.2017.09.024.
69. Mei, D., Lamaka, S.V., Lu, X., Zheludkevich, M.L., Selecting medium for corrosion testing of bioabsorbable magnesium and other metals – A critical review. *Corros. Sci.*, 171, 108722, 2020, https://doi.org/10.1016/j.corsci.2020.108722.
70. Shakir, M., Jolly, R., Khan, M.S., Iram, N.E., Khan, H.M., Nano-hydroxyapatite/chitosan-starch nanocomposite as a novel bone construct: Synthesis and *in vitro* studies. *Int. J. Biol. Macromol.*, 80, 282–292, 2015, https://doi.org/10.1016/j.ijbiomac.2015.05.009.

71. Jaiswal, S., Dubey, A., Lahiri, D., The influence of bioactive hydroxyapatite shape and size on the mechanical and biodegradation behaviour of magnesium based composite. *Ceram. Int.*, 46, 27205–27218, 2020, https://doi.org/10.1016/j.ceramint.2020.07.202.
72. Ratna Sunil, B., Sampath Kumar, T.S., Chakkingal, U., Nandakumar, V., Doble, M., Friction stir processing of magnesium-nanohydroxyapatite composites with controlled *in vitro* degradation behavior. *Mater. Sci. Eng. C*, 39, 315–24, 2014, https://doi.org/10.1016/j.msec.2014.03.004.
73. Dutta, S., Devi, K.B., Mandal, S., Mahato, A., Gupta, S., Kundu, B. *et al.*, *In vitro* corrosion and cytocompatibility studies of hot press sintered magnesium-bioactive glass composite. *Materialia*, 5, 100245, 2019, https://doi.org/10.1016/j.mtla.2019.100245.
74. Esmaily, M., Svensson, J.E., Fajardo, S., Birbilis, N., Frankel, G.S., Virtanen, S. *et al.*, Fundamentals and advances in magnesium alloy corrosion. *Prog. Mater. Sci.*, 89, 92–193, 2017, https://doi.org/10.1016/j.pmatsci.2017.04.011.
75. Iqbal, H.M.N. and Keshavarz, T., The challenge of biocompatibility evaluation of biocomposites. *Biomed. Compos.*, 303–334, 2017, https://doi.org/10.1016/b978-0-08-100752-5.00014-7.
76. Williams, D.F., On the mechanisms of biocompatibility. *Biomaterials*, 29, 2941–2953, 2008, https://doi.org/10.1016/j.biomaterials.2008.04.023.
77. Biological evaluation of medical devices. *BioMed. Saf. Stand*, 26, 7, 54, 1996, https://doi.org/10.1097/00149078-199604150-00011.
78. Michelmore, A., Clements, L., Steele, D.A., Voelcker, N.H., Szili, E.J., Gradient technology for high-throughput screening of interactions between cells and nanostructured materials. *J. Nanomater.*, 1–7, 2012, https://doi.org/10.1155/2012/839053

6

New Frontiers of Bioinspired Polymer Nanocomposite for Biomedical Applications

Sonika[1], Gopikishan Sabavath[2*], Sushil Kumar Verma[3], Ram Swaroop[4] and Arbind Prasad[5]

[1]*Department of Physics, Rajiv Gandhi University, Rono Hills, Doimukh, Itanagar, India*
[2]*Department of Physics, University of Allahabad, Prayagraj, Uttar Pradesh, India*
[3]*Department of Chemical Engineering, Indian Institute of Technology Guwahati, Guwahati, India*
[4]*Department of Physics, Central University of Punjab, Bathinda, Punjab, India*
[5]*Department of Mechanical Engineering, Katihar Engineering College (under DST, Govt. of Bihar), Katihar, Bihar, India*

Abstract
Polymers are very imperative materials and widely used but due to their physical properties having limited applications. Applications of polymers tremendously increased by fillers like magnetic and graphene materials. Exceptional properties like electrical, thermal and mechanical of graphene led to chosen as a filler in polymers. In this chapter, graphene-polymer nano-composites and their biomedical applications presented. Fillers creating opportunities to apply in diversified areas therefore, nanocomposites have been growing continuously. Magnetic-polymer nanocomposites also have attracting with considerable attention due to their amazing dispersion stability, biocompatibility, and magnetism. In this period diverse types of magnetic-polymer nanocomposites also exist. At the end of this chapter, progress in graphene and magnetic polymer nanocomposites, and also discussed future trends of this potential research field.

***Keywords*:** Graphene, magnetic material, polymer nanocomposites, biomedical applications, nano filler, polymer matrix, nanocomposites

**Corresponding author*: gopikishan@allduniv.ac.in

Arbind Prasad, Ashwani Kumar and Manoj Gupta (eds.) Advanced Materials and Manufacturing Techniques for Biomedical Applications, (135–156) © 2024 Scrivener Publishing LLC

6.1 Introduction

A fundamental insight of the interactions between polymers, nanostructures, and cells is necessary for the strategy and synthesis of bio-inspired nanomaterials for biomedical uses. The majority of polymers interacted with various nanomaterials to develop composite structures with the desired properties. The promising potential of polymer nanocomposites for a wide range of electric, magnetic, optical, and biological applications has recently drawn strong scientific interest. The advantages of polymeric nanocomposites combine improved features including lightweight, flexibility, processability, and low cost of the finished product. The emphasis is on polymeric nanocomposites, with basic major strategies all over. First, bioinspired structures with a fibrous design, second, the current state of complex nanocomposites with many functions is evaluated. The importance of medical technology, which has enhanced the standards for biomedical materials, has led to the rise of biomedical applications as an essential research field among the aforementioned domains. Advanced materials with special attributes are in high demand.

6.1.1 Polymers Used in Biomedical Applications

Polymer nanocomposites are very important and attractive materials for biomedical applications. Natural polymers like proteins (fibrin gels, soy, collagen) or polysaccharides (alginate, starch, chitin/chitosan) and synthetics polymers poly (lactic acid), poly (hydroxyl butyrate), poly (ε-caprolactone) and glycolic acid etc. polymers used in biomedical applications. Similarly, water dissolvable polymers poly(vinyl acetate), poly (ethylene glycol), poly (acrylic acid), poly (vinyl acetic acetate) were used for the part of this applications. Conducting polymers polypyrrole (PP), polyaniline (PANI) and poly-vinylperrolidone (PVP) are used. Further, the research community found to add other polymers polypropylene (PP), polyethylene (PE) and polyurethane (PU) for biomedical applications.

6.1.2 Graphene-Polymer Nanocomposites

Graphene is one of the allotropes of the carbon newly discovered by scientists. It has many times greater thermal conductivity than the Copper (Cu) metal. Moreover it has high surface area than the single wall carbon nanotubes (SWNT). Graphene consider as superior form of carbon based on its properties reported as [1].

1. Outstanding electricity
2. It is the newest form of carbon
3. It has huge speculative specific area 2630 m^2/g
4. It is thinnest and lightest material
5. It has extremely excessive thermal conductivity $Wm^{-1} K^{-1}$
6. It has high Young's modulus 1.0PTA
7. It has highest intrinsic mobility $2\times10^5 cm^2 V^{-1}s^{-1}$
8. It has optical transmittance more than 98%
9. It has zero band gap

Therefore, one can consider graphene as rising star for nano-composite materials and foremost optional material for composition. Allotropes of the carbon presented in Figure 6.1.

Polymers enhancing their properties such as thermal or electrical conductivity, stiffness including controlled interfacial interactions with reinforcement loads [2]. Carbon based nanoparticles such as carbon nanotubes (CNT), carbon block (CB), exfoliated graphite have been regularly used as reinforcement to polymer nano-composites. Graphene has exceptional properties like high electronic transport, high thermal conductivity, and advanced mechanical properties therefore attracted industrial and academically interest. Filling of graphene to polymer matrix conductivity enhancement of the polymer leads to apply in the field of health science like biomedical implants, drug distribution, diabetic monitoring and tissue, biosensors, artificial joints, cancer therapy etc. [3, 4]. Surface area of the polymer can be increase by filling titanium(Ti) and carbon (C) which helps to store the drugs [5]. Drugs like herapin, dexamethasone, dopamine, naproxen, 2-ethylhexyl phosphate have been successfully delivered using conducting polymers [6–9].

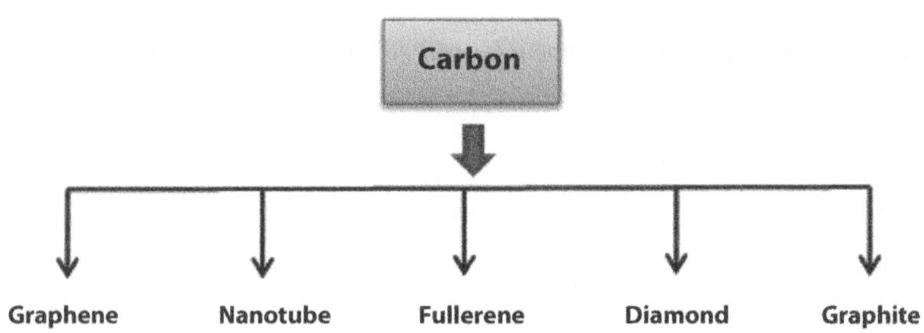

Figure 6.1 Allotropes of carbon [1].

6.2 Methods to Prepare Graphene-Based Polymer Nanocomposites

Basically there are three methods [Figure 6.2]: 1. *In situ* polymerization, 2. Solution blending method, 3. Melt compounding method. *In-situ* polymerization is useful for two types grapheme based polymer nanocomposites without any bonding and with covalent bonding and it is observed that better novel properties of the reinforcement in case of covalent interaction. Schematic diagram of *in-situ* polymerization is presented in Figure 6.3. Various polymers like PMMA, epoxy, nylon, HDPE, polystyrene have been used as mediums graphene polymer nanocomposites [10].

Solution mixing technique is another very imperative technique but research community not frequently applied like *in-situ* polymerization and melts compounding methods. Using this technique more than two or two polymer matrix and nano filler mixed with proper solvent at the desirable temperature and for homogeneously mixing can reaches with stirring. This method frequently applied for insulating and conducting polymers. In this method sample can be casted in the form of film by softening both the polymers in a collective solvent. For instance polypyrrole-clay based nano composite and poly-aniline and graphene were reported [11, 12]. Melt compounding method is most commonly used method. In this method

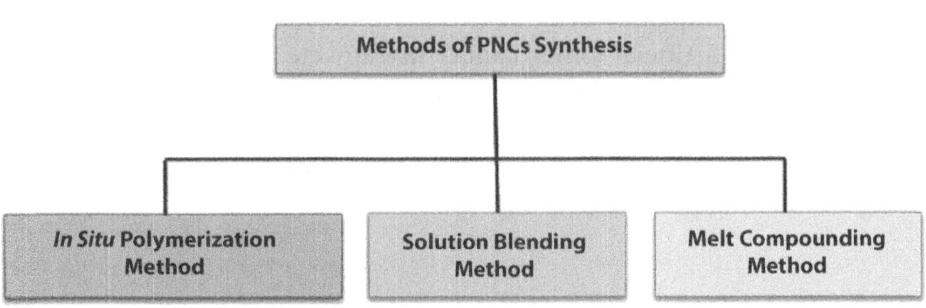

Figure 6.2 Simple methods for graphene based PNCs preparation method [1].

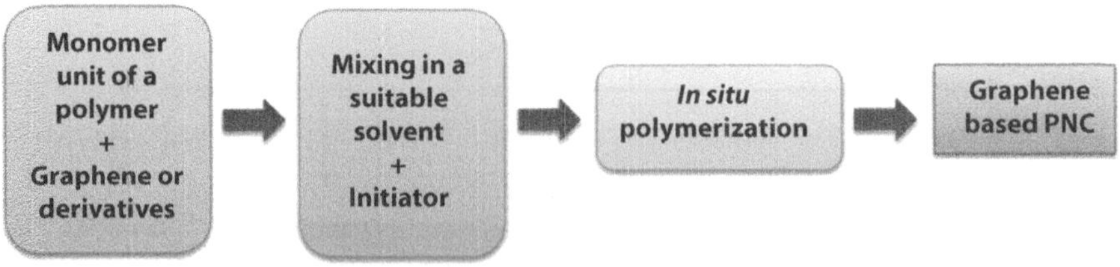

Figure 6.3 Schematic diagram for the *in-situ* polymerization of grapheme (Gr)-based polymer nanocomposites [1].

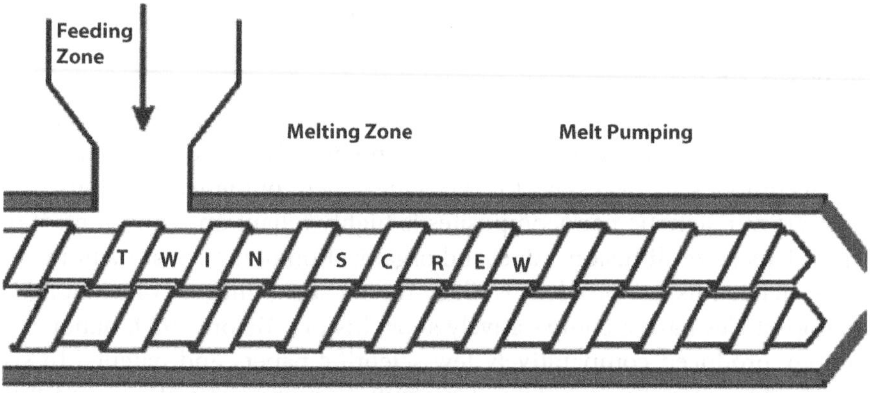

Figure 6.4 Illustrative depiction of extruder used in melts compounding approach [1].

with the help of extruder two or more than two polymer matrix can be filled with nanofiller. Extruder will help to mix the components below their melting point. Melt compounding method is having importance because to produce large scale polymer nanocomposites [13]. The diagram illustration of extruder [Figure 6.4] is shown above.

6.3 Magnetic Material – Polymer Nanocomposites

Materials made of an inorganic magnetic constituent surrounded in an carbon-based polymer, or vice versa, having at minimum one aspect in the nanometer range (1–100 nm), are known as magnetic polymer nanocomposites. Incorporating multiple component elements and their qualities into one substance is the fundamental concept behind a composite. Synergies between organic and inorganic components add additional properties to magnetic polymer nanocomposites that are not possible with either organic or inorganic components alone. Because the charismatic component is nanometer-sized, its compelling characteristics can diverge qualitatively from those of the bulk and change depending on the particle dimension. High processability, intriguing mechanical, optical, and electrical properties are all provided by the polymer. Additionally, the polymer interphase can have a major impact on the magnetic features, and the ability of polymers to be structured can be exploited to regulate the magnetic interactions between particles. In order to reduce functional area, enhance magnetic performance, modulate magnetic characteristics, multi-functionalize, and create materials with distinctive qualities, magnetic nanocomposites offer a wide range of design opportunities. Biomedical applications of polymer nanocomposites have been presented in Figure 6.5.

A deeper ability of the physical science at the nano-scale, the creation of description techniques, and the development of cheap cost manufacturing techniques are necessary for the achievement of these aspirations. The concert of magnetic nanocomposites is highly reliant on a fine control of size, shape, size-diffusion, edge, and organization of the constituent materials, hence this last problem is likely the most significant [14]. Magnetic nanoparticles have a comparatively short history. Before 1989, nanotechnology produced comparatively few scientific papers and patents. In six years, there were three orders of magnitude more publications and two orders of magnitude more patents [15]. The problem then rose to the top of government and business research.

The inherent properties of the materials suggests to progress by characterization tools with the nano scale range (SPM, HRTEM, STEM, SAXS,

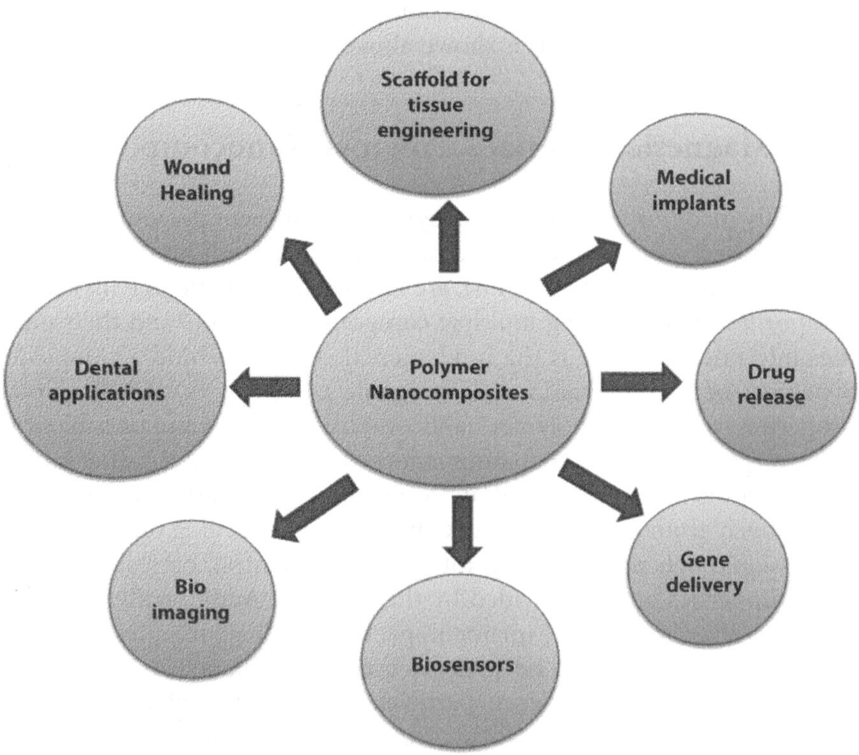

Figure 6.5 Biomedical and biotechnological applications of polymer nanocomposites.

EELS, XMCD), the emergence of synthetic bio-mimetization techniques, the development toward smaller devices, and the identification of novel nano-related occurrences (super-paramagnetism, ferro-antiferrobias exchange, GMR, TMR, quantum tunneling). In magnetic nanocomposites, the polymer can take on a variety of functions. Mangetic core polymer shell works as a covering shield from the atmosphere, an isolator, a separator, or a condensing medium. Due to the integration of the specific features of the cores and shells that achieve cooperatively enhanced performance, magnetic porous materials with core-shell architectures have recently received a lot of attention. Core-shell magnetic nanoparticles are recognized for their exceptional qualities, comprising enhanced stability, the ability to shield the active species from challenging situations, improved physical, chemical, and photoelectric properties, and convenience of surface functionalization, among several other things. The precise coordination of the interaction between their cores and shells, along with their controlled and innovative design, is what gives them all their interesting synergistic properties.

However, the field of magnetic nanocomposites generating the largest activity, actual realizations, and expectations is certainly biomedicine. As divergence agents in magnetic resonance imaging, they were used for the first time in this field(MRI) [16]. As an illustration, the use of nanoparticles has increased the analysis of lymphatic metastasis from 60% to 94%. They now have interest in a wide range of other things: Drug targeting, molecular biology, immunoassays, DNA refining, cell separation and purification, hyperthermia therapy, and diagnostics come next [17]. Magnetic inks, rotating shaft sealing, reciters, shock absorption systems, magnetic sensors, etc., are some further contemporary uses for magnetic ferrofluids. Their applications in radio-electronics, MEMS, improved photo systems [18], and are currently being worked on other industrially relevant technologies include magnetic refrigeration and X-ray switching [19]. Additionally, polymer nano-composites (PNC) are employed as transitional materials in the production of nano-organized inorganic magnetic materials, including spotted media [20–22], and magnetic hollow spheres [23]. They are highly intriguing in basic magnetism and other technical domains, such as life science [24] and ground and global sciences, in addition to their industrial value.

In nature, iron oxides are typically found as nanoparticles. An essential instrument for tracking the evolution of the earth's crust is the study of its magnetic characteristics. For instance, the discovery of magnetite nanoparticles in a Stains meteorite that are very indistinguishable to those formed by magneto tactic microorganisms constitutes pertinent proof of extraterrestrial life [25]. Studying the development of residues, the sea floor, soils, and mainstay foundation also uses magnetic investigations as evidence. Animals that employ them as navigational aids also create iron oxide magnetic nanoparticles [26]. In mammals, they are also used to store iron (Fe), and their development is irregularly linked to illnesses [27].

Magnetic polymer nanocomposites can take many different shapes. Each of these forms is created to produce the best results possible for a certain purpose. The show of composites is associated to their qualities, which are resolute by their configuration and arrangement, important to the growth of methods for their research, following the fundamental chain in materials study in an inductive path. The various magnetic nanocomposites will be given in accordance with this strategy in the following portion of this chapter. Before, in subsequent sections, each component of the materials exploration chain will be reviewed, beginning with an explanation of the processes used to create magnetic nanocomposites and continuing with an explanation of the fundamental methods used to characterize their structure and key physical characteristics. We shall comment on the upcoming trends in this sector before we conclude.

6.3.1 Organization of Magnetic Polymer Nanocomposites

Magnetic polymer-nanocomposites can take the form of tridimensional solids, fibers, films, and multilayers in addition to powders and solutions. According to their respective qualities, they can be employed in an extensive range of applications in each of these forms. Figure 6.6 depicts a scheme for this classification.

6.3.2 Residues and Suspensions

Biomedical uses superparamagnetic nanoparticles which are frequently used in them. A high magnetic moment that enables the recognition, parting, and/or path of biotic objects involved to the particles, as well as their thermal influence in an oscillating magnetic field, which is useful for hyperthermia therapy, are two fundamental magnetic properties that are related to their performance. A magnetic vital, a biocompatible covering, and one or more physiologically active particles (drugs, antibodies, enzymes, etc.) linked to the apparent are the typical components of magnetic biomedical materials.

They are particularly helpful in diagnostics as detecting agents. They have paved the way for new developments in immunoassays, early disease detection, and magnetic biosensors (10–13 mole/L), among other

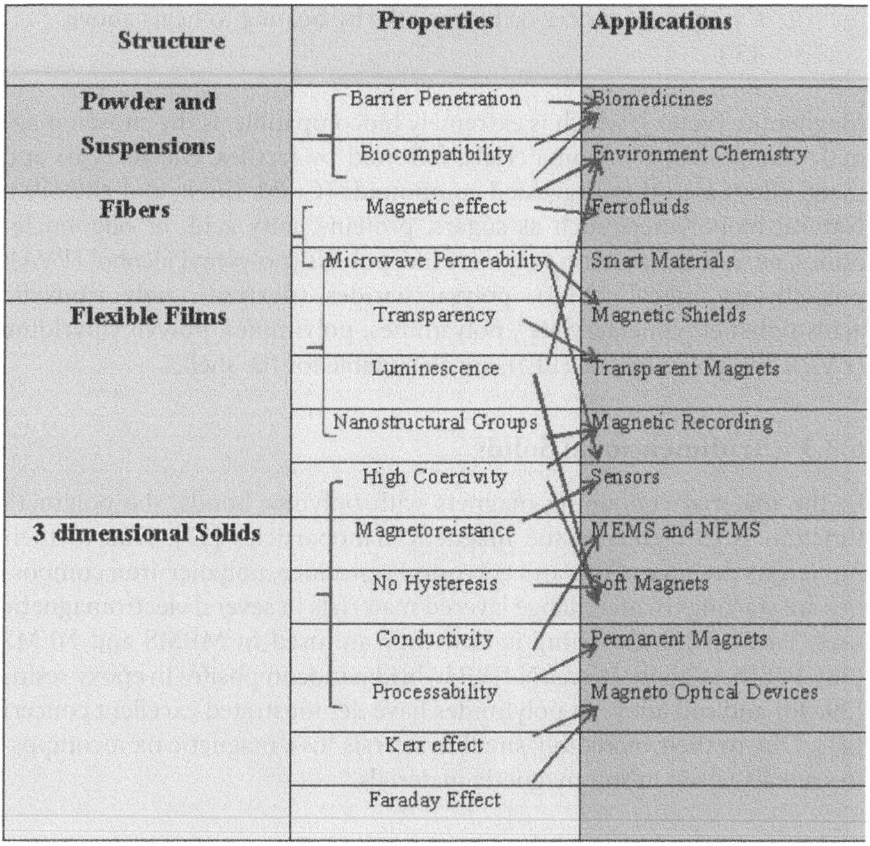

Figure 6.6 Arrangement of magnetic polymer nanocomposites.

fields [28–30]. Their application in MRI enables the concurrent, in vivo visualization of biological methods, such as cell following [31] and DNA manifestation [32], with a determination that is nearly molecular [33]. Currently, they are utilized to diagnose cancer, particularly lymphatic and brain tumors [34]. They have been widely employed as separation tools for blood purification [35], DNA separation, and cell sorting [36]. They are employed as targetable drug carriers to keep medications in small blood plasma irrigation areas and to localize the medications in the damaged extent to prevent adverse properties on the relaxation of the body.

They can perform two functions in hyperthermia:

1. Increasing medicine activity or muscle restoration by heating to heat less 41°C.

2. Cytolysis of cancer or tumor cells by heating to heats above 45°C.

Maghemite (Fe_2O_3), which is extremely biocompatible, is the chosen magnetic component for biomaterials, followed by ferrites. Metals (Co) and metal alloys are other suggested compounds (CoPd, CoNi, and NiCoPd). Natural biopolymers such as sugars, proteins, fatty acid, or oligonucleotides, as well as synthetic polymers like polyols (polyvinyl alcohol (PVA), polyethylene glycol (PEG), polysaccharides (dextran), polycarboxylic acids polymethylmetacrylate , polyamines, polyimines, polyvinylpyridine (PVP), and polyethylene (PE)), are all suitable for the shell.

6.3.3 Tridimensional Solids

In the material containing magnets with polymer bonds, the polymer's function is to condense the magnetic nanoparticles [37]. Due to their numerous design options and corrosion resistance, polymer-iron composites are starting to interchange layered materials in several electromagnetic uses. Particularly intriguing is how they are used in MEMS and NEMS [38]. For these applications, NdFeB hybrid nanocomposites in epoxy resins [39, 40] and soil alloys in polyimides have demonstrated excellent concert [41]. Due to their incredibly small hysteresis loss, magnetic nanocomposites are also used in soft magnetic materials.

6.3.4 High-Permeability Materials for the Microwave

In imperative to enhance the concert of electromagnetic strategies such as small control converters, inductors, and whiskers in arrangements technology [42], magnetic shielding [43], and microwave absorbers for interior and army resolutions, there is a robust request for materials with a high permeability and a low influence loss at high frequencies [44]. The ferromagnetic resonance controls a magnetic material's microwave characteristics. The resonance's frequency and amplitude should both be as high as possible to achieve high microwave permeability [45]. In comparison to composites formed of micron size(μ) particles, nanoparticles exhibit a considerable increase in the real and unreal parts of the absorptivity as well as a narrowing of the magnetic resonance peak [46] that explains the amazing 0.1–18 GHz microwave absorption of metal nanocomposites [47]. For use at extremely high frequencies, or as inductive mechanisms in mobile communications, nanocomposites of ferrites or metals in epoxy resins have been projected [48].

6.3.5 Piezoelectric Materials

High magnetostrictive strain and a high rupture modulus are both displayed by polymer-bonded nanocomposites with oriented nanoparticles in relatively modest magnetic fields. TbDyFeTerfenol-D polymer bound magnets exhibit this behavior [49].

There are also polymer nanocomposites of magnetic nanoparticles in thermo setting or thermoplastic adhesive compounds that can be used for magnetic glueing, as well as nanocomposites of iron (Fe) oxide in a polystyrene (PS) resin that have been projected as magnetic conserving materials. In reaction to imposed magnetic fields, nanocomposite polymer gels exhibit remarkable phenomena, including large deformational properties, high elasticity, anisotropic characteristics, transient support, and rapid reaction to magnetic field [50]. These gels have been created using magnetite and hydroxypropyl cellulose [51].

6.3.6 Multifunctional Materials

Modern sensors and devices are made primarily of multifunctional materials. Following are just a few examples of the numerous ways that magnetism can be combined with other properties in polymer nanocomposites: the creation of sensors and gadgets can benefit from the usage of conductive magnetic materials. They are made up of compelling nanoparticles surrounded in a matrix of different conductive polymer [52]. The material can function as a single electronic system by establishing a charge transfer among the particle surface and the polymer [53]. A few compositions that have been suggested include magnetite-polyaniline, maghemite-polypyrrole(PPy), cobalt ferrite-polypyrrole(PPy), and a number of other metal-polymer combinations [54].

6.3.6.1 Transparent Magnetic Materials

As nanoparticles compared to their bulk counterparts, magnetic oxides are significantly further transparent to visible light [55, 56]. With magnetism that is more than an order of magnitude stronger than that of the durable transparent magnets, magnetic nanocomposites can be produced [57].

6.3.6.2 Luminescent Magnetic Materials

They can be helpful in bioanalytical tests that combine detection and separation. For example, MCF-7 breast cancer cells have been separated and seen using polymer coated Fe_2O_3 cores and CdSe/ZnS shells with restrained anti-cyclin E antibodies [58]. 3D (three dimensional) binary superlattices of self-assembled Fe_2O_3 and PbSe nanoparticles have been future as materials with possibly tunable optical and magnetic characteristics [59].

6.4 Nanostructured Composites

Previous sections have already covered this topic. It does, however, merit some further words due to its importance. The polymeric composite based on bioactive materials and bioresorbable polymers also play a vital role in orthopedic applications [60–65]. There are various advantages such as stress shielding, metal ions leaching and resurgery is not required after fracture healing thus bioresorbable composite in the field of biomedical application shows various worthy applications in these days [66–69]. The polymer (matrix detention, involuntary assembly by molecular deference), the inorganic component (interparticle conjugation, outline directed association), a substrate (substrate production), or a magnetic field can all be used to create nanostructural ordering. There are several different techniques for arranging organic-inorganic composites in a laminar stack. Some of them are two-dimensional perovskites [70], sequential deposition of poly-electrolites [71], and Langmuir-Blodgett films [72]. The preparation of multilayered core-shell nanoparticles can also benefit from layer-by-layer assembly [73, 74]. However, using block copolymers is the most adaptable method for creating ordered magnetic nanostructures. In the solid state, linear (A)n(B)m chain diblock-copolymers can self-assemble structures that can be sustained throughout centimeter-scale length ranges. With different values for n and m, the structures' shapes can be altered from scopes to tubes, gyroids, and lamellae. A variety of nanostructures, comprising hairy nanospheres, shaved nanospheres, porous nanospheres, hollow nanospheres, nanospheres with cross-linked shells, crosslinked polymer clashes, nanofibers, and nanochannels in polymeric thin films, were produced by selectively softening one of the lumps, measured by crosslinking and selective poverty [75]. For the conception of systematic magnetic nanocomposites, block copolymers such as aniline, aminophenol and formaldehyde [76, 77], poly(norbornene-dicarboxylic acid) and poly(norbornene) [78, 79], vinylstyrene-4-vinylpyridineand poly(4-vinylpyridine) [80], poly(styrene-b-2-vinyl pyridine) [81], acrylonitrile-methyl methacryl-divinyl-benzene [82], polystyrene-polymethylmethacrylate [83] etc. A fixed magnetic field can cause structural orientation [84], while irregular magnetic fields have been employed to yield nanocomposites with rotationally liberated particles [85]. It has been said that when it comes to the formation of nanostructures, nature is the master. So why not use live things to make superior magnetic nanocomposites? Several lines are available in this direction:

- Magnetite-protein nanoparticles of various dimensions and shapes are produced by genetically engineering magnetotactic bacteria [86].
- Magnetic nanoparticles with a regulated dimension, shape, and organization have been created using viruses [87].

6.5 Conclusion and Future Trends

Graphene itself is having outstanding properties therefore with different polymer matrices have been applied via various methods for biomedical applications. It was also presented that the graphene and graphene oxides are significant materials to get the stable dispersions and improve the compatibility with polymers. Advantage with graphene-polymer nanocomposite is with desired biomedical applications it is controllable [2]. Additive manufacturing also known as 3D printing based graphene based polymer nanocomposites is extensive focused research area towards the advanced applications in energy storage, electronics, and healthcare. This research area is under investigation and having the challenges like cost, efficiency, practical issues, and mass production. In this method computer-aided design (CAD) models can be fabricating complex structures with relatively low cost and short time. Additive manufacturing applied in biomedical field for tissue engineering to design artificial tissue drug delivery systems, organ transplants and tools to detect orthoses, prostheses, specific medical parameter and medical instruments for surgery and diagnostics [88–90].

Magnetic nanocomposites are primarily being developed in two different directions. One of them is concerned with biomedical applications, while the other is working on techniques for creating nano-organized assemblies and their uses in sensors, electronics, and recording media. These are among the most recent among the major European and American nanotechnology programs. Two hundred magnetic nanoparticles' use for imaging, splitting, guiding, and heating has been well established in the field of biomedicine. Looking ahead, there are numerous biological processes to be observed, illnesses to be identified, preparations to convey or localize, therapies to be enhanced by hyperthermia, materials and cells to be refined or disconnected, and devices for microanalysis to be intended, all of which can make use of magnetic nanoparticles. The concert of the materials in use is, however, far from ideal. Colloidal chemistry (mono-disperse elements), organic-inorganic interaction (stable shells suitable for attaching natural vectors), biochemistry (finding living vectors), hematology and toxicology (biocompatibility evaluates), infections (such as cancer), description and magnetism (structure-properties models), and engineering would all be required (building magnetic devices to produce and extent magnetic fields). The exploration in this area has been fairly split up to now. The conception of multidisciplinary groups that can approach the issue from a worldwide viewpoint is the way of the future. Numerous comprehensive programs have already been started in this approach [91]. We can specifically note the Defense Science Office's Bio-Magnet ICs program, which is focused on the formation of magnetically targeted therapies, magnetic base microanalysis devices, and biosensors [92]. Additionally, a number of businesses are creating and researching magnetic nanoparticles for biomedical uses.

The mechanism of element extent and association by self-assembly, block copolymers, dendrimers, or fluid crystals is the focus of most of the research in the area of nanostructured magnetic nanocomposites. These techniques have piqued the interest of major players in the industry because they make it possible to easily fabricate nanodevices under being setting [93]. The work of the magnetic ingredients assembly at NIST serves as an exemplary example of study in this area [94–96].

High manufacturing costs present a further challenge for other applications, including printing, magnetic shielding, microwave applications, etc. Designing economically cost manufacture techniques for magnetic nanoparticles and films is also another topic for future advancements [97–99].

References

1. Tiwari, S.K., Kumar, V., Huczko, A., Oraon, R., Adhikari, A.D., Nayak, G.C., Magical allotropes of carbon: Prospects and applications. *Crit. Rev. Solid State Mater. Sci.*, 41, 257–317, 2016.
2. Silva, M., Alves, N.M., Paiva, M.C., Graphene-polymer nanocomposites for biomedical applications. *Polym. Adv. Technol.*, 29, 687–700, 2017.
3. Sharma, S., Sudhakara, P., Abdoulhdi, A., Omran, B., Singh, J., Ilyas, R.A., Recent trends and developments in conducting polymer nanocomposites for multifunctional applications. *Polymers*, 13, 2898, 2021.
4. Singh, D.K. and Verma, R.K., Development of reduced graphene oxide modified ultrahigh molecular weight polyethylene (rGO/UHMWPE) based

nanocomposites for biomedical applications. *J. Thermoplast. Compos. Mater.*, 0, 1–36, 2022.

5. Luo, X., Matranga, C., Tan, S., Alba, N., Cui, X.T., Carbon nanotube nano-reservoir for controlled release of anti-inflammatory dexamethasone. *Biomaterials*, 32, 6316–6323, 2011.
6. Sever, E. and Unal, H., Colloidal properties of surface functionalized nano-cube-TiO_2/poly(3-octylthiophene) core/shell conducting nanocomposite. *Appl. Surf. Sci.*, 355, 1028–1036, 2015.
7. Erdönmez, S., Karabul, Y., Kilic, M., Özdemir, Z., Esmer, K., Structural characterization and dielectric parameters of polyindole/WO_3 nanocomposites. *Polym. Compos.*, 42, 1347–1355, 2021.
8. Galembeck, A. and Alves, O., Chemical polymerization of pyrrole on CeO_2 films. *Synth. Met.*, 84, 151–152, 1997.
9. Karimi, A., Husain, S.W., Hosseini, M., Azar, P.A., Ganjali, M.R., Rapid and sensitive detection of hydrogen peroxide in milk by Enzyme-free electrochemiluminescence sensor based on a polypyrrole-cerium oxide nanocomposite. *Sens. Actuators B Chem.*, 271, 90–96, 2018.
10. Ibrahim, A., Klopocinska, A., Horvat, K., Hamid, Z.A., Graphene-based nanocomposites: Synthesis, mechanical properties, and characterizations. *Polymers*, 13, 2869, 2021.
11. Zhao, L., Xu, Y., Qiu, T., Zhi, L., Shi, G., Polyanilineelectrochromic devices with transparent graphene electrodes. *Electrochim. Acta*, 55, 491, 2009.
12. Aldo, J.G., Transparent and conductive thin films of graphene/polyanilinenanocomposites prepared through interfacial polymerization. *Chem. Commun.*, 47, 2592, 2011.
13. Salavagione, H.J., Covalent grapheme polymernanocomposites, in: *Graphene Materials: Fundamentals and Emerging Applications*, vol. 45, p. 149, 2015.
14. Kodama, R.H., Magnetic nanoparticles. *J. Magn. Magn. Mater.*, 200, 359–372, 1999.
15. Meyer, M., Persson, O., Power, Y., Mapping excellence in nanotechnologies. preparatory study, in: *Nanotechnology Expert Group and Eurotech Data*, European Comission, Brussels, 2001, http://europa.eu.int/comm/research/era/.
16. Weissleder, R. and Papisov, M., Pharmaceutical iron oxides for MR imaging. *Rev. Magn. Reson. Med.*, 4, 1–20, 1992.
17. Jordan, A., Scholz, R., Maier-Hauff, K., Johannsen, M., Wust, P., Nadobny, J., Schirra, H., Schmidt, H., Deger, S., Loening, S., Presentation of a new magnetic field therapy system for the treatment of human solid tumors with magnetic fluid hyperthermia. *J. Magn. Magn. Mater*, 225, 118–126, 2001.
18. Chin, T.S., Permanent magnet films for applications in microelectromechanical systems. *J. Magn. Magn. Mater.*, 209, 75–79, 2000.
19. *Advanced magnetic technologies and consulting*, http://amtc.org/Code/product_rd.htm.

20. Park, M., Chaikin, P.M., Register, R.A., Adamson, D.H., Large area dense nanoscale patterning of arbitrary surfaces. *Appl. Phys. Lett.*, 79, 257–259, 2001.
21. Park, M., Harrison, C., Chaikin, P.M., Register, R.A., Adamson, D.H., Block copolymer lithography: Periodic arrays of similar to 10(11) holes in 1 square centimeter. *Science*, 276, 1401–1404, 1997.
22. Sun, Q.H., Lam, J.W.Y., Xu, K.T., Xu, H.Y., Cha, J.A.K., Wong, P.C.L., Wen, G.H., Zhang, X.X., Jing, X.B., Wang, F.S., Tang, B.Z., Nanocluster-containing mesoporousmagnetoceramics from hyperbranched organometallic polymer precursors. *Chem. Mater.*, 12, 2617–2624, 2000.
23. Caruso, F., Spasova, M., Susha, A., Giersig, M., Caruso, R.A., Magnetic nanocomposite particles and hollow spheres constructed by a sequential layering approach. *Chem. Mater.*, 13, 109–116, 2001.
24. Safarik, I. and Safarikova, M., Magnetic nanoparticles and biosciences. *Monatsh. Chem.*, 133, 737–759, 2002.
25. Thomas-Keprta, K.L., Bazylinski, D.A., Kirschvink, J.L., Clemett, S.J., MacKay, D.S., Wentworth, S.J., Vali, H., Gigson, E.K., Romanek, C.S., Elongated prismatic magnetite crystals in ALH84001 carbonate globules: Potential Martian magnetofossils. *Geochim. Cosmochim. Acta*, 64, 4049–4081, 2000.
26. Diebel, C.E., Proksch, R., Green, C.R., Neilson, P., Walker, M.M., Magnetite defines a magnetoreceptor. *Nature*, 406, 299–302, 2000.
27. Dobson, J., Nanoscale biogenic iron oxides and neurodegenerative disease. *FEBS Lett.*, 496, 1–5, 2001.
28. Lange, J., Kotitz, R., Haller, A., Trahms, L., Semmler, W., Weitschies, W., Magnetorelaxometry - a new binding specific detection method based on magnetic nanoparticles. *J. Magn. Magn. Mater.*, 252, 381–383, 2002.
29. Bogoyevitch, M.A., Kendrick, T.S., Ng, D.C.H., Barr, R.K., Taking the cell by stealth or storm? Protein transduction domains (PTDs) as versatile vectors for delivery. *DNA Cell Biol.*, 21, 879–894, 2002.
30. Grossman, H.L., Myers, W.R., Vreeland, V.J., Bruehl, R., Alper, M.D., Bertozzi, C.R., Clarke, J., Detection of bacteria in suspension by using a superconducting quantum interference device. *Proc. Natl. Acad. Sci. U. S. A.*, 101, 129–134, 2004.
31. Bulte, J.W.M., Douglas, T., Witwer, B., Zhang, S., Strable, E., Lewis, B.K., Zywicke, H., Miller, B., Gelderen, P.V., Moskowitz, B.M., Duncan, I.D., Frank, J.A., Magnetodendrimers allow endosomal magnetic labeling and *in vivo* tracking of stem cells. *Nat. Biotech.*, 19, 1141–1147, 2001.
32. Weissleder, R., Moore, A., Mahmood, U., Bhorade, R., Benveniste, H., Chiocca, E.A., Basilion, J.P., *In vivo* magnetic resonance imaging of transgene expression. *Nat. Med.*, 6, 351–354, 2000.
33. Weissleder, R., Scaling down imaging: Molecular mapping of cancer in mice. *Nat. Rev. Cancer*, 2, 11–18, 2002.

34. Hafeli, U., Schutt, W., Teller, J., Zborowski, M., *Scientific and clinical applications of magnetic carriers*, Plenum Press, New York, 1997.
35. Safark, I. and Safarkova, M., Use of magnetic techniques for the isolation of cells. *J. Chromatogr. B*, 722, 33–53, 1999.
36. Willard, M.A., Kurihara, L.K., Carpenter, E.E., Calvin, S., Harris, V.G., Chemically prepared magnetic nanoparticles. *Int. Mater. Rev.*, 49, 125–170, 2004.
37. Fidler, J., Schrefl, T., Hoefinger, S., Hajduga, M., Recent developments in hard magnetic bulk materials. *J. Phys.: Condens. Matter*, 16, S455–S470, 2004.
38. Zahn, M., Magnetic fluid and nanoparticle applications to nanotechnology. *J. Nanopart. Res.*, 3, 73–78, 2001.
39. Pawlowski, B., Schwarzer, S., Rahmig, A., Topfer, J., NdFeB thick films prepared by tape casting. *J. Magn. Magn. Mater.*, 265, 337–344, 2003.
40. Cho, H.J. and Ahn, C.H., Microscale resin-bonded permanent magnets for magnetic micro-electro-mechanical systems applications. *J. Appl. Phys.*, 93, 8674–8676, 2003.
41. Yudin, V.E., Otaigbe, J.U., Bui, T.X., Svetlichnyi, V.M., Polyimide bonded magnets: Processing and properties. *J. Appl. Polym. Sci.*, 88, 3151–3158, 2003.
42. Korenivski, V., GHz magnetic film inductors. *J. Magn. Mater.*, 215, 800–806, 2000.
43. Lagarkov, A.N., Rozanov, K.N., Simonov, A., Starostenko, S.N. (eds.), Microwave permeability of magnetic films, in: *Handbook of Advanced Magnetic Materials*, pp. 1742–1773, Springer, Boston, 2006.
44. Rozanov, K.N., Ultimate thickness to bandwidth ratio of radar absorbers. *IEEE Trans. Ant. Propag.*, 48, 1230–1234, 2000.
45. Nguyen, M.T. and Diaz, A.F., A novel method for the preparation of magnetic nanoparticles in a polypyrrole powder. *Adv. Mater.*, 6, 858–860, 1994.
46. Brosseau, C., Ben Youssef, J., Talbot, P., Konn, A.M., Electromagnetic and magnetic properties of multicomponent metal oxides heterostructures: Nanometer versus micrometer-sized particle. *J. Appl. Phys.*, 93, 9243–9256, 2003.
47. Fujieda, T., Ikeda, S., Suzuki, S., Abe, T., Aono, Y., Microwave absorption properties of Fe-SiO2 nanocomposite powder. *J. Jpn. Inst. Met.*, 66, 135–138, 2002.
48. Brosseau, C., Malle Âgo, S., Que Âffelec, P., Youssef, J.B., Nonreciprocal electro- magnetic properties of nanocomposites at microwave frequencies. *Phys. Rev. B.*, 70, 092401, 2004.
49. Kwon, O.Y., Kwon, Y.D., Kim, D.H., Kim, J.C., Lee, Z.H., Mechanical and magnetic properties of a novel polymer infiltrated directionally solidified RE-Fe giant magnetostrictive materials. *IEEE Trans. Magn.*, 40, 2778–2780, 2004.
50. Varga, Z., Feher, J., Filipcsei, G., Zrinyi, M., Smart nanocomposite polymer gels. *Macromol. Symp.*, 200, 93–100, 2003.

51. Chatterjee, J., Haik, Y., Chen, C.J., Biodegradable magnetic gel: Synthesis and characterization. *Colloid Polym. Sci.*, 281, 892–896, 2003.
52. Gangopadhyay, R. and De, A., Conducting polymer nanocomposites: A brief overview. *Chem. Mater.*, 12, 608–622, 2000.
53. Pokhodenko, V.D., Krylov, V.A., Kurys, Y.I., Posudievsky, O.Y., Nanosized effects in composites based on polyaniline and vanadium or iron oxides. *Phys. Chem. Chem. Phys.*, 1, 905–908, 1999.
54. Grigorev, E.I., Zavyalov, S.A., Chvalun, S.N., The effect of a magnetic field on the conductivity of poly(p-xylylene) films containing iron and nickel nanoparticles. *Polym. Sci. Ser. A.*, 45, 780–784, 2003.
55. Ziolo, R.F., Giannelis, E.P., Weinstein, B.A., Ohoro, M.P., Ganguly, B.N., Mehrotra, V., Russel, M.W., Huffman, D.R., Matris-mediated synthesis of nanocrystalline gamma-Fe2O3-a new optically transparent magnetic material. *Science*, 257, 219–223, 1992.
56. Sohn, B.H. and Cohen, R.E., Processible optically transparent block copolymer films containing superparamagnetic iron oxide nanoclusters. *Chem. Mater.*, 9, 264–269, 1997.
57. Vassiliou, J.K., Mehrotra, V., Otto, J., Wand Dollahon, N.R., Optical absorption tail and super-paramagnetism of gamma-Fe2O3 nanoparticles in a polymer matrix. *Mater. Sci. Forum*, 225, 725–730, 1996.
58. Wang, D.S., He, J.B., Rosenzweig, N., Rosenzweig, Z., Superparamagnetic Fe2O3 Beads-CdSe/ZnS quantum dots core-shell nanocomposite particles for cell separation. *Nano Lett.*, 4, 409–413, 2004.
59. Redl, F.X., Cho, K.S., Murray, C.B., Obrien, C.B., Three-dimensional binary superlattices of magnetic nanocrystals and semiconductor quantum dots. *Nature.*, 423, 968–971, 2003.
60. Bhoi, S., Prasad, A., Kumar, A., Sarkar, R.B., Mahto, B., Meena, C.S., Pandey, C., Experimental study to evaluate the wear performance of UHMWPE and XLPE material for orthopedics application. *Bioengineering*, 9, 676, 2022, https://doi.org/10.3390/bioengineering9110676.
61. Bhoi, S., Kumar, A., Prasad, A., Meena, C.S., Sarkar, R.B., Mahto, B., Ghosh, A., Performance evaluation of different coating materials in delamination for micro-milling applications on high-speed steel substrate. *Micromachines*, 13, 1277, 2022, https://doi.org/10.3390/mi13081277.
62. Gupta, A. *et al.*, Multifunctional nanohydroxyapatite-promoted toughened high-molecular-weight stereocomplex Poly(lactic acid)-based bionanocomposite for both 3D-printed orthopedic implants and high-temperature engineering applications. *ACS Omega*, 2, 7, 4039–52, 2017.
63. Prasad, A., Bioabsorbable polymeric materials for biofilms and other biomedical applications: Recent and future trends. *Mater. Today: Proc.*, 44, 2447–53, 2021a, https://doi.org/10.1016/j.matpr.2020.12.489.
64. Prasad, A., State of art review on bioabsorbable polymeric scaffolds for bone tissue engineering. *Mater. Today: Proc.*, 44, 1391–1400, 2021b, https://doi.org/10.1016/j.matpr.2020.11.622.

65. Prasad, A., Siddharth, B., Vimal, K., Ravi Sankar, M., Biowastes processed hydroxyapatite filled Poly (lactic acid) bio-composite for open reduction internal fixation of small bones. *Mater. Today: Proc.*, 4, 9, 10153–57, 2017, http://linkinghub.elsevier.com/retrieve/pii/S2214785317311410.
66. Prasad, A., Devendar, B., Ravi Sankar, M., Robi, P.S., Micro-scratch based tribological characterization of Hydroxyapatite (HAp) fabricated through fish scales. *Mater. Today: Proc.*, 2, 4–5, 1216–24, 2015, http://linkinghub.elsevier.com/retrieve/pii/S2214785315002795.
67. Prasad, A., Ravi Sankar, M., Vimal, K., State of art on solvent casting particulate leaching method for orthopedic scaffolds fabrication. *Mater. Today: Proc.*, 4, 2, 898–907, 2017, http://dx.doi.org/10.1016/j.matpr.2017.01.101.
68. Prasad, A., Chakraborty, G., Kumar, A., Bio-based environmentally benign polymeric resorbable materials for orthopedic fixation applications, in: *Advanced Materials for Biomedical Applications*, vol. Chapter 15, A. Kumar, Y. Gori, A. Kumar, C.S. Meena, N. Dutt (Eds.), pp. 251–266, CRC Press, Boca Raton, FL, USA, 2022, https://doi.org/10.1201/9781003344810-15.
69. Prasad, A., Biomaterial-based nanofibers scaffolds in tissue engineering application, in: *Functional Biomaterials*, pp. 245–264, Springer, Singapore, 2022.
70. Wortham, E., Zorko, A., Arcon, D., Lappas, A., Organic-inorganic perovskites for magnetic nanocomposites. *Physica B.*, 318, 387–391, 2002.
71. De Cuyper, M. and Joniau, M., Magnetoliposomes ± formation and structural characterization. *Eur. Biophys. J.*, 15, 311–319, 1988.
72. Manne, S. and Aksay, I.A., Thin films and nanolaminates incorporating organic/inorganic interfaces. *Curr. Opin. Solid State Mater.*, 2, 358–364, 1997.
73. Zhu, Y.H., Yang, X.L., Li, P.L., Ying, H., From layer-by-layer assembled core-shell particles to medical/biochemical diagnostics and drug delivery. *Prog. Chem.*, 15, 512–517, 2003.
74. Caruso, F., Susha, A.S., Giersig, M., Mohwald, H., Magnetic core-shell particles: Preparation of magnetite multilayers on polymer latex microspheres. *Adv. Mater.*, 11, 950, 1999.
75. Liu, G., Functional crosslinked nanostructures from block copolymers. *Mater. Sci. Eng. C-Biomim.*, 10, 159–164, 1999.
76. Vadera, S.R., Tuli, A., Kumar, N., Sharma, B.B., Gupta, S.R., Prakash, C., Kishan, P., Preparation and study of finely dispersed magnetic oxide in polymer matrix. *J. Phys. IV.*, 7, 549–550, 1997.
77. Vadera, S.R., Mathur, R., Parihar, M., Kumar, N., Direct synthesis of nanocomposite of gamma-Fe O in the copolymer matrix of aniline-formaldehyde in presence of zinc ions. *Nanostruct Mater.*, 8, 889–898, 1997.
78. Ahmed, S.R., Ogale, S.B., Kofinas, P., Magnetic properties and morphology of block copolymer template ferrimagnetic CoFe2O4 nanoparticles. *IEEE Trans. Magn.*, 39, 2198–2200, 2003.

79. Ahmed, S.R., Ogale, S.B., Papaefthymiou, G.C., Ramesh, R., Kofinas, P., Magnetic properties of CoFe2O4 nanoparticles synthesized through a block copolymer nanoreactor route. *Appl. Phys. Lett.*, 80, 1616–1618, 2002.
80. Chen, L., Yang, W.J., Yang, C.Z., Preparation of nanoscale iron and Fe_3O_4 powders in a polymer matrix. *J. Mater. Sci.*, 32, 3571–3575, 1997.
81. Abes, J.I., Cohen, R.E., Ross, C.A., Block-copolymer-templated synthesis of iron, iron-cobalt, and cobalt-nickel alloy nanoparticles. *Mater. Sci. Eng. C*, 23, 641–650, 2003.
82. Rabelo, D., Lima, E.C.D., Barbosa, D.P., Silva, V.J., Silva, O., Azevedo, R.B., Silva, L.P., Lemos, A.P.C., Morais, P.C., Investigation of magnetic nanoparticles in acrylonitrile-methyl methacrylate-divinylbenzene mesoporous template. *J. Magn. Magn. Mater.*, 252, 13–15, 2002.
83. Thurn-Albrecht, T., Schotter, J., Kastle, G.A., Emley, N., Shibauchi, T., Krusin-Elbaum Guarini, K., Black, C.T., Tuominen, M.T., Russell, T.P., Ultrahigh-density nanowire arrays grown in self-assembled diblock copolymer templates. *Science*, 290, 2126–2129, 2000.
84. Bizdoaca, E.L., Spasova, M., Farle, M., Hilgendorff, M., Caruso, F., Magnetically directed self-assembly of submicron spheres with a Fe3O4 nanoparticle shell. *J.Magn. Magn. Mater.*, 240, 44–46, 2002.
85. Tejada, J., Zhang, X.X., Kroll, E., Bohigas, X., Ziolo, R.F., Solid containing rotationally free nanocrystalline gamma-Fe2O3: Material for a nanoscale magnetic compass? *J. Appl. Phys.*, 87, 8008–8012, 2000.
86. Heyen, U. and Schüler, D., Growth and magnetosome formation by microaerophilic Magnetospirillum strains in an oxygen-controlled fermentor. *Appl. Microbiol. Biotechnol.*, 61, 536–544, 2003.
87. Flynn, C.E., Lee, S.W., Peelle, B.R., Belcher, A.M., Viruses as vehicles for growth, organization and assembly of materials. *Acta Mater.*, 51, 5867–5880, 2003.
88. Culmone, C., Smit, G., Breedveld, P., Additive manufacturing of medical instruments: A state-of-the-art review. *Addit. Manuf.*, 27, 461, 2019.
89. Donate, R., Monzón, M., Ortega, Z., Wang, L., Ribeiro, V., Pestana, D., Oliveira, J.M., Reis, R.L., Comparison between calcium carbonate and β-Tricalcium phosphate as additives of 3D printed scaffolds with Polylactic acid matrix. *J. Tissue Eng. Regen. Med.*, 14, 272, 2020.
90. Silva, M., Pinho, I.S., Covas, J.A., Alves, N.M., Paiva, M.C., 3D printing of graphene-based polymeric nanocomposites for biomedical applications, in: *Functional Composite Materials*, vol. 2, pp. 1–21, 2021.
91. National Nanotechnology Inititative (http://www.nano.gov/), Alexandria.
92. Defense Advanced Research Projects, USA Program, http://www.darpa.mil/dso/thrust/biosci/biomagn.htm.
93. *Some large companies with ongoing research on organized magnetic nanocomposites,* IBM (www.research.ibm.com/), Seagate (www.seagate.com/).
94. Materials Science and Engineering Laboratory: National Institute of Standards and Technology, http://www.metallurgy.nist.gov/.

95. Kumar, A., Gangwar, A.K.S., Kumar, A., Meena, C.S., Singh, V.P., Dutt, N., Prasad, A., Gori, Y., Biomedical study of femur bone fracture and healing, in: *Advanced Materials for Biomedical Applications*, vol. Chapter 14, A. Kumar, Y. Gori, A. Kumar, C.S. Meena, N. Dutt (Eds.), pp. 235–250, CRC Press, Boca Raton, FL, USA, 2022, https://doi.org/ 10.1201/9781003344810-14.
96. Kumar, R., Kumar, A., Kant, L., Prasad, A., Bhoi, S., Meena, C.S., Singh, V.P., Ghosh, A., Experimental and RSM-based process-parameters optimisation for turning operation of EN36B steel. *Materials*, 16, 339, 2023. https://doi.org/10.3390/ma16010339.
97. Prasad, A., Chakraborty, G., Kumar, A., Gajrani, K.K., Introduction to biodegradable polymers, in: *Biodegradable Composites for Packaging Applications*, vol. Chapter 1, A. Prasad, A. Kumar, K.K. Gajrani (Eds.), pp. 1–12, CRC Press, Boca Raton, FL, USA, 2022, https://doi.org/10.1201/9781003227908-1.
98. Chakraborty, G., Prasad, A., Kumar, A., Processing of biodegradable composites, in: *Biodegradable Composites for Packaging Applications*, vol. Chapter 3, A. Prasad, A. Kumar, K.K. Gajrani (Eds.), pp. 33–48, CRC Press, Boca Raton, FL, USA, 2022, https://doi.org/ 10.1201/9781003227908-3.
99. Paul, A.K., Prasad, A., Kumar, A., Review on artificial neural network and its application in the field of engineering. *J. Mech. Eng.: PRAKASH*, 1, 1, 53–61, 2022. https://doi.org/10.56697/JMEP.2022.1107.

7

Nanohydroxyapatite-Based Composite Materials and Processing

Atanu Kumar Paul[1]*, Shasanka Sekhar Borkotoky[2] and Arbind Prasad[3]

[1]Department of Chemical Engineering, National Institute of Technology Andhra Pradesh, Tadepalligudem, West Godavari District, Andhra Pradesh, India
[2]Department of Chemical Engineering, Assam Engineering College, Jalukbari, Guwahati, Assam, India
[3]Department of Mechanical Engineering, Katihar Engineering College (Under Department of Science & Technology, Government of Bihar), Katihar, Bihar, India

Abstract

Biomaterials are used to restore function to traumatized or degenerated connective tissues. In most cases, bone fractures are caused by natural ageing, trauma, tumors, or infections. Without the use of bone substitute materials, restoration is extremely difficult. Limited availability, risk of improper healing at the donor site, and risk of pathogen transfer from allografts have hindered the use of conventional treatments such as autografts and allografts. It is believed that natural bone is a composite material made up of organic and inorganic components, consisting of collagen fibrils that have hydroxyapatite nanocrystals embedded within them. This may offer a resolution to the issues previously mentioned. The biological activity of HAp is greatly affected by its chemical composition, phase purity, and microstructural properties, including its Ca/P ratio, particle size, and shape. The stoichiometric and size controlled HAp nanorod preparation method using a simple sol-gel method will be discussed in this study. An indigenous high temperature and high pressure hydrothermal experimental setup has been designed and built for mass production of nanocrystalline HAp (nHAp). The production of synthetic bone using ceramics, polymers, and metals has advanced significantly in the past few years. Nanohydroxyapatite (nHAp), a common bioceramic material for bone graft alternatives, is biocompatible and osteoconductive. In this chapter, various materials, including polylactic acid, poly(lactic-co-glycolic) acid, polycaprolactone polyamide, polyurethane, polyhydroxybutyrate, and polyvinyl alcohol were examined as substitutes for bone grafts

**Corresponding author*: atanu.kr.paul@gmail.com

Arbind Prasad, Ashwani Kumar and Manoj Gupta (eds.) Advanced Materials and Manufacturing Techniques for Biomedical Applications, (157–184) © 2024 Scrivener Publishing LLC

that use nHAp. The process of generating high-quality biomaterials that resemble natural bone is discussed by producing HAp on both synthetic and natural polymers, such as poly(vinyl alcohol), sodium alginate, and chitosan-gelatin (CG).

Keywords: Biomaterials, collagen, dental implants, fracture fixation, nanohydroxyapatite (nHAp), nanotechnology, polymeric implants, polymethylmethacrylate (PMMA)

7.1 Introduction

Nanotechnology involves the development of technology at the atomic or molecular level, typically within the range of 1-100 nm in length. This provides a basic comprehension of materials and processes at the nanoscale. Another goal of nanotechnology is to develop and apply smallscale systems, technologies, and structures that have unique characteristics and capabilities [1–3]. Nanotechnology is a form of fusion technology since it incorporates several disciplines, including physics, chemistry, materials science, and biotechnology.

Throughout evolution, a wide range of natural materials with exceptional mechanical properties have emerged. This includes materials such as wood, grass, bone, sea shells, and glass sponges. Biological materials are typically composites of various polymers and minerals. They are constructed in a hierarchical manner, allowing the material to be optimized for its function at numerous structural levels [4]. For instance, a collagen-rich matrix and calcium phosphate nanoparticles make up roughly equal proportions of bone.

These elements are combined to create a material with remarkable fracture resistance using a sophisticated hierarchy of fibers and lamellar structures. The basic understanding of bone's hierarchical structure, in connection to its properties, and in relation to its function is improved by study of bone utilizing approaches derived from material physics. However, it might also be useful in understanding how diseases and treatments affect people. Another instance of how a small bit of protein significantly improves the inherent brittleness of glass is in the skeletons of glass sponges. Contrarily, plant cell walls are essentially entirely polymeric composites that can produce significant strains and even complicated motions in response to changing humidity. These actuation capabilities are based on intricate cellulose fiber architectures and hemicellulose water swelling [5]. The fact that these structures, along with other materials systems produced by organisms for mechano-sensing and actuation, do not require the biological system's metabolism to function is particularly intriguing for the materials scientist.

Those composites that can be used in bioengineering are referred to as "bio composites". The composite's component parts maintain their individual identities [6]. Despite working together, they don't totally meld or disappear into one another. The properties of composites, such as the elastic modulus, differ greatly from those of their individual constituents, due to the significant alteration of constituent structures and contents. Composites exhibit anisotropy in their structural makeup. A reinforcement of natural fibers and a matrix (resin) combine to create bio composites (usually derived from plants or cellulose) [7]. Bio composites are formed by mixing natural fibers derived from sources like wood (both hardwood and softwood) as well as non-wood sources such as hemp, flax, wheat, jute, and sisal with polymer matrices that are created from both renewable and non-renewable resources. One of the key elements of bio composites is bio fiber. The term "bio fiber" refers to the fiber-like material obtained from sources such as trees, plants, or shrubs [8]. nHAp, which is the major inorganic constituent of bones, is a chemically synthesized and biocompatible substance that finds use in bone treatments, dental implants, toothpaste additives, and controlled drug release [9]. The wet chemical process in aqueous solution is the most frequent method for preparing nHAp [10]. The composites contain nHAp that has structural similarities to biological apatites [11]. Biocompatibility tests on the nHAp revealed that it is harmless to mesenchymal stem cells (MSC) [12].

The primary aim of this chapter is to provide an overview of the preparation, modification, and use of nanohydroxyapatite in the biomedical sector. Additionally, it discusses the current challenges and research status, thereby establishing a theoretical and practical basis for future studies.

7.2 Biomaterials

The purpose of developing biomaterials is to enhance the quality and duration of human life by creating materials that can interact with biological systems and be used to assess, treat, enhance, or replace tissues, organs, or bodily functions [13]. Alternatively, a biomaterial is an inert material employed in a medical tool that is designed to interact with biological systems. The significance and potential uses of the initial, subsequent, and third generations of biomaterials are presented in Table 7.1 [14].

By creating superior metal, ceramic, polymer, and composite systems, biomaterials have been widely exploited in biomedical applications to enhance human health [15]. The characteristics of biomaterials determine the tissue reactions or interactions that occur when they are introduced

Table 7.1 Classification, properties, and applications of biomaterials.

Period	Category	Applications
1950–1969	First generation	• Fracture fixation • Bone and joint replacement • Dental implants and surgical instruments
1970–2000	Second generation	• Coatings that promote tissue growth into or onto a material • Dental implants • Temporary bone space fillers • Biodegradable polymers and hydrogel
2000–till date	Third generation	Microelectromechanical systems (MEMS) and Nanoelectromechanical systems (NEMS) devices • Blood pressure sensors • DNA array systems Tissue engineering • Tissue repair/regeneration • Cell encapsulation • Bioartificial skin and liver construction Nanostructured biological systems • Biological structure imitations at atomic scales

into the human body. Depending on the response, biomaterials are categorised as bioinert, bioactive, or bioresorbable.

7.3 Types of Biomaterials

New methods of synthesis have been employed to confer desirable chemical, physical, and biological properties on biomaterials. Metals and alloys, ceramics, polymers, and composites have all been investigated for biomedical applications [16–19].

7.3.1 Polymers

Polymers, which consist of repeating units or isomers, are large molecules with long chains [20]. The lengthy chains are formed by either the primary covalent bonding or secondary bonding forces, such as van der Waals, along the backbone chain. Based on the purpose of the repeating units, polymers are structured into linear, branching, or network topologies [21]. Biomaterials can be synthesised from various types of polymers such as polymethylmethacrylate (PMMA) [22, 23], polyethylene (PE) [24, 25], polystyrene (PS) [26, 27], and polyvinylchloride (PVC) [28, 29]. A broad range of applications use these artificial polymers, such as in implantable devices, as device coatings, in tubing, and for injectable drug delivery [20, 30]. Polymeric implants that are bioresorbable or biodegradable offer advantages over metal alloy implants, including avoidance of issues like mechanical properties that exceed those of surrounding tissue, foreign body reactions, and the need for additional procedures to remove the implants. Numerous biodegradable polymers, including collagen, gelatin, chitosan, silk fibroin, and alginate, have been widely studied for various medical applications due to their favorable qualities such as good biocompatibility, controlled biodegradability, and relatively easy processing [31–40]. Polymeric biomaterials offer several benefits compared to metal or ceramic materials, including simpler production, flexibility in further processing, affordability, and the ability to obtain specific mechanical and physical characteristics.

7.3.2 Composites

The term "composite," typically used for materials that have two or more distinct components or phases, refers to a larger scale than that of individual materials. These materials' physical and mechanical properties are significantly altered, making them more advantageous than homogeneous materials. The majority of biological materials, such as human hard tissues, are inorganic-organic nanocomposites [41, 42]. The properties of a composite material are affected by various factors, such as the size and shape of its heterogeneous components, the amount of space they occupy, and the interaction between the different constituents. Among composite materials, ceramics and composites based on polymers are considered to be the most favorable for use as biomaterials. Some of the composites' significant characteristics and applications are outlined in Table 7.2. Because of their unique properties, the biomedical applications of the aforementioned materials have been extensively researched [37, 43].

7.4 Structure of Hydroxyapatite

Hydroxyapatite is the most prevalent bioceramic made from calcium phosphate. It has a stoichiometric chemical formula of $Ca_{10}(PO_4)_6(OH)_2$,

Table 7.2 Some of the properties and applications of various composites.

Sl. No.	Name of the composite	Properties	Application	Reference
1	Nano HAp/collagen	HAp nanocrystals are regularly aligned along collagen fibrils	Synthetic bone materials in medical applications	[44]
2	Nanostructured HAp/ sodium alginate (SA)	HAp nanocrystals are well embedded in alginate matrix, homogeneous microstructure and good mechanical strength	Bone regeneration and implant fixation	[45]
3	HAp/SA nanocomposite beads	The well-swelling, drug loading and controlled release behavior	Drug-controlled release	[46]
4	HAp/alginate composite	Uniformly packed spherical particles with uniform pore size	Infection resistant and drug delivery	[47]
5	Nano HAp/chitosan composite scaffold	Good biocompatibility and sufficient porosity	Bone tissue engineering	[36]
6	Chitosan/nanocrystalline calcium phosphate	Increased proliferation of osteoblast cells and improved mechanical strength	Bone regeneration	[48]

(Continued)

Table 7.2 Some of the properties and applications of various composites. (*Continued*)

Sl. No.	Name of the composite	Properties	Application	Reference
7	HAp/chitosan–gelatin (CG) composite	Enhanced protein and calcium ion adsorption and improved initial cell adhesion and long term growth	Development of human stem cell	[49]
8	Hydroxyapatite/ Chitosan–Gelatin	Good mechanical properties	Bone substitution and bone remodeling	[50]
9	Polyvinyl alcohol (PVA)/ collagen/HAp	HAp nanorods are preferentially oriented parallel to electrospun, porous structure with adjustable pore size and shape and improved rigidity	Bone remodeling	[51]
10	HAp/chitosan/PMMA	No cytotoxicity and more biocompatibility and osteoconductibility	Bone cement	[52]
11	HAp/PMMA	Improved cytocompatibility and physical properties	Dental implant restoration	[53]
12	HAp/collagen/ alginate	Active bone formation around the material and tissue invasion into the material	Bone filler and a drug delivery carrier	[54]

and its atomic calcium to phosphate (Ca/P) ratio is 1.67 [55]. The symmetry space group of hydroxyapatite is $P6_3/m$, and it has a hexagonal structure with lattice parameters $a = b = 9.432$ Å, $c = 6.881$ Å [55, 56]. In each unit cell of hydroxyapatite, which contains 44 atoms, there is one formula unit of $Ca_{10}(PO_4)_6(OH)_2$. The hexagonal primitive cell of HAp is illustrated in Figure 7.1 and has two distinct Ca sites (Ca(I) and Ca(II)), one P site, four different oxygen sites (O(I), O(II), O(III), and O(IV)), and one H site situated along the c axis [57]. In hydroxyapatite, the P atom binds with O(I), O(II), and two O(III) atoms, forming a PO_4 tetrahedron that connects along the c axis through Ca(I) and/or Ca(II). In this configuration, Ca(I) interacts with three O(I) and three O(II) atoms, while Ca(II) interacts with an O(II) atom, four O(III) atoms, and an O(IV) atom. At z = 1/4 and 3/4 (perpendicular to the c axis), Ca(II) atoms create triangles, and

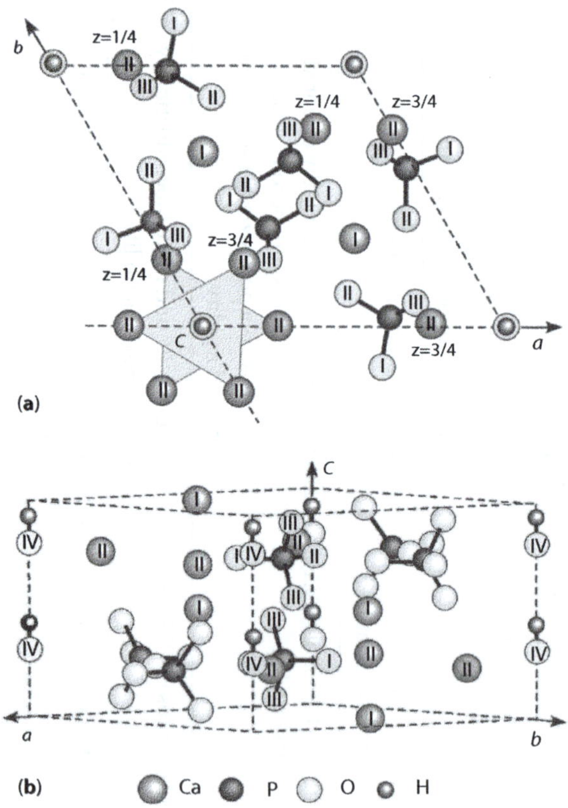

Figure 7.1 This figure shows the unit-cell structure of hexagonal HAp, with (a) shaded triangles representing the Ca (II) triangles at Z = 1/4 and 3/4, and (b) thick lines connecting the P and O atoms forming the tetrahedra, with the fractional z coordinates inserted for Ca (II).

OH groups from O(IV) and H penetrate the centers of the Ca(II) triangles. The hexagonal structure has a $P6_3/m$ space group and mirror planes at z = 1/4 and 3/4 perpendicular to the c axis. There are four potential OH group sites along the c axis, each with a site occupancy of 0.5.

Under neutral to alkaline conditions, hydroxyapatite is the most stable calcium phosphate in fluids and humid conditions [58], and its similarity in structure and composition to the primary inorganic compounds of teeth and bones makes it highly significant [59]. HAp has bioactivity and osteoconductivity, which are crucial in repairing damaged bone [60]. As a result, hydroxyapatite has found widespread use in a variety of biomedical devices, such as dental implants [61], hip implants made of titanium [62, 63], biodegradable scaffolds for tissue engineering and drug delivery [64], and orthopedic implants [65].

HAp has undergone significant research for biomedical uses, including powders, composites, and microscale coatings [66, 67]. Additionally, it has been studied for non-medical applications like column chromatography packing media, gas sensors, and catalysts [68]. Due to the range of applications, the properties of HAp materials must be customized to suit real-world usage.

7.5 Nanohydroxyapatite

When hydroxyapatite (HAp) is engineered at the nano level, it exhibits remarkable functional properties attributed to its larger surface area [69]. Studies have revealed that the rate of osteoblast adhesion and proliferation on nanocrystalline hydroxyapatite (nHAp) is faster compared to conventional HAp [70]. Additionally, Nano HAp demonstrates better sinterability and densification, which lowers the sintering temperature and leads to increased fracture toughness in HAp ceramic [71]. Moreover, nanocrystalline hydroxyapatite (nHAp) can facilitate the adhesion, proliferation, differentiation, and osteointegration of cells while also promoting the deposition of calcium mineral on the surface. These properties enable the formation of new bone tissue within a brief period. Compared to larger crystals, nano HAp exhibits greater bioactivity [72]. As a result, nanophase ceramics are a distinct and promising category of orthopedic/dental implant compositions that possess superior osseointegrative characteristics. nHAp significantly influences the interaction between implants and cells within the human body.

Various techniques have been employed to create nanocrystalline hydroxyapatite (nHAp), such as solid-state reactions, co-precipitation, hydrothermal synthesis, sol-gel synthesis, microemulsion synthesis, and mechanochemical synthesis [73–78]. Among these methods, sol-gel and hydrothermal approaches are effective and eco-friendly ways of producing nHAp. Sol-gel method provides molecular-level mixing of calcium and phosphorus precursors and reduces HAp processing temperature compared to traditional methods [79]. Meanwhile, the hydrothermal technique is one of the most promising methods to produce uniform size, morphology, and composition of highly crystalline and non-agglomerated nHAp crystals [80].

7.5.1 Nanohydroxyapatite/Polymer Composite

To create a hydroxyapatite/collagen bone-like nanocomposite material, Kikuchi *et al.* utilized the self-organization mechanism between the surfaces of HAp and collagen [44]. The resulting HAp/collagen composite had a bone-like structure with the c-axes of HAp nanocrystals and collagen fibrils regularly aligned. According to crystallography, the self-organization between HAp and collagen surfaces was driven by the interaction between Ca^{2+} ions on HAp crystals and dissociated carboxyl residues on collagen molecules. The composite that was produced displayed exceptional biocompatibility and biointegration capabilities that were similar to natural bone and significantly better than other synthetic bone materials.

Rhee *et al.* conducted a study in 2000 to investigate the chemical interaction between collagen and HAp crystals [81]. They submerged a collagen membrane in a supersaturated simulated bodily fluid (1.5 SBF) solution with an ion concentration greater than 1.5 times that of regular simulated body fluid. After four weeks, a few carbonate-containing HAp crystals started to develop near the border of the collagen membrane. The formation of HAp crystals led to a red chemical shift in the carboxylate band of the collagen membrane, which coincided with the reduction in the bond orders of the carboxylate group when chelated with a calcium ion. This first-step nucleation of HAp crystals on the carboxylate group of collagen was observed to be a result of the binding of calcium ion to the carboxylate group of collagen, which is one of the critical factors for HAp crystal nucleation in a 1.5 SBF solution.

7.5.2 Nanohydroxyapatite/Poly (Vinyl Alcohol) Composite

Mollazadeh *et al.* used PVA as a template *in situ* synthesis of HAp particles. The researchers investigated the physical properties of HAp crystals based on the polymer content and molecular weight [82]. They found that the presence of PVA influenced the shape of the crystallite as well as the number of reaction sites. Ca^{2+} ions were responsible for initiating HAp formation by interacting with the negative side groups on the polymer surface. A higher molecular weight PVA resulted in a higher number of HAp nuclei and a smaller crystallite size due to a higher number of reaction sites. XRD patterns and characteristic absorption bands in FTIR spectra confirmed the formation of HAp crystals at room temperature.

Chung *et al.* employed the co-precipitation technique to produce composite powders of hydroxyapatite and PVA, which were then mixed with pure PVA in dimethyl sulfoxide for two hours at 363 K to make wet spinning dopes [83]. In order to determine how HAp particles affect the shear viscosities of the spinning dopes, rheometer analysis was performed on HAp/PVA solutions with HAp concentrations ranging from 0 to 15 wt%. The spinning dopes with pure PVA and a HAp/PVA concentration of 12 wt% exhibited non-Newtonian flow behavior when the HAp content was increased from 1 to 15 wt%. This was due to the interaction between PVA chains, citric acid, and HAp particles, which acted as cross-linkers. PVA fiber breakage was reduced and spinnability was improved by adding citric acid, which reduces the aggregation of HAp particles. This results in producing smooth-surfaced HAp/PVA composite fibers with improved thermal and mechanical properties due to hydrogen bonding between uniformly dispersed HAp particles and PVA chains.

7.5.3 Nanohydroxyapatite/Sodium Alginate Composite

Wang *et al.* synthesised a range of HAp/SA nanocomposites using an *in situ* hybridization method, with alginate concentrations ranging from 10 to 40 wt% [45]. The inorganic phase of the composites was made up of low crystallinity carbonate-substituted HAp, and higher alginate concentrations resulted in a reduction in HAp crystallinity. HAp crystals had a needle-like shape with dimensions of 20 to 50 nm in length and 5 nm in breadth. FTIR analysis revealed a chemical interaction between the mineral phase and the polymer matrix. The inclusion of HAp nanocrystals into the alginate matrix produced composite materials with a more uniform microstructure and greater compactness when compared to pure HAp. The presence of 30-40 wt% of alginate allowed for full fusion of HAp crystallites with the polymer matrix, leading to immobilization of HAp nanoparticles upon implantation. A HAp/SA weight ratio of 70/30 produced a 3D network with higher structural organization. The *in situ* processing technique used has promising potential for developing HAp/SA composites for bone regeneration and implant fixation.

Using the constant composition approach, Malkaj *et al.* [84] examined the impact of SA on the crystal formation of HAp at prolonged supersaturation. With a reduced crystal growth rate of 42–86% at the lower inhibitor doses of $2.1\times10^{-}7–12.6\times10^{-}7$ mol/L, SA was found to inhibit the formation of HAp crystals.

7.5.4 Nanohydroxyapatite/Chitosan Composite

Wilson and Hull found that by aging nanosphase HAp particles in chitosan acetate gel solutions for 30 days, the surface chemistry, colloid stability, and chemical makeup of HAp changed due to adsorption and solubility effects [65]. Chitosan improved the colloid stability of HAp and strongly adsorbed onto HAp particles. Increasing the chitosan content from 0 to 2.5 wt% resulted in a higher TGA mass loss of 3.3–6.5 wt%. CHN analysis revealed that 2.8 to 3.1 wt% of chitosan adsorbed onto HAp. After aging in chitosan acetate gel solutions, the specific surface area of HAp increased from 85 to 160 m^2/g.

Zhang and colleagues developed a new biomimetic nanocomposite nanofiber of HAp/chitosan by combining an *in situ* co-precipitation synthesis method with an electrospinning procedure [85]. The co-precipitation method was used to create a HAp/chitosan nanocomposite with a 30 wt% HAp mass ratio, which had spindle-shaped HAp nanoparticles of approximately 100×30 nm uniformly dispersed in the chitosan matrix. With the addition of a small amount (10 wt%) of ultrahigh molecular weight poly(ethylene oxide) (UHMWPEO) as a fiber-forming additive, continuous HAp/chitosan nanofibers with a diameter of 214 ± 25 nm were successfully generated. *In vitro* biological culture of human fetal osteoblast (hFOB) cells for up to 15 days demonstrated that the addition of HAp nanoparticles to chitosan nanofibrous scaffolds resulted in considerable bone formation.

7.5.5 Nanohydroxyapatite/Gelatin Composite

Using the dispersion polymerization approach [88], Sivakumar and Panduranga created composite microspheres containing coralline HAp and gelatin [86]. The absorption technique was used to include the gentamicin. XRD tests were used to examine the composite microspheres' crystal structure. FTIR analysis confirmed the presence of amide and hydroxyl groups in the composite microspheres. The SEM and optical micrographs showed that the composite microspheres were spherical and porous. Analysis revealed that the average particle size of the composite microspheres was 16 μm. Both TGA and DSC analyses were employed to investigate the thermal behavior of the composite microspheres. The *in vitro* release of gentamicin from the composite microspheres showed a pattern of almost negligible order.

7.5.6 Nanohydroxyapatite/Chitosan-Gelatin Composite

Li and colleagues produced nHAp/chitosan-gelatin (nHAp/CG) films by biomineralizing CG network films in a $Ca(NO_3)_2$-Na_3PO_4 Tris buffer solution under alkaline conditions [87]. In order to create the micro-HAp/chitosan-gelatin (mHAp/CG) films, CG network films were submerged in HAp crystal suspensions with an average size of 5 *μm*. According to the findings, ion/polar interactions were the main driving forces behind the production of nHAp/CG through biomineralization. The formation of mHAp/CG, on the other hand, was largely influenced by hydrogen bonding between the COOH, OH, and $-NH_2$ groups of CG films and the OH groups of HAp crystals. Excellent biocompatibility was shared by both nHAp/CG and mHAp/CG, and mesenchymal stem cells on nHAp/CG exhibit better osteogenic differentiation activity than on mHAp/CG.

Mohamed and Mostafa (2008) developed novel biodegradable HAp-titania/chitosan/gelatin polymeric composites by incorporating HAp filler containing 10% and 30% titania powder [88]. The mechanical properties of the composites were evaluated by conducting compressive strength and hardness tests. Additionally, an *in vitro* study using simulated body fluid (SBF) was performed to assess the composites' bioactivity. The authors expected that the HAp and 10% titania biocomposites could be used in biomedical applications, such as bone substitutes and tissue engineering scaffolds.

Given the aforementioned factors, which include HAp's morphology, crystallite size, purity, stoichiometry, and structure, there is a better chance that its attributes can be changed to satisfy the specifications. As a result, developing the preparation methods to regulate the characteristics of HAp for relevant biomedical applications is of great research interest.

7.5.7 Nanohydroxyapatite/Polylactic Acid Nanocomposites

PLA is frequently used in bone tissue engineering because it has the properties of being able to degrade naturally, being compatible with living tissue, and being able to be molded when heated. The HA and PLA combination would have mechanical, osteoconductive, and osteoinductive capabilities [89–91]. Induced phase separation techniques [92] and other types of methods, such as mechanical mixing, *in situ* precipitation and ultrasound method [93], *in situ* polymerization [94], and melt blending [95], were utilized to prepare the nHAp-PLLA composite scaffold. Rod-shaped particles of nHAp, measuring between 37-65 nm in width and 100-400 nm in length, were utilized in creating the composite scaffold nHAp/PLLA [96]. This resulted in an improvement in the mechanical properties of the HA-PLLA nanocomposite, with the highest values observed in PLLA composites containing 4 wt% of HA [97]. Figure 7.2 depicts the novel and straightforward process of fabricating the composite using Air Jet Spinning (AJS) of nHAp/PLA.

To produce the composite resin nHAp-PDLLA, micro-stereolithography was utilized, which involved combining a liquid photoinitiator with variable quantities of nHAp powder into a photo-crosslinkable PDLLA-diacrylate resin. The elasticity modulus of the composite nHAp-PDLLA network materials rose with increasing nHAp content, as evidenced by the mechanical characteristics, which also demonstrated that strong rigid materials were created [98]. By blending, nHAp-chitosan-PLA nanocomposites have been created. Ternary biocomposite showed a high compressive strength of up to 160 MPa when the HA concentration was between 60–67% [99].

The addition of nHAp enhanced the compressive mechanical properties of the newly fabricated scaffolds. Porous composite scaffolds of nHAp/collagen/PLA/chitosan were created using a technique called thermally induced phase separation. As the proportion of microspheres in the

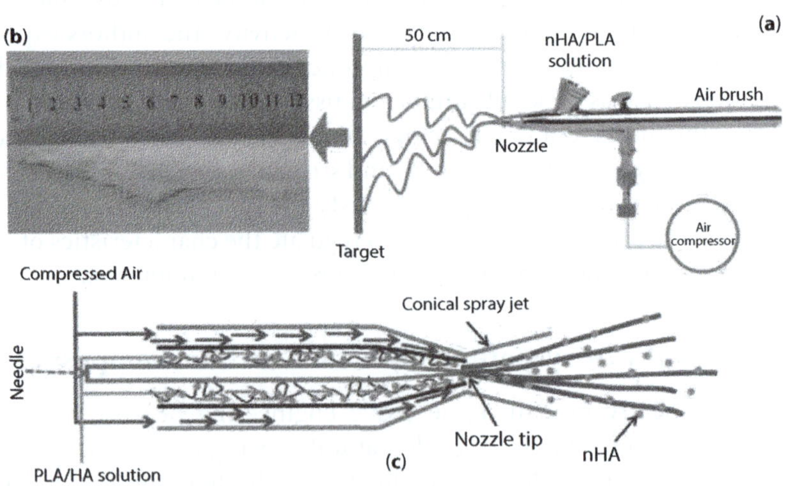

Figure 7.2 Figure (a) displays the schematic diagram of the air jet spray equipment, while Figure (b) shows the mat that was gathered (nHAp-PLA). Figure (c) demonstrates the schematic diagram of the composites process [101].

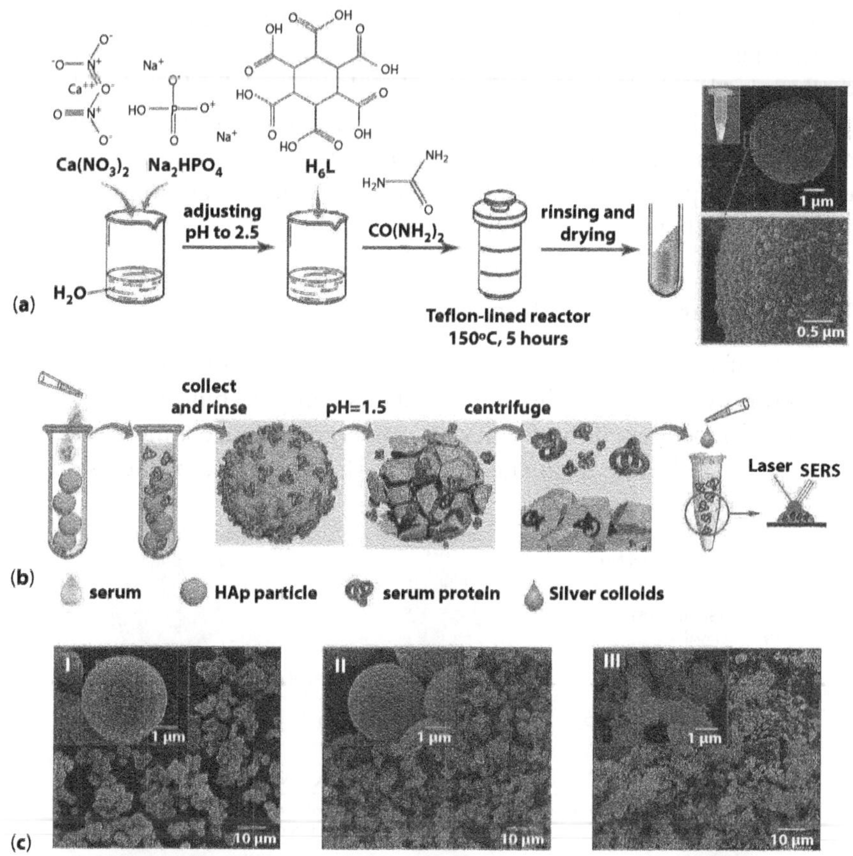

Figure 7.3 (a) The process for producing HAp. (b) A diagram that illustrates how HAp adsorbs and exfoliates serum proteins. (c) Scanning electron microscopy (SEM) images of HAp particles in their raw state (I), HAp particles adsorbed with protein (II), and HAp particles exfoliated with protein (III).

composite increased from 0% to 50%, the compressive modulus of the composite scaffolds increased from 15.4 to 25.5 MPa, and their compressive strength increased from 1.42 to 1.63 MPa [100].

7.6 Cancer Detection and Cell Imaging

Hydroxyapatite has unique properties that enable it to be utilized in the diagnosis of tumors. Upon dispersion in aqueous media, the OH^- on the HAp surface dissolves, exposing Ca^{2+} and creating numerous positively charged sites [102]. Due to this characteristic, HAp is often used for protein adsorption, particularly for acidic proteins such as human serum albumin [103]. There are more negatively charged carboxyl groups on the surface of albumin than on other proteins, and its long peptide chain has greater spatial flexibility, allowing the carboxyl terminus of internal amino acids to be more easily exposed and adsorbed onto the HAp surface. This special technique simplifies the extraction and measurement of albumin in blood, as depicted in Figure 7.3.

7.6.1 Size and Morphology

The size of nanohydroxyapatite is an important factor for delivering it intravenously to cancer cells, as it affects its ability to bind to specific cell membrane receptors, flow within cells, undergo pharmacokinetics, extravasate, and be cleared from the body [104]. When microspheres with a

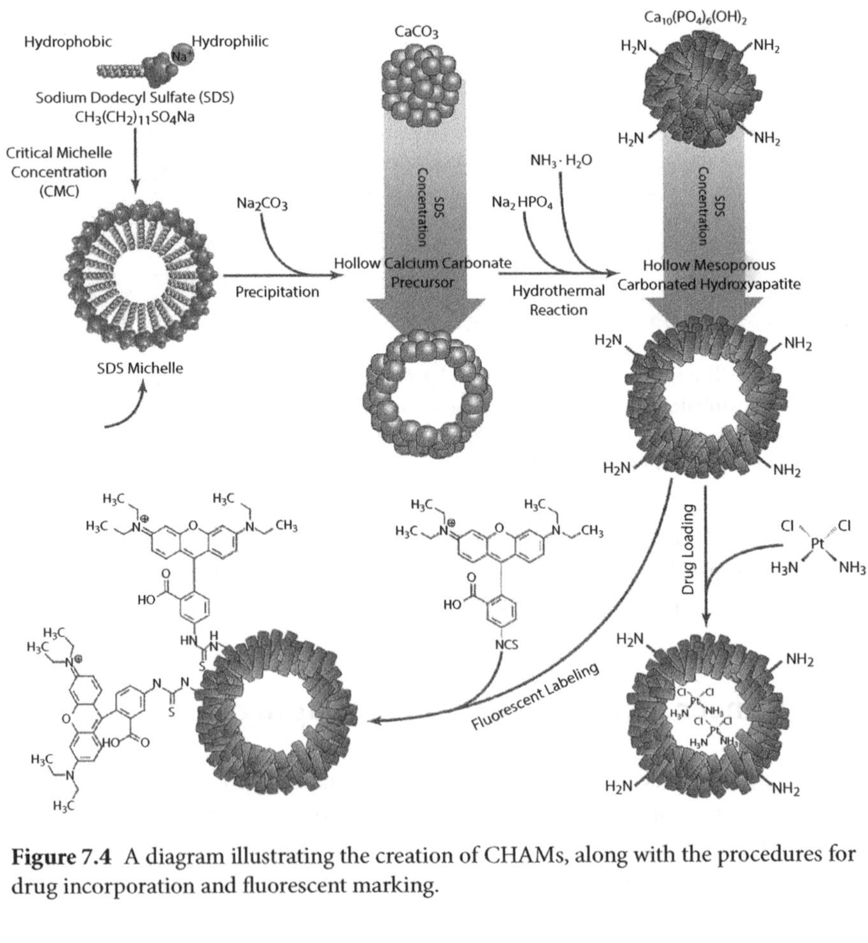

Figure 7.4 A diagram illustrating the creation of CHAMs, along with the procedures for drug incorporation and fluorescent marking.

particle size of approximately 200 nm are injected into the vein, they are rapidly ingested and cleared by macrophages of the reticuloendothelial system. This leads to their accumulation primarily in the liver, spleen, and other tissues [105].

The size of nanohydroxyapatite is crucial for its intravenous delivery to cancer cells and its subsequent clearance by the reticuloendothelial system. Microspheres with an ideal particle size of approximately 200 nm are ingested and cleared by macrophages, primarily in the liver, spleen, and other tissues. Spherical hydroxyapatite, with its large specific surface area, is suitable for drug delivery. Studies have investigated the production of specific morphology microspheres for cancer treatment, including biomimetic hollow mesoporous hydroxyapatite microspheres (Figure 7.4). By regulating the SDS content, hollow microspheres can be produced to provide the highest medicine release [106]. This approach allows for the control of microsphere shape, degradability, and drug loading/release performance during preparation.

Various researchers have also established nano hydroxyapatite for their enhanced biobased polymer composite fabrications for scaffolds, implants, fixation devices and other orthopedic applications. It was observed that the hydroxyapatite not only enhances mechanical strength to the resorbable polymers but also enhanced bioactivity of the composites [107–121].

7.7 Conclusion

The use of nHAp as a biomaterial for bone tissue engineering shows promise, and the most effective way to create artificial bone is by combining nHAp with polymer nanocomposites. When biopolymer and nHAp are combined, a 3D dimensional scaffold that has appropriate bone behavior characteristics including mechanical strength, pore size, and osteoconductivity can be produced as a bone graft alternative. However, to ascertain the biomedical importance of nHAp-based nanocomposites, a complete investigation of the cell-material interaction, mechanical robustness, and biological response is required. In order to ascertain the biophysiochemical consequences, clinical tests on the nHAp-based composite biomaterial must also be carried out.

Hydroxyapatite was first discovered over 50 years ago and was initially used in dentistry for dental restoration. However, with advancements in technology, it is now able to support weight and exhibit anti-infection and anti-tumor properties, making it useful in orthopedics and drug delivery. As a result, this field is rapidly expanding and developing.

Despite modifying hydroxyapatite, there are still several challenges that need to be addressed. For instance, while we know that acidic amino acids can repair and mineralize early-stage dental caries, the mechanisms behind this process are not well understood. Additionally, there is a need to investigate the response parameters for targeted drug delivery and release from drug carriers. Although the problem of brittleness and degradation rate has been partially resolved for bone restoration by incorporating polymer compounds, there is still a significant gap between laboratory research and clinical application. Therefore, further research and development are needed to understand the mechanisms of hydroxyapatite-based composite materials and to advance their clinical use.

References

1. Kim, F., Kwan, S., Akana, J., Yang, P., Langmuir–Blodgett nanorod assembly. *J. Am. Chem. Soc.*, *123*, 4360–4361, **2001**.
2. Weller, H., Rogach, B.A.L., Talapin, D.V., Shevchenko, E.V., Kornowski, A., Haase, M., Weller, H., Organization of matter on different size scales: Monodisperse nanocrystals and their superstructures. *Adv. Funct. Mater.*, 12, 10, 653–664, **2002**.
3. Xia, Y., Yang, P., Sun, Y., Wu, Y., Mayers, B., Gates, B., Yin, Y., Kim, F., Yan, H., One-dimensional nanostructures: Synthesis, characterization, and applications. *Adv. Mater.*, *15*, 353–389, **2003**.
4. Zartner, L., Muthwill, M.S., Dinu, I.A., Schoenenberger, C.A., Palivan, C.G., The rise of bio-inspired polymer compartments responding to pathology-srelated signals. *J. Mater. Chem. B*, *8*, 6252–6270, **2020**.
5. Kaviya, M., Ramakrishnan, P., Mohamed, S.B., Ramakrishnan, R., Gimbun, J., Veerabadran, K.M., Kuppusamy, M.R., Kaviyarasu, K., Sridhar, T.M., Synthesis and characterization of nano-hydroxyapatite/graphene oxide composite materials for medical implant coating applications. *Mater. Today Proc.*, *36*, 204–207, **2021**.
6. Zhang, R., Hu, H., Liu, Y., Tan, J., Chen, W., Ying, C., Liu, Q., Fu, X., Hu, S., Wong, C.P., Homogeneously dispersed composites of hydroxyapatite nanorods and poly(lactic acid) and their mechanical properties and crystallization behavior. *Compos. Part A Appl. Sci. Manuf.*, *132*, 105841, **2020**.
7. Zhou, B.L., Bio-inspired study of structural materials. *Mater. Sci. Eng. C*, *11*, 13–18, **2000**.
8. Farokhi, M., Mottaghitalab, F., Samani, S., Shokrgozar, M.A., Kundu, S.C., Reis, R.L., Fatahi, Y., Kaplan, D.L., Silk fibroin/hydroxyapatite composites for bone tissue engineering. *Biotechnol. Adv.*, *36*, 68–91, **2018**.

9. Zakaria, S.M., Sharif Zein, S.H., Othman, M.R., Yang, F., Jansen, J.A., Nanophase hydroxyapatite as a biomaterial in advanced hard tissue engineering: A review. *Tissue Eng. - Part B Rev.*, *19*, 431–441, **2013**.
10. Mostafa, N.Y., Characterization, thermal stability and sintering of hydroxyapatite powders prepared by different routes. *Mater. Chem. Phys.*, *94*, 333–341, **2005**.
11. Jiang, D. and Zhang, J., Calcium phosphate with well controlled nanostructure for tissue engineering. *Curr. Appl. Phys.*, *9*, S252–S256, **2009**.
12. Cai, Y. and Tang, R., Calcium phosphate nanoparticles in biomineralization and biomaterials. *J. Mater. Chem.*, *18*, 3775–3787, **2008**.
13. Williams, D.F., *The williams dictionary of biomaterials*, Liverpool University Press, Liverpool, United Kingdom, ISBN 9781846314438, 1999.
14. Rajendran, V., Bhandari, S.K., Defence Research & Development Organisation (India), *Bioactive glasses for implant applications*, vol. 1, DESIDOC, India, **2010**.
15. Freed, L.E., Vunjak-Novakovic, G., Biron, R.J., Eagles, D.B., Lesnoy, D.C., Barlow, S.K., Langer, R., Biodegradable polymer scaffolds for tissue engineering. *Bio/Technology*, *12*, 689–693, **1994**.
16. Webster, T.J. and Ejiofor, J.U., Increased osteoblast adhesion on nanophase metals: Ti, Ti6Al4V, and CoCrMo. *Biomaterials*, *25*, 4731–4739, **2004**.
17. Balasundaram, G., Sato, M., Webster, T.J., Using hydroxyapatite nanoparticles and decreased crystallinity to promote osteoblast adhesion similar to functionalizing with RGD. *Biomaterials*, *27*, 2798–2805, **2006**.
18. Thapa, A., Miller, D.C., Webster, T.J., Haberstroh, K.M., Nano-structured polymers enhance bladder smooth muscle cell function. *Biomaterials*, *24*, 2915–2926, **2003**.
19. Sato, M., Slamovich, E.B., Webster, T.J., Enhanced osteoblast adhesion on hydrothermally treated hydroxyapatite/titania/poly(lactide-co-glycolide) sol–gel titanium coatings. *Biomaterials*, *26*, 1349–1357, **2005**.
20. Shastri, V., Non-degradable biocompatible polymers in medicine: Past, present and future. *Curr. Pharm. Biotechnol.*, *4*, 331–337, **2005**.
21. Bronzino, J.D., *The biomedical engineering handbook*, Third Edition, CRC Press, Taylor & Francis Group, Boca Raton, FL, 2006.
22. Socol, G., Macovei, A.M., Miroiu, F., Stefan, N., Duta, L., Dorcioman, G., Mihailescu, I.N., Petrescu, S.M., Stan, G.E., Marcov, D.A. *et al.*, Hydroxyapatite thin films synthesized by pulsed laser deposition and magnetron sputtering on PMMA substrates for medical applications. *Mater. Sci. Eng. B*, *169*, 159–168, **2010**.
23. Oh, S.H., Choi, S.Y., Choi, S.H., Lee, Y.K., Kim, K.N., The influence of lithium fluoride on *in vitro* biocompatibility and bioactivity of calcium aluminate–PMMA composite cement. *J. Mater. Sci. Mater. Med.*, *15*, 25–33, **2004**.
24. Wang, M. and Bonfield, W., Chemically coupled hydroxyapatite-polyethylene composites: Structure and properties. *Biomaterials*, *22*, 1311–1320, **2001**.

25. Eniwumide, J.O., Joseph, R., Tanner, K.E., Effect of particle morphology and polyethylene molecular weight on the fracture toughness of hydroxyapatite reinforced polyethylene composite. *J. Mater. Sci. Mater. Med.*, *15*, 1147–1152, **2004**.
26. Yook, S.W., Kim, H.E., Yoon, B.H., Soon, Y.M., Koh, Y.H., Improvement of compressive strength of porous hydroxyapatite scaffolds by adding polystyrene to camphene-based slurries. *Mater. Lett.*, *63*, 955–958, **2009**.
27. Gong, X.H., Tang, C.Y., Hu, H.C., Zhou, X.P., Xie, X.L., Improved mechanical properties of HIPS/hydroxyapatite composites by surface modification of hydroxyapatite via *in-situ* polymerization of styrene. *J. Mater. Sci. Mater. Med.*, *15*, 1141–1146, **2004**.
28. Kano, J., Zhang, Q., Saito, F., Baron, M., Nzihou, A., Synthesis of hydroxyapatite with the mechanochemical treatment products of PVC and CaO. *Process Saf. Environ. Prot.*, *84*, 309–312, **2006**.
29. Vadukumpully, S., Paul, J., Mahanta, N., Valiyaveettil, S., Flexible conductive graphene/poly(vinyl chloride) composite thin films with high mechanical strength and thermal stability. *Carbon N. Y.*, *49*, 198–205, **2011**.
30. Beiko, D.T., Knudsen, B.E., Watterson, J.D., Denstedt, J.D., Biomaterials in urology. *Curr. Urol. Rep.*, *4*, 51–55, **2003**.
31. Zhang, W., Liao, S.S., Cui, F.Z., Hierarchical self-assembly of nano-fibrils in mineralized collagen. *Chem. Mater.*, *15*, 3221–3226, **2003**.
32. Boccafoschi, F., Habermehl, J., Vesentini, S., Mantovani, D., Biological performances of collagen-based scaffolds for vascular tissue engineering. *Biomaterials*, *26*, 7410–7417, **2005**, doi: 10.1016/j.biomaterials.2005.05.052.
33. Gründer, T., Gaissmaier, C., Fritz, J., Stoop, R., Hortschansky, P., Mollenhauer, J., Aicher, W.K., Bone morphogenetic protein (BMP)-2 enhances the expression of type II collagen and aggrecan in chondrocytes embedded in alginate beads. *Osteoarthr. Cartil.*, *12*, 559–567, **2004**.
34. Bigi, A., Panzavolta, S., Roveri, N., Hydroxyapatite-gelatin films: A structural and mechanical characterization. *Biomaterials*, *19*, 739–744, **1998**.
35. Chang, M.C., Ko, C.C., Douglas, W.H., Preparation of hydroxyapatite-gelatin nanocomposite. *Biomaterials*, *24*, 2853–2862, **2003**.
36. Kong, L., Gao, Y., Cao, W., Gong, Y., Zhao, N., Zhang, X., Preparation and characterization of nano-hydroxyapatite/chitosan composite scaffolds. *J. Biomed. Mater. Res. - Part A*, *75*, 275–282, **2005**.
37. Liao, S., Chan, C.K., Ramakrishna, S., Biomimetic nanocomposites for tissue engineering. *J. Bionanosci.*, *1*, 1–13, **2008**.
38. Furuzono, T., Taguchi, T., Kishida, A., Akashi, M., Tamada, Y., Preparation and characterization of apatite deposited on silk fabric using an alternate soaking process. *J. Biomed. Mater. Res.*, *50*, 344–352, **2000**.
39. Wang, L., Nemoto, R., Senna, M., Changes in microstructure and physicochemical properties of hydroxyapatite-silk fibroin nanocomposite with varying silk fibroin content. *J. Eur. Ceram. Soc.*, *24*, 2707–2715, **200**.

40. Ribeiro, C.C., Barrias, C.C., Barbosa, M.A., Calcium phosphate-alginate microspheres as enzyme delivery matrices. *Biomaterials, 25*, 4363–4373, **2004**.
41. Berman, A., Addadi, L., Weiner, S., Interactions of sea-urchin skeleton macromolecules with growing calcite crystals- a study of intracrystalline proteins. *Nature, 331*, 546–548, **1988**.
42. Berman, A., Hanson, J., Leiserowitz, L., Koetzle, T.F., Weiner, S., Addadi, L., Biological control of crystal texture: A widespread strategy for adapting crystal properties to function. *Science (80-.), 259*, 776–779, **1993**.
43. Evans, S.L. and Gregson, P.J., Composite technology in load-bearing orthopaedic implants. *Biomaterials, 19*, 1329–1342, **1998**.
44. Kikuchi, M., Ikoma, T., Itoh, S., Matsumoto, H.N., Koyama, Y., Takakuda, K., Shinomiya, K., Tanaka, J., Biomimetic synthesis of bone-like nanocomposites using the self-organization mechanism of hydroxyapatite and collagen. *Compos. Sci. Technol., 64*, 819–825, **2004**.
45. Han, Y., Wang, X., Li, S.A., Simple route to prepare stable hydroxyapatite nanoparticles suspension. *J. Nanopart. Res., 11*, 1235–1240, **2009**.
46. Zhang, H. B, Zhou, K. C, Li, Z. Y, Huang, S. P, Plate-like hydroxyapatite nanoparticles synthesized by the hydrothermal method. *J. Phys. Chem. Solids, 70*, 243–248, **2009**.
47. Paul, W. and Sharma, C.P., Infection resistant hydroxyapatite/alginate plastic composite. *J. Mater. Sci. Lett., 16*, 2050–2051, **1997**.
48. Chesnutt, B.M., Viano, A.M., Yuan, Y., Yang, Y., Guda, T., Appleford, M.R., Ong, J.L., Haggard, W.O., Bumgardner, J.D., Design and characterization of a novel chitosan/nanocrystalline calcium phosphate composite scaffold for bone regeneration. *J. Biomed. Mater. Res. - Part A, 88*, 491–502, **2009**.
49. Zhao, F., Grayson, W.L., Ma, T., Bunnell, B., Lu, W.W., Effects of hydroxyapatite in 3-D chitosan-gelatin polymer network on human mesenchymal stem cell construct development. *Biomaterials, 27*, 1859–1867, **2006**.
50. Ji Yin, Y., Zhao, F., Feng Song, X., De Yao, K., Lu, W.W., Chiyan Leong, J., Preparation and characterization of hydroxyapatite/chitosan–gelatin network composite. *J. Appl. Polym. Sci., 77*, 2929–2938, **2000**.
51. Asran, A.S., Henning, S., Michler, G.H., Polyvinyl alcohol–collagen–hydroxyapatite biocomposite nanofibrous scaffold: Mimicking the key features of natural bone at the nanoscale level. *Polymer (Guildf)., 51*, 868–876, **2010**.
52. Kim, S.B., Kim, Y.J., Yoon, T.L., Park, S.A., Cho, I.H., Kim, E.J., Kim, I.A., Shin, J.-W., The characteristics of a hydroxyapatite–chitosan–PMMA bone cement. *Biomaterials, 25*, 5715–5723, **2004**.
53. Zhang, Y. and Lu, J.A., Mild and efficient biomimetic synthesis of rodlike hydroxyapatite particles with a high aspect ratio using polyvinylpyrrolidone as capping agent. *Cryst. Growth Des., 8*, 2101–2107, **2008**.

54. Sotome, S., Uemura, T., Kikuchi, M., Chen, J., Itoh, S., Tanaka, J., Tateishi, T., Shinomiya, K., Synthesis and *in vivo* evaluation of a novel hydroxyapatite/collagen-alginate as a bone filler and a drug delivery carrier of bone morphogenetic protein. *Mater. Sci. Eng. C*, *24*, 341–347, **2004**.
55. Kay, M.I., Young, R.A., Posner, A.S., Crystal structure of hydroxyapatite. *Nature*, *204*, 1050–1052, **1964**.
56. Posner, A.S., Perloff, A., Diorio, A.F., Refinement of the hydroxyapatite structure. *Acta Crystallogr.*, *11*, 308–309, **1958**.
57. Elliott, J.C., Structure and chemistry of the apatites and other calcium orthophosphates. *Studies in Inorganic Chemistry*, 1st Edition, Elsevier Science, United Kingdom, *18*, 94008066–94008066, **1994**.
58. Aoki, H., *Medical applications of hydroxyapatite*, pp. 13–74, Ishiyaku Euro Am, Tokyo, Japan, **1994**.
59. Veis, A.A., Window on biomineralization. *Science (80-.)*, *307*, 1419–1420, **2005**.
60. de Groot, K., Bioceramics consisting of calcium phosphate salts. *Biomaterials*, *1*, 47–50, **1980**.
61. Ichikawa, T., Hirota, K., Kanitani, H., Wigianto, R., Kawamoto, N., Matsumoto, N., Miyake, Y., Rapid bone resorption adjacent to hydroxyapatite-coated implants. *J. Oral. Implantol.*, *22*, 232–235, **1996**.
62. Ozawa, S. and Kasugai, S., Evaluation of implant materials (hydroxyapatite, glass-ceramics, titanium) in rat bone marrow stromal cell culture. *Biomaterials*, *17*, 23–29, **1996**,.
63. Tampieri, A., Celotti, G., Sprio, S., Delcogliano, A., Franzese, S., Porosity-graded hydroxyapatite ceramics to replace natural bone. *Biomaterials*, *22*, 1365–1370, **2001**.
64. Martins, V.C.A. and Goissis, G., Nonstoichiometric hydroxyapatite-anionic collagen composite as support for the double sustained release of gentamicin and norfloxacin/ciprofloxacin. *Artif. Organs*, *24*, 224–230, **2000**.
65. Wilson, O.C. and Hull, J.R., Surface modification of nanophase hydroxyapatite with chitosan. *Mater. Sci. Eng. C*, *28*, 434–437, **2008**.
66. Bunker, B.C., Rieke, P.C., Tarasevich, B.J., Campbell, A.A., Fryxell, G.E., Graff, G.L., Song, L., Liu, J., Virden, J.W., Ceramic thin film formation on functionalized interfaces through biomimetic processing. *Science (80-.)*., *264*, 48–55, **1994**.
67. Riman, R.E., Suchanek, W.L., Byrappa, K., Chen, C.W., Shuk, P., Oakes, C.S., Solution synthesis of hydroxyapatite designer particulates. *Solid State Ion.*, *151*, 393–402, **2002**.
68. Cheng, K., Han, G., Weng, W., Qu, H., Du, P., Shen, G., Yang, J., Ferreira, J.M.F., Sol-gel derived fluoridated hydroxyapatite films. *Mater. Res. Bull.*, *38*, 89–97, **2003**.

69. Uota, M., Arakawa, H., Kitamura, N., Yoshimura, T., Tanaka, J., Kijima, T., Synthesis of high surface area hydroxyapatite nanoparticles by mixed surfactant-mediated approach. *Langmuir*, *21*, 4724–4728, **2005**.
70. Hijón, N., Victoria Cabañas, M., Peña, J., Vallet-Regí, M., Dip coated silicon-substituted hydroxyapatite films. *Acta Biomater.*, *2*, 567–574, **2006**.
71. Wu, Y. and Bose, S., Nanocrystalline hydroxyapatite: Micelle templated synthesis and characterization. *Langmuir*, *21*, 3232–3234, **2005**.
72. Webster, T.J., Ergun, C., Doremus, R.H., Siegel, R.W., Bizios, R., Enhanced functions of osteoblasts on nanophase ceramics. *Biomaterials*, *21*, 1803–1810, **2000**.
73. Young, R.A. and Holcomb, D.W., Variability of hydroxyapatite preparations. *Calcif. Tissue Int.*, *34*, S17–32, **1982**.
74. Bernard, L., Freche, M., Lacout, J.L., Biscans, B., Preparation of hydroxyapatite by neutralization at low temperature - influence of purity of the raw material. *Powder Technol.*, *103*, 19–25, **1999**.
75. Liu, J., Ye, X., Wang, H., Zhu, M., Wang, B., Yan, H., The influence of PH and temperature on the morphology of hydroxyapatite synthesized by hydrothermal method. *Ceram. Int.*, *29*, 629–633, **2003**.
76. Liu, D.M., Troczynski, T., Tseng, W.J., Water-based sol-gel synthesis of hydroxyapatite: Process development. *Biomaterials*, *22*, 1721–1730, **2001**.
77. Lim, G.K., Wang, J., Ng, S.C., Chew, C.H., Gan, L.M., Processing of hydroxyapatite via microemulsion and emulsion routes. *Biomaterials*, *18*, 1433–1439, **1997**.
78. Suchanek, W.L., Shuk, P., Byrappa, K., Riman, R.E., TenHuisen, K.S., Janas, V.F., Mechanochemical-hydrothermal synthesis of carbonated apatite powders at room temperature. *Biomaterials*, *23*, 699–710, **2002**.
79. Sanchez, C., Lebeau, B., Ribot, F., In, M., Molecular design of sol-gel derived hybrid organic-inorganic nanocomposites. *J. Sol-Gel Sci. Technol.*, *19*, 31–38, **2000**.
80. Ashok, M., Kalkura, S.N., Sundaram, N.M., Arivuoli, D., Growth and characterization of hydroxyapatite crystals by hydrothermal method. *J. Mater. Sci. Mater. Med.*, *18*, 895–898, **2007**.
81. Rhee, S.H., Lee, J.D., Tanaka, J., Nucleation of hydroxyapatite crystal through chemical interaction with collagen. *J. Am. Ceram. Soc.*, *83*, 2890–2892, **2000**.
82. Mollazadeh, S., Javadpour, J., Khavandi, A., *In situ* synthesis and characterization of nano-size hydroxyapatite in poly(vinyl alcohol) matrix. *Ceram. Int.*, *33*, 1579–1583, **2007**.
83. Chung, Y.S., Kang, S., II, Kwon, O.W., Shin, D.S., Lee, S.G., Shin, E.J., Min, B.G., Bae, H.J., Han, S.S., Jeon, H.Y. *et al.*, Preparation of hydroxyapatite/poly(vinyl alcohol) composite fibers by wet spinning and their characterization. *J. Appl. Polym. Sci.*, *106*, 3423–3429, **2007**.

84. Malkaj, P., Pierri, E., Dalas, E., The crystallization of hydroxyapatite in the presence of sodium alginate. *J. Mater. Sci. Mater. Med.*, *16*, 733–737, **2005**.
85. Zhang, J., Liao, J., Mo, A., Li, Y., Li, J., Wang, X., Characterization and human gingival fibroblasts biocompatibility of hydroxyapatite/PMMA nanocomposites for provisional dental implant restoration. *Appl. Surf. Sci.*, *255*, 328–330, **2008**.
86. Sivakumar, M. and Panduranga Rao, K., Preparation, characterization and *in vitro* release of gentamicin from coralline hydroxyapatite-gelatin composite microspheres. *Biomaterials*, *23*, 3175–3181, **2002**.
87. Wang, L., Li, Y., Li, C., *In situ* processing and properties of nanostructured hydroxyapatite/alginate composite. *J. Nanoparticle Res.*, *11*, 691–699, **2009**.
88. Mohamed, K.R. and Mostafa, A.A., Preparation and bioactivity evaluation of hydroxyapatite-titania/chitosan-gelatin polymeric biocomposites. *Mater. Sci. Eng. C*, *28*, 1087–1099, **2008**.
89. Yun, Y.P., Kim, S.J., Lim, Y.M., Park, K., Kim, H.J., Jeong, S.I., Kim, S.E., Song, H.R., The effect of alendronate-loaded polycarprolactone nanofibrous scaffolds on osteogenic differentiation of adipose-derived stem cells in bone tissue regeneration. *J. Biomed. Nanotechnol.*, *10*, 1080–1090, **2014**.
90. Zhou, H., Lawrence, J.G., Bhaduri, S.B., Fabrication aspects of PLA-CaP/PLGA-CaP composites for orthopedic applications: A review. *Acta Biomater.*, *8*, 1999–2016, **2012**.
91. Iafisco, M., Palazzo, B., Ito, T., Otsuka, M., Senna, M., Delgado-Lopez, J.M., Gomez-Morales, J., Tampieri, A., Prat, M., Rimondini, L., Preparation of core-shell poly(L-lactic) acid-nanocrystalline apatite hollow microspheres for bone repairing applications. *J. Mater. Sci. Mater. Med.*, *23*, 2659–2669, **2012**.
92. Wei, G. and Ma, P.X., Structure and properties of nano-hydroxyapatite/polymer composite scaffolds for bone tissue engineering. *Biomaterials*, *25*, 4749–4757, **2004**.
93. Han, W., Zhao, J., Tu, M., Zeng, R., Zha, Z., Zhou, C., Preparation and characterization of nanohydroxyapatite strengthening nanofibrous poly(L-lactide) scaffold for bone tissue engineering. *J. Appl. Polym. Sci.*, *128*, 1332–1338, **2013**.
94. Ren, J., Zhao, P., Ren, T., Gu, S., Pan, K., Poly (D,L-lactide)/nanohydroxyapatite composite scaffolds for bone tissue engineering and biocompatibility evaluation. *J. Mater. Sci. Mater. Med.*, *19*, 1075–1082, **2008**.
95. Diao, H., Si, Y., Zhu, A., Ji, L., Shi, H., Surface modified nano-hydroxyapatite/poly(lactide acid) composite and its osteocyte compatibility. *Mater. Sci. Eng. C*, *32*, 1796–1801, **2012**.
96. Nejati, E., Mirzadeh, H., Zandi, M., Synthesis and characterization of nano-hydroxyapatite rods/poly(L-lactide acid) composite scaffolds for bone tissue engineering. *Compos. Part A Appl. Sci. Manuf.*, *39*, 1589–1596, **2008**.

97. Hong, Z., Zhang, P., He, C., Qiu, X., Liu, A., Chen, L., Chen, X., Jing, X., Nano-composite of poly(L-lactide) and surface grafted hydroxyapatite: Mechanical properties and biocompatibility. *Biomaterials*, *26*, 6296–6304, **2005**.
98. Ronca, A., Ambrosio, L., Grijpma, D.W., Preparation of designed poly(D,L-lactide)/nanosized hydroxyapatite composite structures by stereolithography. *Acta Biomater.*, *9*, 5989–5996, **2013**.
99. Zhang, C.Y., Zhang, C.L., Wang, J.F., Lu, C.H., Zhuang, Z., Wang, X.P., Fang, Q.F., Fabrication and *in vitro* investigation of nanohydroxyapatite, chitosan, poly(L -lactic acid) ternary biocomposite. *J. Appl. Polym. Sci.*, *127*, 2152–2159, **2013**.
100. Niu, X., Feng, Q., Wang, M., Guo, X., Zheng, Q., Porous nano-HA/collagen/PLLA scaffold containing chitosan microspheres for controlled delivery of synthetic peptide derived from BMP-2. *J. Control. Release*, *134*, 111–117, **2009**.
101. Abdal-Hay, A., Sheikh, F.A., Lim, J.K., Air jet spinning of hydroxyapatite/poly(lactic acid) hybrid nanocomposite membrane mats for bone tissue engineering. *Colloids Surf. B Biointerfaces*, *102*, 635–643, **2013**.
102. Kandori, K., Hamazaki, H., Matsuzawa, M., Togashi, S., Selective adsorption of acidic protein of bovine serum albumin onto sheet-like calcium hydroxyapatite particles produced by microreactor. *Adv. Powder Technol.*, *25*, 354–359, **2014**.
103. Tanidjaja, I. and Damodaran, S., Influence of amino acids on thermal stability and heat-set gelation of bovine serum albumin. *Food Chem.*, *337*, 127670, **2021**.
104. Khalifehzadeh, R. and Arami, H., Biodegradable calcium phosphate nanoparticles for cancer therapy. *Adv. Colloid Interface Sci.*, *279*, 102157, **2020**.
105. Arami, H., Khandhar, A., Liggitt, D., Krishnan, K.M., *In vivo* delivery, pharmacokinetics, biodistribution and toxicity of iron oxide nanoparticles. *Chem. Soc. Rev.*, *44*, 8576–8607, **2015**.
106. Qiao, W., Lan, X., Tsoi, J.K.H., Chen, Z., Su, R.Y.X., Yeung, K.W.K., Matinlinna, J.P., Biomimetic hollow mesoporous hydroxyapatite microsphere with controlled morphology, entrapment efficiency and degradability for cancer therapy. *RSC Adv.*, *7*, 44788–44798, **2017**.
107. Bhoi, S., Prasad, A., Kumar, A., Sarkar, R.B., Mahto, B., Meena, C.S., Pandey, C., Experimental study to evaluate the wear performance of UHMWPE and XLPE material for orthopedics application. *Bioengineering*, 9, 676, 2022. https://doi.org/10.3390/bioengineering9110676.
108. Bhoi, S., Kumar, A., Prasad, A., Meena, C.S., Sarkar, R.B., Mahto, B., Ghosh, A., Performance evaluation of different coating materials in delamination for micro-milling applications on high-speed steel substrate. *Micromachines*, 13, 1277, 2022. https://doi.org/10.3390/mi13081277.

109. Gupta, A. *et al.*, Multifunctional nanohydroxyapatite-promoted toughened high-molecular-weight stereocomplex poly(lactic acid)-based bionanocomposite for both 3D-printed orthopedic implants and high-temperature engineering applications. *ACS Omega*, 2, 7, 4039–52, 2017.
110. Prasad, A., Bioabsorbable polymeric materials for biofilms and other biomedical applications: Recent and future trends. *Mater. Today: Proc.*, 44, 2447–53, 2021a. https://doi.org/10.1016/j.matpr.2020.12.489.
111. Prasad, A., State of art review on bioabsorbable polymeric scaffolds for bone tissue engineering. *Mater. Today: Proc.*, 44, 1391–1400, 2021b. https://doi.org/10.1016/j.matpr.2020.11.622.
112. Prasad, A., Bhasney, S., Katiyar, V., Ravi Sankar, M., Biowastes processed hydroxyapatite filled poly (lactic acid) bio-composite for open reduction internal fixation of small bones. *Mater. Today: Proc.*, 4, 9, 10153–57, 2017. http://linkinghub.elsevier.com/retrieve/pii/S2214785317311410.
113. Prasad, A., Devendar, B., Ravi Sankar, M., Robi, P.S., Micro-scratch based tribological characterization of hydroxyapatite (HAp) fabricated through fish scales. *Mater. Today: Proc.*, 2, 4–5, 1216–24, 2015. http://linkinghub.elsevier.com/retrieve/pii/S2214785315002795.
114. Prasad, A., Chakraborty, G., Kumar, A., Bio-based environmentally benign polymeric resorbable materials for orthopedic fixation applications, in: *Advanced Materials for Biomedical Applications*, vol. Chapter 15, A. Kumar, Y. Gori, A. Kumar, C.S. Meena, N. Dutt (Eds.), pp. 251–266, CRC Press, Boca Raton, FL, USA, 2022, https://doi.org/10.1201/9781003344810-15.
115. Prasad, A., Biomaterial-based nanofibers scaffolds in tissue engineering application, in: *Functional Biomaterials*, pp. 245–264, Springer, Singapore, 2022.
116. Prasad, A., Ravi Sankar, M., Katiyar, V., State of art on solvent casting particulate leaching method for orthopedic scaffolds fabrication. *Mater. Today: Proc.*, 4, 2, 898–907, 2017. http://dx.doi.org/10.1016/j.matpr.2017.01.101.
117. Kumar, A., Gangwar, A.K.S., Kumar, A., Meena, C.S., Singh, V.P., Dutt, N., Prasad, A., Gori, Y., Biomedical study of femur bone fracture and healing, in: *Advanced Materials for Biomedical Applications*, vol. Chapter 14, A. Kumar, Y. Gori, A. Kumar, C.S. Meena, N. Dutt (Eds.), pp. 235–250, CRC Press, Boca Raton, FL, USA, 2022, https://doi.org/ 10.1201/9781003344810-14.
118. Kumar, R., Kumar, A., Kant, L., Prasad, A., Bhoi, S., Meena, C.S., Singh, V.P., Ghosh, A., Experimental and RSM-based process-parameters optimisation for turning operation of EN36B Steel. *Materials*, 16, 339, 2023. https://doi.org/10.3390/ma16010339.
119. Prasad, A., Chakraborty, G., Kumar, A., Gajrani, K.K., Introduction to biodegradable polymers, in: *Biodegradable Composites for Packaging Applications*, vol. Chapter 1, A. Prasad, A. Kumar, K.K. Gajrani (Eds.), pp. 1–12, CRC Press, Boca Raton, FL, USA, 2022, https://doi.org/10.1201/9781003227908-1.

120. Chakraborty, G., Prasad, A., Kumar, A., Processing of biodegradable composites, in: *Biodegradable Composites for Packaging Applications*, vol. Chapter 3, A. Prasad, A. Kumar, K.K. Gajrani (Eds.), pp. 33–48, CRC Press, Boca Raton, FL, USA, 2022, https://doi.org/ 10.1201/9781003227908-3.
121. Paul, A.K., Prasad, A., Kumar, A., Review on artificial neural network and its application in the field of engineering. *J. Mech. Eng.: PRAKASH*, 1, 1, 53–61, 2022. https://doi.org/10.56697/JMEP.2022.1107.

8

Self-Healing Materials and Hydrogel for Biomedical Application

Arabinda Majhi[1,2], Megha Dhiman[1,2], Partha Roy[3] and Debrupa Lahiri[1,2*]

[1]*Department of Metallurgical and Materials Engineering, Indian Institute of Technology Roorkee, Roorkee, India*
[2]*Centre for Nanotechnology, Indian Institute of Technology Roorkee, Roorkee, India*
[3]*Department of Biosciences & Bioengineering, Indian Institute of Technology Roorkee, Roorkee, India*

Abstract

The materials have the inherent property to recover their physical, chemical, mechanical properties, and rheological behavior through the dissipation of applied external rapturing forces. These materials can be of organic, organometallic, and polymeric origin. This advanced feature of these materials is called self-healing property, which enhances the lifespan of materials by balancing their mechanical integrity. Major factors that affect the self-healing capability of these materials are specific healing mechanisms, rate of propagation, repairing rate, and time of healing. Living species possess self-healing ability against damages up to certain limits, whereas non-living things are not capable of the same. At the current time, scientists are trying to incorporate and mimic nature's self-repairing capabilities into a non-living body. Self-healing materials have a variety of applications, from the construction area to the electronics and biomedical fields. In biomedical fields, polymeric hydrogel plays a significant role as a self-healing material. Hydrogel is a polymeric biomaterial that can keep a huge amount of water through the 3D hydrophilic polymeric network. But hydrogel owes specific issues, like, poor mechanical properties, large swelling ratio, and weakness against environmental stimuli. These critical issues are minimized by developing self-healing smart hydrogel that owes self-repairable ability, leading to superior stability and high strength. Their endowed property of biocompatibility, biodegradability, mechanical integrity, stimuli responsiveness, shear-thinning property, and self-healing property has ensured their prospective utilization in a variety of biomedical domains, e.g., tissue regeneration, drug delivery, wound healing, 3D printing, cell-laden bio-ink, soft robotics.

**Corresponding author*: debrupa.lahiri@mt.iitr.ac.in

Arbind Prasad, Ashwani Kumar and Manoj Gupta (eds.) Advanced Materials and Manufacturing Techniques for Biomedical Applications, (185–210) © 2024 Scrivener Publishing LLC

***Keywords*:** Self-healing, polymer, hydrogel, mechanism, biomedical application, tissue engineering

8.1 Introduction

In biomedical applications, hydrogel plays an essential role as a self-healing material. This chapter mainly emphasizes hydrogel consisting of self-healing properties with their biomedical application. Hydrogels constitute a 3D cross-linked network architecture that has the property to swell in water and store enormous amounts of water while maintaining structural integrity due to the physical or chemical crosslinking of hydrophilic polymeric chains. Hydrogels are frequently referred to as wet and soft materials. They have drawn significant recognition owing to their distinctive properties, like, as huge water-retention ability, biocompatibility, biodegradability, elasticity, etc. In current years, a variety of hydrogels are produced, including supramolecular hydrogels, shape memory, and self-assembled conductive hydrogels, and their properties are explored and tested in a variety of applications [1–3]. Some hydrophilic functional groups, like, hydroxyl (-OH), carboxyl (-COOH), amines ($-NH_2$), and amides ($-CONH_2$) connected to the backbone of polymer provide hydrogels their capability to absorb water, while crosslinks between network chains give them resistance to disintegration. Because of their more water retention ability, network structure, and ability to mimic the extracellular matrix, they acquire a high degree of elasticity, like the native tissues, making them fit candidates for tissue engineering, implant material, wound coverage, and therapeutic drug release. Once these hydrogels are subjected to different external environmental factors (pH, light, heat, etc.), they can respond to the signal by showing events like welling and deswelling.

However, these conventional hydrogels have low mechanical characteristics and are quickly distorted when mechanical forces are applied [4]. Since the materials' irreversible covalent crosslinks prohibit it from healing after the structural damage and such hydrogels are unable to maintain their function and acceptable mechanical qualities. This swelling property of hydrogel tends to restrain their toughening mechanism where the absorption of a large volume of solvent by hydrogel decreases the number of network chains and reduces the fracture energy, as there is less number of chains to break. Solvent molecules increase the separation between network chains, minimize the chance of crystallization and slow down the hardening process [5]. Various properties of hydrogels have assisted them in becoming most suitable in the different biomedical fields, while the weak mechanical properties have impeded their advancement. Therefore, it is of critical importance to develop self-healing hydrogels that are mechanically robust. This would not only increase the lifetime of material but also prolong the reliability and durability of hydrogel for specific applications, by preventing the buildup of cracks or damages.

Current research is going on with the introduction of various other important functionalities, such as, electrical conductivity, magnetic property, transient sol-gel transition, shape memory, bio adhesion, anti-biofouling behavior, anti-freezing ability to the self-healing system. Self-healing behavior of these hydrogels depend on different kind of mechanisms, such as, dynamic covalent bonds (including imine bonds, disulfide bonds, acyl hydrazone, anhydride linkage, Diels Alder reaction, thiol-Michael reaction, boronic ester, boronates, exchange of Si-O bond), metal-ligand interaction, supramolecular interaction (like, host-guest interaction), non-covalent interaction (such as hydrogen bonding, van der Waals interactions, Zwitter-ionic interaction). These mechanisms of self-repairing allow hydrogel to prevent external damages, as well as, gain multiple stimuli-responsive properties. A combination of two or three mechanisms has opened the path for developing more advanced self-healing hydrogel for biomedical applications. Metal-ligand coordination bond-based self-healing hydrogel improves injectability, visco-elastic properties, as well as, physiological functionalities. Hydrogel, having metal ion(s) and polymeric backbone containing pendent groups or chelating groups (such as bipyridine and terpyridine bisphosphonate, catechol, histidine, thiolate, carboxylate, imidazole), form a metal-ligand bond, which manages self-healing behavior during external damage. Dynamic covalent bonding and supramolecular interaction offers excellent physicochemical properties and mimick the dynamic extracellular matrix. It improves the capabilities of shear-thinning, self-healing and shape memorizing. Through supramolecular interaction of hydrogel, cells can be easily encapsulated in the hydrogel matrix via breaking and making of the reversible bond and differentiation can be improved. The shear-thinning and self-healing hydrogels have been used in cell-laden bio-inks for the design of complex structures. Self-healing hydrogel enhances the progress of future generation's tissue regeneration, 3D organ printing, drug delivery studies.

8.2 Self-Healing Hydrogels

Self-healing hydrogels can retrieve and bring back their shape, function and structure entirely or partially after being physically damaged either

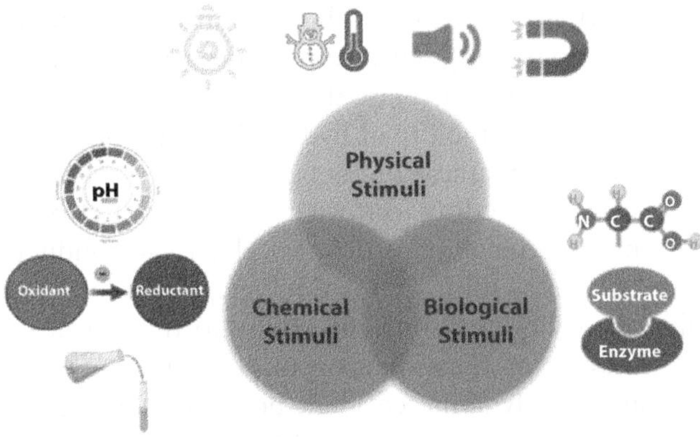

Figure 8.1 Different factors affecting the self-healing hydrogel.

on their own or with the help of external stimuli. The development of the self-healing property arises from the wound healing phenomenon that occurs in the living organism, which has the potential to recover the mechanical strength and morphology of the materials after being repeatedly harmed [6]. These self-healing hydrogels are characterized into two categories: autonomic hydrogels that do not need any external stimuli and non-autonomic hydrogels, where Figure 8.1 shows different physical, biological, and chemical external factors are required to initiate the recovery process, in the form of heat, amino acids, enzymes, light, magnetic field, pH, temperature etc. [7].

8.3 Mechanism of Self-Healing in Hydrogels

Hydrogels with self-healing properties can be fabricated by employing different kinds of design methodologies. The mechanism for healing can be classified into two types: physically and chemically crosslinked self-healing hydrogels (Figure 8.2). Chemically crosslinked hydrogels involve dynamic covalent bonds, including various reactions, like, Diel-Alder reaction, Schiff base, Boronate bonds, Disulfide bonds, etc. In this case, a network is formed through reversible chemical reactions in which bonds can be disrupted and repaired under modest physiological stimuli and conditions. Physically crosslinked self-healing hydrogels include non-covalent interactions, like, host-guest attractions, hydrogen bonding and ionic and hydrophobic interactions. The polymer chains can be introduced with different functional groups of both self-healing hydrogels for favourable chemical reactions and non-covalent physical attractions, which

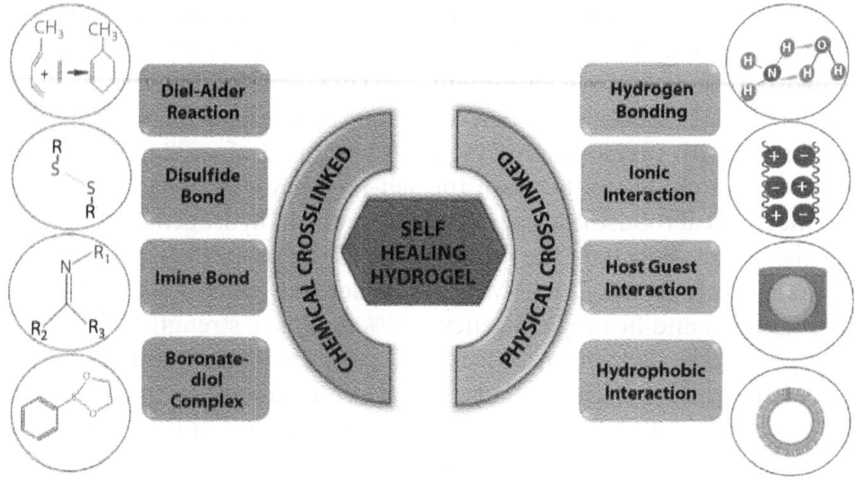

Figure 8.2 Schematic representation of different mechanisms occurring in self-healing hydrogels.

facilitate the process of self-recovery to initiate at the injured region. At the time of the healing process, the dissociation of crosslinks momentarily enhances the diffusion of polymer chains at the damaged site, followed by the creation of new bonds, resulting in network repair and mechanical property recovery [8]. The hydrogels involving covalent interactions in their framework have been reported to be highly stable. In contrast, those having physical non-covalent interactions show a high degree of flexibility, because of their ease in breaking bonds and rebuilding the crosslinks.

8.3.1 Physically Cross-Linked Self-Healing Hydrogels

8.3.1.1 Hydrogen Bonding

Hydrogen bonding is a kind of electrostatic force involving dipole-dipole forces, which is formed when different types of highly electronegative atoms, like, fluorine, oxygen and nitrogen interacts with hydrogen atom. Hydrogels, formed incorporating hydrogen bonding, can self-heal based on the reversible crosslinks formed between different moieties having donor and acceptor groups. 2-ureido-4[1H]-pyrimidinone (UPy) based self-healing hydrogels are studied extensively due to the quadruple hydrogen bond formation unit that provides them stability through reversible supramolecular assembly due to multiple hydrogen bonds [9]. However, the dynamic reversible hydrogen bonding makes the hydrogel less stable and decreases the self-recovery capacity. To surmount this limitation and to make a more robust hydrogel, hydrophobic groups can be grafted onto the polymer backbone. The polypropylene glycol-2-amino-5-(2-hydroxyethyl)-6-methylpyrimidin-4-ol (PPG-mUPy) self-healing polyurethane (P.U.) elastomer has been described by Cheng *et al.* by combining PPG motif with ureidopyrimidone (UPy) segment of explicit design. To control the equilibrium between the self-recovery capability of hydrogel and mechanical robustness, they used the switch of hydrogen bonding that was thermally triggered. The sample showed the proper balance between mechanical properties (20.62 MPa of tensile strength and 100.49 MJ m^{-3} of toughness) and healing efficiency (93% in tensile strength) after being damaged and healed within 24 h at 80°C [10]. Zhao *et al.* created the poly (N-vinylpyrrolidone)/gallic acid (PVP/GA) hydrogel by interacting a hydroxyl group of G.A. with the tertiary amine group of PVP via hydrogen bonding. The hydrogel, thus formed, comprised of properties, like, self-healing, conductivity, self-adhesive, biocompatibility, and stretchability etc. All these properties allowed this hydrogel to be used as wearable electronic sensors and electronic skin, as it can attach with the skin and observe physiological activities with outstanding reliability and sensitivity. The hydrogel showed excellent self-healing property within 30 seconds without any external stimuli. The variation in the brightness intensity of a blue light-emitting diode (LED) bulb in the circuit was used to evaluate

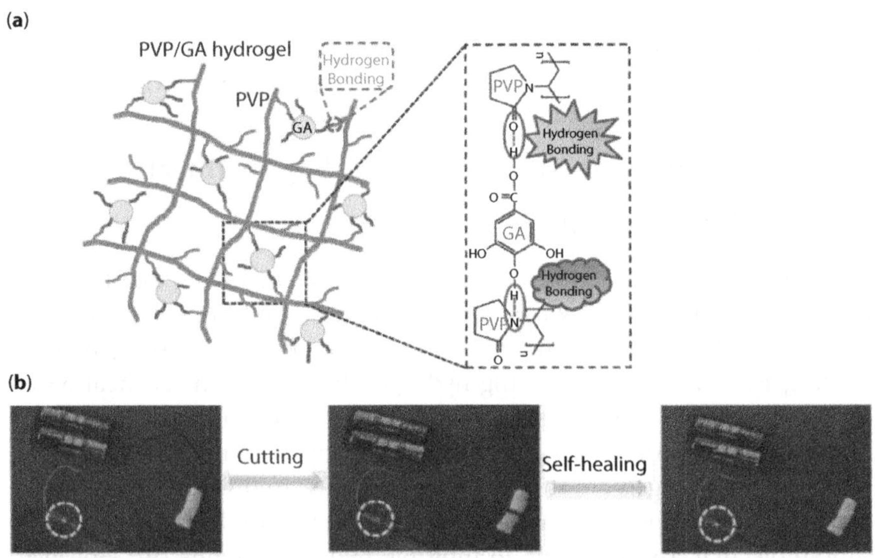

Figure 8.3 (a) Diagrammatic representation of hydrogen bonding (b) Illuminous change of self-healing process. Reproduced with permission from ref [11] Copyright 2020, Elsevier.

the electrical conductivity of the hydrogel during the process of cutting and contacting operation. The conductivity of hydrogel was because of the presence of ionized carboxylate, phenoxide and hydride ions present in the backbone of hydrogel that conducted electricity and lit the LED bulb. The bulb turned off when the hydrogel was sliced apart, and the bulb illuminated again when the hydrogel was re-contacted, suggesting the ability to self-heal [11]. Figure 8.3 shows the hydrogen bonding in the hydrogel between PVP and G.A. and the illuminance variance at the time of self-healing.

8.3.1.2 Ionic Interactions

Ionic interaction is the electrostatic attraction that occurs when two oppositely charged groups of polymer interact with each other and facilitate the crosslinking during the formation of a hydrogel. Self-healing occurs when free ions migrate to the non-crosslinked or damaged portion of a polymeric chain that results into a reversible reaction, involving electrostatic interaction between free ions and polymeric chains. Zhong *et al.* have shown the synthesis of self-healing hydrogels with better mechanical properties, by the formation of ionic bonds when carboxylic group of polyacrylic acid (PAA) and Fe^{3+} ions interacts. These hydrogels can be employed as scaffolds in tissue engineering because of their astonishing features, such as, robustness, stretchability and self-healing ability [12]. Xu *et al.* synthesized the self-healable metallo-hydrogel constructed on the particular crosslinking interaction between Ni^{2+} and polymeric ligand formed when 2,6-diethynyl pyridine was added in PEG-diazide. The hydrogenation was specifically initiated by the addition of Ni^{2+} and its metal-ligand coordination with the ligand moieties. Changes in the metal/ligand ratio and polymer concentration could alter the gelation and mechanical properties. The reversibility and dynamic nature of the ionic interaction imparted hydrogel its excellent self-healing capability that, upon the damage, was completely healed within 2 minutes, after simply reuniting the fragmented parts on removing the strain. This self-healing behavior may have used in the repair of metal coatings [13]. The hydrogels based on such ionic interactions have conductive properties. Sun *et al.* synthesized the hydrogel sensors having high mechanical strength, excellent conductivity, and self-repairable property and demonstrated their application in wearable devices. A double network ionically conductive hydrogel was engineered by incorporating polyacrylamine (PAAm) and gelatin in sodium citrate solution. This soaking method not only endowed hydrogel with an enhanced mechanical and electrochemical property, but increased conductivity also as compared to the reported sensors. On immersing the hydrogel in sodium citrate solution, ionic interactions formed between citrate ion and gelatin, and more diffusion of ions in hydrogel increased the conductivity. The hydrogel showed 1.66 MPa tensile strength, 849% elongation with 1.5 Sm-1 conductivity. This sensor can monitor the body's movement based on the sensitive pressure and strain graph [14].

8.3.1.3 Host–Guest Interactions

Host guest interactions is a defining field of supramolecular chemistry that occurs when a guest molecule of suitable size is incorporated into the host molecule and is held together with the help of non-covalent interactions. Crown ethers [15], cyclodextrin [16], calix [n] arene [17], cucurbit [n] and arenes [18] are host molecules that are employed for the synthesis of these inclusion complexes having self-recovering capabilities, owing to the reversible nature of these interactions. Rui Yu *et al.* synthesized injectable self-healing hydrogel by using aromatic amino acids of silk having host guest interaction with acryloyl-β-cyclodextrin via a facile method of photopolymerization using U.V. radiation and demonstrated improved effectiveness as a wound dressing as compared to the commercial dressing. Figure 8.4 shows the process of fabricating the hydrogel and the intrinsic self-healing property of the S.F./Ac-CD hydrogel, while it was divided into different pieces and immersed as one within seconds and with no external stimuli. Because of the presence of various hydrophobic groups in cyclodextrin, the hydrophobic pocket was used to entrap the curcumin drug molecule. It demonstrated the controlled and sustained release profile of the drug [19].

Highley and his group have synthesized self-healing hydrogel by adding hyaluronic acid (H.A.) modified cyclodextrin (CD) and adamantane (Ad) consecutively that forming a host-guest complex with the help of supramolecular host-guest bonding and used this hydrogel as ink and scaffold matrix in 3D printing technology to show excellent shear-thinning and self-healing capacity. This hydrogel was able to return to 100% storage modulus in 30 seconds under 3.5 cycles of 0.5 percent and then 250 percent strain testing circumstances [20].

8.3.1.4 Hydrophobic Interactions

Hydrophobic bonding is a non-covalent interaction that arises when the non-polar hydrophobic moieties form aggregates in aqueous conditions. The hydrophobic structures, thus formed, are easy to create and can quickly reform, when disrupted due to physical crosslinking. Therefore, using the monomers having the self-assembling mechanism in an aqueous

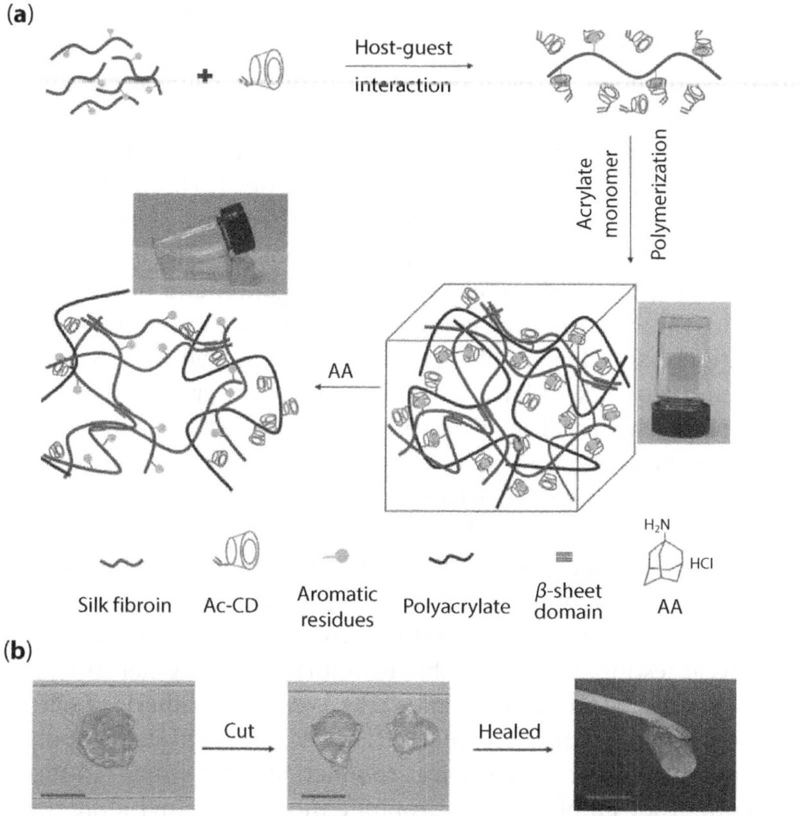

Figure 8.4 (a,b) Process to fabricate the hydrogel and the intrinsic self-healing capability of the S.F./Ac-CD hydrogel Reproduced with permission from ref [19] Copyright 2021, Elsevier.

medium, this property might be used to create physically self-healing hydrogels. Micelle copolymerization is most widely utilized strategies for incorporating self-healing capabilities into hydrogels. This mechanism involves the stabilized hydrophobic group being incorporated into hydrogels without compromising their hydrophilic character. The hydrophobic moiety must be dissolved in the micelle, made by the surfactant, for micelle copolymerization to occur. The surfactant takes on various forms, including rod, multilayer, and spherical. In an aqueous solution, the hydrophobic monomer group copolymerizes with the hydrophilic moieties that entangle tightly in micelles. When external forces are applied to hydrogels, the micelles operate as different crosslinking points, preventing the disrupting of hydrogel by dispersing the stress. In contrast, the hydrophobic chemical polymeric chains, interwined within the micelles, unravel or slide to dispel vast amount of energy within the structure that immensely enhances the toughness of hydrogel. Liu *et al.* firstly synthesized hydrophobic association (H.A.) hydrogel using acrylamide (Am) and octylphenolpolyethoxyether as hydrophilic and hydrophobic monomers, respectively, to generate associated micellar structures that serve as physical crosslinking sites in an HA-gel network [21]. This hydrogel compressed higher mechanical property and self-healing capability. The effect of various substituents, including hydrophobic, acrylamide and SDS, was also studied, which showed increased elongation and decrease in elastic modulus on increasing the content of SDS, owing to the constrained conformation shape changes of chains caused by the higher SDS content [22]. Rong and co-workers have fabricated a physical crosslinked double network (D.N.) hydrogel involving xanthan gum (X.G.) chains, crosslinked with Ca^{2+} as a primary network that provided reversible sacrificial bonding and energy dissipation to the hydrogel and polyacrylamide (PAM) as a hydrophobically crosslinked secondary network that gave elasticity to this D.N. hydrogel. Figure 8.5 shows the self-healing process occurring in the hydrogel. This double crosslinked method provided hydrogel with unique mechanical properties. After continuous loading and unloading compressive cycle testing, they were able to return their original forms, proving their exceptional fatigue resistance [23]. Wei *et al.* synthesized self-healable polymeric ionic liquid hydrogels (PIL) via the micellar copolymerization method with high electrical conductivity and enhanced mechanical strength. They employed a hydrophilic monomer, vinyl ionic liquid, and stearyl methacrylate with 18 carbons as hydrophobic moiety copolymerized in a micellar solution of sodium dodecyl sulphate (SDS) to impart the hydrogel improved self-healing characteristics through hydrophobic interactions. The hydrogel exhibited a self-healing efficiency of 85% without external stimuli, mechanical strength (5.8 MPa), and electrical conductivity (1.258 S/m). Because of their exceptional characteristics, the resulting PIL self-healable hydrogels are exciting candidates for various disciplines, including soft electrical devices and biological applications [24].

8.3.2 Chemically Self-Healing Hydrogels

8.3.2.1 Imine Bond

Imine bond (known by the name Schiff base) is a covalent dynamic reversible in nature interaction that can form hydrogel with self-healing property by crosslinking between polymers having amine and ketone or aldehyde group, resulting in loss of water under different physiological conditions.

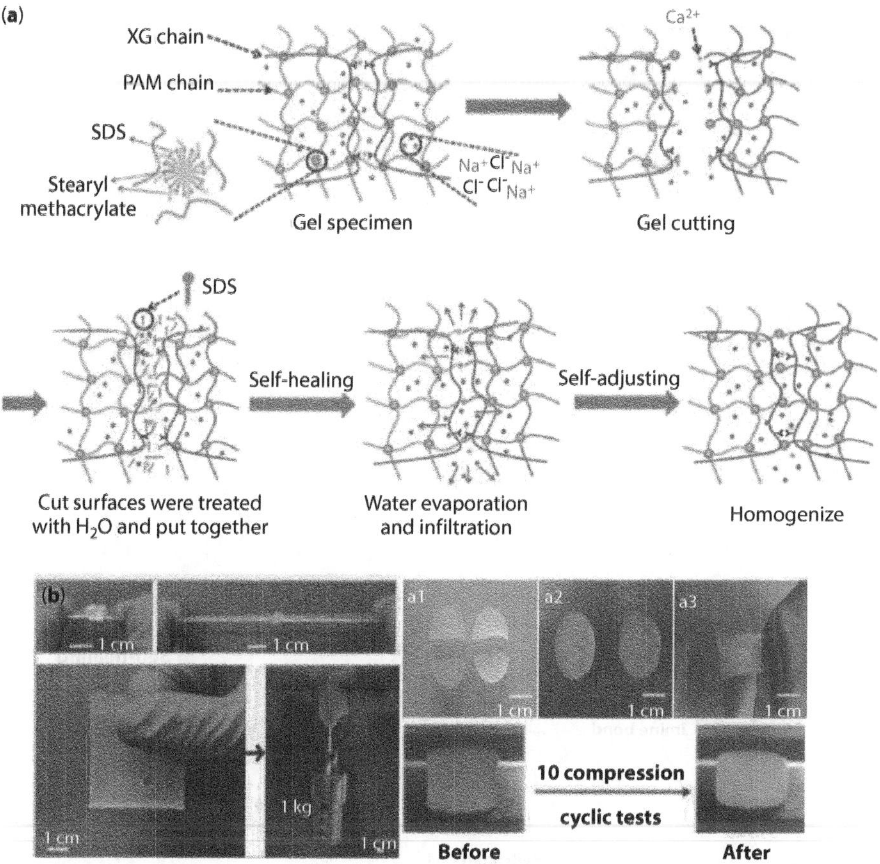

Figure 8.5 (a) The schematic of PAM/XG hydrogel healing process (b) Notch insensitivity self-healing, and fatigue resistance properties of the hydrogel. Reproduced with permission from Ref. [23] Copyright 2016, ACS.

Aryl imine is relatively more stable than aliphatic imine, with hydrolysis because of the delocalization of electrons across the π system. Zhang *et al.* have synthesized hybrid collagen chitosan (COL-CS) hydrogel using dibenzaldehyde-modified PFG that acted as a telechelic cross-linker, demonstrating dynamic self-healing properties within 15 minutes of damage and completely disappeared within 45 minutes [25] (Figure 8.6).

The hydrogels, based on imine linkages, possess many properties, other than self-healing. pH-responsive hydrogel can also be synthesized because of the dynamic property of the Schiff base. pH-responsive injectable self-healing hydrogel of N carboxyethyl chitosan (CEC) and dibenzaldehyde modified PEG has demonstrated the sustained drug release

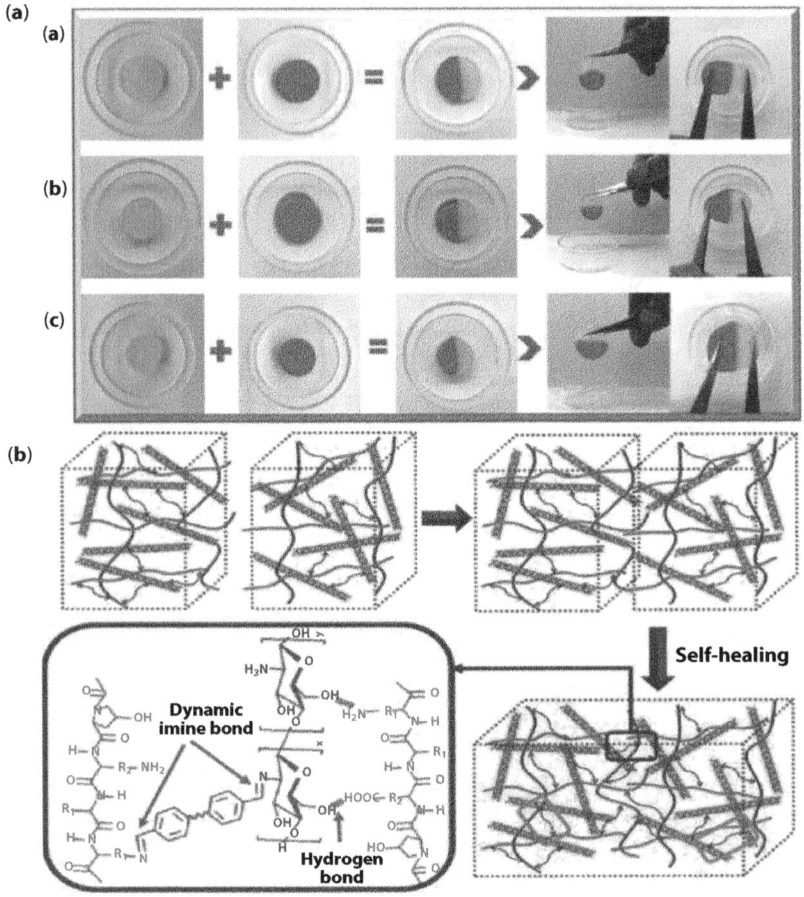

Figure 8.6 (a) Self-healing capability of Collagen-Chitosan hydrogel with different ratio: (a) 1/1 (b) 1/2 (c) chitosan (b) Self-healing through dual crosslinking (imine and hydrogel bond). Reproduced with permission from ref [25] Copyright 2020, ACS.

of DOX for hepatocellular carcinoma therapy [26]. Since cancer tissues exhibit low extracellular pH as compared to the bloodstream and normal tissues, this drug-loaded hydrogel demonstrated that the DOX release rate at the acid pH groups (6.8, 5.5, and 4.0) was much faster than the physiological pH (7.4) group. The protonated CEC amino groups at acidic pH may be the reason of this behavior, that causes the loss of the reversible Schiff base bond inside the hydrogel and a quicker disintegration rate [26]. Another study demonstrated the degradability property by using dextran and ε-poly-L-lysine (EPL) from dynamic Schiff linkage having antimicrobial, self-healing properties and can take the shape of a wound. Both these components degraded completely *in-vivo*, and the hydrogel's deterioration could be due to the oxidized dextran degradation [27]. Wang *et al.* reported the antimicrobial property of the hydrogel formed through the creation of a covalent reversible imine bond between ABA copolymer and polyethyleneimine (PEI). The electrostatic interaction of partially positive charge of amine group present on the polyethyleneimine and quaternary amine groups of triblock polymer ABA exposed on the surface of hydrogel with the negatively charged microorganism's cell membrane gave rise to the antimicrobial property of the hydrogel that results in the death and lysis of cell [28]. Kim *et al.* demonstrated the printable self-healing hydrogel combining adipic acid dihydrazide (ADH) and glycol chitosan (G.C.), added to oxidized hyaluronate (OHA) through imine and hydrazone bond formation, with application in cell-laden construct for cartilage tissue engineering [29].

8.3.2.2 Diel-Alder Reaction

Diel-Alder reaction is a conjugate addition reaction that occurs between the 4π electrons of diene and 2π electrons of dienophile. The driving factor behind the reaction is the creation of new bonds, which are energetically more stable than the bonds. Many literature reports have shown the use of diel-alder reaction to synthesize self-healing hydrogels with different properties. The covalent linkage, formed in the reaction, possess numerous qualities, like, reversibility at different physiological conditions, reaction does not require a catalyst, higher efficiency, and no hazardous products. The thermoresponsive hydrogel was prepared by using furan and maleimide as diene and dienophile, respectively, showing the role of temperature in shifting the equilibrium between diel- alder and retro diel-alder reaction [30]. A thermosensitive hydrogel was synthesized, using modified chitin and polyethylene glycol (PEG) as a cross-linker, to form dually crosslinked hydrogel via diel-alder reaction and used it for 3D cell culture. Compared to the simple physically crosslinked thermosensitive chitin hydrogel, the chitin-modified hydrogel that were dually crosslinked had much enhanced structural integrity properties and decreased biodegradation rate. The cells, encapsulated in the synthesized hydrogel, showed higher viability, enhanced proliferation and self-assembled into 3D spheroids [31]. Li *et al.* have fabricated hybrid hydrogel for drug delivery of pectin/chitosan, having self-healing and pH responsiveness properties via diel-alder reaction. The resultant hydrogel showed exceptional self-healing capacity in normal conditions, which is due to the reversible property of the D-A reaction within a few hours. Moreover, the pH-responsive release of 5 fluorouracil drug was also demonstrated. On varying the pH from 6.8 to 7.4 of the medium, hydrogel showed different release process due to different intraparticle diffusion mechanism of anion drug within the pores of hydrogel illustrating that changing the crosslinking density of hydrogel carriers can control the drug release mechanism [32].

8.3.2.3 Disulphide Bond

Disulphide linkages have become evident as a feasible technique for creating crosslinked self-healing degradable hydrogels, because these linkages are susceptible to diverse stimuli, like, heat, photon radiation, redox potential, pH, etc. Li Song *et al.* have synthesized multi-responsive hydrogel consisting of a single component, a thermoresponsive polymer, Pluronic F127, and Lipoic acid (L.A.). With the help of the disulphide exchange reaction of L.A. components induced by the U.V. radiation, crosslinking of L.A. components occurred. Then, in the aqueous solution, the synthesized F127-LA copolymer was self-assembled into micelles and formed the hydrogel. This hydrogel offered a good self-healing capacity, owing to the disulfide exchange reactions. The drug release was also monitored by adding a reducing agent that was controlled by manipulating the surrounding temperature [33]. Jiao and his team have reported the synthesis of redox-sensitive poly(N isopropyl acrylamide) hydrogel, using *N, N'*-methylene bisacrylamide (BIS) as a permanent cross-linker and *N, N'*-bis(acryloyl)cystamine (BAC) as a reversible cross-linker within a microfluidic device.. They have used this microfluidic device incorporated via photopolymerization as a simple and rapid method for reversibly capturing and releasing proteins and drugs with the help of disulfide-thiol exchange. The crosslinking densities of the hydrogel dots were altered due to the process of reformation and breakage of these disulfide linkages, causing them to swell or shrink in microfluidic devices. It showed the release efficiency of 83% for the bovine serum albumin (BSA) reversibly and can be repeated up to three times within the device. This showed that such reproducible devices have great capability to be used as integrated lab-on-chip devices for delivery and fast detection [34].

8.3.2.4 Boronate-Diol Complexation

Boronate-diol complexes are reversible boronate ester bonds, formed by the covalent linkage between the boronic acid complex and diol-containing moieties. Boronate bond was combined with the nanofilaments M13 virus to synthesize hydrogel, having stimuli-responsive, good self-healing ability and mechanical properties. Phenylboronic acid (PBA) was first conjugated with filamentous M13 virus, and polyvinyl alcohol (PVA) was then added as a cross-linker that resulted in to a three-dimensional structure by the creation of a boronate-ester bond between boronic acid and diol. The gel was cut into two pieces to monitor the self-healing capability, and they self-healed within 1 hour. It arises because of the formation of reversible dynamic covalent crosslinks of boronate bonds. This hydrogel was also used for a controlled insulin release study, showing good sugar responsiveness to insulin at physiological pH. To evaluate the glucose behavior, the FTIC labeled insulin was injected inside the hydrogel. The burst insulin release was more in the presence of glucose than in the absence. The reason behind such behavior is that glucose partially substitutes the PVA diols as it diffuses inside hydrogel that results in the manufacture of a complex with the free dangling PBA bond of the viral chain, causing the gel to swell and facilitates the insulin release [35]. Amaral *et al.* used PBA-modified laminarin, microalga food reserve and PVA to synthesize hydrogel due to the properties, like, antioxidant, anti-inflammatory, biodegradability and ease of chemical modification. The synthesized hydrogel was tested for self-healing and cytocompatibility. After 48 hours of direct interaction with preosteoblasts, hydrogels showed no toxicity, making them a suitable platform for tissue engineering and medication delivery [36]. Hong *et al.* have reported the synthesis of pH-triggered mussel mimetic alginate-boronic acid (Alg-BA) bioadhesive glue that can be used as tissue sealant, hemostatic agent and wound dressing. The catechol cis diol groups, present in mussels, have similar adhesive properties to diols present in boronic acid. The deprotonation of catechol diol groups in mussel adhesive proteins is activated by basic pH, which causes catechol moieties to cure. Boronic acid can also be fixed in a weakly basic state, reinforcing cis-diol binding from alginate backbones and boronic acid. In boronic acid-mediated hydrogen bonding with gel polymers, Alg-BA chains are productively dispersed and intertwined with polymeric hydrogel chains before cure. They represented the hydrogel-based model that can mimic the *in vivo* conditions [37].

8.4 Application of Self-Healing Hydrogel in Biomedical Application

8.4.1 Drug Delivery

Self-healing hydrogel plays an essential role in drug delivery applications. Self-healing hydrogel perform their drug delivery action through a various mechanism, such as, supramolecular interaction, metal-ligand interaction, host-guest chemistry and reversible chemical bonding triggered by temperature, pH and light. In the case of site-specific and local drug delivery applications, self-healing hydrogel efficiently works due to its injectability [38]. After being injected by a self-healing mechanism, it forms a site-specific shaped scaffold with extended sustainable drug-releasing capability. Various biocompatible materials, such as, polysaccharides, peptides, chitosan, pectin, catechol functionalized, hyaluronic acid, gelatine, collagen and alginate are used for developing self-healing hydrogel. Injectable hydrogels, with biodegradability and self-healing property, are excellent for drug loading and release, in biomedical applications.

Jianglei Qin *et al.* developed a biodegradable pectin-based injectable self-healing hydrogel for anticancer drug delivery therapy [39]. In this research work, biobased pectin bearing aldehyde group and a thermo-responsive copolymer of poly(N-isopropyl acrylamide) (PNIPAM) holding acyl hydrazide (AH)groups formed a hydrogel through acyl hydrazone connection, which provides better biocompatibility, biodegradability, as well as, sustainable drug release property (Figure 8.7). Synergetic treatment effect had seen on CT26 tumors, when hydrogel coloaded with DOX and CA4 during *in vivo* study. This biodegradable self-healing hydrogel effectively worked in antitumor treatment. Poly(N-isopropyl acrylamide) and its derivatives are widely used polymers for biomedicine, because of their lower critical solution temperature (LCST), which is close to the human body temperature, showing efficient and fast temperature response at the range of human body temperature. Near the LCST, PNIPAM-based hydrogel act as smart hydrogel by swelling and shrinking. The hydrogels' self-healing behaviors and mechanical characteristics are depends on the concentration ratio of P(NIPAM-stat-AH) to pectin. the injectable hydrogels can be loaded with anticancer drugs, doxorubicin hydrochloride (DOX·HCl) and combretastatin A4 disodium phosphate (CA4) and injected at around the tumor site for sustained drug release which improves the therapeutic effect by decreasing the drug diffusion and long-term toxicity to normal healthy tissues. The hydrogel could simultaneously deliver anticancer drugs with multiple mechanisms, through sustainable drug delivery mechanism that significantly increased the antitumor therapy with reduced cytotoxicity. cyclodextrin (CD) and adamantane (AD) grafted quaternized chitosan and cyclodextrin grafted graphene oxide (GO-CD) polymer b) quaternized chitosan-graphene oxide based supramolecular hydrogel, (c) Characterization of supramolecular hydrogel and wound healing application. Figure 8.8 shows the outline of drug delivery system of pectin based self-healing hydogel.

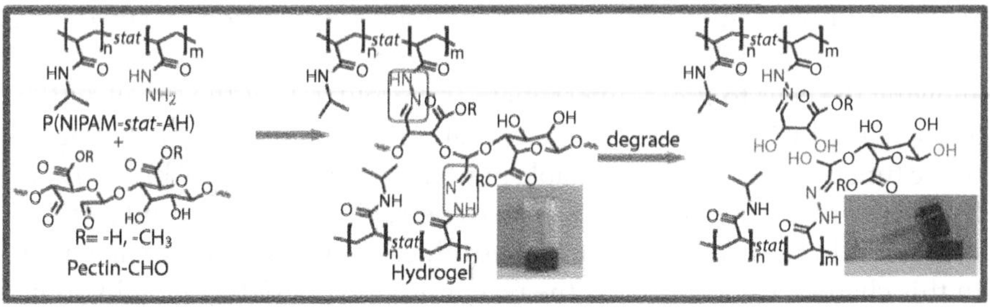

Figure 8.7 Synthesis of pectin-based hydrogel. Reproduced with permission from Ref [39].

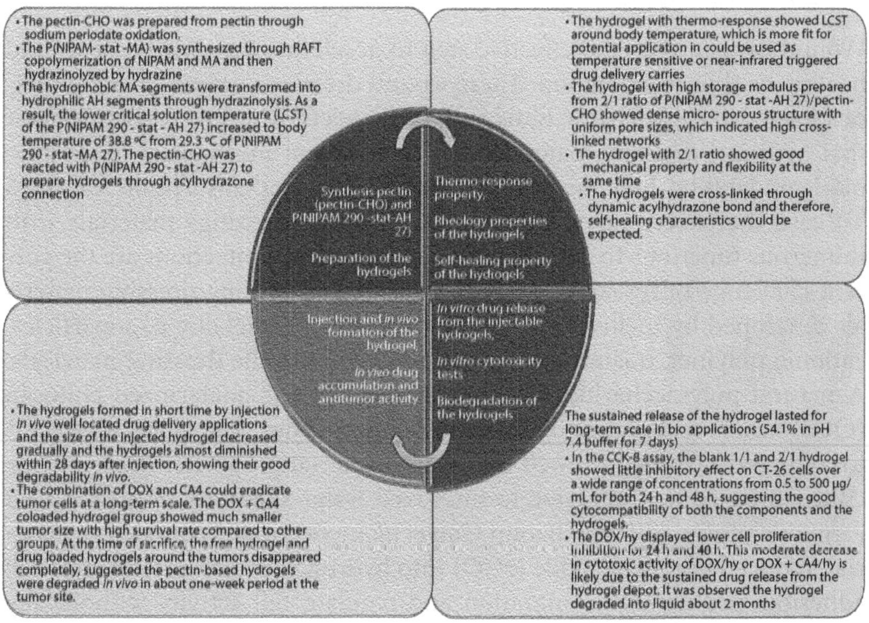

Figure 8.8 Outline of drug delivery system of self-healing hydrogel. Reproduced with permission from Ref. [39].

8.4.2 Tissue Engineering Application

Tissue engineering is an interdisciplinary research field, where biology and medicine meet with the engineering domain to solve the problem of the healthcare sector by providing practical solutions in terms of regeneration, reconstruction of damaged tissue and organs, regenerative medicine, and sustainable drug delivery carriers, etc. Polymeric hydrogels offer an excellent opportunity for research in tissue regeneration by fabricating a scaffold, encapsulating drugs, and biomolecules into the scaffold, and allowing the cells to grow, differentiate and proliferate at the site of action. Self-healing hydrogels offer fundamental properties that mimic biological tissue functionality, such as, tunable mechanical property, self-healing ability, flexibility, external trigger responsiveness, biomimicry, modularity and specificity, like, electroactive by the introduction of electroactive materials. In this chapter, we will discuss the recent progress of self-healing hydrogel in the skin, neural, cartilage and bone tissue regeneration.

8.4.2.1 Wound Healing

Wound dressings are an effective step to accelerate the process of wound healing or skin regeneration. These wound dressings normally are in various forms, like, hydrogels, films and foam sheets. Multiple reports suggest that hydrogel dressing materials exhibit excellent mechanical properties, biodegradability, biocompatibility and high water absorption capacity. To ease the successful wound healing, wound dressings materials with numerous functions to create resistance to bacterial infections are the current challenge of research. Antibacterial wound healing dressing materials developed by inclusion of antibiotic molecule, metallic nanoparticles, cationic polymer, quaternary ammonium salt into the dressing materials. A natural polysaccharide, chitosan (CS) and their quaternized chitosan (QCS) have a wide range of antibacterial properties. The quaternized chitosan based hydrogel showed better antibacterial properties than pristine chitosan based hydrogel due to improved water solubility of quaternized form of chitosan [40, 41]. Baolin Guo *et al.* incorporated conductive material, reduced graphene oxide (rGO) into hydrogel, which can enhance cell adhesion, proliferation, migration and regulate cell activity without any electrical stimulation. Therefore, it can support wound healing as a new type of excellent antibacterial conductive material [42]. In this research work, an injectable self-healing supramolecular conductive hydrogel was prepared by introduction of the host-guest interaction between cyclodextrin (CD) and adamantane (AD), by special modification of QCS and rGO (Figure 8.9). The QCS-CD-AD/GO supramolecular hydrogel retained its all properties, such as, electrical, swelling, rheological and excellent photo-thermal properties. During the *in-vitro* study, hydrogels showed excellent antibacterial capability against all types bacterial growth, as well as, showed blood compatibility and cytocompatibility against co-culturing with mouse fibroblasts (L929 cells) and red blood cells (RBC). A mouse

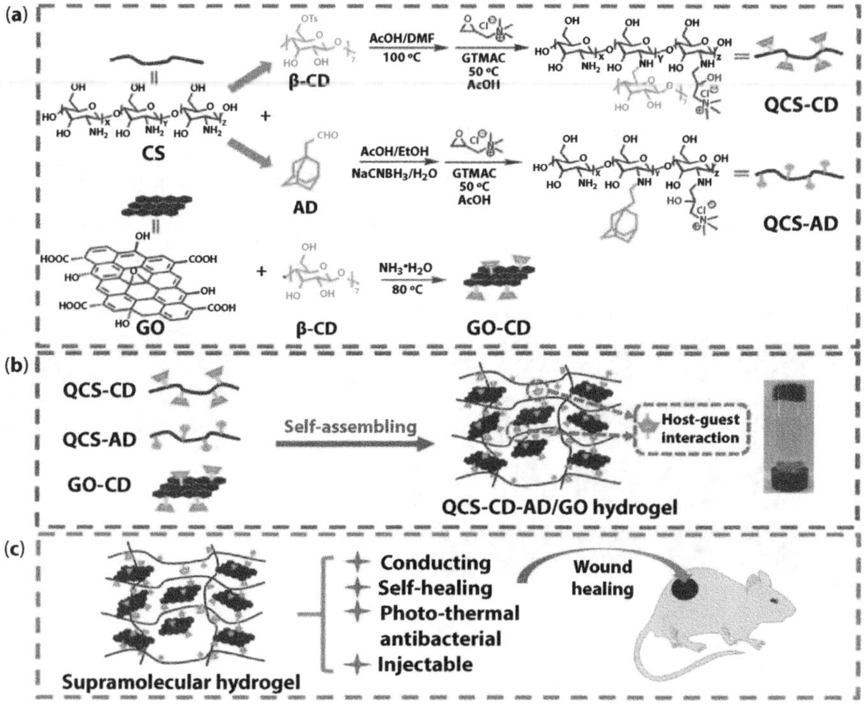

Figure 8.9 (a) Schematic representation of synthesis of cyclodextrin (CD) and adamantane (AD) grafted quaternized chitosan and cyclodextrin grafted graphene oxide (GO-CD) polymer (b) quaternized chitosan-graphene oxide based supramolecular hydrogel, (c) Characterization of supramolecular hydrogel and wound healing application Reproduced with permission from ref [42] Copyright 2020, Elsevier.

wound healing model showed that these self-healing supramolecular conductive hydrogels have potential as good candidates for wound healing.

8.4.2.2 Neural Tissue Engineering

Keeping neural tissue regeneration in mind, a chitosan–cellulose nanofiber (CS–CNF) composite self-healing hydrogels was synthesized [42]. The hydrogel offered tunable self-healing property and nearly the same modulus (approximately 2 kPa) by adding (0.06–0.15) wt% of CNFs in chitosan (CS) through the reversible dynamic Schiff base reaction (Figure 8.10). *In vitro* study suggest that stem cells embedded in the CS–CNF hydrogel with self-healing properties showed better neural differentiation and significant improvement in oxygen metabolism. In the stem cell embedded self-healing hydrogel, cells metabolic changes were highly correlated with

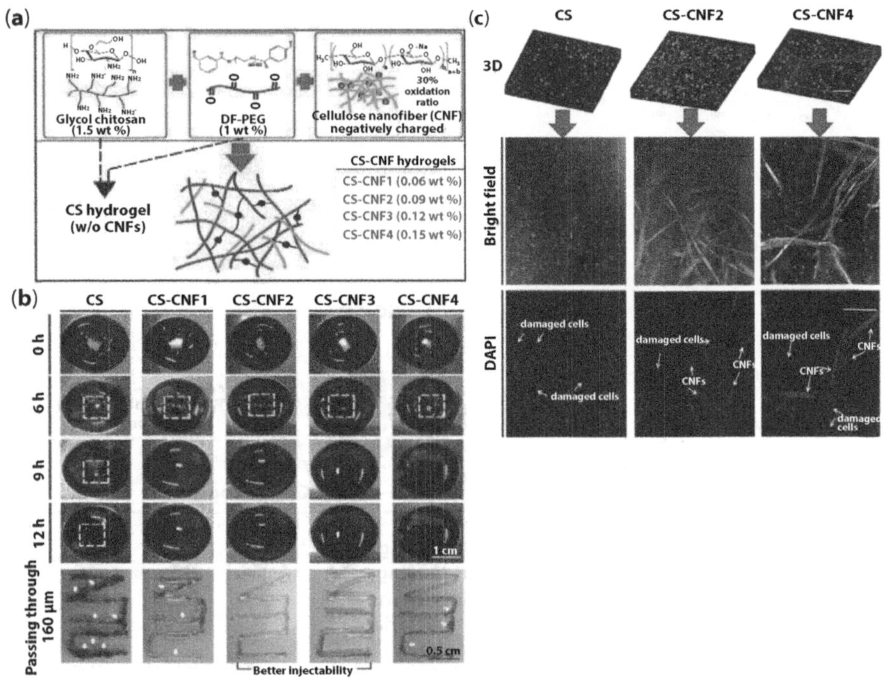

Figure 8.10 (a) Synthetic scheme of self-healing hydrogels of CS-CNF (b) observation of self-healing by disappearance of the central holes in the hydrogels created by160 μm diameter needle (30G), (c) Cell viability study of the hydrogel through confocal microscopy. Creative commons cc by license Ref [42].

differentiation. *In vivo* study of the zebrafish brain injury model revealed that CS-CNF with 0.09 wt% CNFs hydrogel optimized neural regeneration with the best self-healing properties over pristine CS hydrogel. This new self-healing hydrogel provide importance of self-healing behavior in the biological performance of hydrogels with better injectability and tissue regeneration potential.

8.4.2.3 Bone Tissue Engineering

Generally, bone defect treatment uses autogenous, allogeneic, and prosthetic materials. However, these materials are limited by their associated problems and limitations. Therefore, craniofacial bone repair encourages alternative biomaterials that lead to bone defect repair. Recently, two types of biomaterials have been used for bone tissue engineering, ceramic-based biomaterials and polymer-based on natural polymer or synthetic or mixer of synthetic and natural polymer, as well as, hybrids of both ceramic and polymer-based materials [43, 44]. Polymer or polymer ceramic composite-based bone scaffolds have been widely used for bone regeneration, due to their easy chemical modification, tunable biodegradable property and bio-compatibility. Specifically, polymeric hydrogels are favourable for medical application, because of their injectability. Injectable hydrogel scaffolds can easily achieve the proper shape of the damaged site of bone by minimizing surgical operational time and without arises of any significant medical complication of surrounding tissue damage.

Mingzhu Liu *et al.* developed a cytocompatible, injectable, self-healing tuneable hydrogel by implementing DA click chemistry and dynamic acyl hydrazone bond crosslinking to form chondroitin sulfate (ChS) hydrogels [45]. DA click chemistry and acyl hydrazones bond crosslinking affect the covalent crosslink density by optimizing the ratio of reactive functional groups, leading towards a tuneable hydrogel. Hydrogel injectability and self-healing ability depends on the dynamic acyl hydrazone bonds, which broke and formed simultaneously. The copolymer of F127@ChS/furan have been synthesised by modification of pluronic F127 to N-maleoyl alanine grafted F127 (F127-AMI) as dienophile and ChS to ChS–furan as diene. Adipic dihydrazide (ADH) conjugated F127@ChS/furan synthesized by EDC (1-Ethyl-3-(3-dimethylaminopropyl)carbodiimide)/NHS (N-hydroxysuccinimide) carbodiimide coupling reaction between ADH and F127@ChS/furan. Reaction of F127@ChS/furan/ADH with N-maleoyl alanine grafted PEG (PEG-AMI) form a hydrogel, F127@ChS–PEG–OChS hydrogel The hydrogel transformed into a gel by both DA reaction and dynamic acyl hydrazone bond crosslinking. The crosslinking density and the hydrogel concentration manipulate the gelation time of the hydrogel in the range of seconds to minutes. In the present study showed that injectable hydrogel formed a gel within 70 s, which gives suitable time window for *in vivo* study of injectable hydrogel. Within 72 h, cell viability increases to 144.4% ± 6.7%, when loaded with rat mesenchymal stem cells. These above results support that hydrogel is cytocompatible and allowed cell proliferation in suitable manner. Prepared hydrogel can be tuned with physicochemical properties to support bone repairs and delivering BMP-4 growth factors for bone regeneration. Bone growth rate speed up in case of BMP-4 loaded hydrogel treated group of rat. According to report, the bone is formed through step by step mechanism, as following: (1) migration of the progenitor cells and adhering to the hydrogel matrix through chemotactic activity of the BMP-4 gradient; (2) the hydrogel system allowed to pass the cell and differentiate into bone cells; (3) resulting angiogenesis and bone marrow-like tissue formation by production collagen fibers;

(4) mineralization occurs and bone forms from tissue over time and slowly replaces the hydrogel with bone.

8.5 Conclusion and Future Prospects

Self-healing materials open new domains in the field of biomedical applications, from drug delivery systems to tissue regeneration to organ printing through 3D bioprinting. In recent times, biomedical scientists have tried to incorporate various multifunctional properties into hydrogel systems, leading to a new generation of self-healing hydrogel with antimicrobial,antifouling,electroconductiveability,shapememorycapacity,nonequilibrium dynamics system, vascularisation designing ability and sustainable drug-releasing mechanism over the long period system. With these functionalities, self-healing hydrogels can be easily explored in the newly emerging fields of bioacutarors, nano energy harvesting hydrogels, soft robotics, and injectable bioelectronic hydrogels. However, with promising design, development of such hydrogel creates new challenges that have to addressed to accelerate the regeneration process of the damaged tissue. It is crucial to control the hydrogels' biodegradability, biocompatibility, and noncytotoxity to facilitate tissue regeneration and cell migration, enabling the hydrogels to be sustainable in physiological environments. Otherwise, the hydrogel may be quickly degraded and lose its mechanical integrity and fail the long-term performance of the devices within the human body. Tissue regenerative performance of hydrogel will be degraded over if the engineered hydrogels scaffold should not enhance the differentiation, cell spreading, and proliferation. Various self-healing hydrogels perform well in their mechanical behavior, as well as, *in vitro* study. Still, these materials have not been tested in *in vivo* or animal models. So we can say that their biocompatibility in the human body system is still unclear. With implications of combined knowledge from various interdisciplinary fields, such as, chemistry, mathematics, physics, engineering, and medicine, we can solve the hydrogel's problems and narrow the gap between laboratory and clinical application.

References

1. Yin, J., Liu, Q., Zhou, J., Zhang, L., Zhang, Q., Rao, R., Liu, S., Jiao, T., Self-assembled functional components-doped conductive polypyrrole composite

hydrogels with enhanced electrochemical performances. *RSC Adv.*, 10, 18, 10546–10551, 2020.

2. Cui, Y., Li, D., Gong, C., Chang, C., Bioinspired shape memory hydrogel artificial muscles driven by solvents. *ACS Nano*, 15, 8, 13712–13720, 2021.
3. Mihajlovic, M., Fermin, L., Ito, K., Van Nostrum, C.F., Vermonden, T., Hyaluronic acid-based supramolecular hydrogels for biomedical applications. *Multifunct. Mater.*, 4, 3, 032001, 2021.
4. Girirajan, S., Campbell, C., Eichler, E., HHS public access. *Physiol. Behav.*, 176, 5, 139–148, 2011.
5. Naficy, S., Brown, H.R., Razal, J.M., Spinks, G.M., Whitten, P.G., Progress toward robust polymer hydrogels. *Aust. J. Chem.*, 64, 8, 1007–1025, 2011.
6. Fan, F., Saha, S., Hanjaya-Putra, D., Biomimetic hydrogels to promote wound healing. *Front. Bioeng. Biotechnol.*, 9, 1–24, 2021.
7. Williams, K.A., Dreyer, D.R., Bielawski, C.W., The underlying chemistry of self-healing materials. *MRS Bull.*, 33, 8, 759–765, 2008.
8. Amaral, A.J.R. and Pasparakis, G., Stimuli responsive self-healing polymers: Gels, elastomers and membranes. *Polym. Chem.*, 8, 42, 6464–6484, 2017.
9. Söntjens, S.H.M., Sijbesma, R.P., Van Genderen, M.H.P., Meijer, E.W., Stability and lifetime of quadruply hydrogen bonded 2-Ureido-4[1H]-pyrimidinone dimers. *J. Am. Chem. Soc.*, 122, 31, 7487–7493, 2000.
10. Fan, C.J., Huang, Z.C., Li, B., Xiao, W.X., Zheng, E., Yang, K.K., Wang, Y.Z., A robust self-healing polyurethane elastomer: From H-bonds and stacking interactions to well-defined microphase morphology. *Sci. China Mater.*, 62, 8, 1188–1198, 2019.
11. Zhao, M., Tang, Z., Zhang, X., Li, Z., Xiao, H., Zhang, M., Liu, K., Ni, Y., Huang, L., Chen, L., Wu, H., A self-healing, stretchable, and conductive Poly(N-vinylpyrrolidone)/gallic acid composite hydrogel formed via hydrogen bonding for wearable electronic sensors. *Compos. Sci. Technol.*, 198, 108294, 2020.
12. Zhong, M., Liu, X.Y., Shi, F.K., Zhang, L.Q., Wang, X.P., Cheetham, A.G., Cui, H., Xie, X.M., Self-healable, tough and highly stretchable ionic nanocomposite physical hydrogels. *Soft Matter*, 11, 21, 4235–4241, 2015.
13. Xu, X., Jerca, V.V., Hoogenboom, R., Self-healing metallo-supramolecular hydrogel based on specific Ni2+ coordination interactions of Poly(ethylene glycol) with bistriazole pyridine ligands in the main chain. *Macromol. Rapid Commun.*, 41, 4, 6–11, 2020.
14. Sun, X., Yao, F., Wang, C., Qin, Z., Zhang, H., Yu, Q., Zhang, H., Dong, X., Wei, Y., Li, J., Ionically conductive hydrogel with fast self-recovery and low residual strain as strain and pressure sensors. *Macromol. Rapid Commun.*, 41, 13, 1–6, 2020.
15. Li, Z., Davidson-Rozenfeld, G., Vázquez-González, M., Fadeev, M., Zhang, J., Tian, H., Willner, I., Multi-triggered supramolecular DNA/Bipyridinium dithienylethene hydrogels driven by light, redox, and chemical stimuli for

shape-memory and self-healing applications. *J. Am. Chem. Soc.*, 140, 50, 17691–17701, 2018.
16. Cai, T., Huo, S., Wang, T., Sun, W., Tong, Z., Self-healable tough supramolecular hydrogels crosslinked by poly-cyclodextrin through host-guest interaction. *Carbohydr. Polym.*, 193, 54–61, 2018.
17. Granata, G., Petralia, S., Forte, G., Conoci, S., Consoli, G.M.L., Injectable supramolecular nanohydrogel from a micellar self-assembling calix[4]arene derivative and curcumin for a sustained drug release. *Mater. Sci. Eng. C*, 111, 110842, 2020.
18. Liu, Y., Zhou, M., Liu, Y., Han, X., Zhang, X., Liu, S., Host–guest interaction-mediated fabrication of aggregation-induced emission supramolecular hydrogel for use as aqueous light-harvesting systems. *Supramol. Chem.*, 32, 8, 445–451, 2020.
19. Yu, R., Yang, Y., He, J., Li, M., Guo, B., Novel supramolecular self-healing silk fibroin-based hydrogel via host–guest interaction as wound dressing to enhance wound healing. *Chem. Eng. J.*, 417, 128278, 2021.
20. Highley, C.B., Rodell, C.B., Burdick, J.A., Direct 3D printing of shear-thinning hydrogels into self-healing hydrogels. *Adv. Mater.*, 27, 34, 5075–5079, 2015.
21. Jiang, G., Liu, C., Liu, X., Zhang, G., Yang, M., Liu, F., Construction and properties of hydrophobic association hydrogels with high mechanical strength and reforming capabilitya. *Macromol. Mater. Eng.*, 294, 12, 815–820, 2009.
22. Jiang, H., Duan, L., Ren, X., Gao, G., Hydrophobic association hydrogels with excellent mechanical and self-healing properties. *Eur. Polym. J.*, 112, 660–669, 2019.
23. Yuan, N., Xu, L., Wang, H., Fu, Y., Zhang, Z., Liu, L., Wang, C., Zhao, J., Rong, J., Dual physically cross-linked double network hydrogels with high mechanical strength, fatigue resistance, notch-insensitivity, and self-healing properties. *ACS Appl. Mater. Interfaces*, 8, 49, 34034–34044, 2016.
24. He, X., Sun, X., Meng, H., Deng, S., He, T., Zang, H., Wei, D., Self-healing polymeric ionic liquid hydrogels with high mechanical strength and ionic conductivity. *J. Mater. Sci.*, 56, 17, 10231–10248, 2021.
25. Ding, C., Tian, M., Feng, R., Dang, Y., Zhang, M., Novel self-healing hydrogel with injectable, pH-Responsive, strain-sensitive, promoting wound-healing, and hemostatic properties based on collagen and chitosan. *ACS Biomater. Sci. Eng.*, 6, 7, 3855–3867, 2020.
26. Qu, J., Zhao, X., Ma, P.X., Guo, B., pH-responsive self-healing injectable hydrogel based on N-carboxyethyl chitosan for hepatocellular carcinoma therapy. *Acta Biomater.*, 58, 168–180, 2017.
27. Gao, L., Chen, J., Feng, W., Song, Q., Huo, J., Yu, L., Liu, N., Wang, T., Li, P., Huang, W., A multifunctional shape-adaptive and biodegradable hydrogel with hemorrhage control and broad-spectrum antimicrobial activity for wound healing. *Biomater. Sci.*, 8, 24, 6930–6945, 2020.

28. Wang, W., Xiang, L., Gong, L., Hu, W., Huang, W., Chen, Y., Asha, A.B., Srinivas, S., Chen, L., Narain, R., Zeng, H., Injectable, self-healing hydrogel with tunable optical, mechanical, and antimicrobial properties. *Chem. Mater.*, 31, 7, 2366–2376, 2019.
29. Kim, D.Y., Park, H., Kim, S.W., Lee, J.W., Lee, K.Y., Injectable hydrogels prepared from partially oxidized hyaluronate and glycol chitosan for chondrocyte encapsulation. *Carbohydr. Polym.*, 157, 1281–1287, 2017.
30. Gandini, A., The furan/maleimide Diels-Alder reaction: A versatile click-unclick tool in macromolecular synthesis. *Prog. Polym. Sci.*, 38, 1, 1–29, 2013.
31. Bi, B., Ma, M., Lv, S., Zhuo, R., Jiang, X., *In-situ* forming thermosensitive hydroxypropyl chitin-based hydrogel crosslinked by diels-alder reaction for three dimensional cell culture. *Carbohydr. Polym.*, 212, 368–377, 2019.
32. Li, D.Q., Wang, S.Y., Meng, Y.J., Guo, Z.W., Cheng, M.M., Li, J., Fabrication of self-healing pectin/chitosan hybrid hydrogel via diels-alder reactions for drug delivery with high swelling property, pH-responsiveness, and cytocompatibility. *Carbohydr. Polym.*, 268, 118244, 2021.
33. Song, L., Zhang, B., Gao, G., Xiao, C., Li, G., Single component pluronic F127-lipoic acid hydrogels with self-healing and multi-responsive properties. *Eur. Polym. J.*, 115, 346–355, 2019.
34. Jiao, C., Obst, F., Geisler, M., Che, Y., Richter, A., Appelhans, D., Gaitzsch, J., Voit, B., Reversible protein capture and release by redox-responsive hydrogel in microfluidics. *Polym. (Basel)*, 14, 2, 1–16, 2022.
35. Zhi, X., Zheng, C., Xiong, J., Li, J., Zhao, C., Shi, L., Zhang, Z., Nanofilamentous virus-based dynamic hydrogels with tunable internal structures, injectability, self-healing, and sugar responsiveness at physiological pH. *Langmuir*, 34, 43, 12914–12923, 2018.
36. Amaral, A.J.R., Gaspar, V.M., Mano, J.F., Responsive laminarin-boronic acid self-healing hydrogels for biomedical applications. *Polym. J.*, 52, 8, 997–1006, 2020.
37. Hong, S.H., Shin, M., Park, E., Ryu, J.H., Burdick, J.A., Lee, H., Alginate-boronic acid: pH-triggered bioinspired glue for hydrogel assembly. *Adv. Funct. Mater.*, 30, 26, 1–8, 2020.
38. Chen, J., Nichols, B.L.B., Norris, A.M., Frazier, C.E., Edgar, K.J., All-polysaccharide, self-healing injectable hydrogels based on chitosan and oxidized hydroxypropyl polysaccharides. *Biomacromolecules*, 21, 10, 4261–4272, 2020.
39. An, H., Yang, Y., Zhou, Z., Bo, Y., Wang, Y., He, Y., Wang, D., Qin, J., Pectin-based injectable and biodegradable self-healing hydrogels for enhanced synergistic anticancer therapy. *Acta Biomater.*, 131, 149–161, 2021.
40. Chen, Y., Wang, Q., Li, D., Mensah, A., Qiu, Y., Ke, H., Wei, Q., Mussel-inspired double crosslinked hydrogels with desirable mechanical properties, strong tissue-adhesiveness, self-healing properties and antibacterial properties. *Mater. Sci. Eng. C*, 120, 111690, 2021.

41. Qu, J., Zhao, X., Liang, Y., Zhang, T., Ma, P.X., Guo, B., Antibacterial adhesive injectable hydrogels with rapid self-healing, extensibility and compressibility as wound dressing for joints skin wound healing. *Biomaterials*, 183, 185–199, 2018.
42. Zhang, B., He, J., Shi, M., Liang, Y., Guo, B., Injectable self-healing supramolecular hydrogels with conductivity and photo-thermal antibacterial activity to enhance complete skin regeneration. *Chem. Eng. J.*, 400, 125994, 2020.
43. Chen, J., Peng, Q., Peng, X., Han, L., Wang, X., Wang, J., Zeng, H., Recent advances in mechano-responsive hydrogels for biomedical applications. *ACS Appl. Polym. Mater.*, 2, 3, 1092–1107, 2020.
44. Pathan, N. and Shende, P., Strategic conceptualization and potential of self-healing polymers in biomedical field. *Mater. Sci. Eng. C*, 125, 112099, 2021.
45. Lü, S., Bai, X., Liu, H., Ning, P., Wang, Z., Gao, C., Ni, B., Liu, M., An injectable and self-healing hydrogel with covalent crosslinking: *In vivo* for cranial bone repair. *J. Mater. Chem. B*, 5, 20, 3739–3748, 2017.

Section II

ADVANCED MANUFACTURING TECHNIQUES FOR BIOMEDICAL APPLICATIONS

9

Biomimetic and Bioinspired Composite Processing for Biomedical Applications

Hemant Kumar[1], Purnima Justa[3], Nancy Jaswal[3], Balaram Pani[2] and Pramod Kumar[3]*

[1]Department of Chemistry, Ramjas College, University of Delhi, Delhi, India
[2]Bhaskaracharya College of Applied Sciences, Department of Chemistry, University of Delhi, Delhi, India
[3]Department of Chemistry& Chemical Science, School of Physical & Material Sciences, Central University of Himachal Pradesh, Dharamshala, India

Abstract

An emerging discipline of research called biomimetics fine-tunes engineering and design biomaterials to replicate biological structure and functions by adapting natural operating principles. The main goal of the application is to make medical implants that can replace both hard and soft tissues. To accomplish the necessary biological function with a minimally adverse biological reaction, the design as well as development of the multifunctional smart biomaterials that are compatible with human physiology is essential. To increase life expectancy and improve the quality of life, several biosensors, biomimetic medication delivery systems, and medical bio implants have been tried. We critically evaluate this method's benefits above simple biomimicry and how it differs from bio-integrated design as a systematic tool for presenting creative solutions to difficult situations. Using examples from the area of biomedicine, where most of the inspiration is still more tightly focused on imitation or bio-integration, the extent of the applicability of the nature-inspired method is shown. The tools, techniques, and applications within these domains that may be helpful in biomedical engineering and human health will be reviewed in this chapter. Examples include genome editing and the creation of quorum-sensing systems. We'll talk about notable developments, such as those in cell therapy, cancer diagnostics and treatment, microbiome engineering, and vaccine development. Finally, there will be discussion of these disciplines' difficulties and prospects.

**Corresponding author*: pramodgang03@hpcu.ac.in, pramodgang03@gmail.com

Arbind Prasad, Ashwani Kumar and Manoj Gupta (eds.) Advanced Materials and Manufacturing Techniques for Biomedical Applications, (213–240) © 2024 Scrivener Publishing LLC

Keywords: Biomimetic, bioinspired, drug delivery, biomaterials, cancer diagnostics, microbiome engineering

9.1 Introduction

Natural selection had also produced a variety of materials, structures, models, systems, and processes that have been specifically adapted for a number of uses. Learning from nature's design principles is a key component of biomimicry or bio inspiration, which entails creating rapidly improving models at different length scales and using richness of the information to address some of humanity's most pressing problems. When you consider the variety of tasks that natural substances and systems have evolved to perform, such as providing structural support, catalysis, sensing, actuation, gating, light-harvesting, trafficking, charge transfer, self-assembly, self-organization, molecular recognition and self-replication and combinations of two or more of these tasks, you can say that nature has pretty much written the "solution manual." Food, water, public health, homeland security, and clean energy are all very important global issues. This comprehensive solution handbook could be a very effective way to deal with these issues, as well as food, public health, water, homeland security, and clean energy [1, 2].

A new wave of recent advances in biology and engineering is giving the scientific subject of biomimetics a boost in popularity. The area focuses on cutting-edge technologies created by transferring functionality from biological systems. Bioinspired design and biomimetics have attracted a lot of attention in recent years. The application of scientific and technical ideas in the design of engineering elements and structures that are motivated by biological systems is known as "bioinspired design." In contrast, engineering structures and parts that mimic biological processes are called "biomimetics." Airfoils and airplane wings are therefore instances of bioinspired design that are motivated by the flight of birds but controlled by the laws of lift and drag from aerodynamics. On the other hand, biomimetics is based on the simple idea of copying nature without fully understanding the scientific principles that make natural systems work the way they do. This is an example of an early idea for an airplane with wings that flap [3].

This is precisely achieved by fusing biopolymers and nanoparticles into a nanostructure to create a hierarchical hybrid material. In light of this, this chapter presents research on the use of biomimetic, bioresponsive, and bioactive nanomaterials in the context of several scientific fields. Due to the many biological interactions and architectural nanoscale structures, nanocomposites are proving to be excellent biomimetic, bioresponsive, and bioactive nanomaterials. As a result, many researchers have now been investigating nanocomposites for their use in bio applications. A variety of configurations, including organic-inorganic, bioinorganic, and bioinorganic nanomaterials, may be found in nanocomposites, which are made of ceramic, metallic, and polymeric materials. This enables a broad range of features, such as magnetic characteristics, MR imaging comparison, or drug administration, as well as a combination of these sorts of qualities, to be included in nanocomposite materials [4]. The bulk of earlier studies focused on bone regeneration, while more recent advancements have opened up opportunities for soft tissue body areas. Despite the most recent technical developments, more investigation is still required to fully comprehend the effects of using nanoparticles inside a human body on its long-term biocompatibility.

A number of metals' unique structural, mechanical, chemical, and other properties also contribute to their suitability for the manufacture of a range of implanted biomedical items. Metals may be used in load-bearing and medical applications because of many factors, most notably their great strength and rigidity. Therefore, metals must be used in orthopedic and dental applications. Titanium or cobalt-chromium alloys, along with stainless steels are often preferred. These goods are often intended to sustain the high level of stress with low level of strain, and only practical answer is provided by the necessary dimensional and geometrical requirements of metal-based items. Metals also provide a number of other biological aspects. Metals are widely used in biomedical fields, and various well-known metal-based products have been used often in a wide range of situations. Many current studies focus on improving and modifying metal surfaces to lessen or even avoid certain tissue responses. Any biomaterial used for structural purposes should typically have strong mechanical properties such as stiffness, resistance to fatigue, superior resistance to degradation, high biocompatibility, etc. Learning from nature has sparked a lot of study in recent years in a variety of sectors, including robotics [5], bioelectronics, medicine, catalysis, self-cleaning, and energy.

Regenerative medicine and tissue engineering aim to regenerate the transplantable tissues. Biomaterial scaffolds offer three-dimensional tissue regeneration templates and synthetic extracellular matrix habitats. Simulating natural extracellular matrix, developmental, or wound-healing processes is advantageous for scaffolds. Tissue engineering uses biomimetic materials [6]. Regenerative medicine, which replaces damaged tissues and organs, relies on biomaterials. Biomaterial scaffolds facilitate neotissue growth and organization. In order to stimulate natural developmental or wound healing processes and replicate extracellular matrix (ECM) at the nanoscale scale for tissue regeneration. Here we discuss the creation and modification of nanofibrous, nanocomposite, and nanostructured drug-delivering scaffolds. ECM-mimicking nanostructured biomaterials govern cell adhesion, proliferation, differentiation, and matrix deposition. Drug delivery systems on a nanoscale may improve tissue regeneration in porous 3D scaffolds. Nanostructured biomaterials are an intriguing and fast-growing scientific topic that is enabling regenerative medicine [7]. Directly repairing bone abnormalities or manufacturing bone tissue for transplantation is intriguing with bone tissue engineering. Biomaterials provide a scaffold along with the extracellular environment for regeneration of cells and tissue. Several signaling signals affect cellular activity, tissue formation, and repair. Multiple studies and trials have revealed the potential of tissue engineering; however, practical applications of bone tissue engineering research have been restricted owing to a lack of effective signal-delivery devices. Biomedical engineers are encouraged to design biomimetic drug delivery devices that imitate release of signaling molecule or presentation by the adjacent extracellular matrix during the healing. Engineered biomimetic drug delivery systems manage signal molecule position, timing, and release kinetics based on drug physiochemical characteristics and biological processes. This chapter discusses biomimetic signaling mechanisms of delivery for bone tissue engineering, not chemicals. Recent study shows the importance and promise of the biomimetic delivery technologies for bone tissue synthesis [8]. Naturally inspired structures and methods have been used to produce several drug delivery systems (DDS) to treat serious illnesses in recent years. Smart DDSs are constructed with biologically inspired materials and stimuli-responsive properties. To avoid failure or rejection, biobased biomedical constructions may mimic biological structures to generate comparable responses or synthesis using a biologically inspired approach. Several DDS have shown promise in treating illnesses by targeting sick tissue with improved medication effectiveness, bioavailability, and biocompatibility. Depending on the application, nanocarrier-based DDSs may deliver biopharmaceuticals. We discussed nanoscale DDS biomimicry here. Nanoparticles, nanofibers, micelles, liposomes, and hydrogel-based DDS are highlighted. The action mechanisms of the DDS, as well as their use as new carriers for biological purposes or drug delivery, are all biomimetic [9–12].

9.2 Synthesis of Biomimetic and Bioinspired Composite

9.2.1 3D (Three-Dimensional) Printing

3D printing of composites has turned up as one of the optimistic techniques for the synthesis of bioderived materials. 3D printing of composites requires two components i.e., matrix and reinforcement. Matrix material includes polymers, mortar, and paste whereas particles, fibers, and plates ranging in several nanometers to microns and sometimes even millimeters are suitable candidates for reinforcements. It was discovered that a magnetically composed DLP 3D printing method could align fibers to produce ordered structures. The suggested technique showed the prepared sample's tensile, deformation, and friction properties with various fiber contents and angles. According to the findings, the directionally reliant anisotropy of the properties of composites with aligned fibers, the mechanical properties of composites are best when the fiber is parallel to the loading direction, and the friction properties are best when the fiber is placed perpendicular to the sliding direction [13]. 3D printing opens new avenues for tissue regeneration by using single or multiple cells distributed in biocompatible materials compared to traditional methods where the fabrication of different artificial tissues is challenging because of the difficulty in achieving homogenous distribution of seed cells in biomaterials. Advanced bioprinting technologies such as magnetic bioprinting, inkjet-based bioprinting, and stereolithographic bioprinting have proved to be efficient in the successful fabrication of various complex artificial tissues such as lungs, skin, heart, bone, and cartilage [14]. A 3D-printed hydrogel scaffold has been synthesized containing human articular chondrocytes and human mesenchyme stem cells in order to attain integrated regeneration of bone and cartilage defects [15].

9.2.2 Synthesis of Bioinspired Nanomaterials

An important area of biomimetics is the synthesis of materials, and among the numerous materials, the synthesis of the nanomaterials is fastest-growing or most promising subject. Process biomimetic synthesis (PBS) and Functional biomimetic synthesis (FBS) are two categories into which biomimetic synthesis can be divided (Figure 9.1). The former uses a variety of materials and techniques to imitate certain characteristics of natural materials structures and systems while the latter mimics the "methods" i.e., aims to create different desirable nanomaterials/structures by imitating the synthesis routes, processes, or procedures of natural substances, materials, and structures [16]. It has been reported that the selective clearance of disease-causing antibodies can be achieved by using biodegradable PEGylated nanoparticles manufactured of poly(lactide-coglycolide) that are encased within intact RBC membranes [17]. Water-rich structures called hydrogels are created when hydrophilic monomers are crosslinked. The hydrophilic three-dimensional network architectures are widely used in the biomedical industry because they closely resemble the extracellular matrix of real soft tissue. However, their limited applicability in the biomedical industry is due to their low mechanical strength and also lack of structural stability. The hydrogel's shear-thinning properties have been improved, making nanocomposite hydrogel a novel biomaterial that has recently attracted significant research interest for biomedical applications. Soft polymer matrixes can be filled with various nanomaterials to enhance the hydrogel's physical, chemical, and biological properties. As a result, nanocomposite hydrogels with improved chemical and biological properties are produced [18]. Syzygiumcumini seed extract was used to simultaneously reduce chloroauric acid and graphene oxide, resulting in bioinspired gold nanoparticles coated with reduced graphene oxide. Elaeagnus umbellate extract was used to reduce silver nitrate to silver nanoparticles, and a bimodal catalytic peptide surfactant stabilized nanoemulsion template method was used to create bioinspired iron oxide nanoparticles [19–21].

9.3 Biomaterials for Biomedical Applications

9.3.1 Biomaterials-Based Cell Therapy

Cell therapy has already significantly improved healthcare and given patients access to new medications for diseases that were previously incurable. Examples include the use of mesenchymal stem cells for tissue regeneration, islet transplantation for the treatment of diabetes, and T cell therapy for cancer immunotherapy. Biomaterials may enhance the therapeutic effects of cell therapies by serving as carriers that enable 3D organization, cell viability, and function [22]. In order to encourage cell adhesion and proliferation since conserving the genotype and phenotype of target cells, biomaterials can be used during the cell-production process. Various biomaterials such as chitosan, Poly(β-aminoesters), PDMAEMA/DNA, PLL, and PLGA have been used for nonviral gene delivery offering advantages like enhanced cell uptake, high transfection efficiency, high biological activity, and efficient delivery of peptides and nucleic acids [23–27].

One of the major issues nowadays is cardiovascular disease, which has a high mortality rate worldwide. There are few mechanical support therapies for end-stage heart failure, but the only effective therapy is heart transplantation, which itself is nearly impossible due to donor and surgical limitations. Because necrotic or ischemic myocardium cannot be reversed by any non-cellular medicinal treatments, percutaneous coronary intervention, or surgical methods, stem cell applications to regenerate injured myocardium to restore cardiac function are currently a reliable and focusable

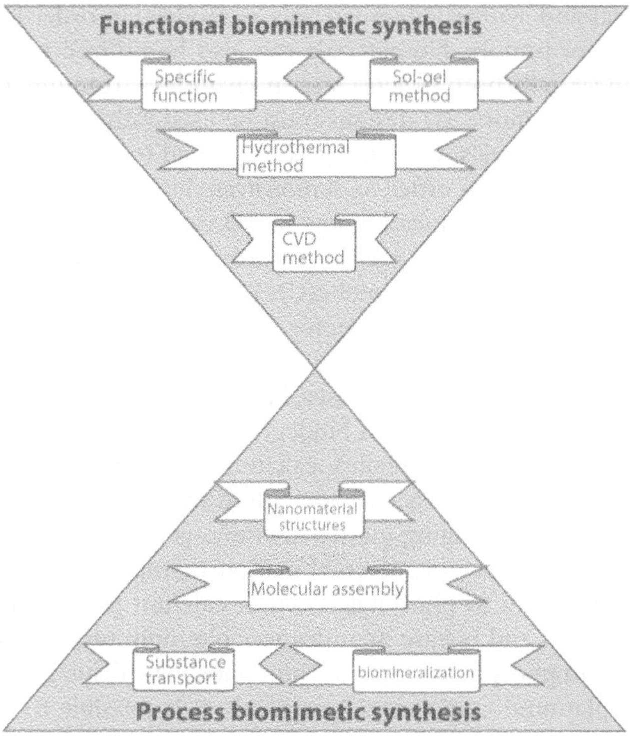

Figure 9.1 Categories of biomimetic synthesis.

technology. However, the use of stem cells is constrained by a number of factors, including poor tissue integration, low retention, the brief survival rate of cells in the wounded area, and even immunological sensitivity. As a result of the incorporation of stem cells into biomaterials and organic scaffolds, advancements in tissue engineering and regenerative medicine have exploded. Not only can biomaterials protect stem cells from harm and maintain their ability to proliferate, differentiate, and self-renew, but in some circumstances, they can also create the microenvironment of stem cells by imitating the extracellular matrix present naturally [28].

9.3.2 Biomaterials for Cancer Diagnostics

Our knowledge of carcinogenesis and our capacity to locate and seize tumor cells *in vitro* and *in vivo* have both greatly benefited from the usage of biomaterials. Key elements of the tumor microenvironment, including its structural, mechanical, and biological roles, can be reproduced *in vitro* using natural and synthetic biomaterials as models. In order to track the development of tumors, biomaterials can also be utilized to collect and identify cancer cells both *in vitro* and *in vivo* [29]. To mimic the micro-environment of tumors, 3D designs must be used rather than 2D ones to promote morphogenesis. Collagen gels were used to replicate *in vivo* cell morphologies *in vitro* in order to demonstrate how normal mouse mammary epithelial cells produce lumens in 3D rather than as a monolayer on 2D substrates [30]. In the medical industry, intensive research is being done to create new carbon nanotube(CNT) biomaterials that will be used in disease treatment and diagnosis. for instance, the use of CNTs in cancer treatment and diagnosis, including *in vivo* imaging, hyperthermia, and drug delivery systems for diseases, has been studied [31]. Due to their exceptional biocompatibility and biodegradability, calcium-based biomaterials such as calcium phosphates, calcium carbonates, calcium silicate, and calcium fluoride have been widely used for a variety of therapeutic approaches such as imaging-guided therapy, gene therapy, chemotherapy, photodynamic therapy, or combination therapy [32]. According to studies, the Calcium to phosphate ratio of calcium phosphate is crucial to the adsorption and release of anticancer platinum bisphosphates, with higher Calcium phosphate ration calcium phosphate displaying a stronger affinity for platinum complexes. As a result, it is possible to modify the characteristics of calcium phosphate biomaterials to create calcium-phosphate drug conjugates for certain therapeutic uses [33]. Cancer-associated fibroblasts(CAFs), vascular space-related cells, and other innate adaptive immune are present in the extracellular matrix scaffold where cancer cells are embedded. It is attempted to employ markers to distinguish CAFs from their precursors since CAFs exhibit preinvasive and pro-metastatic activities that are distinct from those of their precursor cells. In several cancer forms, an elevated level of CAFs is known to be a poor prognostic predictor. Tumor environment biomimetics significantly advances our comprehension. CAFs can be used to either trap cancer cells in order to stop metastasis or to analyze the functionality of circulating tumor cells. The ongoing collaboration of tissue engineers, biomaterials specialists, and cancer researchers opens up new possibilities for cancer diagnosis and treatment by allowing for the biomimicry of the tumor environment [34].

9.3.3 Biomaterials for Vaccine Development

The use of biomaterial delivery systems for vaccines is not new; approximately 20 years ago, scientists showed that PLGA microparticles loaded with antigens improved immunization effectiveness [35]. Biomaterials serve as promising candidate adjuvants to increase vaccine effectiveness. We can extend the use of vaccinations to illnesses like AIDS, malaria, and cancer through adjuvant design. The degradable polymer PLGA and its derivatives, particularly those that contain innate activators, have by far been the most widely used biomaterials in adjuvant research. The polymer has received approval for a number of clinical purposes, and both its transport and degradation properties have undergone extensive research [36]. In general, methods that integrate material science, engineering, and immunogen design have become a potential approach for many vaccination platforms. By encasing cargos in lipid or polymer carriers, tethering signals to the surfaces of spherical carriers (like polystyrene beads or gold nanoparticles), or allowing cues to self-assemble into nano- or micro-complex, virus-like particles, or polyelectrolyte multilayers, biomaterials can aid in the achievement of this goal. Several toll-like receptors have been administered with antigen using material-based methods as well. Several researchers have proven applications to adorn the surface of a particular carrier or to construct a multivalent scaffold to display multiple copies of an antigen, in addition to researching biomaterials platforms to encapsulate signals [37]. In a cluster of pneumonia cases in Wuhan Province, China in December 2019, the SARS CoV-2 was first discovered. Since then, the virus has spread at an astounding rate throughout the world. COVID-19 shows that the biomaterials community concentrated its research efforts on fungus and bacteria, with minimal attention paid to viruses. Numerous biomaterial-based solutions have been offered, including nanoparticles, liposomes, scaffolds, and microneedles. Specific immune cell targeting, antigen cargo protection, and biomimetic antigen presentation are all possible using nanoscale structures. Modification in nanoparticle size, shape, surface chemistry (particle charge, hydrophobicity), and material composition can be used to achieve this. Biomaterials have been used to inactivate viruses, and they hold a lot of promise for halting the spread of SARS-CoV-2 and other viral outbreaks in the future. Innovative filtration devices capable of collecting and neutralizing SARS-CoV-2 and other viruses can be created using the biomaterials method [38].

9.4 Bioinspired Materials

In the course of billions of years of evolution, nature has created biomaterials with remarkable qualities or smart behaviors. The preparation of man-made materials inspired by nature has been one of the frontiers of materials science and engineering. These biomaterials are highly admired by material scientists for their exceptional performances as well as for another reason that is even more significant i.e., they are typically synthesized and processed in aqueous circumstances at room temperature with very little chemical resource availability for living organisms. Bioinspired materials can be categorized as one-dimensional, two-dimensional, and three-dimensional on the basis of low to high-dimensional structures [39].

9.4.1 One-Dimensional Bioinspired Material

The connection between surfaces that may carry liquid droplets and people's daily lives and business operations has recently drawn a lot of interest. Surfaces that may carry liquid droplets are common in the natural world. Butterfly wings, goose feathers, and rice leaves are a few common examples. The directional arrangement of the micro and nanostructures on the surface of these organisms is shown to explain this particular characteristic. One-dimensional (1D) materials have recently attracted the attention of scientists due to their potential applications in microfluidics, textile dyeing, filtration, and smog removal. Examples of such materials include spider silk and cactus spine [40]. 1D photonic crystals are a desirable material for full-color color detection due to their ability to combine structural and water vapor sensitivity. This draws inspiration from the many different animals found in nature, many of which exhibit an iridescent color due to the presence of photonic crystal structures on the surface of their bodies. It has been discovered that by adjusting the incident angles, periods, and number of layers in spin-coated titania and PHEMA-co-PGMA films, bioinspired organic/inorganic hybrid 1D photonic crystals can have their optical properties tailored. These demonstrated reversible water vapor sensitivity, which makes them an appealing option for affordable humidity sensors and exhibits vibrant detection opportunities in both chemical and biological fields [41]. The construction of stimuli-responsive Janus wettability films using 1D photonic crystals was inspired by the Janus wettability of lotus leaves and Bragg stacks of bottles. Poly(N-Isopropyl acrylamide Bis-acrylamide-Acrylic Acid) P(NiPAAm-bis-AA) hydrogel and TiO2 were spin-coated alternately on a silicon wafer to create the 1DPCs.The two components' layer thicknesses could be adjusted to produce films with a striking structural color and adjustable wettability. Contrary to the above-mentioned film, which offers an intelligent monitor for practical applications such as water quality monitoring, microfluidics and artificial hydrogels techniques used to create Janus wettability surfaces inspired by lotus leaves exhibit simple function and complex operation [42].

9.4.2 Two-Dimensional (2D) Bioinspired Materials

Tunable chemical characteristics and controllable structures of bio-inspired nanochannels have helped them to receive a great attention in recent years. Nanotechnology and material science advancements also led to the emergence of bioinspired nanochannels that imitated the perceptive capabilities of biological ion channels. By taking inspiration from the layered microstructure of nacre, 2D materials-based bionic nanochannels have been developed possessing advantages of easy expansion, modification, tunable channel size and length, and facile preparation, *etc.*, and also showing great potential in membrane separations, biomimetic microsensors, and energy conversion [43]. In order to create nacre-like materials, graphene oxide and chemically converted graphene have currently emerged as the most desirable building components. Larger molecular targets, such as DNA, proteins, and nanoparticles, cannot often be transported by GO membranes because of their unique channel width, which is typically too tiny (1nm). However, it has been demonstrated that the interlayer distance can be finely tuned in a wide range from sub-nanometer to more than ten nanometers [44].

9.4.3 Three Dimensional (3D) Bioinspired Materials

The use of supercapacitors in high power delivery or energy harvesting is promising as a key application for energy storage. Nature is an inspiration for better electrode surface design. The unique functionality of organisms is governed by both their micro-and nano-architectures as well as their material properties. Given that electrochemical reactions frequently take place in the nanometer region, problems with length scale have prevented the mimicked spider web structure from being used for energy storage applications. A 3D carbon network and zeolite imidazole framework were used in the creation of spider web bionic nanomaterials, which have great potential for electrolyte-based energy storage applications and can provide energy much more quickly in supercapacitors [45]. The rapid development of modern electrical equipment towards miniaturization and high power has established strict criteria for mechanical durability, dielectric property, and heat resistance of electric insulating materials. Mica-based nano paper was developed using a sol-gel-film transformation method and was inspired by natural nacre. Its matrix is a 3D aramid nanofiber framework. An impressive aspect of this bioinspired structure is how well it endows the mica nanopaper with exceptional mechanical qualities, such as excellent strength, extremely large ductility, and extremely high toughness. Due to its incredible multi-property combination and capacity for continuous production, mica-based nanopaper is a very promising electrical insulating material in miniature high-power electrical equipment [46].

9.5 Biomimetic Drug Delivery Systems

Circulatory cells have been used as sources for drug delivery applications because they can cross biological barriers by using the function of membrane proteins while avoiding negative immune responses [47, 48]. Among all the cells, Leukocytes can act as powerful sources to develop biomimetic DDS as they are able to overcome the barriers of inflamed arteries as well as the BBB. Leukolike vectors were synthesised by modifying leukocyte membrane with nonporous silica nanoparticles. Electrostatic interactions as well as chemical conjugation helps the leukocyte call membranes to bound on the nanoparticles. LLVs were shown to considerably pass through an inflammatory endothelium monolayer as compared to non-modified nanoparticles, but not a healthy monolayer [49, 50].

9.5.1 Cell Membrane-Based Drug Delivery System

Neutrophil cellular membranes coated nanoparticles were developed to treat an inflammatory disease i.e. rheumatoid arthritis. The nanoparticles were synthesised by modifying the polymeric cores of poly(lactic-co-glycolic acid) with the plasma membranes of neutrophil. These neutrophil-imitating nanoparticles get accumulated in inflamed vasculature as well as lesioned area results into the inhibition of endogenous neutrophil influx. Thus, suppress the symptoms of arthritis [51]. Also, cancer metastasis can be treated by encapsulating the drugs with neutrophil-coated PLGA nanoparticles. The resultant can efficiently target metastatic as well as circulating tumor cells by exerting the mimicking function of neutrophil [52]. The aforementioned reports collectively suggest that nanoparticle surface modification with leukocyte membranes or membrane proteins may be a useful method to achieve targeted drug delivery to inflamed disease sites by actively navigating through endothelial cell barriers (Figure 9.2).

9.5.2 Lipoprotein-Based Drug Delivery System

The nanocarriers can endure the immune system response while evading the mononuclear phagocyte system (MPS), for a prolonged period by mimicking the lipoproteins. Agents with diverse physiochemical properties

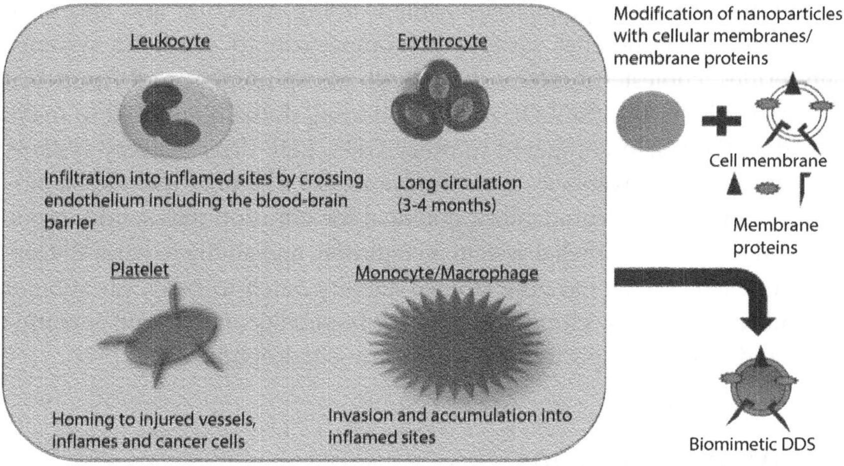

Figure 9.2 Representation of efficient applications of leukocyte, erythrocyte, platelet in drug delivery system [53].

can be carried by lipoprotein nanoparticles. Lipoprotein, an endogenous molecule, has the ability to evade MPS clearance and prevent an immunological response. Additionally, it has high biocompatibility and is fully biodegradable. The amphiphilic and hydrophobic drugs can be introduced into the membrane and core of lipoprotein Drug delivery system respectively whereas the hydrophilic molecules can be introduced into the surface of rHDL, or they can be wrapped further into core of rHDL with the assistance of specific hydrophobic components, such as calcium phosphate [54, 55]. ApoE-rHDL a high density lipoprotein which mimics is a well-known example of a biomimetic nanocarrier for Alzheimer's disease therapy as it has high BBB permeability [56, 57]. Biomimetic lipoprotein based drug delivery system has many biomedical applications in CNS diseases. During the treatment of Brain tumor the main challenges that the drug has to overcome is tumor cell targeting, to cross BBB, intratumoral heterogeneity as well as immunosuppressive environment. Lipoprotein based drug delivery system holds a good efficiency to overcome these challenges as it has BBB penetrating property, specific targeting ability, long half-life circulation, controlled release of drug, reduced immunogenicity as well as delivery of hydrophobic drugs [58].

Among the four main classifications of lipoproteins for lipid transport *in vivo*, low-density lipoprotein (LDL) is becoming an alluring vehicle for the targeted delivery of theranostic drugs. LDL particles have inherent biocompatibility and biodegradability, lower immunogenicity, and the ability to naturally target malignant cells while evading detection and clearance by the reticuloendothelial system, in comparison to synthetic systems. Triglycerides, cholesterol esters, cholesterol, phospholipids, and ApoB-100 or short peptides with the LDL receptor-binding domain are used to make LMPs directly. When compared to other DDS, LDL drug delivery systems have a number of advantages and disadvantages. The pros include its biocompatibility and biodegradability, potential for diffusion into solid tumors, decreased reticuloendothelial system recognition, and ability to naturally target tumors. However, it has poor storage stability, batch-to-batch variation in sizes and compositions, a limited supply, the possibility of pathogen contamination, and a limited capacity for encapsulating hydrophilic drugs [59].

9.6 Artificial Organs

The application of artificial organs (AO) in the rehabilitation of organ failure has become essential. Each organ requires a tailored membrane because each has distinct functions of its own. It is necessary to highlight that the main hindrance in membrane-based artificial organ technology is hemocompatibility not the perm selectivity of membrane at blood-material junction. The approach to improve hemocompatibility must also be specifically designed for each organ, as per needs of the kidney and lungs. Additionally, the most crucial parameter, especially for artificial lungs, is to enhance the mass transfer coefficient.

9.6.1 Artificial Kidney

To create dialysis membranes, a variety of polymers are currently being used. These polymers can be categorised into five groups: cellulose derivatives, polysulfone derivatives (PSU), polyacrylontrile (PAN), polymethylmethacrylate (PMMA), and ethyl-vinyl-acetate copolymer (EVAL). Cellulose materials were used to synthesize hemodialysis membranes mainly. Due to some concerns over the biocompatibility of these membranes the research for new biomaterials begins. Hence, Cellulosic membranes have been superseded by synthetic membranes with improved biocompatibility in the modern era [60] Polysulfone based dialysis membranes had better clinical outcomes and higher biocompatibility, according to research. Long-term studies have demonstrated that to achieve sufficient biocompatibility, a precise balance between the hydrophilic as well as hydrophobic qualities is required. PSU polymers are said to be hydrophobic intrinsically; thus, researchers have used polyvinylpyrrolidone (PVP) additives to boost the

PSU membrane's hydrophilicity. This has been clinically demonstrated to have a good impact on biocompatibility [61]. Polyacrylonitrile along with highly biocompatible hydrophilic sodium meth allyl sulfonate additives were used to synthesize AN69 [62]. Polymethylmethacrylate membranes exhibit excellent adsorption properties for molecules at the outer walls as well as inner surfaces of the membranes [63]. Ethyl-vinyl-acetate copolymer shows good blood compatibility hence proven to be good candidate for dialysis material [64]. More subsequently, 3D printing technology has been employed in conjunction with microfluidic technology to mimic a kidney-like structure [65].

9.6.2 Artificial Liver

Non-biological artificial system that enhances the role of liver in detoxification of blood and bio-artificial liver with the hepatocyte cells are the two main categories of modern liver-support system. Non-biological artificial systems focus to eliminate the toxins that are bound to albumin in the blood stream as it is the main cause of liver failure. Molecular absorbent recirculation system has been employed for dialysis of albumin. The concentration gradient of unbound albumin across the membrane acts as the driving force for the permeability of bilirubin [66]. Molecular absorbent recirculation system (50-60kDa) along with a typical hemodialysis membrane is utilized for this procedure to be done. As a result, this membrane prevents albumin from passing through while enabling bilirubin (MW of 584 Da) to pass through easily. Fractionated plasma separation and adsorption is another technique of artificial liver. This strategy differs from the MARS in using the permeable polysulfone membrane having molecular weight 250 kDa [67]. The membrane enables the toxins that are bound to albumin proteins in the blood to selectively penetrate rather than requiring albumin dialysate. Bio-artificial liver systems along with the hepatocytes cells can perform biological functions of the liver. It is extremely difficult to produce these bio-artificial systems since so many factors must be taken into account at once, including the hepatocyte supply, bioreactor design, immunological response as well as blood homeostasis. The currently used BAL membranes are made of relatively standard components including polysulfone, cellulose, and cellulose acetate [68, 69].

9.6.3 Artificial Pancreas

Artificial pancreas system constitutes of islet cells wrapped in a capsule semipermeable membrane which acts as a barrier against immune system.

This capsule provides essential nutrients to the islet cells as well as protects them from immunoglobin. The molecular weight of the smallest immunoglobin is 150 kDa. Therefore, 50-150kDa is supposed to be the ideal membrane to accomplish the objectives of artificial pancreatic system. This system is further classified into three categories based on the size of islet cells: microencapsulation and microencapsulation. Mainly polymeric membranes have been used to encapsulate these cells [70–72].

9.6.4 Artificial Lung

Polypropylene and poly-4-methylpentene polymers have been mainly used for synthesizing artificial lung membranes [73]. The polypropylene-based microporous membranes are highly porous having great gas permeability. These membranes have a drawback of having short life time period. To overcome this poly-4-methylpentene-based lung membrane is developed with a longer lifetime period. On comparing both the membranes poly-4-methylpentene have a thin, dense layer that may effectively stop the leakage of plasma. Additionally, PMP have high hydrophobicity, which inhibits blood plasma from unintentionally hydrating the membrane. Due to these benefits, poly-4-methylpentene membranes can be used for continuous long-term treatment, for more than 42 days [74]. The emergence of thrombosis and an inflammatory reaction are the main problems with artificial lung membranes. Especially, thrombosis can obstruct the blood flow and significantly reduce the effectiveness of oxygenation and protein adsorption is the key step of blood coagulation. Modern artificial lung membranes are so advanced that they can persist for a few days until blood clots (thrombosis) negatively impact patients. Furthermore, hydrophobic qualities are necessary in artificial lung membranes to prevent pore hydration and plasma passage. Exactly for this reason, the majority of modern artificial lung membranes are made of hydrophobic substances like Polypropylene and poly-4-methylpentene polymers. Minimizing surface energy is another approach to avoiding protein adsorption, notably for artificial lung membranes. It has been demonstrated that the membrane surface energy and the proportion of protein adsorption are inversely related. The current Polypropylene and poly-4-methylpentene polymer materials could be replaced with fluorinated polymers in the future. The development of the implantable and wearable artificial lungs is still in its beginnings. The hemocompatibility of the material presents the biggest obstacle. A thorough knowledge of the interaction between blood and material is needed in order to achieve the safety for longer term [75–79] (Figure 9.3).

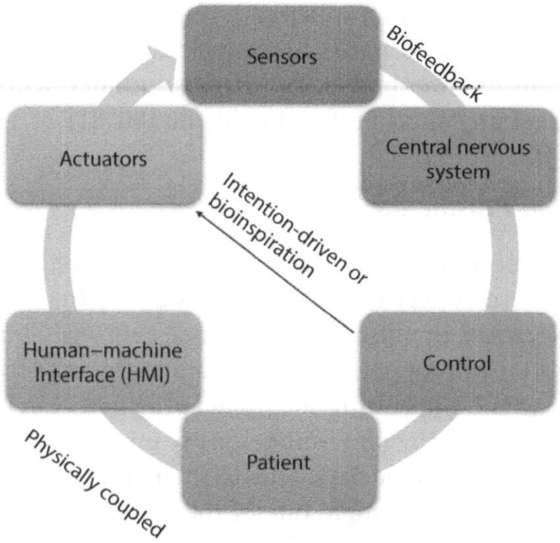

Figure 9.3 Mechanism of biomechanotronics.

9.7 Neuroprosthetics

Neuroprosthetics nowadays are offered as replacements for the sensory or cognitive modality losses brought on by an injury or an illness and were created by combining neuroscience with biomedical engineering techniques. Novel hybrid biomaterials that can meet the requirements for optimum neuroprosthetics have emerged as a result of the merging of biomedical engineering with neural prosthetics.

9.7.1 Sensory Prosthetics

9.7.1.1 Auditory Prosthetics

The cause of deafness is loss of sensory hair cells present in the inner ear. Using electrical current to stimulate remaining neurons enables the hair cells bypass. Auditory prosthetic is most extensively used as restorative implant ranging from basic to advance level. Speech processor used in an auditory prosthesis transforms the electric signals delivered via microphone to array of electrodes(placed in the cochlea's scala tympani close to osseous spiral lamina and the spiral ganglion. There are several types of cochlear implants that can be positioned along cochlear spiral. The characteristics of electrode such as geometry, mechanical properties vary based on the location. Spatial hearing, noisy milieu as well as pitch recognition are the major drawbacks after implantation of cochlear. Cochlear nucleus complex are stimulated by auditory implants in lower brainstem whereas auditory neurons are stimulated by midbrain implants in inferior colliculu [80, 81].

9.7.1.2 *Visual Prosthetics*

Utilizing photoreceptors, that are specialized neuron cells situated in retina, photons are converted to electrical signals. Image sensing is the outcome of electrical impulses of the retinal neurons. Stimulating optic neurons allows for interpretation of sent data to the implants. Depending on the technique and implanting site, visual prosthesis can be divided into 2 groups: (1) retina nearby photodiodes array implants, and (2) telemetry technology using external devices. Due to their convenient qualities, photodiodes are preferred; although, one of the shortcomings is the absence of energy for retinal stimulation. The most challenging issue of implanting electrodes on retina is the mechanical stability. Reason behind this is the site of implantation i.e. suprachoroidal, epiretinal as well as subretinal. Owing to the movement of eyes i.e. around 125000 per day in this situation causes mechanical tension in the electrodes. In addition to having poor mechanical stability, epiretinal electrodes have also demonstrated fixation issues and size restrictions. The subretinal region is the ideal location for electrode implantation since it contains the majority of the visual neurons. Moreover, the thickness of the electrodes should be less than 400 m due to surgical challenges. Even though the suprachoroidal site has been utilized frequently due to the simpler surgical procedure and mechanical stability of the region, a larger stimulating energy is required in contrast to epiretinal and subretinal locations due to greater distance to the retina [82–84].

9.7.2 Motor Prosthetics

Motor impairment is a result of spinal cord damage. The limitations of tetraplegia can be restored by motor neuroprostheses. Electrical signals are sent through motor neuron in order to stimulate contraction of the muscle. The muscle may be regulated and trained using this system. In order to restore paralyzed tissue, brain-computer interface interfaces i.e. (BCIs) have now been widely used as a tool for decoding brain signals like electromyograms (EMGs), electro-oculograms (EOGs), kinematics, and force. Controlling the hand, and particularly grabbing, in non-intrusive ways, is challenging task. Action potentials in efferent neurons are produced when paralyzed muscles are electrically stimulated using surface electrodes.

Four different muscle groups should be activated for grasping: the finger and thumb, opening and closing the hand, flexing of thumb, and the stabilizing muscle. As a non-invasive technique, electrodes are placed adjacent to the motor site of the muscles. For recording the sequence of muscle activity, electroencephalogram electrodes based on the motor cortex are used to control the hand and elbow [85, 86] (Table 9.1).

9.7.3 Cognitive Prosthetics

Cognitive neural prostheses have helped amputees. Cortical neural prostheses, particularly cognitive ones that use implanted electrodes in the cortex, translate and communicate decisions made by paralyzed patients that have an impact on neural activity to external assistance devices such robotic limbs using a decoding algorithm. Since cognitive prosthesis is placed inside the brain, biocompatibility represents one of the most crucial considerations. A battery that can be installed in the prosthesis can provide power. The implanted circuit can be optimized to change power consumption. A surgical procedure is required to replace a new battery after the battery has exhausted. Rechargeable batteries are ones that don't require surgery and are able to charged wirelessly using electromagnetic energy. Rechargeable batteries, particularly self-recharging ones, seem to be more favorable to ordinary batteries than they are for neural prostheses because

Table 9.1 Shows the materials that can be used in neuroprosthetics [89].

<table>
<tr><th>Type</th><th>Material</th><th>Mechanism</th></tr>
<tr><td rowspan="8">Polymer</td><td>PEDOT</td><td rowspan="8">Faradic</td></tr>
<tr><td>PEDOT-OH</td></tr>
<tr><td>PPY-Cl</td></tr>
<tr><td>PEDOT-PSS</td></tr>
<tr><td>PEDOT-COOH</td></tr>
<tr><td>PPY-PU</td></tr>
<tr><td>PANI</td></tr>
<tr><td>PPY-PLGA</td></tr>
<tr><td rowspan="4">Metal</td><td>Pt</td><td>Faradic</td></tr>
<tr><td>TIN</td><td>Capacitive</td></tr>
<tr><td>Au</td><td rowspan="2">Faradic</td></tr>
<tr><td>IrOx</td></tr>
<tr><td>Carbon family</td><td>Carbon nanotube</td><td>Capacitive</td></tr>
</table>

of their long-term biocompatibility. The other issue is the brain's sensitivity to heat; as a result of this tolerance, power and devices shouldn't produce heat over what the brain can tolerate [87, 88].

9.8 Conclusion

Nature inspired the thought endeavor to mimic nature. The goal of biomedical engineering is to closely mimic body behavior and treat disabilities. The research is relocating toward the design of novel materials with excellent performance due to the increasing demand for lightweight engineering materials with the best strength-toughness balance. When it comes to structural applications, composites frequently offer the best solution due to their excellent mechanical properties and lightweight. The scientific field of biomimetics is enjoying increased prominence due to a new wave of recent developments in biology and engineering. The entire potential of taking cues from nature hasn't yet been achieved, but materials scientists are constantly creating new biomimetic and nature-inspired materials with superior qualities. Natural elements like molluscan nacre, spider web, and minerals like calcium have already significant contributions to the development of innovative tissue engineering technologies. The composition of biological membranes, cells, and harmful microbes has been used as a model for developing drug delivery technology. The use of biomimetics has been expanded for the highly promising creation of artificial organs and neuroprosthetics. While biomimetic nanostructures are fascinating and present new opportunities for designing novel nanomaterials with distinctive properties and applications, a different line of inquiry is connected to the growing interest and emphasis on developing precise and more sophisticated methods for the preparation of nanomaterials using bio-templates by imitating biological processes. Future developments in bioinspired and biomimetic systems hold great promise for overcoming current challenges in drug and gene delivery. Although medical technology is currently advancing and innovating, there are still many obstacles that must be overcome before such bioinspired materials may be successfully and often used for therapeutic applications that serve many purposes.

References

1. Naik, R.R. and Singamaneni, S., Introduction: Bioinspired and biomimetic materials. *Chem. Rev.*, *117*, 20, 12581–12583, 2017.

2. Lepora, N.F., Verschure, P., Prescott, T.J., The state of the art in biomimetics. *Bioinspir. Biomim.*, *8*, 1, 013001, 2013.
3. Salifu, A.A., Mannoor, M.S., Soboyejo, W., 1 Bioinspired and biomimetic design of multilayered and multiscale structures. *Bioinspir. Struct. Des.*, 1, 2020.
4. Zan, G. and Wu, Q., Biomimetic and bioinspired synthesis of nanomaterials/nanostructures. *Adv. Mater.*, *28*, 11, 2099–2147, 2016.
5. Pfeifer, R., Lungarella, M., Iida, F., Self-organization, embodiment, and biologically inspired robotics. *Science*, *318*, 5853, 1088–1093, 2007.
6. Ma, P.X., Biomimetic materials for tissue engineering. *Adv. Drug Deliv. Rev.*, *60*, 2, 184–198, 2008.
7. Wei, G. and Ma, P.X., Nanostructured biomaterials for regeneration. *Adv. Funct. Mater.*, *18*, 22, 3568–3582, 2008.
8. Dang, M., Saunders, L., Niu, X., Fan, Y., Ma, P.X., Biomimetic delivery of signals for bone tissue engineering. *Bone Res.*, *6*, 1, 1–12, 2018.
9. Rasheed, T., Nabeel, F., Raza, A., Bilal, M., Iqbal, H.M.N., Biomimetic nanostructures/cues as drug delivery systems: A review. *Mater. Today Chem.*, *13*, 147–157, 2019.
10. Sabu, C., Rejo, C., Kotta, S., Pramod, K., Bioinspired and biomimetic systems for advanced drug and gene delivery. *J. Controlled Release*, *287*, 142–155, 2018.
11. Sheikhpour, M., Barani, L., Kasaeian, A., Biomimetics in drug delivery systems: A critical review. *J. Controlled Release*, *253*, 97–109, 2017.
12. Yang, G., Liu, Y., Jin, S., Zhao, C.X., Development of core-shell nanoparticle drug delivery systems based on biomimetic mineralization. *ChemBioChem*, *21*, 20, 2871–2879, 2020.
13. Ma, Y., Wu, Q., Duanmu, L., Wu, S., Liu, Q., Li, B., Zhou, X., Bioinspired composites reinforced with ordered steel fibers produced via a magnetically assisted 3D printing process. *J. Mater. Sci.*, *55*, 32, 15510–15522, 2020.
14. Li, T., Chang, J., Zhu, Y., Wu, C., 3D printing of bioinspired biomaterials for tissue regeneration. *Adv. Healthc. Mater.*, *9*, 23, 2000208, 2020.
15. Idaszek, J., Costantini, M., Karlsen, T.A., Jaroszewicz, J., Colosi, C., Testa, S., Święszkowski, W., 3D bioprinting of hydrogel constructs with cell and material gradients for the regeneration of full-thickness chondral defect using a microfluidic printing head. *Biofabrication*, *11*, 4, 044101, 2019.
16. Zan, G. and Wu, Q., Biomimetic and bioinspired synthesis of nanomaterials/nanostructures. *Adv. Mater.*, *28*, 11, 2099–2147, 2016.
17. Gagliardi, M., Biomimetic and bioinspired nanoparticles for targeted drug delivery. *Ther. Deliv.*, *8*, 5, 289–299, 2017.
18. Das, S.K., Parandhaman, T., Dey, M.D., Biomolecule-assisted synthesis of biomimetic nanocomposite hydrogel for hemostatic and wound healing applications. *Green Chem.*, *23*, 2, 629–669, 2021.

19. Ali, S., Perveen, S., Ali, M., Jiao, T., Sharma, A.S., Hassan, H., ... Chen, Q., Bioinspired morphology-controlled silver nanoparticles for antimicrobial application. *Mater. Sci. Eng.: C, 108*, 110421, 2020.
20. Kadiyala, N.K., Mandal, B.K., Ranjan, S., Dasgupta, N., Bioinspired gold nanoparticles decorated reduced graphene oxide nanocomposite using Syzygiumcumini seed extract: Evaluation of its biological applications. *Mater. Sci. Eng.: C, 93*, 191–205, 2018.
21. Wilson, R.J., Hui, Y., Whittaker, A.K., Zhao, C.X., Facile bioinspired synthesis of iron oxide encapsulating silica nanocapsules. *J. Colloid Interface Sci., 601*, 78–84, 2021.
22. Facklam, A.L., Volpatti, L.R., Anderson, D.G., Biomaterials for personalized cell therapy. *Adv. Mater.*, *32*, 13, 1902005, 2020.
23. Zhang, C.G., Zhu, W.J., Liu, Y., Yuan, Z.Q., Yang, S.D., Chen, W.L., ... Zhang, X.N., Novel polymer micelle mediated co-delivery of doxorubicin and P-glycoprotein siRNA for reversal of multidrug resistance and synergistic tumor therapy. *Sci. Rep.*, *6*, 1, 1–12, 2016.
24. Van de Wetering, P., Cherng, J.Y., Talsma, H., Crommelin, D.J.A., Hennink, W.E., 2-(Dimethylamino) ethyl methacrylate based (co) polymers as gene transfer agents. *J. Controlled release*, *53*, 1-3, 145–153, 1998.
25. Männistö, M., Vanderkerken, S., Toncheva, V., Elomaa, M., Ruponen, M., Schacht, E., Urtti, A., Structure–activity relationships of poly (l-lysines): Effects of pegylation and molecular shape on physicochemical and biological properties in gene delivery. *J. Controlled Release*, *83*, 1, 169–182, 2002.
26. Akinc, A., Anderson, D.G., Lynn, D.M., Langer, R., Synthesis of poly (β-amino ester)s optimized for highly effective gene delivery. *Bioconjugate Chem.*, *14*, 5, 979–988, 2003.
27. McNeer, N.A., Chin, J.Y., Schleifman, E.B., Fields, R.J., Glazer, P.M., Saltzman, W.M., Nanoparticles deliver triplex-forming PNAs for site-specific genomic recombination in CD34+ human hematopoietic progenitors. *Mol. Ther.*, *19*, 1, 172–180, 2011.
28. Hashemzadeh, M.R., Yazdi, M.E.T., Amiri, M.S., Mousavi, S.H., Stem cell therapy in the heart: Biomaterials as a key route. *Tissue Cell, 71*, 101504, 2021.
29. Beri, P., Matte, B.F., Fattet, L., Kim, D., Yang, J., Engler, A.J., Biomaterials to model and measure epithelial cancers. *Nat. Rev. Mater.*, *3*, 11, 418–430, 2018.
30. Bissell, M.J., Hall, H.G., Parry, G., How does the extracellular matrix direct gene expression? *J. Theor. Biol.*, *99*, 1, 31–68, 1982.
31. Saito, N., Haniu, H., Usui, Y., Aoki, K., Hara, K., Takanashi, S., Endo, M., Safe clinical use of carbon nanotubes as innovative biomaterials. *Chem. Rev.*, *114*, 11, 6040–6079, 2014.
32. Qi, C., Lin, J., Fu, L.H., Huang, P., Calcium-based biomaterials for diagnosis, treatment, and theranostics. *Chem. Soc. Rev.*, *47*, 2, 357–403, 2018.

33. Iafisco, M., Palazzo, B., Martra, G., Margiotta, N., Piccinonna, S., Natile, G., ... Roveri, N., Nanocrystalline carbonate-apatites: Role of Ca/P ratio on the upload and release of anticancer platinum bisphosphonates. *Nanoscale*, *4*, 1, 206–217, 2012.
34. De Vlieghere, E., Verset, L., Demetter, P., Bracke, M., De Wever, O., Cancer-associated fibroblasts as target and tool in cancer therapeutics and diagnostics. *VirchowsArchiv*, *467*, 4, 367–382, 2015.
35. Jones, K.S., Biomaterials as vaccine adjuvants. *Biotechnol. Prog.*, *24*, 4, 807–814, 2008.
36. Eldridge, J.H., Staas, J.K., Meulbroek, J.A., Tice, T.R., Gilley, R.M., Biodegradable and biocompatible poly (DL-lactide-co-glycolide) microspheres as an adjuvant for staphylococcal enterotoxin B toxoid which enhances the level of toxin-neutralizing antibodies. *Infect. Immun.*, *59*, 9, 2978–2986, 1991.
37. Gebre, M.S., Brito, L.A., Tostanoski, L.H., Edwards, D.K., Carfi, A., Barouch, D.H., Novel approaches for vaccine development. *Cell*, *184*, 6, 1589–1603, 2021.
38. Chakhalian, D., Shultz, R.B., Miles, C.E., Kohn, J., Opportunities for biomaterials to address the challenges of COVID-19. *J. Biomed. Mater. Res. Part A*, *108*, 10, 1974–1990, 2020.
39. Zhao, N., Wang, Z., Cai, C., Shen, H., Liang, F., Wang, D., ... Xu, J., Bioinspired materials: From low to high dimensional structure. *Adv. Mater.*, *26*, 41, 6994–7017, 2014.
40. Ju, J., Zheng, Y., Jiang, L., Bioinspired one-dimensional materials for directional liquid transport. *Accounts Chem. Res.*, *47*, 8, 2342–2352, 2014.
41. Wang, Z., Zhang, J., Xie, J., Li, C., Li, Y., Liang, S., Yang, B., Bioinspired water-vapor-responsive organic/inorganic hybrid one-dimensional photonic crystals with tunable full-color stop band. *Adv. Funct. Mater.*, *20*, 21, 3784–3790, 2010.
42. Liu, C., Zhang, L., Zhang, X., Jia, Y., Di, Y., Gan, Z., Bioinspired free-standing one-dimensional photonic crystals with janus wettability for water quality monitoring. *ACS Appl. Mater. Interfaces*, *12*, 36, 40979–40984, 2020.
43. Hao, J., Wang, W., Zhao, J., Che, H., Chen, L., Sui, X., Construction and application of bioinspired nanochannels based on two-dimensional materials. *Chin. Chem. Lett.*, *33*, 5, 2291–2300, 2022.
44. Feng, Y., Zhu, W., Guo, W., Jiang, L., Bioinspired energy conversion in nanofluidics: A paradigm of material evolution. *Adv. Mater.*, *29*, 45, 1702773, 2017.
45. Deng, X., Zhu, S., Li, J., He, F., Liu, E., He, C., ... Zhao, N., Bio-inspired three-dimensional carbon network with enhanced mass-transfer ability for supercapacitors. *Carbon*, *143*, 728–735, 2019.

46. Zeng, F., Chen, X., Xiao, G., Li, H., Xia, S., Wang, J., A bioinspired ultratough multifunctional mica-based nanopaper with 3D aramid nanofiber framework as an electrical insulating material. *ACS nano, 14*, 1, 611–619, 2019.
47. Fang, R.H., Jiang, Y., Fang, J.C., Zhang, L., Cell membrane-derived nanomaterials for biomedical applications. *Biomaterials, 128*, 69–83, 2017.
48. Fang, R.H., Kroll, A.V., Gao, W., Zhang, L., Cell membrane coatingnanotechnology. *Adv. Mater., 30*, 23, 1706759, 2018.
49. Parodi, A., Quattrocchi, N., Van De Ven, A.L., Chiappini, C., Evangelopoulos, M., Martinez, J.O., Brown, B.S., Khaled, S.Z., Yazdi, I.K., Enzo, M.V., Isenhart, L., Synthetic nanoparticles functionalized with biomimetic leukocyte membranes possess cell-like functions. *Nat. Nanotechnol, 8*, 1, 61–68, 2013.
50. Palomba, R., Parodi, A., Evangelopoulos, M., Acciardo, S., Corbo, C., De Rosa, E., Yazdi, I.K., Scaria, S., Molinaro, R., Furman, N.E., You, J., Biomimetic carriers mimicking leukocyte plasma membrane to increase tumor vasculature permeability. *Sci. Rep., 6*, 1, 1–11, 2016.
51. Zhang, Q., Dehaini, D., Zhang, Y., Zhou, J., Chen, X., Zhang, L., Fang, R.H., Gao, W., Zhang, L., Neutrophil membrane-coated nanoparticles inhibit synovial inflammation and alleviate joint damage in inflammatory arthritis. *Nat. Nanotechnol., 13*, 12, 1182–1190, 2018.
52. Kang, T., Zhu, Q., Wei, D., Feng, J., Yao, J., Jiang, T., Song, Q., Wei, X., Chen, H., Gao, X., Chen, J., Nanoparticles coated with neutrophil membranes can effectively treat cancer metastasis. *ACS Nano, 11*, 2, 1397–1411, 2017.
53. Fukuta, T. and Kogure, K., Biomimetic nanoparticle drug delivery systems to overcome biological barriers for therapeutic applications. *Chem. Pharm. Bull., 70*, 5, 334–340, 2022.
54. Ma, X., Song, Q., Gao, X., Reconstituted high-density lipoproteins: Novel biomimetic nanocarriers for drug delivery. *Acta Pharm. Sin. B, 8*, 1, 51–63, 2018.
55. Rajora, M.A., Ding, L., Valic, M., Jiang, W., Overchuk, M., Chen, J., Zheng, G., Tailored theranostic apolipoprotein E3 porphyrin-lipid nanoparticles target glioblastoma. *Chem. Sci., 8*, 8, 5371–5384, 2017.
56. Song, Q., Huang, M., Yao, L., Wang, X., Gu, X., Chen, J., Chen, J., Huang, J., Hu, Q., Kang, T., Rong, Z., Lipoprotein-based nanoparticles rescue the memory loss of mice with Alzheimer's disease by accelerating the clearance of amyloid-beta. *ACS nano, 8*, 3, 2345–2359, 2014.
57. Song, Q., Song, H., Xu, J., Huang, J., Hu, M., Gu, X., Chen, J., Zheng, G., Chen, H., Gao, X., Biomimetic ApoE-reconstituted high density lipoprotein nanocarrier for blood–brain barrier penetration and amyloid beta-targeting drug delivery. *Mol. Pharm., 13*, 11, 3976–3987, 2016.
58. Chen, Y.X., Wei, C.X., Lyu, Y.Q., Chen, H.Z., Jiang, G., Gao, X.L., Biomimetic drug-delivery systems for the management of brain diseases. *Biomater. Sci., 8*, 4, 1073–1088, 2020.

59. Zhu, C. and Xia, Y., Biomimetics: Reconstitution of low-density lipoprotein for targeted drug delivery and related theranostic applications. *Chem. Soc. Rev.*, *46*, 24, 7668–7682, 2017.
60. Henderson, L.W. and Chenoweth, D.E., Cellulose membranes—Time for a change? in: *Disputed Issues in Renal Failure Therapy*, vol. 44, pp. 112–126, 1985.
61. Streicher, E. and Schneider, H., Polysulphone membrane mimicking human glomerular basement membrane. *Lancet*, *322*, 8359, 1136, 1983.
62. Thomas, M., Moriyama, K., Ledebo, I., AN69: Evolution of the world's first high permeability membrane. *High-performance Membrane Dialyzers*, *173*, 119–129, 2011.
63. Fabbrini, P., Sirtori, S., Casiraghi, E., Pieruzzi, F., Genovesi, S., Corti, D., Brivio, R., Gregorini, G., Como, G., Carati, M.L., Viganò, M.R., Polymethylmethacrylate membrane and serum free light chain removal: Enhancing adsorption properties. *Blood Purif.*, *35*, Suppl. 2, 52–58, 2013.
64. Nakano, A., Ethylene vinyl alcohol co-polymer as a high-performance membrane: An EVOH membrane with excellent biocompatibility. *High-performance Membrane Dialyzers*, *173*, 164–171, 2011.
65. Sochol, R.D., Gupta, N.R., Bonventre, J.V., A role for 3D printing in kidney-on-a-chip platforms. *Curr. Transplant. Rep.*, *3*, 1, 82–92, 2016.
66. Siami, G.A. and Siami, F.S., Membrane plasmapheresis in the United States: A review over the last 20 years. *Ther. Apheresis*, *5*, 4, 315–320, 2001.
67. Di Paola, L., Artificial kidney: A chemical engineering challenge, in: *Current Trends and Future Developments on (Bio-) Membranes*, pp. 1–20, Elsevier, Amsterdam, Netherlands, 2020.
68. Sivaraman, B. and Latour, R.A., The relationship between platelet adhesion on surfaces and the structure versus the amount of adsorbed fibrinogen. *Biomaterials*, *31*, 5, 832–839, 2010.
69. Sperling, C., Fischer, M., Maitz, M.F., Werner, C., Blood coagulation on biomaterials requires the combination of distinct activation processes. *Biomaterials*, *30*, 27, 4447–4456, 2009.
70. Schweicher, J., Nyitray, C., Desai, T.A., Membranes to achieve immunoprotection of transplanted islets. *Front. Biosci. (Landmark edition)*, *19*, 49, 2014.
71. Iacovacci, V., Ricotti, L., Menciassi, A., Dario, P., The bioartificial pancreas (BAP): Biological, chemical and engineering challenges. *Biochem. Pharmacol.*, *100*, 12–27, 2016.
72. Li, R.H., Materials for immunoisolated cell transplantation. *Adv. Drug Delivery Rev.*, *33*, 1-2, 87–109, 1998.
73. Federspiel, W.J. and Henchir, K.A., Lung, artificial: Basic principles and current applications. *Encycl. Biomat. Biomed. Eng.*, *9*, 910, 2004.
74. Awad, J.A., Cloutier, R., Fournier, L., Major, D., Martin, L., Masson, M., Guidoin, R., Pumpless respiratory assistance using a membrane oxygenator

as an artificial placenta: A preliminary study in newborn and preterm lambs. *J. Invest. Surg.*, *8*, 1, 21–30, 1995.
75. LaFayette, N.G., Schewe, R.E., Montoya, J.P., Cook, K.E., Performance of a MedArray silicone hollow fiber oxygenator. *ASAIO J.*, *55*, 4, 382–387, 2009.
76. Motlagh, N.V., Birjandi, F.C., Sargolzaei, J., Shahtahmassebi, N., Durable, superhydrophobic, superoleophobic and corrosion resistant coating on the stainless steel surface using a scalable method. *Appl. Surface Sci.*, *283*, 636–647, 2013.
77. Philipp, A., Foltan, M., Gietl, M., Reng, M., Liebold, A., Kobuch, R., Keyl, C., Bein, T., Müller, T., Schmid, F., Interventionelle extrakorporale Lungenunterstützung (ILA) mittels arterio-venösem Shunt und einem neu entwickelten low resistance Lung Assist Device (LAD). *Kardiotechnik*, *1*, 7–13, 2003.
78. Park, A., Song, Y., Yi, E., Duy Nguyen, B.T., Han, D., Sohn, E., Park, Y., Jung, J., Lee, Y.M., Cho, Y.H., Kim, J.F., Blood oxygenation using fluoropolymer-based artificial lung membranes. *ACS Biomaterials Sci. Eng.*, *6*, 11, 6424–6434, 2020.
79. Duy Nguyen, B.T., Nguyen Thi, H.Y., Nguyen Thi, B.P., Kang, D.K., Kim, J.F., The roles of membrane technology in artificial organs: Current challenges and perspectives. *Membranes*, *11*, 4, 239, 2021.
80. Zeng, F.G., *Cochlear implants: Auditory prostheses and electric hearing*, vol. 20, Springer Science & Business Media, New York, USA, 2004.
81. Shannon, R.V., Advances in auditory prostheses. *Curr. Opin. Neurol.*, *25*, 1, 61, 2012.
82. Shepherd, R.K., Shivdasani, M.N., Nayagam, D.A., Williams, C.E., Blamey, P.J., Visual prostheses for the blind. *Trends Biotechnol.*, *31*, 10, 562–571, 2013.
83. Maghami, M.H., Sodagar, A.M., Lashay, A., Riazi-Esfahani, H., Riazi-Esfahani, M., Visual prostheses: The enabling technology to give sight to the blind. *J. Ophthalmic Vis. Res.*, *9*, 4, 494, 2014.
84. Barriga-Rivera, A., Guo, T., Yang, C.Y., Abed, A.A., Dokos, S., Lovell, N.H., Morley, J.W., Suaning, G.J., High-amplitude electrical stimulation can reduce elicited neuronal activity in visual prosthesis. *Sci. Rep.*, *7*, 1, 1–13, 2017.
85. Sarasola-Sanz, A., Irastorza-Landa, N., López-Larraz, E., Bibián, C., Helmhold, F., Broetz, D., Birbaumer, N., Ramos-Murguialday, A., A hybrid brain-machine interface based on EEG and EMG activity for the motor rehabilitation of stroke patients, in: *2017 International Conference on Rehabilitation Robotics (ICORR)*, London, UK, pp. 895–900, IEEE, 2017.
86. Müller-Putz, G.R., Plank, P., Stadlbauer, B., Statthaler, K., Uroko, J.B., 15 years of evolution of non-invasive EEG-based methods for restoring hand & arm function with motor neuroprosthetics in individuals with high spinal cord injury: A review of grazbci research. *J. Biomed. Sci. Eng.*, *10*, 6, 317–325, 2017.

87. Andersen, R.A., Hwang, E.J., Mulliken, G.H., Cognitive neural prosthetics. *Annu. Rev. Psychol.*, *61*, 169, 2010.
88. Harrison, R.R., Kier, R.J., Chestek, C.A., Gilja, V., Nuyujukian, P., Ryu, S., Greger, B., Solzbacher, F., Shenoy, K.V., Wireless neural recording with single low-power integrated circuit. *IEEE Trans. Neural Syst. Rehabil. Eng.*, *17*, 4, 322–329, 2009.
89. Zarrintaj, P., Saeb, M.R., Ramakrishna, S., Mozafari, M., Biomaterials selection for neuroprosthetics. *Curr. Opin. Biomed. Eng.*, 6, 99–109, 2018.

10

3D Printing in Drug Delivery and Healthcare

B. Mahesh Krishna, Francis Luther King M.*, G. Robert Singh and A. Gopichand

Department of Mechanical Engineering, Swarnandhra College of Engineering and Technology, Narsapur, Andhra Pradesh, India

Abstract

Personalized medicine for people has been around for a long time, but it has gained a lot of notification recently. It is anticipated that 3D printing will be a game-changer in the pharmaceutical industry. Prosthetics, bio-printing, and pharmaceutics are being studied using 3DP. 3DP technologies in drug carriers have expanded rapidly because to the potential advantages of customizing drugs in individually personalized doses. It produces pharmacological combinations on demand with various release and dose behaviors based on shape and size. 3DP systems are inkjet, extrusion, or laser-based depending on material stacking. Several medicine delivery methods use these methods. Drug administration routes include intrauterine, oral, transdermal, and pulmonary. 3D printing's potential in healthcare is explored, along with ways to overcome integration issues. Dentistry, engineered tissue samples, tissue engineering, regenerative medicine, medical devices, anatomical modeling, and pharmaceutical formulation are some healthcare applications being explored. This chapter summarizes current original research on 3DP, focusing on bio-printing and other kinds of 3DP on drug delivery applications, advantages and future possibilities in this sector.

Keywords: 3D printing, bio-printing, computer-aided design (CAD), drug delivery, customized medicine, pharmaceutics

**Corresponding author*: francis.mech@gmail.com

Arbind Prasad, Ashwani Kumar and Manoj Gupta (eds.) Advanced Materials and Manufacturing Techniques for Biomedical Applications, (241–274) © 2024 Scrivener Publishing LLC

10.1 Introduction

Today's constantly changing healthcare business is abandoning "one size fits all" therapeutic techniques. NHS England showed that mass-producing drugs fails 70% of patients [1]. Novel patient-specific medicines have resulted. Conventional manufacturing processes are too inflexible in dose and duration to create personalized drug-delivery systems. Thus, the healthcare business must embrace new platforms to generate personalized treatments. 3D printing lets scientists mimic anything from Neanderthal skulls to protein folding. 3D printing may possibly play a part in personalized drug delivery, considering the many options. The first 3D-printed pharmaceutical product, Spritam®, treats epilepsy, and scientists want to apply this technique to create other drugs. This breakthrough will change the pharmaceutical business from mass manufacturing to dosage form individualization [2]. "Personalized medicine" is a common way to tailor treatment to an individual's requirements. Technology may enable patient-specific therapy! Customized therapy depends on biological differences, lifestyle adjustments, drug-herb and drug-food combinations, and illness types. Thus, a universal method for developing formulations that release medications with the right profile for specific purposes is essential. 3D printing can now manage and distribute low-volume therapeutic doses for personalized therapy. Since the FDA authorized the first 3D-printed pill, levetiracetam under the brand name Spritam, there has been extensive research into its drug delivery potential [3, 4]. The American Society for Testing and Materials (ASTM) defines seven kinds of additive manufacturing equipment, although only some may legally make pharmaceutical formulations. Stereolithography, digital light processing, 2-photon polymerization, continuous liquid interface production, selective laser sintering, fused deposition modeling, pneumatic extraction/syringe extrusion, material jetting, and binder jetting have been used in pharmaceutical research [5]. 3D printing tablets allows for a wide range of release profiles and exquisite micro-control. 3D bioprinting technology's precision may allow it to manufacture artificial "organs" or "organs-on-a-chip" that mimic the human body's functions. Drug testing may use these "organs" or organs-on-a-chip. 3D printing can manufacture painless microneedle patches, which might improve patient compliance. 3D printing in medicine may save costs, boost production with fewer workers, improve accessibility, and promote collaboration [6]. Bioprinting tissues and organs, designing customized implants and prostheses, creating anatomical models for pre-operative surgery planning, and studying medication discovery, administration, and dosage forms are among 3D printing's most significant medical uses [7–9]. 3D printing has been used to create drug-eluting implants, medicinal devices, and customized solid oral dosage forms [10–16] individualized. Thus, 3D-printed biological tissues at the point-of-care have been examined for intra-individual therapy response detection [17]. 3DP technology can produce personalized, individually dosed pharmaceuticals for the pharmaceutical industry [18]. 3D printing uses polyvinyl alcohol (PVA), poly (lactic acid) (PLA), poly (caprolactone) (PCL), gelatin methacryloyl (GelMA), nanocellulose, chitosan, alginate, pectin, gelatin, and chocolate [19].

10.2 3D Printing in Healthcare Technologies

Clinicians have employed 3D printing for several medicinal needs [20]. This technology allows biomedical therapy to be personalized [21]. 3D printing in medical also allows for complicated implants, prostheses, surgical aids, and models. Due to material science advances, subsequent 3D printers could accurately and durably create metal items. Powder, filament, and bio-inks are just some of the feedstock options for today's 3D printers, which can also handle ceramic and metal materials. Most of the problems connected with conventional medical device production procedures have been alleviated thanks to the advent of 3D printing technology, which may assist in the manufacture of patient-specific implants [21].

10.3 Four Dimensions Printing (4D)

Four-dimensional printing is a part of three-dimensional printing where Functional aspects and intelligent induced materials are utilized in a 3D printing method to create 4D printed objects that can react to changes in their environment, such as temperature or pH [22]. It has several medical uses [23], such as in tissue engineering, medication administration, and implantable organ replacement therapy. It was originally believed that 4D printing was a means of connecting AM and time. In fact, the technique proposed in the early investigations in 2013 included the creation of a 3D printed element that undergoes a regulated form change. A definition of 4D printing was published by Momeni *et al.* [24–26]. This method was characterized as the systematic improvement of the form, function, and utility of 3D printed objects [27]. According to the Atlantic Council of the United States, "AM of objects with the capacity to self-transform in function or form when treated to a pre-specified stimulus" such as heat, osmotic pressure, UV light, electricity, or other energy sources constitutes 4D printing. This means that 4D printing is the programmable part of a 3D printed area that may alter the performance of readily available materials or material hybrids designed for self-assembly at certain shapes and locations [28]. 4D bio-printing is a medical science and technology application of 4D printing that is linked to the development of the included cells [29]. Cells mature by a process of self-organization that makes use of the fourth dimension of environmental cues. Lipids, biopolymers, and hydrogels are only some of the biomedically relevant materials that have been made possible by 4D printing [30]. Hydrogels that respond to external stimuli, either synthetically or naturally, are among the most promising materials for application in 4D printing. Because of their high interconnectedness and adaptable porosity, hydrogels are useful for 4D printing [31].

10.4 Transformation Process and Materials

The 3D printing process alters the material's consistency at several points in the making of the model. The potential for geometry collapse, model destruction, or lost features is high [32, 33]. To address the issue of how to print complicated models at high resolutions, the field of Additive Manufacturing (AM) has advanced. The contrast between hard and soft materials that define the human body is the most important. Rigid tissue includes human bones, whereas soft tissue includes ligaments and articular cartilage. Because the bulk of the materials used in 3D printing are stiff, bones are the simplest and easiest biological tissue to manufacture. Acrylonitrile butadiene styrene (ABS), plaster powder [34], and hydroquinone [35] are some of the materials utilized in 3D printing to represent the bone structure.

There is a significant disparity between the anatomical model generated on a 3D printer and the real thing when it comes to soft tissues, thus further research is needed. The bulk of 3D printing materials aren't realistic enough to reliably mimic human biological tissue. In order to soften the printed structures, postprocessing may be required. The regeneration of cartilaginous tissues [36], arteries for practicing valve replacement [37], hepatic segment [38], and hearts [39] are some examples. An exciting example of a useful tool for planning the operational strategy in the treatment of congenital cardiac illness is the 3D-printed brain aneurysm [40] constructed from the flexible TangoPlusTM photopolymer. Urethane and rubber-like polymers combined with a hard photopolymer are used because they have a Shore value and elastic characteristics that are similar to the physiological one [41, 42].

10.4.1 3D Bioprinting

Bioprinting "uses computer-aided transfer technologies with the objective of creating and patterning both living and non-living materials by a specific 3D or 2D organization" to develop bioengineered structures for medical, regenerative medicine, and basic cell biology research [43]. It takes precise and closely controlled biomaterial and live cell generation to duplicate the intricate 3D architecture of human tissues and organs. In order to meet these demands, scientists have developed four separate 3D bioprinting techniques based on biological release. Methods such as extrusion, ink-jet, laser-assisted, and stereolithography-based bioprinting technologies are examples of this [44, 45].

One possible use for this technology in the future is printing functioning organs. Organs, which are composed of many different types of tissue, have a very intricate structure. "Organ printing," first used in 2003, it is a fast prototyping using computer-aided 3D printing process based on stack deposition of cells using a 3D gel with progressive maturation of the printed construct into vascularized and perfused biological organs or tissue." It's still common to hear the term "organ printing" used, particularly in popular fiction [46–49]. Catalysis, biotechnology, sensing, and medicine are just a few examples of the many fields that utilize the phrase "bio-fabrication" to represent both natural and artificial processes. According to the definition, biofabrication is "the automated manufacturing of biologically functional products with the structural organization from bioactive molecules, living cells, cell aggregates such as micro-tissues, biomaterials, or hybrid cell-material constructs, through bio-assembly or bio-printing and subsequent tissue maturation processes [50–52]." Biofabrication can be accomplished through bio-assembly or bio-printing and subsequent tissue maturation processes.

10.4.1.1 Bioinks

Bioinks are cell formulations that can be handled by automated bio-fabrication processes. Furthermore, it has the potential to include physiologically active biomaterials and components [49]. Bio inks may include a wide variety of cell types and cell structures, including single cells, cellular rods, spheroids, aggregates, organoids, mini-tissues, cells with a thin layer of material coating, cells contained in specialized colloidal microenvironments, or cells seeded onto microcarriers. In addition, bio-inks could or might not include bioactive substances such DNA, miRNA, cytokines, exosomes [53–57]. Researchers are working to improve upon the properties of FDM 3D printing by switching to a modified extrusion process, which has the added benefit of lowering production costs [58].

10.4.2 Bioceramics

Orthopedic tissue replacement is one area where bio-ceramics have facilitated medical advancement. It's no secret that ceramics like alumina, calcium phosphate, bioglass, and zirconia have several applications [45]. More interconnectivity was found inside the planned macro-pores after microwave sintering of both pure tri-calcium phosphate (TCP) and SrO-MgO-doped TCP scaffolds (3DP) [59]. SrO and MgO dopants were more effective in speeding up osteogenesis and vasculogenesis than TCPd alone. Z-3D Corporation's printing equipment was used to create a matrix of CaHPO4H2O, DCPD, brushite, and unreacted TCP from Ca3(PO4)2 and TCP powder in the presence of phosphoric acid mixture. The printed samples had compressive strengths between 0.9 and 8.7 MPa, which were effectively improved upon by further hardening to reach 22 MPa [53]. Ceramics are a great material to utilize in hospitals because of their bioactivity and durability. Ceramics are osteoinductive, meaning they promote bone formation, and they are also biocompatible.

10.4.3 Synthetic Biopolymers

Polymeric materials' biodegradability and simplicity of manufacture make them popular in medicine and science. Natural and synthetic polymers exist. Depending on their function, polymeric scaffolds must consider biological and technical considerations. Tissue engineering uses poly(lactic acid), poly(glutamic acid), and poly(l-lactic acid) polymers. Tissue engineering scaffolds also use poly-capro-lactone (PCL). Poly(-hydroxyethyl) ethers (a-hydroxy esters) alter PCL and form copolymers with desired characteristics. Two examples are poly(-lactic acid-co—capro-lactone) (PLLACL) and poly(-hydroxyethyl) ethers (a-hydroxyethyl) [60]. SLS printing creates porous 3D scaffolds using PCL. In another work, chloroform was utilized as a natural binder to blend biodegradable polymer powder consisting of PGLA and PLLA in a 75%:25% ratio. The liver construct was then used to imitate liver tissue [54]. In 2013, Bose *et al.* employed 3D-printed PLLA/PLGA scaffolds to create an *in vitro* liver organogenesis process and a host vein and artery network [61]. The researchers printed 3D cell-containing constructs using PCL, PEG, and other biomaterials [62].

Scaffolds are often fabricated from biomaterials like biopolymers and biological molecules. Scaffolds for tissue engineering (TE) may be made from a wide variety of materials, such as alginate-based substrates, collagen, chitosan, and proteoglycans. Complications with biocompatibility, biodegradability, or both may arise from only one synthetic process in composite scaffold production. Hydrolyzed collagen is mostly gelatin. When temperatures are low, the extracellular matrix acts as a natural buoyancy aid, keeping cells from sinking into gel [63].

Scanning laser micromachining (SLM) was utilized to create starch-cellulose and starch-cellulose acetate scaffolds, with polymer particle size, laser scan speed, and laser intensity all having an impact on the characteristics of the scaffolds [64]. Following the fabrication of the scaffold, samples with tiny particle sizes were developed, exhibiting desirable mechanical characteristics and porosity for drug transport and TE. Another work [65] created a multi-layered PCL (core)-alginate-PM (shell) scaffold using bio-plotting, simple coating, and core-shell nozzle techniques. Scientists have 3D-printed a microfluidic system to transport blood. The device maintained normoglycemic limits (4–6 mM) in the storage solution, allowing cells to release ATP just as they would have if they hadn't been stored [66].

10.5 3D Printing's Pharmaceutical Potentials

Traditional pharmaceutical operations such as milling, granulation, compression, and mixing may result in erratic final product quality, which is dependent on parameters such as drug release, drug loading, drug stability, and pharmaceutical dose. As a strong tool technology, three-dimensional printing offers competitive benefits such as better safety, increased R&D efficiency, drug accessibility, and effectiveness. The primary benefits of three-dimensional printing that make it more appealing are as follows [67–69]:

10.5.1 Personalization

The term "personalized medicine" refers to the practice of tailoring a patient's medical care to their specific traits, preferences, and needs. It entails careful diagnosis, treatment, and monitoring. Personalized medicine may be expanded to include pre-emptive medicine aimed at lowering the risk of illnesses for which a patient has demonstrated predisposition by adjusting the patient's food, habits, and lifestyle, as well as recommending the use of certain pharmaceuticals or supplements.

10.5.2 Personalized Therapy

Surgeons were the first medical professionals to make use of 3D printing in the field of medicine. Initially, the purpose of this technology was to assist in the creation of 3D models of patients in order to assist in the visualization of their anatomy. This was particularly helpful for patients who had certain abnormalities or traits that required complex operations. Because it permits the production of items that perfectly fit the anatomical features of the patient according to diagnostic imaging techniques, it has partially superseded more established methods. These products may be made from biocompatible material. Scaffolds are three-dimensional constructs whose goal is to transport active compounds, fill up gaps, and drive cell growth towards target tissues.

a) Customized Medicine

Customizing medicine to the needs of the patient or the substance at hand is another use of 3D printing, like 3D printing food. This method may be employed in the case of pediatric dosages, which have a wide range of doses. Similarly, the shape of the dose form may be altered for individuals who have difficulty swallowing. 3D printing is a straightforward and versatile method for changing the shape of a dosage form.

b) Creating Complicated Forms

As low as 10–12 mol per tablet, 3D printing makes it possible to produce precise medication dosages or active pharmaceutical components, which helps to reduce negative effects associated with excessive dosing. In compared to traditional methods, where complex geometries were impossible to accomplish, 3D printing makes it simple. In addition, the size and form of the release device provide a unique release profile. There is a different release, a distinct chemical loading, and the treatment's flavor is hidden when a more complex version is used.

c) Long-Acting

3D printing enables for precise control and targeting of medication release. Matrix powdered layers can be printed using a binder, making it possible to use. As a result, a barrier between the API levels is formed, allowing for change in the release profile.

d) One-of-a-Kind Dosage Form

One possible use of 3D printing in the pharmaceutical industry is the creation of customized dosage forms. A new dosage model was made via 3D printing.

e) Mini Dispenser Unit

The 3D printer configuration takes up very little area, allowing these printers to fit into any location while being cost effective. A computer-aided design is three-dimensional printing. It implies that 3D printing can be managed through computer software and a network. Moreover, this technologies allow for the personalization of medication, functions as a mini-dispenser based on these features, possibly bringing tablet manufacture closer to patients.

f) Healthcare Network

Based on the patient's requirements, pharmacists and doctors may adjust the next dosage or medication combinations. Remote control of 3D printers is possible. As a result, patients have easy access to 3D printing. As a result, patient compliance improves, and the time it takes for a clinician to respond to a patient's requirements is reduced.

g) Rapid Disintegration

When comparing 3D printing to powder compaction, there is a notable difference in the disintegration process. Powder agglomeration follows a unique pattern with both contemporary and older methods. Increased rapid tablet disintegration speed is achieved by higher binding strength at the outer region and lower binding strength in the middle while using 3D printing powder. While Aprecia's Zip Amount® provides a high dose of piracetam, it has been discovered that it may dissolve in under 10 seconds (1000 mg).

h) Tool-Free

3D printing may remove the necessity for tool fabrication. As a result, expenses, lead times, and related labor are all decreased.

i) Sustainable/Environmentally Friendly

3D printing is a developing and unique technology that is energy efficient and environmentally friendly. It makes use of up to 90% standard materials, resulting in less wastes. When compared to conventionally made items, it features a powerful model that has a lower carbon impact.

j) Reduced Manufacturing Time

3D printers are time efficient, which reduces the length of product development design cycles.

k) Production Process

There is a simpler and more cost-effective production process available. Improved equipment shortens the processing time, lessens waste, streamlines setup assembly, and shortens the lead time via the coordination and integration of components.

l) Engineering and Maintenance

3D printers have more flexible setup and maintenance operations. It is a low-cost industrial engineering solution.

m) Logistic

Compared to traditional manufacturing methods, 3D printing has several benefits, including less time spent on stock and inventory management, lower shipping costs, and lower overall costs.

10.6 Drug Administration Routes

One possible remedy of transdermal medication distribution is the use of AM to create the complex and individually shaped geometries required by pharmaceutical drugs and medical devices. Successful local and systemic API delivery has been shown using 3D-printed implants, microneedles, masks, and patches tailored to individual patients' requirements.

10.6.1 Transdermal Route

Drugs applied topically don't move through the liver's first-pass effect (PM) like those taken orally. Patients comply more. SLA produced a salicylic acid nasal patch to treat acne. A portable 3D scanner scanned a volunteer's head and face to create a patient-specific nose patch. SLA administered drug-loaded PEGDA nasal patch salicylic acid (anti-acne). Microneedle-embedded transdermal patches are fascinating (MN). These MNs may generate superficial pores in the skin that enable large molecules like insulin, melanostatin, and erythropoietin to enter. A patient-specific microneedle patch may be made using a DLP printer and 3DM-Cast. For maximum penetration, curved microneedles must be perpendicular to the skin [70]. A high molecular weight polymer's structure may have reduced drug release patterns in films. Expectedly, PEO-based systems maintained consistency. High-molecular-weight PEO films have better bio-adhesion. Above, HME developed multilayer films with reservoir and backing layers. These multi-layer films may provide a diffusion control layer to the drug-loaded film to enhance its mechanical properties [71]. PEG 400 concentrations between 0.75 and 1.5 percent may extrude HPC films. The film cannot be extruded at concentrations below 0.75 percent owing to inadequate plasticization, while concentrations exceeding 1.5 percent may cause polymer powder agglomeration and irregular hopper flow [72].

10.6.2 Ocular Route

One possible approach is to utilize contact lenses to transport drugs directly to the cornea. An *in vitro* ocular model was manufactured utilizing a mold made using 3D printing technology to investigate the kinetics of ciprofloxacin and moxifloxacin release from commonly used daily disposable contact lenses. These 3D printed polycarbonate-acrylonitrile-but adiene-styrene molds comprised a corneal/scleral portion of an eyelid component, and the eyes were connected to a microfluidic syringe pump to simulate tear flow and volume control. To study the kinetics of drug release, we soaked contact lenses in ciprofloxacin and moxifloxacin solutions and then placed them in an eye model. As a result, using an eye mold allows for a more accurate simulation of *in vivo* circumstances [70]. In a study, an ocular prosthesis was made with the use of CAD software and fast manufacturing equipment, and it was compared to a traditional custom-made prosthesis in terms of fabrication time, weight, cosmesis, comfort, and motility (CMP). The production of the CAD prosthesis took an average of 2.5 hours, and it weighed 2.9 grams. CMP, on the other hand, needed an average of 10 hours of production time and weighed 4.4 grams. Patients also reported that the CAD prosthesis was more comfortable [73].

10.6.3 Rectal and Vaginal Routes

Drugs delivered vaginally or rectally might have local or systemic effects. Suppositories, pessaries, intrauterine devices, surgical meshes, vaginal stents, and dilators are delivery methods. 3D-printed, customized delivery devices improve efficacy, patient compliance, and surgical outcomes including hematoma [70]. 3D-printed intrauterine medicine delivery devices and implants. Custom-sized 3D-printed devices allow intrauterine delivery of local and systemic APIs. These devices will control API delivery. Dutcher *et al.* fused deposition modeled a T-shaped intrauterine device. Poly-caprolactone 3D printed devices released indomethacin quicker than extruded filament [74].

Similar researchers employed fused deposition modeling (FDM)-based 3D printing to produce intrauterine systems (IUS) and subcutaneous rods from ethylene vinyl acetate (EVA) (SR). Over 30 days, customized, 3D-printed T-shaped prototypes enhanced drug release by 30%. This technology uses 3D extrudable polymer ethylene vinyl acetate (EVA) to evaluate drug-loaded implanted devices [75]. Fu *et al.* [76] FDM 3D printed progesterone vaginal rings. The PLA and polycaprolactone filament was made using HME. Drug-loaded filament creates O, Y, and M vaginal rings. Due to its larger surface area and geometry, O released drugs better than Y and M. 3D printing creates customized contraception [77] Created non dissolving anti-inflammatory suppositories and pessaries using FDM 3D printer molds on 3D castable resin. Elastic polymers make rectal and vaginal delivery pleasant. Another study [78] employed a cerclage pessary to prevent preterm deliveries. Another unique pessary employed biodegradable PCL and PLA. Hormone delivery discs, IUDs, subdermal rods, surgical mesh, and several pessaries were created. [79] created vaginal stents to deliver antibiotics or hormones, control bleeding, ease pain, and prevent hematoma following surgery in a study on children and adolescents. Laarhoven *et al.* [80] created a vaginal ring contraceptive from polyethylene vinyl acetate copolymers, etonogestrel, and ethinyl estradiol via melt extrusion. HME molds thermoplastics. Intravaginal rings are HME implants. Water-insoluble polymers release ethinylestradiol slowly, improving patient compliance. Polyurethanes are versatile polymers that perform well for implant extrusion. Kavita *et al.* observed that dapivirine supplied consistently for 30 days in polyurethane intravaginal rings. Polyurethane's unusual form caused a zero-order release of dapivirine from the ring.

10.6.4 Pulmonary Drug Delivery

3D printing medical equipment and models is used to treat respiratory diseases. 3D-printed lungs may help doctors diagnose and treat respiratory organ diseases. These methods will aid inhalable drug research via 3D printing.

Morrison *et al.* demonstrated the usefulness of 3D printing in pulmonary medicine by constructing bioresorbable airway splints for pediatric tracheobronchomalacia (TBM). Thanks to tailored devices for life-threatening conditions, 3D printed airway splints may prevent airway collapse and provide a root map [81]. Zopf *et al.* [82] developed a 3D-printed tracheobronchial bioresorbable airway splint that reduces airway collapse in pig models of tracheobronchomalacia and increases survival.

3D-printed medical equipment has helped asthma and other respiratory disorders. 3D-printed "sneezometers" measure airflow and sneeze speed [83]. This spirometer is faster and more precise. This gadget uses 3D printing to assess airflow rate in chronic and other breathing patients and generates personalized inhalers for better therapy. A 3D-printed ergonomic inhaler was created to encourage asthmatics to utilize it [84]. This 3D-printed device makes any asthma inhaler more effective, convenient, and user-friendly. Quinone's team 3D printed a lung tumor motion simulator to better comprehend lung cancer treatment. This device detected tumor movement while breathing and enabled lung cancer radiation therapy [85].

10.7 Custom Design 3D Printed Pharmaceuticals

Since the United States launched its precision medicine program in 2015 [86–89], there has been a rise in interest in the customization of treatment interventions. Individual variations exist, such as genetics, physical appearance, living environment, and eating habits [90]. Distinct people have different pharmacological needs, even for the same condition. Most people get the intended effect based on the formulation used, however traditional pharmaceutical techniques place stringent limitations on the qualities of the medicine, including its size, shape, and method of release. This may lead to a typical issue with medication usage, in which the tablets are successful in some people but not in others. Because every patient's illness, genetic makeup, and physical make-up are different, there is no universal set of guidelines that can be used for treatment [91–93]. Most drugs, for instance, are meant for adults, so doctors often have to split pills in half before giving them to kids. Production of medicines using 3D printing is a state-of-the-art method [94, 95].

3D printing technology has a promising future in personalized medicine [96] since traditional manufacturing methods are inadequate for making individualized drugs and are unable to make specialized dosage forms with modified release profiles and highly intricate geometries. Due to the adaptability of 3D printing, it is now possible to tailor the dosage regimen to the specific needs of each individual patient by adjusting factors such as pill size and fill percentage [97, 98]. If doctors have access to low-cost desktop 3D printers in the future, they may be able to print out medicines on the spot in the exact quantity needed by each patient. In humans, certain investigations have shown that 3D-printed tablets are feasible. Excipients for 3D printing are chosen based on the technology employed, the process's unique needs, and the dose form. Flexible medicine delivery strategies are particularly significant for the pediatric population because of the wide range of dosing needs. Furthermore, the technology permits a wider range of dosage form shapes, which is very helpful for those with swallowing difficulties. As 3D printing puts manufacturing closer to the patient, dosing strategies may be modified and therapeutic options expanded [23].

10.8 Excipient Selection for 3D Printing Custom Designs

FFF extrusion printing, also known as fused deposition modeling, is common in 3D printing today (FDM). FFF printers employ solid polymeric filaments instead of extrusion. Rapid-acting acetaminophen contains methylene blue, colloidal silicon dioxide, and polyvinylpyrrolidone K30 (PVP K30). PVP's quick FDT wetting and disintegration and great wettability/water-absorbability make it a desirable prototyping material. PVP, mannitol, lactose, and distilled water (cross-linked polyvinylpyrrolidone). Methylene blue (just a color print marker for easy observation) (0.5 percent w/v) and PVP K30 (5.0 percent w/v) were mixed in a 75 percent ethanol-water solution to form the binder liquid, while the powder combination consisted of the following materials and their weights: 40:20:9. 5:30:0.5. Scientists are advancing polymer science to synthesize novel polymers and change the physicochemical properties of natural and manmade polymers for new DDS [99].

Researchers developed printable dosage forms for poorly soluble drugs using piroxicam, polyethylene glycol 400 (PEG), ethanol, acetone, 2-propanol, glycerol, propylene glycol (PG), and water. Edible icing sheets and papers were printed on. Edible paper typically contains corn starch, glucose, cellulose, glycerine, sugar, vegetable oil, gum Arabica, polysorbate 80, vanilla (vanillin, piperonal), titanium dioxide, and citric acid. Inkjet printing utilized 30:70, 40:60, and 50:50 PEG:ethanol formulations, whereas flexography used 100% PEG-400. To adjust solvent viscosity, cellulose derivatives, PVA, PEG, or PVP may be added. Surfactants in the aqueous API solution may lower surface tension and make the ink suitable for printing. Nozzle clogging induced by water evaporation in aqueous formulations may be mitigated with the use of moisturizing additives like PEG, PG, or glycerol. These solvents have the potential to function as viscosity modifiers, which might limit the loss of solvent during the drying process of alcohol-based formulations. Moisturizing agents, on the other hand, are not required in flexographic printing since wide-aperture cells are employed. If the drug's solubility in any acceptable solvent is a concern, suspension formulation should be investigated. Inkjet printers may be used to print nanosuspensions. To minimize aggregation and precipitation of the API nanoparticles, a tight control over particle size is required. As a result, selecting suspension stabilizing agents wisth care is critical. For instance, inkjet printable formulations may consist mostly of water, ethanol, dimethylacetamide, and isopropanol as the solvent. Viscosity may be adjusted by additives including PEG, PG, glycerol, PVP, PLGA, cellulose derivatives, and PVA. Surfactants, stabilizers, and solubilizes like Tween 20 and PVP are correct. PEG 8000, glycerol, dimethylacetamide, and PG are all emollients. Water, ethanol, PEG 400, and PG are the main solvents in the flexographically printed composition. Several different substances may be used to alter a solution's viscosity, including polyethylene glycol, polyglycerol, glycerol, cellulose derivatives, polyvinylpyrrolidone, and starch. It is not necessary to use moisturizing or surfactant chemicals here [100].

10.9 Customized Medicating of Drugs

The theophylline tablet created by Pietrzak *et al.* has a specific drug loading capacity and an easy-to-swallow tablet shape, making it suitable for both immediate and extended-release. They utilized a technique called Fused Deposition Modeling to print the parts. Cost-effectiveness, digital controllability, and compatibility with polymers such as hydroxypropyl cellulose, Eudragit RS, RL, and We were all shown by the method. They attained a % dose precision with a suitable *in-vitro* release rate and weight accuracy. Because of the printer's highly flexible nature, as well as its compact size and simplicity of use, the approach provides the possibility for future customized treatment [101].

Khaled *et al.* [102] 3D extruded a polypill with five functions. The tablet was modeled in 3D. Rapid release (RR) and slow release (SR) are its stages (SR). Aspirin and hydrochlorothiazide were found in the IR layer, whereas pravastatin, atenolol, and ramipril were separated into three SR chambers for release. For individuals with hypertension, diabetes, or chronic renal failure, the polypill simplifies multiple drug administration. Tagami and colleagues 3D printed polyvinyl alcohol curcumin tablets and examined the effects of various printing conditions. Polymer flow rate and printing temperature impact tablet color and curcumin content [103]. Clark *et al.* used piezo-activated inkjet to make 3D ropinirole hydrochloride tablets. This is the first published study on tablet-based 3D printing using a UV inkjet printer. Low-oxygen photoinitiated cross-linked polyethylene glycol diacrylate hydrogel matrix encapsulated the medication [104]. Gioumouxouzis *et al.* [105] combined metformin and glimepirea in a double-layer antidiabetes dosage form encapsulated in Eudragit®RL sustained release and PVA release layers. Their findings propose 3D printing solid combination drug dose formulations. This method may offer an accurate, personalized dose of many medications in a single tablet, improving patient compliance for chronic illnesses. All of the research above suggests they can be mass-produced and commercialized. Clinical studies are employing 3D-printed drug dosages from many pharmaceutical companies. Thus, this approach has decreased the distance between research and clinical usage of customized drugs, which seems promising. Wang and colleagues used stereolithographic 3D printing to make and test drug-loaded tablets with different release profiles. They created tablets with an expected loading and extended drug release profile using 4-aminosalicylic acid and paracetamol as model medicines [106]. The Tan *et al.* capsule has an open-side coating, a drug-containing surface-eroding matrix, and a drug-free matrix [107]. They assembled the three 3D-printed mold components. The exposed side destroyed the medication layer by layer, and the release curve followed the deteriorating matrix. They could potentially create several release patterns using surface-eroding matrices. Katstra and colleagues 3D-printed chlorpheniramine maleate delayed-release pills. 3D printing using fluorescein as a medication delivery agent produced dosage forms with drug concentrations as low as 10 moles per tablet. Friability and hardness were typical of commercial medications [108].

Liang *et al.* 3D-printed variable-release oral medication delivery devices disguised as dental braces [109]. Field testing showed that 3D-printed medicine delivery systems may be used in people. Yu 3D printed custom-packaged acetaminophen tablets with zero-order medication release. Matrix tablets had drug-free release-barrier layers on the base and radial material gradients. Drug dissolution exhibited a two-dimensional surface erosion process. Printing release retardation agents onto tablets allowed them to quickly make tablets with large doses and desired drug release [110]. To develop controlled-release tablets of extended-release prednisolone, Skowyra and colleagues explored the possibility of using fused deposition modeling in conjunction with a 3D printer. By controlling the volume of the design, they were able to adjust the mass of the printed tablet, establishing a close correlation between the desired and realized doses. The produced tablets were of a regular elliptical form. The study [111] suggests that 3D printing using fused deposition modeling might be utilized to create and regulate the dose of extended-release tablets, providing a highly flexible, economically viable, relatively tiny, and digitally controlled platform for making patient-tailored pharmaceuticals. Nayan *et al.* [112] hot-melt extruded haloperidol tablets and tested several amorphous polymers for prolonged drug release. Okwuosa *et al.* created crystalline dipyridamole and theophylline tablets with high talc fillings. They used 3D printing and hot-melt extrusion with thermostable filler talc to make low-temperature immediate-release tablets. The pills are quick-release, high-quality, and low-variability [113].

10.10 Devices for Personalized Topical Treatment

3D printing has created fascinating new drug delivery methods. Gastro retentive floating pulsatile drug delivery, transdermal drug delivery, micro-porous bioceramics, a variety of tablets, a microfluidic pump, and more are recent advances. 3D printing technology has demonstrated promising results in personalized pharmaceutical distribution, in addition to its widespread usage in formulation research. Personalized pharmaceutical dosage, complicated drug-release profiles, and tailored topical therapeutic devices provide patient-centric design (custom manufacturing), real-time dose form analysis, enhanced logistics, and reduced waste. Technology has transformed the pharmaceutical distribution system, making it possible to meet the demands of the old, young, and visually impaired [114, 115].

Since the FDA approved the first 3D-printed pill, Spritam, in August 2015, pharmaceutical companies and universities are increasingly interested in using 3D printing to make prescription drugs. 3D printing might build medicine serves with complex architectural designs and personalized drugs in new locations. 3DP's widespread use in the pharmaceutical industry is due to its potential benefits, including increased productivity, complicated drug release patterns, multiple dosing, a low-cost single-step method, and drug delivery customization/personalization. This cutting-edge technology helps customize drug administration and fill prescriptions appropriately. 3D printing has a once-in-a-lifetime chance to help cure many diseases with individualized treatment.

10.10.1 Oral Solid Dosage Forms

Using a 3D printing technique, Yu and colleagues created multi-layered medication delivery devices in the form of a doughnut containing acetaminophen. Furthermore, the dose may be changed. As a result, this research found that the 3D approach may be utilized to construct drug delivery devices with complicated and desirable drug release characteristics [116]. Using CAD models, Yu and coworkers developed a drug delivery system with novel internal structural features that disintegrates rapidly. There was a 98.5 % discharge within 2 minutes [117], and the products were determined to have high pharmaco-technical properties and toughness. Bone infections caused by Staphylococcus aureus implants may be treated with rifampin and vancomycin-loaded calcium phosphate scaffolds, which Inzana *et al.* Results showed that medication greatly reduced the metabolic burden of bacteria [118].

Giri *et al.* employed HME to manufacture drug-loaded filaments for 3D printing gastro-retentive floating tablets (GRFT). GRFT's capacity to remain beyond the stomach's typical content may allow it to avoid oral dosage form issues such unexpected gastrointestinal transit and emptying periods and metabolic degradation. Theophylline-loaded filaments were made using HME and HPC. Infill density and shell thickness varied. Medication content, dissolving rates, physicochemical properties, and floating behavior of the tablets were measured. Zero-order release stabilized FDM-printed GRFTs for 10 hours. 3DP pills release more accurately than medication tablets. HME and 3D printing can develop excellent oral dosage forms of GRFTs, enabling adaptability and efficiency above traditional approaches [119].

10.10.2 Semisolid Extrusion (EXT) and Inkjet Printing

Patient-specific doses of the anticoagulant medicine warfarin were created using semisolid extrusion 3D printing (EXT) and inkjet printing, both of which were assessed by Blom *et al.* at the Helsinki University Hospital (HUS) Pharmacy, the largest hospital pharmacy in Finland (IJP). Oral powders in unit dose sachets weren't as consistent in quality as the printed orodispersible films' (ODFs), which were thicker and more malleable (OPSs). After being tested for stability in a month's time, the OPSs and ODFs were deemed safe for nasogastric administration. To provide more details on the dosage form and cut down on medication administration errors, a Quick Response (QR) code was printed onto the ODFs using inkjet printing technology. The study confirmed that printing technologies are practical resources for developing customized dosage forms for individual patients [120].

Curcumin, cisplatin, and 5-fluorouracil are just a few examples of the anti-cancer agents that may be coated onto arrays of metal microneedles using a novel inkjet printing process for transdermal drug administration [121].

To produce paclitaxel microparticles with controlled morphologies, Lee and colleagues used a piezoelectric inkjet printing method. It was found that production using piezoelectric techniques was reliable, consistent, and well-suited to mass production. It was shown that the release rate was related to the shape and surface area of the produced microparticles [122], with an initial burst followed by a steady release. Calcium sulphate powders were developed by Zhou and colleagues for use in tissue generated bone scaffolds. Excellent printing accuracy and sufficient compressive strength were achieved in both solid and porous scaffolds [123].

10.10.3 Stencil Printing

As an innovative printing process, Wickstrom *et al.* propose its use in the pharmaceutical business. Haloperidol (HAL) discs were created using a prototype stencil printer and polyester as the stencil medium. The aperture area and the height of the stencil determined the dosage. The mass and content homogeneity parameters were reached in the development of the therapeutic HAL for use in treating children aged 6 to 17 years. The HAL dose was achieved by combining 1% lactic acid with 16% hydroxypropyl methylcellulose (HPMC). After printing, the medicine maintained its amorphous structure and a pH greater than 4. Disintegration experiments showed that the orodispersible discs produced had a half-life of less than 30 seconds. The results of these analyses led to the conclusion that the novel batchwise stencil printing technology might be used as a practical strategy for pharmaceutical manufacture [124].

10.10.4 Implants

Implants made of non-biodegradable polymers must be surgically removed after usage. 3DP creates patient-specific dose forms using biodegradable polymers that break down into readily excreted compounds. HME filaments were used to 3D-print hollow implants based on commercial implants. Hollow implants with PVA windows were made using PLA and PVA filaments. Models' makeup compressed the implant powder. *In vitro* research showed that implant dose forms were flexible based on materials and design. Stewart *et al.* used coatings to lengthen implant release profiles in this study. Drug delivery devices may be manufactured on demand and applied quickly in a clinical context [125].

3D printing has advanced medical implants using a number of methods and materials. Implants are surgically transplanted into the body to improve, help, or maintain function or enhance/alter shape [125, 126]. Implants might be synthetic like hip prostheses or soft like blood vascular grafts [127]. 3D printing may also create drug-loaded devices for each patient. Goyanes *et al.* developed a nose-shaped salicylic acid mask for acne vulgaris. The patient's nose was 3D scanned. Stereo lithography printing outperformed fused deposition modeling printing in resolution, drug loading, drug release, and degradation. This research found that 3D printing can create patient-specific medicine delivery devices of various forms and sizes [128]. The use of 3D printing has opened the door to the development of implantable medication delivery systems with tunable drug release profiles. A bone infection, for instance, needs antibiotics to be given directly to the infected region as soon as possible. Treatment for osteoarticular TB also includes closing the surgical defect [129]. With the use of 3D printing technology, Zhu *et al.* developed an implant that combines the positive aspects of osseous regeneration with a local combination therapy to lessen the likelihood of drug resistance and adverse consequences. The 3D-printed scaffold comprised a macro/mesoporous composite filled with potent antifungal drugs like isoniazid and rifampin. Since the drug release from the composite scaffolds was more sustained, the necessary drug level was maintained at the defect's peripheral tissues while only a very little amount entered the bloodstream [130].

Huang *et al.*'s levofloxacin-releasing implants. Using a predefined release profile, implants released levofloxacin pulsatilely and steadily. 3D printing might be utilized to manufacture pharmaceutical implants with intricate micro- and macro-architecture in a single device, which would speed up development and manufacturing [131]. Wu and colleagues showed that after 30 days *in vitro*, their 3D-printed slow-release isoniazid implant had a drug concentration greater than the effective bacteriostatic concentration. The implant supplied antibiotics, validating the approach for complicated implants [132].

Long-term pharmacological patients may benefit from implants with effective medications in a sustained release matrix. An embed, a dosage structure with dynamic pharmaceuticals in a supported supply conveyance grid, may benefit long-term pharmaceutical patients. Levofloxacin micro-structured embeds have more complicated dispersion patterns. This embed revealed a bimodal, pulsing profile. (5–25 days), steady-state medication release (25–50 days), and return to beat discharge (50–80 days) [133]. Wu *et al.* created a multi-drug embed for 3D printing bone TB therapy. Isoniazid and rifampicin, two TB drugs, were fused into each layer to form a cluster, creating a multilayered concentric chamber [134]. Wu *et al.* [135] developed a 3DP-based multi-drug implant for chronic osteomyelitis using tobramycin (TOB) and levofloxacin (LVFX) with multilayered scaffolds (layers 0.4 cm3). This study on polycaprolactone (PCL)-based embed coatings aims to delay hydrophilic mixture release from implanted devices created utilizing added substance manufacturing (AM) [136]. Chaudhari *et al.* used FDM-based HME to construct skin patches with quercetin-PVP to address medication delivery difficulties [137]. These research demonstrate that 3D printing can develop novel dosage forms such microparticles, implants, solid dispersions, nanocapsules, patches, and more. Anticancer patches, implants, and other devices have been used in many successful trials.

10.10.5 Tissue Engineering

Injury, inflammation, and trauma may cause organ and joint dysfunction [138, 139]. Invasive surgery is needed to heal injured tissues, organs, and joints [140, 141]. Transplanting healthy organs or joints to replace damaged ones causes several medical complications. Scientists and doctors developed tissue engineering (TE) procedures to repair damaged tissues and organs. Tissue engineering uses bioactive scaffolds and human cells to rebuild damaged or diseased tissues in 3D [142–145]. B-TE scaffolds should mimic the damaged tissue's geometrical shape, have a tunable pore size 100 to 600 mm 3D porous structure with a controlled architecture, degrade in a way that allows cells to migrate and grow, and have sufficient mechanical strength properties to prevent from any deformation and preserve structure integrity [146]. B-TE uses bioactive material scaffolds with precise mechanical alterations to insert living cells into artificial organs or tissues. The scaffold supports the replacement tissue before cells form an extracellular lattice. Bone rejuvenation—dividing, proliferating, differentiating, and creating new tissues—can repair damaged organs.

10.10.6 Regenerative Medicine

Regenerative medicine studies and develops methods for tissue and organ repair and regeneration. The ideal scaffold was biocompatible, biodegradable, biomimetic, three-dimensionally porous, and biomechanically compatible with the host tissue. The scaffolds assist produce extracellular matrix (ECM), although they are only meant for short-term usage until updated. Metals, ceramics, polymers, hydrogels, and composites are employed primarily for bone repair and regeneration [147]. Chou *et al.* (2013) [148] printed biodegradable iron-manganese scaffolds using ink-jet 3DP for craniofacial applications. Titanium silica composite scaffolds were made using 3DP and SLS [149]. Ceramic composites, amorphous glasses, and crystalline ceramics are potential materials for B-TE applications [150]. HA and "betatri-calcium phosphate (b -TCP)" bioceramic were used to make porous bone-healing scaffolds [151]. Indirect 3DP and freeze-drying created collagen-based open porosity scaffolds with interconnected channels. Maxillofacial surgery encompasses dental implant, mandibular, orthognathic, and mid-face reconstruction [152, 153].

10.10.7 Scaffoldings

Researchers have been paying close attention to skin therapy and rejuvenation using 3DP methods. Laser-based cell printing technique was used to print skin tissue [154]. Laser-assisted bio-printing technique was used to manufacture multicellular skin grafts consisting of keratinocytes, fibroblasts, and collagen [155]. The researchers also engineered a skin with a bilayer structure, consisting of 20 layers of collagen transporting fibroblasts and 20 layers of collagen carrying keratinocytes. *In-vivo*, the bilayer skin was applied to full-thickness lesions in nude mice's dorsal skin chamber [156]. After 11 days, the skin was discovered to be totally linked to the mice and functioning normally. The use of 3D bio-printing was recently utilized to generate bilayer functional skin that resembles human skin [157].

10.10.8 Organ Printing

Natural-based materials (bio-inks, hydrogels, and hence mixes) have seen extensive use for 3D printing of organs. In addition, bio-printing may sustain the viability of cells by feeding them from the matrix in which they are embedded [158, 159]. A 3D bioprinting process employing hybrid hydrogel was used to create and build the Trileaflet aortic valve. After seven days of *in vivo* growing settings, the valve was sustained, and highly visible cells with a high ability for remodeling were identified [160]. The anatomy of the liver is complicated, with eight separate parts. Organovo™ was the first to use the 3DP technology to create a vascularized liver. High-density hepatocytes, endothelial cells, and hepatic stellate cells were used by Nguyen *et al.* (2015) [161] in a 3D printing process to create a structure similar to that of liver lobules in a healthy liver. The designed 3D printed model of the liver allows for precise tumor site identification and planning, as well as expertise with laparoscopy or hepatic resections.

10.11 Conclusion

4D pharmaceutical printing may happen sooner than expected. Smart pharmaceuticals might revolutionize medication delivery and healthcare by targeting and personalizing therapy for each patient. This strategy may be implemented due to rising interest in and non-expert use of pharmacological delivery systems. However, the Food and Drug Administration (FDA) encourages complicated dosage forms and risk-based manufacturing procedures, which may speed up healthcare's adoption of such breakthrough technology. 3D printing's major obstacles are technical and quality assurance issues. Once an optimum printing platform is built, 3D printers are expected to swiftly replace existing pharmaceutical distribution routes, ushering in a new digital health age.

It's clear that 3D printing has improved healthcare and medicine. New 3D printing methods are replacing old production methods because they provide better standardization, dimensional precision, and patient-specific devices. However, printing with hard or viscous feedstock may need setting adjustments or component replacement. Commercialization of the highlighted inventions is unknown. Healthcare devices that haven't been FDA-approved may only be used for study.

3DP might be used to add pharmaceuticals to FDM-created injectable devices or mesh implants using HME, or to customize a 3D-printed tablet's release profile and dose in CAD. The end products must be stable and reproducible to combine 3D printing results. It's hard to predict how far 3D printing can go in medicine. Although medical, biological, and healthcare 3D printing applications and products have advanced, additional research is required to assure their efficacy, efficiency, and environmental friendliness. Due to their financial support, connected corporations should promote scholarly advances in this domain.

References

1. NHS, *Improving outcomes through personalised medicine*, National Health Service (NHS), England, 2016, 2019 Aug 15, Available from: https://www.england.nhs.uk/wp-content/uploads/2016/09/improving-outcomes-personalised-medicine.pdf.
2. Alomari, M., Mohamed, F.H., Basit, A.W., Gaisford, S., Personalised dosing: Printing a dose of one's own medicine. *Int. J. Pharm.*, 494, 2, 568–77, 2015.
3. Hamburg, M.A. and Collins, F.S., The path to personalized medicine. *N. Engl. J. Med.*, 363, 4, 301–4, 2010.
4. Arain, F.A., Kuniyoshi, F.H., Abdalrhim, A.D., Miller, V.M., Sex/gender medicine. The biological basis for personalized care in cardiovascular medicine. *Circ. J.*, 73, 10, 1774–82, 2009.
5. Lim, S.H., Kathuria, H., Tan, J.J.Y., Kang, L., 3D printed drug delivery and testing systems — A passing fad or the future? *Adv. Drug Deliv. Rev.*, 132, 139–168, 2018.
6. Ursan, I.D., Chiu, L., Pierce, A., Three-dimensional drug printing: A structured review. *J. Am. Pharm. Assoc.*, 53, 2, 136–44, 2013.
7. Cui, X., Boland, T., D'Lima, D.D., Lotz, M.K., Thermal inkjet printing in tissue engineering and regenerative medicine. *Recent Pat. Drug Deliv. Formul.*, 6, 2, 149–55, 2012.
8. Schubert, C., van Langeveld, M.C., Donoso, L.A., Innovations in 3D printing: A 3D overview from optics to organs. *Br. J. Ophthalmol.*, 98, 2, 159–61, 2014.
9. Banks, J., Adding value in additive manufacturing: Researchers in the United Kingdom and Europe look to 3D printing for customization. *IEEE Pulse*, 4, 6, 22–6, 2013.

10. Muwaffak, Z., Goyanes, A., Clark, V. *et al.*, Patient-specific 3D scanned and 3D printed antimicrobial polycaprolactone wound dressings. *Int. J. Pharm.*, 527, 1, 161–170, 2017.
11. Boetker, J., Water, J.J., Aho, J. *et al.*, Modifying release characteristics from 3D printed drug-eluting products. *Eur. J. Pharm. Sci.*, 90, 47–52, 2016.
12. Trenfield, S.J., Goyanes, A., Telford, R. *et al.*, 3D printed drug products: Non-destructive dose verification using a rapid point-and-shoot approach. *Int. J. Pharm.*, 549, 1, 283–292, 2018.
13. Khaled, S.A., Alexander, M.R., Wildman, R.D. *et al.*, 3D extrusion printing of high drug loading immediate release paracetamol tablets. *Int. J. Pharm.*, 538, 1, 223–230, 2018.
14. Genina, N., Boetker, J.P., Colombo, S. *et al.*, Anti-tuberculosis drug combination for controlled oral delivery using 3D printed compartmental dosage forms: From drug product design to *in vivo* testing. *J. Control Release*, 268, 40–48, 2017.
15. Awad, A., Trenfield, S.J., Gaisford, S. *et al.*, 3D printed medicines: A new branch of digital healthcare. *Int. J. Pharm.*, 548, 1, 586–596, 2018.
16. Vuddanda, P.R., Alomari, M., Dodoo, C.C. *et al.*, Personalisation of warfarin therapy using thermal ink-jet printing. *Eur. J. Pharm. Sci.*, 117, 80–87, 2018.
17. Awad, A., Trenfield, S.J., Goyanes, A. *et al.*, Reshaping drug development using 3D printing. *Drug Discov. Today*, 23, 8, 1547–1555, 2018.
18. Khatri, P., Shah, M.K., Vora, N., Formulation strategies for solid oral dosage form using 3D printing technology: A mini-review. *J. Drug Deliv. Sci. Technol.*, 46, May, 148–55, 2018.
19. Aguilar-de-Leyva, A., Linares, V., Casas, M., Caraballo, I., 3D printed drug delivery systems based on natural products. *Pharmaceutics*, 12, 7, 620, 2020.
20. Durfee, W.K. and Iaizzo, P.A., Medical applications of 3D printing, in: *Engineering in Medicine*, pp. 527–543, Academic Press, 2019.
21. Hoang, D., Perrault, D., Stevanovic, M., Ghiassi, A., Surgical applications of three-dimensional printing: A review of the current literature & how to get started. *Ann. Transl. Med.*, 4, 1–19, 2016.
22. González-Henríquez, C.M., Sarabia-Vallejos, M.A., Rodriguez-Hernandez, J., Polymers for additive manufacturing and 4D-printing: Materials, methodologies, and biomedical applications. *Prog. Polym. Sci.*, 94, 57–116, 2019.
23. Sinha, S.K., Additive manufacturing (AM) of medical devices and scaffolds for tissue engineering based on 3D and 4D printing, in: *3D and 4D Printing of Polymer Nanocomposite Materials*, pp. 119-160. Elsevier, 2020.
24. Momeni, F., Mehdi, S.M., Hassani, N., Liu, X., Ni, J., A review of 4D printing. *Mater. Des.*, 122, 42–79, 2017.
25. Lui, Y.S., Sow, W.T., Tan, L.P., Wu, Y., Lai, Y., Li, H., 4D printing and stimuli-responsive materials in biomedical application. *Acta Biomater.*, 92, 19–36, 2019.
26. Miao, S., Castro, N., Nowicki, M., Xia, L., Cui, H., Zhou, X., Zhu, W., Jun Lee, S., Sarkar, K., Vozzi, G., Tabata, Y., Fisher, J., Zhang, L.G., 4D printing of polymeric materials for tissue and organ regeneration. *Mater. Today*, 20, 577–591, 2017.

27. Le Fer, G. and Becker, M.L., 4D printing of resorbable complex shape memory poly (propylene fumarate) star scaffolds. *ACS Appl. Mater. Interfaces*, 12, 22444–22452, 2020.
28. Grinberg, D., Siddique, S., Le, M.Q., Liang, R., Capsal, J.F., Cottinet, P.J., 4D printing based piezoelectric composite for medical applications. *J. Polym. Sci. Part B Polym. Phys.*, 57, 109–115, 2019.
29. Gao, B., Yang, Q., Zhao, X., Jin, G., Ma, Y., Xu, F., 4D bioprinting for biomedical applications. *Trends Biotechnol.*, 34, 746–756, 2016.
30. Lee, J.Y., An, J., Chua, C.K., Fundamentals and applications of 3D printing for novel materials. *Appl. Mater. Today*, 7, 120–133, 2017.
31. He, Y., Yang, F., Zhao, H., Gao, Q., Xia, B., Fu, J., Research on the printability of hydrogels in 3D bioprinting. *Sci. Rep.*, 6, 1–13, 2016.
32. Helguero, C.G., Mustahsan, V.M., Parmar, S. *et al.*, Biomechanical properties of 3D-printed bone scaffolds are improved by treatment by CRFP. *J. Orthop. Surg. Res.*, 12, 1, 195, 2017.
33. Tan, L., Zhu, W., Zhou, K., Recent progress on polymer materials for additive manufacturing. *Adv. Funct. Mater.*, 30, 43, Article 2003062, 2020, 10.1002/adfm.202003062.
34. Asadi-Edyvand, M., Solati-Hasjin, M., Farzad, A., Osman, N.A.A., Effect of technical parameters on porous structure and strength of 3D printed calcium sulfate prototypes. *Robot. Comput. Integr. Manuf.*, 37, 57–67, 2015.
35. Garcia, J., Yang, Z., Mongrain, R., Leask, R.L., Lachapelle, K., 3D printing materials and their use in medical education: A review of current technology and trends for the future. *BMJ Simul. Technol. Enhanc. Learn.*, 4, 1, 27–40, 2017.
36. Mannoor, M.S., Jiang, Z., James, T. *et al.*, 3D printed bionic ears. *Nano Lett.*, 13, 6, 2634–2639, 2013.
37. Vukievic, M., Mosadegh, B., Little, J.K., Little, S.H., Cardiac 3D printing and its future directions. *JACC: Cardiovasc. Imaging*, 10, 2, 171–184, 2017.
38. Soon, D.S.C., Chae, M.P., Pilgrim, C.H.C., Rozen, W.M., Spychal, R.T., Hunter-Smith, D.J., 3D haptic modelling for preoperative planning of hepatic resection: A systematic review. *Ann. Med. Surg.*, 10, 1–7, 2016.
39. Abudayyeh, I., Gordon, B., Ansari, M.M., Jutzy, K., Stoletniy, L., Hilliard, A., A practical guide to cardiovascular 3D printing in clinical practice: Overview and examples. *J. Interv. Cardiol.*, 31, 3, 375–383, 2017.
40. Cantinotti, M., Valverde, I., Kutty, S., Eckert, J., Three-dimensional printed models in congenital heart disease. *Int. J. Cardiovasc. Imaging*, 33, 1, 137–144, 2016.
41. Stratasys, Case study, *Dante pazzanese cardiology institute uses 3D printed arteries to diagnose and treat diseases.* https://www.stratasys.com/fr/resources/case-studies/dante-pazzanese/.
42. Stratasys, Case study, *Shaping young hearts. Stratasys helps a surgeon save more kids.* https://www.stratasys.co.in/resources/case-studies/nicklaus/.
43. Zhai, Y., Galarraga, H., Lados, D.A., Microstructure evolution, tensile properties, and fatiguedamage mechanisms in Ti-6Al-4V alloys fabricated by two additive manufacturing techniques. *Proc. Eng.*, 114, 658–666, 2015.

44. Zocca, A., Colombo, P., Gomes, C.M., Günster, J., Additive manufacturing of ceramics: Issues, potentialities, and opportunities. *J. Am. Ceram. Soc.*, 98, 7, 1983–2001, 2015.
45. Gul, H., Khan, M., Khan, A.S., Bioceramics: Types and clinical applications, in: *Handbook of Ionic Substituted Hydroxyapatites*, pp. 53–83, Woodhead Publishing, 2020.
46. Tarafder, S., Davies, N.M., Bandyopadhyay, A., Bose, S., 3D printed tricalcium phosphate scaffolds: Effect of SrO and MgO doping on *in vivo* osteogenesis in a rat distalfemoral defect model. *Biomater. Sci.*, 1, 12, 1250–1259, 2013.
47. Gbureck, U., Vorndran, E., Müller, F.A., Barralet, J.E., Low temperature direct 3D printed bioceramicsand biocomposites as drug release matrices. *J. Controlled Release*, 122, 2, 173–180, 2007.
48. Tuan, T.N.A., Abdullah, A.M., Md, A.H., Mohamad, D., Rajion, Z.A., Preparation and characterization of a newly developed polyamide composite utilising an affordable 3D printer. *J. Reinf. Plast. Compos.*, 34, 19, 1628–1638, 2015.
49. Greco, A., Licciulli, A., Maffezzolim, A., Stereolitography of ceramic suspensions. *J. Mater. Sci.*, 36, 1, 99–105, 2011.
50. Park, C.H., Rios, H.F., Jin, Q., Sugai, J.V., Padial-Molina, M., Taut, A.D., Giannobile, W.V., Tissue engineering bone-ligament complexes using fiber-guidingscaffolds. *Biomaterials*, 33, 1, 137–145, 2012.
51. Seliktar, D., Dikovsky, D., Napadensky, E., Bioprinting and tissue engineering: Recent advances and future perspectives. *Israel J. Chem.*, 53, 9-10, 795–804, 2013.
52. Travitzky, N., Bonet, A., Dermeik, B., Fey, T., Filbert-Demut, I., Schlier, L., Greil, P., Additive manufacturing of ceramic-based materials. *Adv. Eng. Mater.*, 16, 6, 729–754, 2004.
53. Germain, L., Fuentes, C.A., van Vuure, A.W., Des Rieux, A., Dupont-Gillain, C., 3D-printed biodegradable gyroid scaffolds for tissue engineering applications. *Mater. Des.*, 151, 113–122, 2018.
54. Griffith, L.G., Wu, B., Cima, M.J., Powers, M.J., Chaignaud, B., Vacanti, J.P., *In vitro* organogenesis of liver tissue. *Ann. N. Y. Acad. Sci.*, 831, 1, 382, 1997.
55. Bose, S., Vahabzadeh, S., Bandyopadhyay, A., Bone tissue engineering using 3D printing. *Mater. Today*, 16, 12, 496–504, 2013.
56. Pati, F., Shim, J.H., Lee, J.S., Cho, D.W., 3D printing of cell-laden constructs for heterogeneous tissue regeneration. *Manuf. Lett.*, 1, 1, 49–53, 2013.
57. Richards, D.J., Tan, Y., Jia, J., Yao, H., Mei, Y., 3D printing for tissue engineering. *Isr. J. Chem.*, 53, 805–814, 2013.
58. Salmoria, G.V., Klauss, P., Paggi, R.A., Kanis, L.A., Lago, A., Structure and mechanical properties of cellulose based scaffolds fabricated by selective laser sintering. *Polym. Test.*, 28, 648–652, 2009.
59. Tarafder, S., Dernell, W.S., Bandyopadhyay, A., Bose, S., SrO- and MgO-doped microwave sintered 3D printed tricalcium phosphate scaffolds: Mechanical properties and *in vivo* osteogenesis in a rabbit model. *J. Biomed. Mater. Res. Part B: Appl. Biomater.*, 103, 3, 679–690, 2013.

60. Gbureck, U., Hölzel, T., Klammert, U., Würzler, K., Müller, F.A., Barralet, J.E., Resorbable dicalcium phosphate bone substitutes prepared by 3D powder printing. *Adv. Funct. Mater.*, 17, 18, 3940–3945, 2007.
61. Bose, S., Vahabzadeh, S., Bandyopadhyay, A., Bone tissue engineering using 3D printing. *Mater. Today*, 16, 12, 496–504, 2013.
62. Pati, F., Shim, J.H., Lee, J.S., Cho, D.W., 3D printing of cell-laden constructs for heterogeneous tissue regeneration. *Manuf. Lett.*, 1, 1, 49–53, 2013.
63. Richards, D.J., Tan, Y., Jia, J., Yao, H., Mei, Y., 3D printing for tissue engineering. *Isr. J. Chem.*, 53, 805–814, 2013.
64. Salmoria, G.V., Klauss, P., Paggi, R.A., Kanis, L.A., Lago, A., Structure and mechanical properties of cellulose based scaffolds fabricated by selective laser sintering. *Polym. Test.*, 28, 6, 648–652, 2009.
65. Billiet, T., Gevaert, E., De Schryver, T., Cornelissen, M., Dubruel, P., The 3D printing of gelatin methacrylamide cell-laden tissue-engineered constructs with high cell viability. *Biomaterials*, 35, 1, 49–62, 2014.
66. Gmeiner, R., Deisinger, U., Schonherr, J., Lechner, B., Detsch, R., Boccaccini, A.R., Stampfl, J., Additive manufacturing of bioactive glasses and silicate bioceramics. *J. Ceram. Sci. Technol.*, 6, 75–86, 2015.
67. Vithani, K., Goyanes, A., Jannin, V., Basit, A.W., Gaisford, S., Boyd, B.J., An overview of 3D printing technologies for soft materials and potential opportunities for lipid-based drug delivery systems. *Pharm. Res.*, 36, 1–20, 2019.
68. Agrawal, A. and Gupta, A.K., 3d printing technology in pharmaceuticals and biomedical: A review. *J. Drug Deliv. Ther. Open*, 9, 1–4, 2019.
69. Liang, K., Brambilla, D., Leroux, J.C., Is 3D printing of pharmaceuticals a disruptor or enabler? *Adv. Mater.*, 31, 1–4, 2019.
70. Lim, S.H., Kathuria, H., Tan, J.J.Y., Kang, L., 3D printed drug delivery and testing systems — A passing fad or the future? *Adv. Drug Deliv. Rev.*, 132, 139–168, 2018.
71. Lang, B., McGinity, J.W., Williams RO, I.I.I., Hot-melt extrusion—Basic principles and pharmaceutical applications. *Drug Dev. Ind. Pharm.*, 40, 9, 1133–55, 2014.
72. Repka, M.A., Gerding, T.G., Repka, S.L., McGinity, J.W., Influence of plastcosers and drugs on the physico-mechanical properties of hydropxypropylcellulose films prepared by hot melt extrusion. *Drug Dev. Ind. Pharm.*, 25, 625-633, 625–33, 1999.
73. Alam, M.S., Sugavaneswaran, M., Arumaikkannu, G., Mukherjee, B., An innovative method of ocular prosthesis fabrication by bio-CAD and rapid 3-D printing technology: A pilot study. *Orbit*, 36, 4, 223–7, 2017.
74. Hollander, J. *et al.*, Three-dimensional printed PCL-based implantable prototypes of medical devices for controlled drug delivery. *J. Pharm. Sci.*, 105, 9, 2665–76, 2016.
75. Genina, N., Hollander, J., Jukarainen, H., Makila, E., Salonen, J., Sandler, N., Ethylene vinyl acetate (EVA) as a new drug carrier for 3D printed medical drug delivery devices. *Eur. J. Pharm. Sci.*, 90, 53–63, 2016.

76. Fu, J., Yu, X., Jin, Y., 3D printing of vaginal rings with personalized shapes for controlled release of progesterone. *Int. J. Pharm.*, 539, 1–2, 75–82, 2018.
77. Sun, Y., Ruan, X., Li, H., Kathuria, H., Du, G., Kang, L., Fabrication of non-dissolving analgesic suppositories using 3D printed moulds. *Int. J. Pharm.*, 513, 1-2, 717–24, 2016.
78. Tan, M., Lipman, S., Lee, H., Sie, L., Carvalho, B., Daniels, K., Impact of the electronic medical record on nurse's time allocation during cesarean delivery [25R]. *Obstetrics & Gynecology*, 127, p.154S, 2016.
79. Hakim, J., Dietrich, J.E., Smith, P.A., Buskmiller, C., Baylor College of Medicine, 2021. Vaginal stents, vaginal dilators, and methods of fabricating the same. U.S. Patent 10,898,699.
80. van Laarhoven, J.A.H., Kruft, M.A.B., Vromans, H., *In vitro* release properties of etonogestrel and ethinyl estradiol from a contraceptive vaginal ring. *Int. J. Pharm.*, 232, 1-2, 163–73, 2002.
81. Morrison, R.J. *et al.*, Mitigation of tracheobronchomalacia with 3D-printed personalized medical devices in pediatric patients. *Sci. Transl. Med.*, 7, 285, 285ra64–285ra64, 2015.
82. Zopf, D.A., Flanagan, C.L., Wheeler, M., Hollister, S.J., Green, G.E., Treatment of severe porcine tracheomalacia with a 3-dimensionally printed, bioresorbable, external airway splint. *JAMA Otolaryngol. Neck Surg.*, 140, 1, 66–71, 2014.
83. *The University of Surrey's 3D printed diagnostic tool is nothing to sneeze at*, https://3dprint.com/122419/3d-printed-sneezometer (accessed Oct. 30, 2021).
84. *3D printed ergonomic & friendly asthma inhaler*, https://hackaday.io/project/27002-3d-printed-ergonomic-friendly-asthma-inhaler, (accessed Oct. 30, 2021).
85. Quinones, D.R. *et al.*, Open source 3D printed lung tumor movement simulator for ~ radiotherapy quality assurance. *Mater. (Basel)*, 11, 8, 1317, 2018.
86. Collins, F.S. and Varmus, H., A new initiative on precision medicine. *N Engl. J. Med.*, 372, 9, 793–795, 2015.
87. Hussein, N., Voyer-Nguyen, P., Portnoy, S., Peel, B., Schrauben, E., Macgowan, C., Yoo, S.J., Simulation of semilunar valve function: Computer-aided design, 3D printing and flow assessment with MR. *3D Print. Med.*, 6, 1, 2, 2020.
88. Vidal, L., Kampleitner, C., Brennan, M.A., Hoornaert, A., Layrolle, P., Reconstruction of large skeletal defects: Current clinical therapeutic strategies and future directions using 3D printing. *Front. Bioeng. Biotechnol.*, 8, 61, 2020.
89. McCormack, A., Highley, C.B., Leslie, N.R., Melchels, F.P.W., 3D printing in suspension baths: Keeping the promises of bioprinting afloat. *Trends Biotechnol.*, 38, 6, 584–593, 2020.
90. Horwitz, R.I., Cullen, M.R., Abell, J., Christian, J.B., Medicine. (De)personalized medicine. *Science*, 339, 6124, 1155–1156, 2013.
91. Alomari, M., Mohamed, F.H., Basit, A.W., Gaisford, S., Personalised dosing: Printing a dose of one's own medicine. *Int. J. Pharm.*, 494, 2, 568–577, 2015.

92. Hamburg, M.A. and Collins, F.S., The path to personalized medicine. *N Engl. J. Med.*, 363, 4, 301–304, 2010.
93. Florence, A.T. and Lee, V.H., Personalised medicines: More tailored drugs, more tailored delivery. *Int. J. Pharm.*, 415, 1-2, 29–33, 2011.
94. Goole, J. and Amighi, K., 3D printing in pharmaceutics: A new tool for designin customized drug delivery systems. *Int. J. Pharm.*, 499, 1–2, 376–394, 2016.
95. Pillay, V. and Choonara, Y.E., 3D printing in drug delivery formulation: You can dream it, design it and print it. How about patent it? *Recent Pat. Drug Deliv. Formul.*, 9, 3, 192–193, 2015.
96. Preis, M. and Öblom, H., 3D-printed drugs for children-are we ready yet? *AAPS PharmSciTech*, 18, 2, 303–8, 2017.
97. Goyanes, A., Buanz, A.B., Basit, A.W., Gaisford, S., Fused-filament 3D printing (3DP) for fabrication of tablets. *Int. J. Pharm.*, 476, 1–2, 88–92, 2014.
98. Allahham, N., Fina, F., Marcuta, C., Kraschew, L., Mohr, W., Gaisford, S., Basit, A.W., Goyanes, A., Selective laser sintering 3D printing of orally disintegrating printlets containing ondansetron. *Pharmaceutics*, 12, 2, 110, 2020.
99. Yu, D.G., Branford-White, C., Yang, Y.C., Zhu, L.M., Welbeck, E.W., Yang, X.L., A novel fast disintegrating tablet fabricated by three dimensional printing. *Drug Dev. Ind. Pharm.*, 35, 12, 1530–6, 2009.
100. Raijada, D., Genina, N., Fors, D. *et al.*, A step toward development of printable dosage forms for poorly soluble drugs. *J. Pharm. Sci.*, 102, 10, 3694–704, 2013.
101. Pietrzak, K., Isreb, A., Alhnan, M.A., A flexible-dose dispenser for immediate and extended release 3D printed tablets. *Eur. J. Pharm. Biopharm.*, 96, 380–7, 2015.
102. Khaled, S.A., Burley, J.C., Alexander, M.R., Yang, J., Roberts, C.J., 3D printing of fivein-one dose combination polypill with defined immediate and sustained release profiles. *J. Control Release*, 217, 308–314, 2015.
103. Tagami, T., Fukushige, K., Ogawa, E., Hayashi, N., Ozeki, T., 3D printing factors importantfor the fabrication of polyvinylalcohol filament-based tablets. *Biol. Pharm. Bull.*, 40, 3, 357–64, 2017.
104. Clark, E.A., Alexander, M.R., Irvine, D.J., Roberts, C.J., Wallace, M.J., 3D printing of tablets using inkjet with UV photo initiation. *Int. J. Pharm.*, 529, 1–2, 523–30, 2017.
105. Gioumouxouzis, C.I., Baklavaridis, A., Katsamenis, O.L., Markopoulou, C.K., Bouropoulos, N., Tzetzis, D., Fatouros, D.G., A 3D printed bilayer oral solid dosage form combining metformin for prolonged and glimepiride for immediate drug delivery. *Eur. J. Pharm. Sci.*, 120, 40–52, 2018.
106 Wang, J., Goyanes, A., Gaisford, S., Basit, A.W., Stereolithographic (SLA) 3D printing of oral modified-release dosage forms. *Int. J. Pharm.*, 503, 1-2, 207–12, 2016.

107. Tan, Y.J.N., Yong, W.P., Kochhar, J.S., Khanolkar, J., Yao, X., Sun, Y., Ao, C.K., Soh, S., On-demand fully customizable drug tablets via 3D printing technology for personalized medicine. *J. Control Release*, 322, 42–52, 2020.
108. Katstra, W.E., Palazzolo, R.D., Rowe, C.W., Giritlioglu, B., Teung, P., Cima, M.J., Oral dosage forms fabricated by three dimensional printing. *J. Control Release*, 66, 1, 1–9, 2000.
109. Liang, K., Carmone, S., Brambilla, D., Leroux, J.C., 3D printing of a wearable personalized oral delivery device: A first-in-human study. *Sci. Adv.*, 4, 5, eaat2544, 2018.
110. Yu, D.G., Yang, X.L., Huang, W.D., Liu, J., Wang, Y.G., Xu, H., Tablets with material gradients fabricated by three-dimensional printing. *J. Pharm. Sci.*, 96, 9, 2446–56, 2007.
111. Skowyra, J., Pietrzak, K., Alhnan, M.A., Fabrication of extendedrelease patient-tailored prednisolone tablets via fused deposition modelling (FDM) 3D printing. *Eur. J. Pharm. Sci.*, 68, 11–7, 2015.
112. Nayan, G.S., Tahsin, Md, Ankita, V.S., Abu, T.M., Formulation of 3D printed tablet for rapid drug release by fused deposition modeling: Screening polymers for drug release, drug-polymer miscibility and printability. *J. Pharm. Sci.*, 107, 1, 390–401, 2018.
113. Okwuosa, T.C., Stefaniak, D., Arafat, B., Isreb, A., Wan, K.W., Alhnan, M.A., A lower temperature FDM 3D printing for the manufacture of patient-specific immediate release tablets. *Pharm. Res.*, 33, 11, 2704–12, 2016.
114. Seoane-Viano, I., Januskaite, P., Alvarez-Lorenzo, C., Basit, A.W., Goyanes, A., Semi- solid extrusion 3D printing in drug delivery and biomedicine: Personalised solutions for healthcare challenges. *J. Contr. Release*, 332, 367–438, 2021.
115. Karavasili, C., Eleftheriadis, G.K., Gioumouxouzis, C., Andriotis, E.G., Fatouros, D.G., Mucosal drug delivery and 3D printing technologies: A focus on special patient populations. *Adv. Drug Deliv. Rev.*, 176, 113858, 2021.
116. Yu, D.G., Branford-White, C., Ma, Z.H., Zhu, L.M., Li, X.Y., Yang, X.L., Novel drug delivery devices for providing linear release profiles fabricated by 3DP. *Int. J. Pharm.*, 370, 1-2, 160–6, 2009.
117. Yu, D.G., Shen, X.X., Branford-White, C., Zhu, L.M., White, K., Yang, X.L., Novel oral fast-disintegrating drug delivery devices with predefined inner structure fabricated by Three-Dimensional Printing. *J. Pharm. Pharmacol.*, 61, 3, 323–9, 2009.
118. Inzana, J.A., Trombetta, R.P., Schwarz, E.M., Kates, S.L., Awad, H.A., 3D printed bioceramics for dual antibiotic delivery to treat implant associated bone infection. *Eur. Cell Mater.*, 30, 232–47, 2015.
119. Giri, B.R., Song, E.S., Kwon, J., Lee, J.-H., Park, J.-B., Kim, D.W., Fabrication of intragastric floating, controlled release 3D printed theophylline tablets using hot-melt extrusion and fused deposition modeling. *Pharmaceutics*, 12, 77, 2020.

120. Öblom, H., Sjöholm, E., Rautamo, M., Sandler, N., Towards printed pediatric medicines in hospital pharmacies: Comparison of 2D and 3D-printed orodispersible warfarin films with conventional oral powders in unit dose sachets. *Pharmaceutics*, 11, 334, 2019.
121. Uddin, M.J., Scoutaris, N., Klepetsanis, P., Chowdhry, B., Prausnitz, M.R., Douroumis, D., Inkjet printing of transdermal microneedles for the delivery of anticancer agents. *Int. J. Pharm.*, 494, 593–602, 2015.
122. Lee, B.K., Yun, Y.H., Choi, J.S., Choi, Y.C., Kim, J.D., Cho, Y.W., Fabrication of drug-loaded polymer microparticles with arbitrary geometries using a piezoelectric inkjet printing system. *Int. J. Pharm.*, 427, 2, 305–10, 2012.
123. Zhou, Z., Buchanan, F., Mitchell, C., Dunne, N., Printability of calcium phosphate: Calcium sulfate powders for the application of tissue engineered bone scaffolds using the 3D printing technique. *Mater. Sci. Eng. C*, 38, 1–10, 2014.
124. Wickström, H., Koppolu, R., Mäkilä, E., Toivakka, M., Sandler, N., Stencil printing—A novel manufacturing platform for orodispersible discs. *Pharmaceutics*, 12, 33, 2020.
125. Adamovic, D., Ristic, B., Zivic, F., Review of existing biomaterials-method of material selection for specific applications in orthopaedics. *Biomater. Clin. Pract. Adv. Clin, Res. Med. Devices*, 47–99, 2017, Springer International Publishing.
126. Khosravi, F., Khorasani, S.N., Ghomi, E.R., Kichi, M.K., Zilouei, H., Farhadian, M., Neisiany, R.E., A bilayer GO/nanofibrous biocomposite coating to enhance 316L stainless steel corrosion performance. *Mater. Res. Express.*, 6, 8, 086470, 2019.
127. Khosravi, F., Khorasani, S.N., Khalili, S., Neisiany, R.E., Ghomi, E.R., Ejeian, F., Das, O., Nasr-Esfahani, M.H., Development of a highly proliferated bilayer coating on 316L stainless steel implants. *Polym. (Basel)*, 12, 1–13, 2020.
128. Goyanes, A., Det-Amornrat, U., Wang, J., Basit, A.W., Gaisford, S., 3D scanning and 3D printing as innovative technologies for fabricating personalized topical drug delivery systems. *J. Control Release*, 234, 41–8, 2016.
129. Prasad, L.K. and Smyth H, I.I.I., 3D Printing technologies for drug delivery: A review. *Drug Dev. Ind. Pharm.*, 42, 7, 1019–31, 2016.
130. Zhu, M., Li, K., Zhu, Y., Zhang, J., Ye, X., 3D-printed hierarchical scaffold for localized isoniazid/rifampin drug delivery and osteoarticular tuberculosis therapy. *Acta Biomater.*, 16, 145–55.0, 2015.
131. Huang, W., Zheng, Q., Sun, W., Xu, H., Yang, X., Levofloxacin implants with predefined microstructure fabricated by three-dimensional printing technique. *Int. J. Pharm.*, 339, 1-2, 33–8, 2007.
132. Wu, G., Wu, W., Zheng, Q., Li, J., Zhou, J., Hu, Z., Experimental study of PLLA/INH slow release implant fabricated by three dimensional printing technique and drug release characteristics *in vitro*. *BioMed. Eng. Online*, 13, 97, 2014.

133. Huang, W., Zheng, Q., Sun, W., Xu, H., Yang, X., Levofloxacin implants with predefined microstructure fabricated by three-dimensional printing technique. *Int. J. Pharm.*, 339, 33–38, 2007.
134. Wu, W., Zheng, Q., Guo, X., Sun, J., Liu, Y., A programmed release multi-drug implant fabricated by three-dimensional printing technology for bone tuberculosis therapy. *Biomed. Mater.*, 4, 065005, 2009.
135. Wu, W., Ye, C., Zheng, Q., Wu, G., Cheng, Z., A therapeutic delivery system for chronic osteomyelitis via a multi-drug implant based on three-dimensional printing technology. *J. Biomater. Appl.*, 31, 250–260, 2016.
136. Cui, M., Pan, H., Li, L., Fang, D., Sun, H., Qiao, S., Li, X., Pan, W., Exploration and preparation of patient-specific ciprofloxacin implants drug delivery system via 3D printing technologies. *J. Pharm. Sci.*, 21, S0022–S3549403–412, 2021 Aug 6.
137. Chaudhari, V.S., Malakar, T.K., Murty, U.S., Banerjee, S., Extruded filaments derived 3D printed medicated skin patch to mitigate destructive pulmonary tuberculosis: Design to delivery. *Expet. Opin. Drug Deliv.*, 18, 301–313, 2021.
138. O'brien, F.J., Biomaterials & scaffolds for tissue engineering. *Mater. Today*, 14, 3, 88–95, 2011.
139. Zhang, S., Chen, Y., Lu, B., Liu, J., Shao, J., Xu, C., Lithographically-generated 3D lamella layers and their structural color. *Nanoscale*, 8, 17, 9118–9127, 2016.
140. Derakhshanfar, S., Mbeleck, R., Xu, K., Zhang, X., Zhong, W., Xing, M., 3D bioprinting for biomedical devices and tissue engineering: A review of recent trends and advances. *Bioact. Mater.*, 3, 2, 144–156, 2018.
141. Turnbull, G., Clarke, J., Picard, F., Riches, P., Jia, L., Han, F., Shu, W., 3D bioactive composite scaffolds for bone tissue engineering. *Bioact. Mater.*, 3, 3, 278–314, 2018.
142. Han, F., Wang, J., Ding, L., Hu, Y., Li, W., Yuan, Z., Zhao, Z., Tissue engineering and regenerative medicine: Achievements, future, and sustainability in asia. *Front. Bioeng. Biotechnol.*, 8, 83, 2020.
143. Mabrouk, M., Beherei, H.H., Das, D.B., Recent progress in the fabrication techniques of 3D scaffolds for tissue engineering. *Mater. Sci. Eng.: C*, 110, 110716, 2020.
144. Leberfinger, A.N., Dinda, S., Wu, Y., Koduru, S.V., Ozbolat, V., Ravnic, D.J., Ozbolat, I.T., Bioprinting functional tissues. *Acta Biomaterialia*, 95, 32–49, 2019.
145. Choudhury, D., Tun, H.W., Wang, T., Naing, M.W., Organ-derived decellularized extracellular matrix: A game changer for bioink manufacturing? *Trends Biotechnol.*, 36, 8, 787–805, 2018.
146. Garreta, E., Oria, R., Tarantino, C., Pla-Roca, M., Prado, P., Fernandez-Aviles, F., Campistol, J.M., Samitier, J., Montserrat, N., Tissue engineering by decellularization and 3D bioprinting. *Mater. Today*, 20, 1, 166–178, 2017.

147. Willerth, S.M. and Sakiyama-Elbert, S.E., Combining stem cells and biomaterial scaffolds for constructing tissues and cell delivery. *Stem J.*, 1, 1, 1–2, 2019.
148. Chou, D.T., Wells, D., Hong, D., Lee, B., Kuhn, H., Kumta, P.N., Novel processing of iron–manganesealloy-based biomaterials by inkjet 3DP. *Acta Biomaterialia*, 9, 10, 8593–8603, 2013.
149. Liu, F.H., Lee, R.T., Lin, W.H., Liao, Y.S., Selective laser sintering of bio-metal scaffold. *Procedia Cirp*, 5, 83–87, 2013.
150. Bonfield, W., Hydroxyapatite-reinforced polyethylene as an analogous material for bone replacementa. *Ann. N. Y. Acad. Sci.*, 523, 1, 173–177, 1988.
151. Ishack, S., Mediero, A., Wilder, T., Ricci, J.L., Cronstein, B.N., Bone regeneration in critical bone defects using three-dimensionally printed b-tricalcium phosphate/hydroxyapatite scaffolds is enhanced by coating scaffolds with either dipyridamole or BMP-2. *J. Biomed. Mater. Res. Part B: Appl. Biomater.*, 105, 2, 366–375, 2017.
152. Ayoub, A.F., Rehab, M., O'neil, M., Khambay, B., Ju, X., Barbenel, J., Naudi, K., A novel approach for planning orthognathic surgery: The integration of dental casts into three-dimensional printed mandibular models. *Int. J. Oral. Maxillofac. Surg.*, 43, 4, 454–459, 2014.
153. Jirman, R., Horak, Z., Mazanek, J., Reznícek, J., Individual replacement of the frontal bone defect: Case report. *Prague Med. Rep.*, 110, 1, 79–84, 2009.
154. Schiele, N.R., Corr, D.T., Huang, Y., Raof, N.A., Xie, Y., Chrisey, D.B., Laser-based direct-write techniques for cell printing. *Biofabrication*, 2, 3, 032001, 2010.
155. Koch, L., Deiwick, A., Schlie, S., Michael, S., Gruene, M., Coger, V., Chichkov, B., Skin tissue generation by laser cell printing. *Biotechnol. Bioeng.*, 109, 7, 1855–1863, 2012.
156. Michael, S., Sorg, H., Peck, C.T., Koch, L., Deiwick, A., Chichkov, B., Reimers, K., Tissue engineered skin substitutes created by laser-assisted bioprinting form skin-like structures in the dorsal skin fold chamber in mice. *PLoS One*, 8, 3, e57741, 2013.
157. Cubo, N., Garcia, M., del Cañizo, J.F., Velasco, D., Jorcano, J.L., 3D bioprinting of functional human skin: Production and *in vivo* analysis. *Biofabrication*, 9, 1, 01, 2016.
158. Zhang, Y., Yu, Y., Chen, H., Ozbolat, I.T., Characterization of printable cellular micro-fluidic channels for tissue engineering. *Biofabrication*, 5, 2, 025004, 2013.
159. Yu, Y., Zhang, Y., Martin, J.A., Ozbolat, I.T., Evaluation of cell viability and functionality in vessel-like bioprintable cell-laden tubular channels. *J. Biomech. Eng.*, 135, 9, 091011, 2013.
160. Duan, B., Kapetanovic, E., Hockaday, L.A., Butcher, J.T., Three-dimensional printed trileaflet valve conduits using biological hydrogels and human valve interstitial cells. *Acta Biomater.*, 10, 5, 1836–1846, 2014.

161. Nguyen, D., Robbins, J., Crogan-Grundy, C., Gorgen, V., Bangalore, P., Perusse, D., Creasey, O., King, S., Lin, S., Khatiwala, C., Halberstadt, C., Functional characterization of three-dimensional (3d) human liver tissues generated by an automated bioprinting platform. *FASEB J.*, 29, 424, 2015.

11

3D Printing in Biomedical Applications: Techniques and Emerging Trends

Gourhari Chakraborty* **and Atanu Kumar Paul**

Department of Chemical Engineering, National Institute of Technology Andhra Pradesh, Tadepalligudem, Andhra Pradesh, India

Abstract

3D printing technique is the cutting-edge technology with the paradigm shifting ability in the context of patient specific biomedical applications. Creation of complex objects, tissue objects, drug delivery systems, and facile system design are the parts of the process which can give the liberty to develop more efficient and application specific product. Inkjet based 3D-printing and extrusion based 3D printing are the two broad techniques which are practiced for development of products. Hydrogels melt cured polymers and metals are utilized for fabrication 3D printing elements. It has profound prospects in biomedical applications like development of tissue, sensor, drug delivery, artificial organ, blood vessel printing. In this chapter basic 3D printing techniques will be discussed in corroboration with recent developments in printing technology. Current polymeric materials, hydrogels, metal powder, and their technology oriented up-gradations will be incorporated into it. Finally, emerging trends and works in different bio medical applications like bone, skin, tissue, blood vessel, drug delivery applications will be thoroughly reviewed.

***Keywords*:** 3D-printing, hydrogel, bone, tissue, bio printing

11.1 Introduction

3D printing is the rapid prototyping technology which constructs three dimensional objects through layer by layer deposition which was initiated in 1986. Design of object in the form of CAD file is required prior

**Corresponding author*: gourhari115@gmail.com

Arbind Prasad, Ashwani Kumar and Manoj Gupta (eds.) Advanced Materials and Manufacturing Techniques for Biomedical Applications, (275–302)

to printing. In comparison to the conventional manufacturing technologies 3D printing of product is more design specific so requirement specific products can be made which leads to more benefit in application [1]. In recent times 3D printing technique has emerged in all the fields of biomedical applications such as tissue culture, bone regeneration, drug delivery, skin application, cartilage making, scaffold making etc. 3D printing techniques varied based on the requirement or design of the products and the nature of the printing material. Inkjet based printing, extrusion based printing, selective laser melting, bio plotting, fused deposition modeling, and stereolithography are the techniques adopted by different groups for 3D printing [1–4]. Bioprinting is the modified version of 3D printing which reduces the inertness of objects towards cell growth. High degree of motion, easy handling, provision of multilayer, high resolution and accuracy, automation etc. are the advantages of bio printing from bio ink [2, 3]. Polymeric materials such as poly lactic acid (PLA), poly ethylene glycol (PEG), cellulose, chitosan, silk, gelatine are used for 3D printing in the form of filament or gel. Metallic substances like Magnesium, Titanium are utilized as 3D printing ingredient. In some cases, composite systems are also employed for making 3D printing filament [4]. 3D printing in biomedical field can be classified in four different levels. Structure without requirement of biocompatibility is the first one and development of non-degradable product with biocompatibility is second. Fabrication of biocompatible and degradable product, bio printing of bio mimicking living tissues are the level three and four, respectively. 3D printing can be utilized for bone regeneration, scaffold making using suitable design and suitable polymer. PLA, Cellulose, silk and chitosan are majorly utilized polymers for 3D printing in biomedical field [5–7]. In case of healthcare applications 3D printed products added the advantages of consumer specificity, biocompatible nature, tunable porosity, mechanical stability etc. Developments in the printing technology, materials and modification techniques can together going resolve many complicated medical issues such as cancer therapy, bio sensing, bone regeneration, nerve regeneration etc. [8, 9].

In this chapter different aspects of 3D-printing techniques are discussed considering biomedical applications. Printing technologies are reviewed in corroboration to recent advancements in technologies. Materials for 3D printing are also one of the basic requirements for successful product development. Filament, hydrogel, powder are different forms of materials used for 3D-printing. In this regard materials for 3D printing are thoroughly discussed and included. Diverse applications of 3D printed products in biomedical domain are thoroughly enlightened and included in view of the recent developments. Challenges and opportunities of 3D-Printing also included before the outlook of the chapter. Thus, through different sections all arrears like technologies, materials and applications are suitably reviewed and documented.

11.2 3D Printing Technologies

3D-Printing can be useful for various complex geometries and the processes proceeds based on layer by layer printing on top of the initial layer. This process was initiated in 1986 with stereolithography and developed over the years with emerging techniques like inkjet printing, extrusion based printing, selective laser melting, bio plotting, fused deposition modeling etc. [10]. Material properties like mechanical property, melting property, processability, biocompatibility, sensitivity to light governs the material of selection for particular printing technique [11]. Figure 11.1 describes the schematic of 3D printing technology for biomedical applications. In case of each printing starting point is digital model and then printing with suitable technique. In this regard digital modeling, inkjet based 3D printing, extrusion based 3D printing, selective laser melting, bio plotting and fused deposition modeling are discussed as follows.

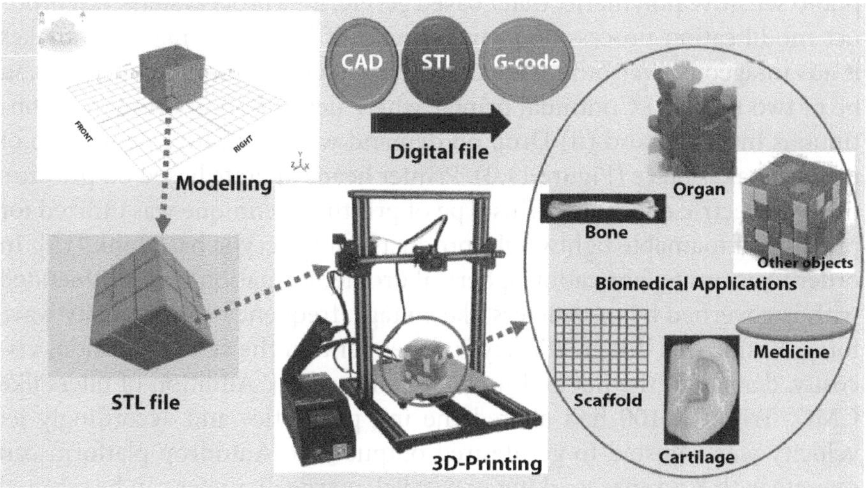

Figure 11.1 Basic flow chart of 3D-printing from modeling to products.

11.2.1 Digital Model

Digital file of printable object is known as computer-aided design (CAD) model. CAD file can be obtained using modeling software and scanner based on existing objects. For example, printing of prostheses can be done from the intraoral scanning data of patients where scanning will allow 3D modeling. CAD file is editable and based on the user requirement variation in design and printing parameters can be done. Software can be open source or professional packages. Each CAD file can record geometry in different file format however conversion of the same in stereolithography file (STL) should be there for printability. STL files uses series of linked triangles to depict the surface of the solid object. Number of triangles determines the resolution of the image. Chord height and angle are two parameters which defines the resolution of the file. Recommended value of the chord height is between 1/20 of the layer thickness to 0.001 mm. STL files are converted into 2D layers using 'slicing' software. User can set the algorithm for printing of each layer with particular fill pattern (fully filled to sparsely filled). Numerical programming language G-code is utilized most of the cases to control the motion of the 3D printer [12, 13].

11.2.2 Inkjet-Based 3D Printing

This is emerging contact-free printing technique operates based on digital image data. It operates by deposition of functional materials precisely on substrate. The functional material can be conductive, semi conductive, photo sensitive polymeric, water based gel etc. Less process steps, fast product modification process and involvement of less harsh materials makes it advantageous over other similar printing techniques. Inkjet printer can be of two types (i) Continual printer where deposition of droplets is continuous in nature and (ii) Drop on demand where selective deposition of material takes place (Figure 11.2). Printer heads operate based on pressure or piezoelectric effect [14]. This type of printing technique was utilized for printing of foamable lightweight products using acrylic based ink [15]. In order to accurate fabrication spherical droplet formation is recommended with is governed by parameters like voltage, frequency and viscosity wave form. Beside this design efficacy is dependent on the surface tension, viscosity, density of the ink and nozzle diameter [16]. Addition of fillers like CMC, Triton X-100 had altered the ink properties and accordingly jet velocity was adjusted to get desired output [17]. Autodrop platform can operate with 8 nozzles working in parallel, nozzle diameter can be selected within 30-100 μm range and drop frequency can be up to 2000/sec per

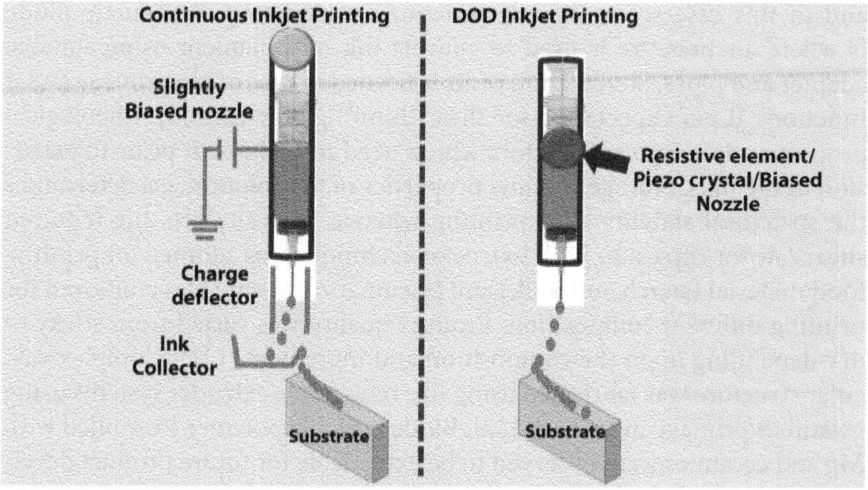

Figure 11.2 Different types of Inkjet printers [17–19].

nozzle. Print head can be heated up to 150°C for this type of ink jet printer. A positional accuracy of 10 ± 3 μm can be maintained in XY direction for the same [18]. Polymer, water based ink, hydrogel, powder mixed matrix can be used as ink for this type of printers. Living cells like protein can be printed using this type of printers for making different products for biomedical applications. Inkjet based bio printing is one of the major route for development of various application oriented products in developing scaffold, complicated structures, bones, medicine etc. [18, 19].

11.2.3 Extrusion-Based 3D Printing

It is also known as direct writing where ink is extruded and dispensed applying mechanical or pneumatic force. In case of extrusion based 3D-printing viscosity of the ink maintained between 30–10^7 mPas range based on the material. During extrusion based 3D printing substrate kept below the syringe and material deposited directly on it. In this process a wide range of material can be used as ink because of operation compatibility. Hydrogel, biodegradable polymers like PLA, PCL were utilized for ink preparation. Depending on the nature of mechanical force screw and piston type arrangements are noticed in commercial extrusion based 3D-printing (Figure 11.3 (a)) [3, 5]. Piston driven extrusion is more general in case of application. Piston motion is controlled using guiding motor and piston operates linearly. In case of screw driven extrusion more volumetric control can be done. In case of more viscous bioinks it is utilized and in this case screw directly functions as conveyer. Pneumatic mode is where air pressure is used to convert ink into filament using suitable adapter and pipes. Sterilization of air is needed in case of bio printing and it functions at per expectation for shear thinning polymers [5]. Rheological properties are important factors which need to be known prior to extrusion of the inks. Storage and loss properties of the solution, gel determines the structural stability after printing whereas viscosity sets the required shear rate for extrusion [20]. Extrusion technique was adopted for printing food material (starch and milk) and pneumatic pressure was employed for printing different composition. Product quality was varied from sticky to dry depending upon the composition and ingredients [21]. Complex vascular structure was fabricated using five retractable extruder systems using gelatin as primary ingredient [22]. Biodegradable polymer PLA filled with Mg and ceramics were observed to be extrudable for future product development [23, 24]. In case of bio printing use of extrusion based 3D printer is growing over the days. Mesenchymal stem cells were successfully printed in alginate-gelatin hydrogel with 65% printing accuracy [25]. Even though extrusion based printing has number of advantages however structural failure and deformation in shape are the prime difficulties associated with this process [26].

11.2.4 Laser-Based 3D Printing

Laser is used to fabricate 3D objects from CAD file using metal powder. Selective laser melting (SLM) and selective laser sintering (SLS) are the methods generally practiced for 3D printing. In 1980, Deckard and Beaman discovered SLS and it came in to use around 2011. In case of SLS based on software control powder material was heated below the melting temperature and they selectively sintered to make prototype (Figure 11.3 (b)). In case of polymeric material heating was continued up to glass transition temperature. SLS process requires more powder and not efficient for small dimension substances. Spherical morphology materials with 10-150 μm size is preferred for SLS [27, 28]. In case of SLM high power laser source is used to melt powder material and molten metal spread on the flat surface. Similar procedure is followed for layer by layer deposition. Laser power varies based on the source such as 170 W for 316L, 180 W for In718 and 125 W for Cu10Sn. Multiple materials suitably deposited on same building layer using SLM technique [29]. An SLS machine, EOS P 396 was utilized for 3D printing of polyamide 12 (PA12) based objects. Denser printed parts with better mixing was noticed for SLS printing of carbon black incorporated matrix [30]. In case of implant fabrication SLM technique has high application. Ti6Al4V based implants were fabricated for bio medical applications through SLM technique [31, 32]. A surface roughness was recorded 17.6 ± 3.7 μm for the fabricated object through SLM [31]. Nickel based alloys were also fabricated in 3D object by SLM technique [33]. SLS was utilized for 3Dprinting of ondansetron containing orodispersable printlets. It was observed that the 3D printed products releases drug within 15 s and about 90% release was noticed within 5 min [34].

11.2.5 Bioplotting

In case of tissue engineering another 3D printing technique explored for biomedical application is Bioplotting. Tubular or spheroid materials are extruded using syringe and provision of multiple syringes involving different cells depends on the requirement of the object. UV and photo curing of polymers sometimes helps in bio plotting [3]. 'Bioink' with bio functionality and biocompatibility enhances the effectivity towards biomedical applications (Figure 11.3 (c)). Predefined objects are used for construction of complex cell-laden constructs [35]. Application of bioplotting is significant in cartilage tissue engineering. Alginate and gelatine were mostly used for bioplotting [36]. Chondrocytes was printed using methyl cellulose and alginate as precursor. Impact of different sterilization methods was investigated towards the effective printing of object [37]. Polyethylenimine (PEI) coating emerged as potential material in developing tissue. Pneumatically controlled three-axis bioplotter was employed for deposition of porous scaffold on stationary substrate. Effectivity of the porous scaffold was examined through mechanical testing, protein release study etc. and high cell viability was recorded for the bioplotted objects [38].

11.2.6 Fused Deposition Modeling (FDM)

FDM is the 3D printing technique where printing is carried out using continuous filament of thermoplastic polymer. In addition to automotive application, aeronautical application and household goods making FDM is emerging in the biomedical application like bone, scaffolds, stents etc. The operation of FDM governed by majorly by material feed, liquefier print head, gantry and build surface (Figure 11.3 (d)). Filament is fed continuously in liquefier where heating arrangement converts the material into molten form. Now using suitable nozzle, molten polymer is deposited on the surface of substrate where the object printing is controlled by x-y directional movement of the nozzle and z-directional movement of stage based on the CAD design [39, 40]. FDM is also referred as fused filament fabrication (FFF) and it is second common 3D printing technique after stereolithography. The mechanical properties of FDM objects varies based on material, structural parameters and operation variables. Structural parameters include rasters angle, printing orientation, sequence of stacking and infill density. Operational variables cover speed of printing, liquefier temperature, extrusion rate, layer time, transverse speed of nozzle and bed temperature [41]. Polyamide and blends were successfully explored for FDM based 3D printing by several groups [42]. Biodegradable polymers like poly caprolactone (PCL), PLA because of thermoplastic nature widely used for making filament. In some cases, at per need suitable modification with reinforcement was carried out [43, 44].

In bioanalytical technology like lab-on-chip use of FDM is increasing gradually [45]. In case of drug delivery application use of FDM is varied based on the temperature requirement. Thermolabile drug printing requires low temperature FDM and it was noticed that Kollidon VA64 and Kollidon 12PF were two low melting polymers which has potential to be used as low-temperature FDM printing material where Ramipril was model drug [46].

In addition to this stereolithography also sometimes practiced for printing of 3D objects from resin bath using UV/laser beam [3]. Biomedical applications are the thrust area where above mention techniques suitably used to improve efficacy of application.

11.3 Materials for 3D Printing

In future engineering systems development of functional materials and structures will open up the complicated areas of biomedical engineering and 3D printing will be the lead among the fabrication technologies. Printing route, strategy and modification of pathway are dependent on the properties of the aimed material. In other way continuous upgrade of materials, bio materials, functional materials etc. enhancing the application domain of the 3D printing. Different materials like polymer, ceramic, metal and composite were utilized for fabrication of objects. Living organisms are also bioplotted for specific applications. In this section materials utilized for 3D printing for biomedical purposes is discussed on the basis of nature of printable inks like hydrogel, polymer solution, metallic and others (ceramic and living organisms) [47].

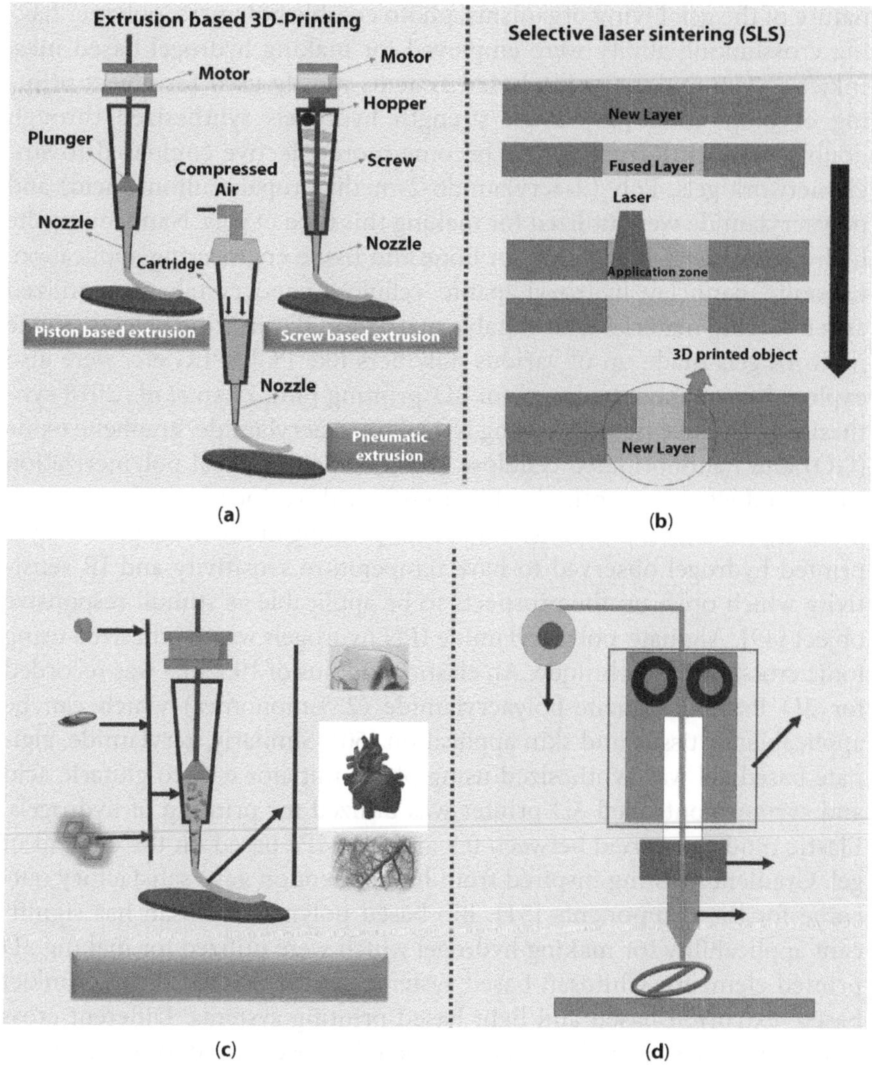

Figure 11.3 (a) Different types of extrusion based 3D-printers [20–22], (b) Selective laser sintering process [30], (c) Bioplotting [36] and (d) Fused deposition modeling based 3D-printing [39, 43].

11.3.1 Hydrogel

Hydrogels with printable nature has become attractive materials for printing of biomedical objects (Table 11.1). Hydrogel strength, viscosity and formulation easily can be tuned depending on the requirement of the ink. Hydrogels are formulated based on the application of the object and nature of the gel. Living organisms, photo curable polymers, polymer having crosslinking ability were employed for making hydrogel based inks. Inkjet system and extrusion based systems mostly used for object printing or bioplotting [48]. High strength hydrogels synthesized through double network formation has become more effective compared to single network gels. Poly (2-acrylamido-2-methylpropanesulfonic acid) and polyacrylamide were utilized for making this type of gels. Nanocomposite hydrogels were also explored for bone and tissue engineering application. Laponite, nanoclay, hydroxyl apatite, cellulose nanocrystals were utilized as reinforcing material with suitable matrix for 3D-printing objects. Single network gels made up of various polymers like PCL, PEG etc. were also explored for making hydrogel for 3D-printing [48]. Zhao *et al.*, 2018 synthesized printable hydrogel using N-isopropyl acrylamide, graphene oxide (GO) and nanofibrillated cellulose (NFC). In situ radical polymerization were conducted for synthesis of hydrogels with various formulations. Lab made stainless steel mold was utilized for printing of the hydrogels. Bilayer printed hydrogel observed to have temperature sensitivity and IR sensitivity which open up the prospects to be applicable as stimuli responsive object [49]. Alginate-polyacrylamide IPN hydrogels was synthesized using ionic crosslinking technique. An elastic modulus of 19.4 kPa was recorded for 3D Printed alginate-polyacrylamide (23%monomer) which can be applicable for tissue and skin application [50]. Similarly, acrylamide, alginate based gel was synthesized using photo initiator α-Keto glutaric acid and syringe controlled 3D printer was utilized for printing of hydrogels. Elastic modulus varied between 0.5 and 2.5 MPa based on the content of gel. Gradient printing inspired from human tendon gave satisfactory outcome for the components [51]. Bio based polymer chitosan has significant applicability for making hydrogel which were utilized for making 3D printed elements. Chitosan based systems can be printed through inkjet based, extrusion based and light based printing systems. Different cross linkers like physical and chemical cross linkers were used for making gels for various applications [52]. Nano cellulosic systems were utilized by several groups for making hydrogel for 3D printing. Biocompatibility, printability, structural nature promotes the application of cellulosic materials for 3D printing applications [53].

11.3.2 Polymers (Melt Cured)

FDM is growing over the times for development of the 3D printed objects in biomedical applications (Table 11.1). Melt cured polymers with thermoplasticity has wide application for the purpose. Polyamide was used for making filament for 3D printing. As filament is prepared through extrusion so melt rheological behavior like melt strength, viscosity, melting temperature, stability of the polymer are the governing parameters for successful printing. Other materials like shape memory polymer, functional polymer, biodegradable polymers were employed for 3D printing of biomedical objects [54]. Poly ether ether ketone (PEEK) of various compositions with fillers like HAP, TiO2 were explored for 3D printing of orthopedic, skull, bone. FDM and SLM were used for printing of PEEK based objects [55]. Different polymers and polymeric composites were utilized as filament for 3D printing. ABS polymer in combination with MWCNT, graphene, clay, carbon fiber, glass fiber, metals were explored for FDM process [56]. Biodegradable polymers like PLA, PCL and different copolymers were subjected for filament fabrication by several groups. Extrusion of PLA-Stainless steel was carried out at 200–230°C with bed temperature 60–70°C and layer thickness 0.18–0.30 mm. PLA based object observed to have toughness of 18 kJ/m2 and has an ultimate tensile strength of 69.0288 MPa at 45° raster orientation of the print. The bioactivity of composite is similar like PLA [57]. Printing nature and design stability is dependent on the nature of the polymer [58]. PCL was explored for FDM based 3D printing by several groups in combination with fillers like calcium carbonate, GO, graphene for biomedical applications [43]. Incorporation biofiller in PLA can improve the mechanical properties and biocompatibility. CNC reinforcement in PLA has improve the mechanical strength of the 3D printed specimen [44].

11.3.3 Metallic Substances

In order to fabricate orthopedic implants metal based 3D printing was also carried out by several groups. SLM, SLS, selective electron beam melting (SEBM), laser induced forward transfer (LIFT) and Inkjet based printing of metal based objects. Ti based alloys were mostly used for fabrication of 3D objects. Ti based alloys have biocompatibility, mechanical strength, corrosion resistance which converts it into prime material for bone fixation, orthopedic implant and dental related printing [59]. Ti6Al4V (Ti-64) is most widely used alloy for orthopedic implant fabrication [31, 32, 59]. Tantalum, chromium alloy, magnesium, iron, zinc and nickel are some of the metals employed for fabrication of biomedical objects through 3D printing. In some cases, liquid metal like Ga was also explored for 3D printing. Ti based metals were explored successfully in combination with biomolecules collagen, morphogenetic protein etc. for biomedical application [60].

11.3.4 Ceramic Substances

Ceramic materials like hydroxyapatite (HAP), alumina, zirconia, phosphates and bioglass were explored in making objects for biomedical applications through 3D printing (Table 11.1). Application of ceramic materials are prominent for making scaffold and bone regeneration materials. Vat polymerization, SLS, FDM, extrusion and FDM techniques were widely utilized for ceramic object printing [61]. HAP with chemical structure Ca10(PO4)6(OH)2 has similar functionality like bone and have biocompatibility, bioactivity, osteointegrity and excellent osteoconductivity. Stereolithographic, extrusion based and inkjet printing of HAP based materials were carried out by several researchers. Tri calcium phosphate (TCP), CaSO4 etc. were incorporated along with HAP for inkjet based printing where as for FDM thermoplastic polymer was used along with HAP [62]. Zirconia and HAP composite based scaffold was found to be porous and bioactive material [63]. Spray dried HA and PLA based filament was extruded using twin screw extruder with thermal profile between 175°C and 200°C. Fused deposition modeling based 3D-printer (3DPRN LAB 3D) was used for printing of 3D object in the form of prism. Porosity of the scaffold was increased up to 53.9% in inclusion of HAP in PLA system. This opens up the possibility to use for drug delivery application [64]. Calcium phosphate and PLA based filament was used for fabrication of 3D printed scaffold through fused deposition modeling technique [65]. Utilization of alumina through stereolithographic 3D printing was reported [66]. Calcium and silicate based bio ceramic materials like akermanite was also explored for making biomedical applications and became attractive material for 3D printing of bone and orthopedic objects [67].

11.3.5 Living Cells

In case of tissue engineering and regenerative medicine 3D bioprinting of living cells are required. Cell based biomaterials are deposited in predetermined structure based on requirement in order to fabricate desired object. Structure are in the form of tissue or organ which can be further characterized based on the application. Bioinks are printed in two different methods i) cell-scaffold based approach and ii) scaffold-free-cell based approach. In case of the former scaffold is made up of biodegradable polymers and gradually cells occupy the space of the scaffold with the degradation. In case of the later living cells are printed directly without any support of scaffold and the growth is normal embryonic in nature. Agarose based bioinks, alginate based bioinks, collagen based bioinks and hyaluronic acid-based bioinks were mostly employed for printing by various groups [68]. Bacterial spores were printed with agarose based system and found that *Bacillus subtilis* can survive up to 1 month and germinated upon rehydration. The material observed to have antibacterial properties [69]. Pluronic F127-DA is one of the widely used ink for bioprinting [70]. In case of bone mimicking application ceramic compound based 3d printing was conducted using hydroxyl apatite and tricalcium phosphate. It gave cell viability 95% after several week of incubation [71]. CaP was explored for drug delivery, scaffold and orthopedic applications in combination with different cells and reagents [72]. Living cells like human embryonic kidney cells were printed using droplet based 3D printer. The printed cell showed 90% cell viability [73]. Now a day's 3D printing is transforming to bio-printing to enhance and control the effectivity of the product in biomedical applications like bone regeneration, tissue engineering, bio mimicking drug delivery, bio sensing etc. In this respect single or multiple syringe based printers are developed and multilayer deposition of materials are also carried out in order to meet the complex structure.

Table 11.1 Materials used for biomedical applications.

Material	Nature of ink	Method of printing	Application	Ref.
N-isopropyl acrylamide, GO, NFC	Hydrogel	Mould based 3D printing	Stimuli responsive object	[49]
Alginate-polyacrylamide	IPN-Hydrogel	Multi syringe system	Skin and Tissue	[50]
Acrylamide, alginate	Hydrogel	Syringe controlled 3D printing	Human Tendon	[51]
PLA-CNC	Filament	FDM	Biomedical	[44]
PLA-Stainless steel	Filament	FDM	Orthopedic	[57]
Ti6Al4V	Bulk metal	SLM	Biomedical	[31]
Inconel 718 Ni-base alloy	Bulk metal	SLM	Biomedical	[33]
Zirconia and HAP	Powder	Digital light processing	Scaffold	[63]
HAP and PLA	Filament	FDM	Porous scaffold	[64]
Bacterial spores	Bioink	FDM	Antibacterial	[69]
Pluronic F127-DA	Bioink	Inkjet	Skin	[70]

11.4 Biomedical Applications: Recent Trends of 3D-Printing

Biomedical application of 3D printed substances grown over the time and became more effective and patient specific. In recent times many instrument molds, surgery items, medicine, implants etc. are fabricated through 3D printing. In the domain of biomedical application printing of pacemakers, hips, knee prostheses, dental implants, artificial kidney, artificial heart need patient specific design and 3D printing has emerged as possible solution to this. 3D printing also showing prospects in cancer treatment, sensing of biomolecules, advanced drug release objects [74, 75]. In this section current works related to use of 3D printing for biomedical applications like skin, bone, tissue, drug delivery is discussed and some of them can be seen in Table 11.2.

Table 11.2 3D printing and biomedical applications.

Material	Application	Method of printing	Key features	Ref.
Biomimetic bioink (GelMA/ HA-NB/LAP)	Skin	DLP	Rapid printing of regenerated skin material	[81]
Photopolymer	Skin cancer	Stereolithography	Cisplatin release 80-90% within 1 h	[82]
Titanium alloy	Bone	SLM	Mechanical strength 142.8 MPa	[86]
Yttria (3% mol)-stabilized zirconia powder	Dentistry	Robocasting	High cell attachment	[92]
PLA	Tissue	Extrusion 3D printing	Yield strength up to 28 MPa	[96]
HAMA	Cartilage Tissue	Robotic assisted bioprinter	Cartilage defect repaired within 12 weeks	[97]
PEG	Drug delivery	SLA	Printable gel was obtained	[104]
Magnetic particle loaded Chitosan	Drug delivery	Two-photon direct laser writing	Release behavior of drug can be controlled using light	[106]
Carbon loaded PLA	Electrochemical sensing	FDM	Caffeine detection (0-90 mg/L)	[111]

11.4.1 Skin

Skin is one of the complex organ in the body and three layers i.e. epithelium, dermis and hypodermis constructs the skin. In case of skin damage existing techniques like autografts and allografts are practiced which are based on donor and recipient system. 3D-printing of bioinks in the form of skin can reduce some of the complicacies and steps of skin care or skin replacement operations. Silk based systems were observed to be promising in some of the skin applications [76, 77]. In case of skin bioprinting first appropriate material selection, cell selection and design of the replaceable skin is required. Then object printing need to be done using suitable techniques described in section 11.2. Sometimes post treatment of ski is required using suitable bioreactor prior to characterization [78]. Multifunctional skin with different sensor elements also printed through 3D printing techniques which is transforming disease diagnosis more effective and easy [79]. ECG sensor, metabolic sensor, motion sensor and transdermal drug delivery sometimes associated with the E-skin elements. E-skin elements are fabricated such a way that it should have stretchability, self-healing nature, biocompatibility, transparency and comfortability [80]. Zhou *et al.*, 2020 used a newly designed newly designed biomimetic bioink (GelMA/HA-NB/LAP) for development of skin. The ink had the rapid gelation time with promising cell adhesion and biocompatibility. DLP-based 3D printing coupled with a biomimetic hydrogel bioink and functional cells gave significant outcome [81]. Microneedles fabricated through stereolithography technique using biocompatible photopolymer resin. Inkjet dispensing of cisplatin formulations on the needle surface was carried out. It was observed through tomography analysis that 3D printed microneedles have excellent piercing capacity with an 80% of penetration depth. Cisplatin release rate was recorded to be rapid in nature with 80-90% release within 1 h [82]. Technological development in printing type and materials leading to end use products like wearable sensing skin [83].

11.4.2 Bone and Dentistry

Bioprinting is utilized for bone and dental applications and products are patient specific in nature. Metallic and polymeric substances were utilized for making implants and scaffold by various groups. In case of bone tissue printing vascularized 3D printing technique was adopted by different groups [84]. Hydroxyapatite/PLA based 3D printed bone was casted successfully by Zare *et al.*, 2021 using hybrid 3D printing-milling machine. HAP and PLA ratio was varied in different formulations in range 95:5 to

50:50. It was observed that 10:90 loading of HA and PLA gave best mechanical property [85]. SLM technique was adopted to make porous scaffold of titanium based alloys and compressive strength was achieved in the range of 114.9 MPa and 142.8 MPa for scaffolds specimens with Truss(S4) and Cube (S5) pore unit [86]. Similarly bone scaffolds using HAP, bioceramic, calcium phosphates were printed successfully incorporating in polymers like PLA, Polydopamine [87–89].

In dentistry application of 3D printing also increasing with times particularly for oral surgery, implantology, maxillofacial prosthesis and prosthodontics. SLM, stereolithography, FDM and DLP are techniques mostly practiced for printing of dentistry elements [90, 91]. Branco *et al.*, 2020 printed Yttria (3% mol) -stabilized zirconia powder using robocasting technique. It was observed that mechanical properties were at per the requirement of dental replacements and specimens were non-toxic and denser in structure [92].

11.4.3 Tissue

Tissue engineering is one of the area where multiple diseases can be addressed using suitable techniques. Scaffold making can originate the tissue into 3D form and can be used for surgery. Sodium alginate is one of the leading chemical used along with different cells for tissue growth purpose. In view of the printing of bioinks different user specific modifications of the 3D-printing unit are carried out in recent periods [93]. Polymer employment for fabrication of scaffold is very wide. Depending upon the requirement of mechanical property of the tissues different polymer and printing techniques were adopted by several groups. In polymeric scaffold fabrication different gradient techniques like mechanical, biochemical and physical gradient were adopted in several occasions [94]. Sucrose and glucose based sugar syrup was subjected to extrusion based for making of objects for tissue engineering [95]. PLA based scaffold can be made with yield strength of 28 MPa setting suitable printing temperature, extrusion speed and layer thickness for tissue culture [96]. Cartilage regeneration, and potential application in clinical application is one of area where 3D printing is playing a vital role for upgrading the design and property. Hyaluronic acid (HA) was modified by reacting with the methacrylic acid (MA) to make bioink HAMA. HAMA with other modifiers was 3D printed using robotic assisted bioprinter system. In order to conduct *in-vivo* experiment in rabbit a lateral para-patellar incision was created on the surface of articular region and a cylindrical defect site with 5 mm dia and 4 mm height was created in the center of the trochlear groove. The defects were directly repaired by implantation of hydrogel in the affected site using 6-DOF robot. Bioprinting process was completed within 60s and highest modulus was reported to be 10.02 GPa. ICRS grade IV cartilage defect was repaired within 12 weeks [97]. In similar to that polyurethane and hyaluronic acid based 3D printed system was successfully studied for cartilage making [98]. 3D printing is explored in various ways in the field and growing from all perspectives like material, printing technique *in-vivo* practice etc. [99–101].

11.4.4 Drug Delivery

3D printing technology has opened up the flexibility of drug product printing from the perspective of design, control on formulation and release profile. It was observed that parameters like printlet geometry, loading of infill and type of binder like polymer. Vat photo polymerization is one of the leading technique where different monomers and oligomers such as methacrylate or acrylic esters were fed for printing. SLA, DLP, two photon polymerization etc. were employed for vat polymerization. Oral dosage form can be different depending upon the system [102]. In case of regenerative medicine application PEG, chitosan, alginate etc. were employed by several groups [103]. PEG based hydrogel filled with ibuprofen was successfully printed using SLA technique and it was observed at 30% w/w water in PEG hydrogel and 10% ibuprofen gave best printability [104]. Drug loading can be done using different forms like cylindrical and geometric tablets, two pulse oral drug carrying capsule, pellets and radiator dosage using 3D print techniques like SLA and DLP [105]. Chitosan based double-helical micro swimmer loaded with a chemotherapeutic drug, doxorubicin was printed using photo initiator. It showed controlled release behavior in presence of light stimulus [106]. Application of 3D printing also observed for transdermal drug delivery [107]. Hydrogel with stimuli response behavior was employed for drug delivery application where release nature can be tuned using stimulus like pH, magnetic field etc. Photo cross-linked acrylic acid loaded with sulforhodamine B (fluorescent property) showed pH dependent release property [108]. In future more use of 3D printing for drug delivery and regenerative medicine will increase with control release behavior which will transform the process into more cost effective and prominent [109, 110].

11.4.5 Other Applications

3D printing application is also covers electrochemical and bio sensing application of disease diagnosis. Carbon loaded PLA filament was fabricated for sensing of caffeine by FDM technique and sensing range was noticed to be 0-90 mg/L [111]. 3D printed PLA/graphene composite through extrusion technique showed excellent property towards biosensing of uric acid [112]. In associated to disease detection 3D printing also applied for therapy. In case of cancer therapy functional scaffold was utilized [113]. Artificial organ printing also carried out by some of the groups. 3D printing for pediatric application like prostheses making was carried out by different groups [114].

11.5 Challenges and Opportunities

3d printing technology dependent on the upgrade of material, upgrade on imaging and digital file development and product oriented technology modification. Printability of material (Hydrogel, polymer) changes depending upon the composition and reinforcement. Viscosity, conductive property, crosslinking ability, time of curing etc. governs the 3D printing ability of the inks. Biocompatibility of the FDM based objects still need to be improved. Use of biodegradable polymers like PCL, PLA need to be explored more. Proper design of the scaffold, incorporation of filler, maintenance of rheological properties need to be done to get more updated and error free objects. In some cases, like regenerative medicine, bone scaffold, cancer therapy stimuli response behavior of the product can improve the applicability. In this regard more emphasis on pH, temperature, magnetic field, electric field controllable product development can be studied. Bioink upgrade and printability with proper mechanical stability and biocompatibility need to examined more [115–117]. Use of crosslinking agent, employment of filler, modification on surface can further improve the ink properties. Technology development needs more product oriented modifications which are specific depending upon the printing technique. More close to desired design is the universal motto for all the processes.

11.6 Conclusion

In this chapter general 3D printing techniques which are commonly used for biomedical items are discussed. Materials like hydrogel, melt cured polymers, metallic and ceramic substances and living cells which are in use as printing ink are discussed with suitable examples. Some emerging techniques, materials related to printing for biomedical applications like skin, bone, drug delivery etc. are discussed in corroboration with current literatures. Finally, shortcomings and possible modification techniques for betterment of 3D printing of biomedical substances are described.

Acknowledgements

The authors are thankful to The National Institute of Technology Andhra Pradesh, for providing the necessary facilities to prepare this book chapter.

References

1. Szymczyk-Ziółkowska, P., Łabowska, M.B., Detyna, J., Michalak, I., Gruber, P., A review of fabrication polymer scaffolds for biomedical applications using additive manufacturing techniques. *Biocybern BioMed. Eng.*, 40, 2, 624–638, 2020.
2. Ozbolat, I.T., Moncal, K.K., Gudapati, H., Evaluation of bioprinter technologies. *Addit. Manuf.*, 13, 179–200, 2017.
3. Mobaraki, M., Ghaffari, M., Yazdanpanah, A., Luo, Y., Mills, D.K., Bioinks and bioprinting: A focused review. *Bioprinting*, 18, e00080, 2020.
4. Vanaei, S., Parizi, M.S., Salemizadehparizi, F., Vanaei, H.R., An overview on materials and techniques in 3D bioprinting toward biomedical application. *Eng. Regen.*, 2, 1–18, 2021.
5. Gu, Z., Fu, J., Lin, H., He, Y., Development of 3D bioprinting: From printing methods to biomedical applications. *Asian J. Pharm. Sci.*, 15, 5, 529–557, 2020.
6. Serra, T., Planell, J.A., Navarro, M., High-resolution PLA-based composite scaffolds via 3-D printing technology. *Acta Biomater.*, 9, 3, 5521–5530, 2013.
7. Xu, W., Wang, X., Sandler, N., Willfor, S., Xu, C., Three-dimensional printing of wood-derived biopolymers: A review focused on biomedical applications. *ACS Sustain. Chem. Eng.*, 6, 5, 5663–5680, 2018.
8. Ahangar, P., Cooke, M.E., Weber, M.H., Rosenzweig, D.H., Current biomedical applications of 3D printing and additive manufacturing. *Appl. Sci.*, 9, 8, 1713, 2019.
9. Tetsuka, H. and Shin, S.R., Materials and technical innovations in 3D printing in biomedical applications. *J. Mater. Chem. B*, 8, 15, 2930–2950, 2020.
10. Ngo, T.D., Kashani, A., Imbalzano, G., Nguyen, K.T., Hui, D., Additive manufacturing (3D printing): A review of materials, methods, applications and challenges. *Compos. B. Eng.*, 143, 172–196, 2018.

11. Chen, Z., Zhao, D., Liu, B., Nian, G., Li, X., Yin, J., Yang, W., 3D printing of multifunctional hydrogels. *Adv. Funct. Mater.*, 29, 20, 1900971, 2019.
12. Awad, R.H., Habash, S.A., Hansen, C.J., 3D printing methods, in: *3D Printing Applications in Cardiovascular Medicine*, pp. 11–32, Academic Press, Elsevier, United States, 2018.
13. Jeong, Y.G., Lee, W.S., Lee, K.B., Accuracy evaluation of dental models manufactured by CAD/CAM milling method and 3D printing method. *J. Adv. Prosthodont.*, 10, 3, 245–251, 2018.
14. Fischer, A.C., Mäntysalo, M., Niklaus, F., Inkjet printing, laser-based micromachining, and micro–3D printing technologies for MEMS, in: *Handbook of Silicon Based Mems Materials and Technologies*, pp. 531–545, Elsevier, United States, 2020.
15. Wagner, A., Kreuzer, A.M., Göpperl, L., Schranzhofer, L., Paulik, C., Foamable acrylic based ink for the production of light weight parts by inkjet-based 3D printing. *Eur. Polym. J.*, 115, 325–334, 2019.
16. Wang, J., Liu, Y., Fan, Z., Wang, W., Wang, B., Guo, Z., Ink-based 3D printing technologies for graphene-based materials: A review. *Adv. Compos. Hybrid Mater.*, 2, 1, 1–33, 2019.
17. Brindha, J., Edwina, R.G.P., Rajesh, P.K., Rani, P., Influence of rheological properties of protein bio-inks on printability: A simulation and validation study. *Mater. Today: Proc.*, 3, 10, 3285–3295, 2016.
18. De Gans, B.J., Duineveld, P.C., Schubert, U.S., Inkjet printing of polymers: State of the art and future developments. *Adv. Mater.*, 16, 3, 203–213, 2004.
19. Habib, M.A. and Khoda, B., Development of clay based novel bio-ink for 3D bio-printing process. *Proc. Manuf.*, 26, 846–856, 2018.
20. Duty, C., Ajinjeru, C., Kishore, V., Compton, B., Hmeidat, N., Chen, X., Kunc, V., What makes a material printable? A viscoelastic model for extrusion-based 3D printing of polymers. *J. Manuf. Process.*, 35, 526–537, 2018.
21. Lille, M., Nurmela, A., Nordlund, E., Metsä-Kortelainen, S., Sozer, N., Applicability of protein and fiber-rich food materials in extrusion-based 3D printing. *J. Food Eng.*, 220, 20–27, 2018.
22. Fitzsimmons, R.E., Aquilino, M.S., Quigley, J., Chebotarev, O., Tarlan, F., Simmons, C.A., Generating vascular channels within hydrogel constructs using an economical open-source 3D bioprinter and thermoreversible gels. *Bioprinting*, 9, 7–18, 2018.
23. Ferrández-Montero, A., Lieblich, M., Benavente, R., González-Carrasco, J.L., Ferrari, B., Study of the matrix-filler interface in PLA/Mg composites manufactured by material extrusion using a colloidal feedstock. *Addit. Manuf.*, 33, 101142, 2020.
24. Faes, M., Valkenaers, H., Vogeler, F., Vleugels, J., Ferraris, E., Extrusion-based 3D printing of ceramic components. *Proc. Cirp*, 28, 76–81, 2015.
25. Koch, F., Tröndle, K., Finkenzeller, G., Zengerle, R., Zimmermann, S., Koltay, P., Generic method of printing window adjustment for extrusion-based

3D-bioprinting to maintain high viability of mesenchymal stem cells in an alginate-gelatin hydrogel. *Bioprinting*, 20, e00094, 2020.
26. Wolfs, R.J.M. and Suiker, A.S.J., Structural failure during extrusion-based 3D printing processes. *Int. J. Adv. Manuf. Technol.*, 104, 1, 565–584, 2019.
27. Duan, B. and Wang, M., Selective laser sintering and its biomedical applications, in: *Laser Technology in Biomimetics*, pp. 83–109, Springer, Berlin, Heidelberg, 2013.
28. Duan, B. and Wang, M., Selective laser sintering and its application in biomedical engineering. *MRS Bull.*, 36, 12, 998–1005, 2011.
29. Wei, C., Li, L., Zhang, X., Chueh, Y.H., 3D printing of multiple metallic materials via modified selective laser melting. *CIRP Ann.*, 67, 1, 245–248, 2018.
30. Cai, C., Tey, W.S., Chen, J., Zhu, W., Liu, X., Liu, T., Zhou, K., Comparative study on 3D printing of polyamide 12 by selective laser sintering and multi jet fusion. *J. Mater. Process. Technol.*, 288, 116882, 2021.
31. Vaithilingam, J., Prina, E., Goodridge, R.D., Hague, R.J., Edmondson, S., Rose, F.R., Christie, S.D., Surface chemistry of Ti6Al4V components fabricated using selective laser melting for biomedical applications. *Mater. Sci. Eng. C*, 67, 294–303, 2016.
32. He, B., Wu, W., Zhang, L., Lu, L., Yang, Q., Long, Q., Chang, K., Microstructural characteristic and mechanical property of Ti6Al4V alloy fabricated by selective laser melting. *Vacuum*, 150, 79–83, 2018.
33. Xia, M., Gu, D., Yu, G., Dai, D., Chen, H., Shi, Q., Selective laser melting 3D printing of Ni-based superalloy: Understanding thermodynamic mechanisms. *Sci. Bull.*, 61, 13, 1013–1022, 2016.
34. Allahham, N., Fina, F., Marcuta, C., Kraschew, L., Mohr, W., Gaisford, S., Goyanes, A., Selective laser sintering 3D printing of orally disintegrating printlets containing ondansetron. *Pharmaceutics*, 12, 2, 110, 2020.
35. Ouyang, L., Yao, R., Zhao, Y., Sun, W., Effect of bioink properties on printability and cell viability for 3D bioplotting of embryonic stem cells. *Biofabrication*, 8, 3, 035020, 2016.
36. Angelats Lobo, D. and Ginestra, P., Cell bioprinting: The 3D-bioplotter™ case. *Materials*, 12, 23, 4005, 2019.
37. Hodder, E., Duin, S., Kilian, D., Ahlfeld, T., Seidel, J., Nachtigall, C., Lode, A., Investigating the effect of sterilisation methods on the physical properties and cytocompatibility of methyl cellulose used in combination with alginate for 3D-bioplotting of chondrocytes. *J. Mater. Sci.: Mater. Med.*, 30, 1, 1–16, 2019.
38. You, F., Wu, X., Zhu, N., Lei, M., Eames, B.F., Chen, X., 3D printing of porous cell-laden hydrogel constructs for potential applications in cartilage tissue engineering. *ACS Biomater. Sci. Eng.*, 2, 7, 1200–1210, 2016.
39. Varma, M.V., Kandasubramanian, B., Ibrahim, S.M., 3D printed scaffolds for biomedical applications. *Mater. Chem. Phys.*, 255, 123642, 2020.

40. Pu'ad, N.M., Haq, R.A., Noh, H.M., Abdullah, H.Z., Idris, M., I, Lee, T.C., Review on the fabrication of fused deposition modelling (FDM) composite filament for biomedical applications. *Mater. Today: Proc.*, 29, 228–232, 2020.
41. Cano-Vicent, A., Tambuwala, M.M., Hassan, S.S., Barh, D., Aljabali, A.A., Birkett, M., Serrano-Aroca, Á., Fused deposition modelling: Current status, methodology, applications and future prospects. *Addit. Manuf.*, 47, 102378, 2021.
42. Zhang, X., Fan, W., Liu, T., Fused deposition modeling 3D printing of polyamide-based composites and its applications. *Compos. Commun.*, 21, 100413, 2020.
43. Prasad, A. and Kandasubramanian, B., Fused deposition processing polycaprolactone of composites for biomedical applications. *Polym.-Plast. Technol. Mater.*, 58, 13, 1365–1398, 2019.
44. Kumar, S.D., Venkadeshwaran, K., Aravindan, M.K., Fused deposition modelling of PLA reinforced with cellulose nano-crystals. *Mater. Today: Proc.*, 33, 868–875, 2020.
45. Salentijn, G., I, Oomen, P.E., Grajewski, M., Verpoorte, E., Fused deposition modeling 3D printing for (bio) analytical device fabrication: Procedures, materials, and applications. *Anal. Chem.*, 89, 13, 7053–7061, 2017.
46. Kollamaram, G., Croker, D.M., Walker, G.M., Goyanes, A., Basit, A.W., Gaisford, S., Low temperature fused deposition modeling (FDM) 3D printing of thermolabile drugs. *Int. J. Pharm.*, 545, 1-2, 144–152, 2018.
47. Zhu, Y., Joralmon, D., Shan, W., Chen, Y., Rong, J., Zhao, H., Li, X., 3D printing biomimetic materials and structures for biomedical applications. *Bio-Des. Manuf.*, 4, 2, 405–428, 2021.
48. Xu, C., Dai, G., Hong, Y., Recent advances in high-strength and elastic hydrogels for 3D printing in biomedical applications. *Acta Biomater.*, 95, 50–59, 2019.
49. Zhao, Q., Liang, Y., Ren, L., Qiu, F., Zhang, Z., Ren, L., Study on temperature and near-infrared driving characteristics of hydrogel actuator fabricated via molding and 3D printing. *J. Mech. Behav. Biomed. Mater.*, 78, 395–403, 2018.
50. Bootsma, K., Fitzgerald, M.M., Free, B., Dimbath, E., Conjerti, J., Reese, G., Sparks, J.L., 3D printing of an interpenetrating network hydrogel material with tunable viscoelastic properties. *J. Mech. Behav. Biomed. Mater.*, 70, 84–94, 2017.
51. Bakarich, S.E., Gorkin III, R., Gately, R., Naficy, S., in het Panhuis, M., Spinks, G.M., 3D printing of tough hydrogel composites with spatially varying materials properties. *Addit. Manuf.*, 14, 24–30, 2017.
52. Rajabi, M., McConnell, M., Cabral, J., Ali, M.A., Chitosan hydrogels in 3D printing for biomedical applications. *Carbohydr. Polym.*, 260, 117768, 2021.
53. Sultan, S., Siqueira, G., Zimmermann, T., Mathew, A.P., 3D printing of nano-cellulosic biomaterials for medical applications. *Curr. Opin. Biomed. Eng.*, 2, 29–34, 2017.

54. Shie, M.Y., Shen, Y.F., Astuti, S.D., Lee, A.K.X., Lin, S.H., Dwijaksara, N.L.B., Chen, Y.W., Review of polymeric materials in 4D printing biomedical applications. *Polymers*, 11, 11, 1864, 2019.
55. Singh, S., Prakash, C., Ramakrishna, S., 3D printing of polyether-ether-ketone for biomedical applications. *Eur. Polym. J.*, 114, 234–248, 2019.
56. Rahim, T.N.A.T., Abdullah, A.M., Md Akil, H., Recent developments in fused deposition modeling-based 3D printing of polymers and their composites. *Polym. Rev.*, 59, 4, 589–624, 2019.
57. Sakthivel, N., Bramsch, J., Voung, P., Swink, I., Averick, S., Vora, H.D., Investigation of 3D-printed PLA–stainless-steel polymeric composite through fused deposition modelling-based additive manufacturing process for biomedical applications. *Med. Devices Sens.*, 3, 6, e10080, 2020.
58. Melnikova, R., Ehrmann, A., Finsterbusch, K., 3D printing of textile-based structures by fused deposition modelling (FDM) with different polymer materials, in: *IOP Conference Series: Materials Science and Engineering*, vol. 62, IOP publishing, p. 012018, 2014.
59. Ni, J., Ling, H., Zhang, S., Wang, Z., Peng, Z., Benyshek, C., Khademhosseini, A., Three-dimensional printing of metals for biomedical applications. *Mater. Today Bio.*, 3, 100024, 2019.
60. Guo, A.X., Cheng, L., Zhan, S., Zhang, S., Xiong, W., Wang, Z., Cao, S.C., Biomedical applications of the powder-based 3D printed titanium alloys: A review. *J. Mater Sci. Technol.*, 125, 252–264, 2022.
61. Varma, M.V., Kandasubramanian, B., Ibrahim, S.M., 3D printed scaffolds for biomedical applications. *Mater. Chem. Phys.*, 255, 123642, 2020.
62. Han, Y., Wei, Q., Chang, P., Hu, K., Okoro, O.V., Shavandi, A., Nie, L., Three-dimensional printing of hydroxyapatite composites for biomedical application. *Crystals*, 11, 4, 353, 2021.
63. Cao, Y., Shi, T., Jiao, C., Liang, H., Chen, R., Tian, Z., Shen, L., Fabrication and properties of zirconia/hydroxyapatite composite scaffold based on digital light processing. *Ceram. Int.*, 46, 2, 2300–2308, 2020.
64. Corcione, C.E., Gervaso, F., Scalera, F., Padmanabhan, S.K., Madaghiele, M., Montagna, F., Maffezzoli, A., Highly loaded hydroxyapatite microsphere/PLA porous scaffolds obtained by fused deposition modelling. *Ceram. Int.*, 45, 2, 2803–2810, 2019.
65. Sahmani, S., Khandan, A., Esmaeili, S., Saber-Samandari, S., Nejad, M.G., Aghdam, M.M., Calcium phosphate-PLA scaffolds fabricated by fused deposition modeling technique for bone tissue applications: Fabrication, characterization and simulation. *Ceram. Int.*, 46, 2, 2447–2456, 2020.
66. Zhang, K., Xie, C., Wang, G., He, R., Ding, G., Wang, M., Fang, D., High solid loading, low viscosity photosensitive Al2O3 slurry for stereolithography based additive manufacturing. *Ceram. Int.*, 45, 1, 203–208, 2019.
67. Zadehnajar, P., Mirmusavi, M.H., Soleymani Eil Bakhtiari, S., Bakhsheshi-Rad, H.R., Karbasi, S., RamaKrishna, S., Berto, F., Recent advances on

akermanite calcium-silicate ceramic for biomedical applications. *Int. J. Appl. Ceram. Technol.*, 18, 6, 1901–1920, 2021.
68. Gopinathan, J. and Noh, I., Recent trends in bioinks for 3D printing. *Biomater. Res.*, 22, 1, 1–15, 2018.
69. González, L.M., Mukhitov, N., Voigt, C.A., Resilient living materials built by printing bacterial spores. *Nat. Chem. Biol.*, 16, 2, 126–133, 2020.
70. Liu, X., Yuk, H., Lin, S., Parada, G.A., Tang, T.C., Tham, E., Zhao, X., 3D printing of living responsive materials and devices. *Adv. Mater.*, 30, 4, 1704821, 2018.
71. Romanazzo, S., Molley, T.G., Nemec, S., Lin, K., Sheikh, R., Gooding, J.J., Roohani, I., Synthetic bone-like structures through omnidirectional ceramic bioprinting in cell suspensions. *Adv. Funct. Mater.*, 31, 13, 2008216, 2021.
72. Trombetta, R., Inzana, J.A., Schwarz, E.M., Kates, S.L., Awad, H.A., 3D printing of calcium phosphate ceramics for bone tissue engineering and drug delivery. *Ann. BioMed. Eng.*, 45, 1, 23–44, 2017.
73. Graham, A.D., Olof, S.N., Burke, M.J., Armstrong, J.P., Mikhailova, E.A., Nicholson, J.G., Bayley, H., High-resolution patterned cellular constructs by droplet-based 3D printing. *Sci. Rep.*, 7, 1, 1–11, 2017.
74. Rajendran, A. and Pattanayak, D.K., Bioactive and antimicrobial macro-/micro-nanoporous selective laser melted Ti–6Al–4V alloy for biomedical applications. *Heliyon*, 8, 3, e09122, 2022.
75. Bozkurt, Y. and Karayel, E., 3D printing technology; methods, biomedical applications, future opportunities and trends. *J. Mater. Res. Technol.*, 14, 1430–1450, 2021.
76. Yu, J.R., Navarro, J., Coburn, J.C., Mahadik, B., Molnar, J., Holmes IV, J.H., Fisher, J.P., Current and future perspectives on skin tissue engineering: Key features of biomedical research, translational assessment, and clinical application. *Adv. Healthc. Mater.*, 8, 5, 1801471, 2019.
77. Wang, Q., Han, G., Yan, S., Zhang, Q., 3D printing of silk fibroin for biomedical applications. *Materials*, 12, 3, 504, 2019.
78. Yan, W.C., Davoodi, P., Vijayavenkataraman, S., Tian, Y., Ng, W.C., Fuh, J.Y., Wang, C.H., 3D bioprinting of skin tissue: From pre-processing to final product evaluation. *Adv. Drug Deliv. Rev.*, 132, 270–295, 2018.
79. Dahiya, R., E-skin: From humanoids to humans [point of view]. *Proc. IEEE*, 107, 2, 247–252, 2019.
80. Ma, Z., Li, S., Wang, H., Cheng, W., Li, Y., Pan, L., Shi, Y., Advanced electronic skin devices for healthcare applications. *J. Mater Chem. B.*, 7, 2, 173–197, 2019.
81. Zhou, F., Hong, Y., Liang, R., Zhang, X., Liao, Y., Jiang, D., Ouyang, H., Rapid printing of bio-inspired 3D tissue constructs for skin regeneration. *Biomaterials*, 258, 120287, 2020.
82. Uddin, M.J., Scoutaris, N., Economidou, S.N., Giraud, C., Chowdhry, B.Z., Donnelly, R.F., Douroumis, D., 3D printed microneedles for anticancer therapy of skin tumours. *Mater. Sci. Eng. C.*, 107, 110248, 2020.

83. Gao, Y., Yu, G., Shu, T., Chen, Y., Yang, W., Liu, Y., Xuan, F., 3D-printed coaxial fibers for integrated wearable sensor skin. *Adv. Mater. Technol.*, 4, 10, 1900504, 2019.
84. Adepu, S., Dhiman, N., Laha, A., Sharma, C.S., Ramakrishna, S., Khandelwal, M., Three-dimensional bioprinting for bone tissue regeneration. *Curr. Opin. Biomed. Eng.*, 2, 22–28, 2017.
85. Zare, R.N., Doustkhah, E., Assadi, M.H.N., Three-dimensional bone printing using hydroxyapatite-PLA composite. *Mater. Today: Proc.*, 42, 1531–1533, 2021.
86. Zhao, L., Pei, X., Jiang, L., Hu, C., Sun, J., Xing, F., Zhang, X., Bionic design and 3D printing of porous titanium alloy scaffolds for bone tissue repair. *Compos. B. Eng.*, 162, 154–161, 2019.
87. Mondal, S., Nguyen, T.P., Hoang, G., Manivasagan, P., Kim, M.H., Nam, S.Y., Oh, J., Hydroxyapatite nano bioceramics optimized 3D printed poly lactic acid scaffold for bone tissue engineering application. *Ceram. Int.*, 46, 3, 3443–3455, 2020.
88. Sahmani, S., Khandan, A., Esmaeili, S., Saber-Samandari, S., Nejad, M.G., Aghdam, M.M., Calcium phosphate-PLA scaffolds fabricated by fused deposition modeling technique for bone tissue applications: Fabrication, characterization and simulation. *Ceram. Int.*, 46, 2, 2447–2456, 2020.
89. Wang, G., Influence of polydopamine/polylactic acid coating on mechanical properties and cell behavior of 3D-printed calcium silicate scaffolds. *Mater. Lett.*, 275, 128131, 2020.
90. Prasad, S., Kader, N.A., Sujatha, G., Raj, T., Patil, S., 3D printing in dentistry. *J. 3D Print. Med.*, 2, 3, 89–91, 2018.
91. Zaharia, C., Gabor, A.G., Gavrilovici, A., Stan, A.T., Idorasi, L., Sinescu, C., Negruțiu, M.L., Digital dentistry-3D printing applications. *J. Interdiscip. Med.*, 2, 1, 50–53, 2017.
92. Branco, A.C., Silva, R., Santos, T., Jorge, H., Rodrigues, A.R., Fernandes, R., Figueiredo-Pina, C.G., Suitability of 3D printed pieces of nanocrystalline zirconia for dental applications. *Dent. Mater.*, 36, 3, 442–455, 2020.
93. Ozbolat, I.T., Chen, H., Yu, Y., Development of 'Multi-arm Bioprinter'for hybrid biofabrication of tissue engineering constructs. *Robot Comput. Integr. Manuf.*, 30, 3, 295–304, 2014.
94. Bracaglia, L.G., Smith, B.T., Watson, E., Arumugasaamy, N., Mikos, A.G., Fisher, J.P., 3D printing for the design and fabrication of polymer-based gradient scaffolds. *Acta Biomat.*, 56, 3–13, 2017.
95. Bégin-Drolet, A., Dussault, M.A., Fernandez, S.A., Larose-Dutil, J., Leask, R.L., Hoesli, C.A., Ruel, J., Design of a 3D printer head for additive manufacturing of sugar glass for tissue engineering applications. *Addit. Manuf.*, 15, 29–39, 2017.
96. Baptista, R., Guedes, M., Pereira, M.F., Maurício, A., Carrelo, H., Cidade, T., On the effect of design and fabrication parameters on mechanical performance of 3D printed PLA scaffolds. *Bioprinting*, 20, e00096, 2020.

97. Ma, K., Zhao, T., Yang, L., Wang, P., Jin, J., Teng, H., Wang, X., Application of robotic-assisted *in situ* 3D printing in cartilage regeneration with HAMA hydrogel: An *in vivo* study. *J. Adv. Res.*, 23, 123–132, 2020.
98. Shie, M.Y., Chang, W.C., Wei, L.J., Huang, Y.H., Chen, C.H., Shih, C.T., Shen, Y.F., 3D printing of cytocompatible water-based light-cured polyurethane with hyaluronic acid for cartilage tissue engineering applications. *Materials*, 10, 2, 136, 2017.
99. Mehrotra, D., Dwivedi, R., Nandana, D., Singh, R.K., From injectable to 3D printed hydrogels in maxillofacial tissue engineering: A review. *J. Oral. Biol. Craniofac Res.*, 10, 4, 680–689, 2020.
100. Marques, C.F., Diogo, G.S., Pina, S., Oliveira, J.M., Silva, T.H., Reis, R.L., Collagen-based bioinks for hard tissue engineering applications: A comprehensive review. *J. Mater. Sci.: Mater. Med.*, 30, 3, 1–12, 2019.
101. Saroia, J., Yanen, W., Wei, Q., Zhang, K., Lu, T., Zhang, B., A review on biocompatibility nature of hydrogels with 3D printing techniques, tissue engineering application and its future prospective. *Bio-Des. Manuf.*, 1, 4, 265–279, 2018.
102. Xu, X., Awad, A., Robles-Martinez, P., Gaisford, S., Goyanes, A., Basit, A.W., Vat photopolymerization 3D printing for advanced drug delivery and medical device applications. *J. Control Release.*, 329, 743–757, 2021.
103. Costa, R.M., Rauf, S., Hauser, C.A., Towards biologically relevant synthetic designer matrices in 3D bioprinting for tissue engineering and regenerative medicine. *Curr. Opin. Biomed. Eng.*, 2, 90–98, 2017.
104. Martinez, P.R., Goyanes, A., Basit, A.W., Gaisford, S., Fabrication of drug-loaded hydrogels with stereolithographic 3D printing. *Int. J. Pharm.*, 532, 1, 313–317, 2017.
105. Trenfield, S.J., Awad, A., Madla, C.M., Hatton, G.B., Firth, J., Goyanes, A., Basit, A.W., Shaping the future: Recent advances of 3D printing in drug delivery and healthcare. *Expert Opin. Drug Deliv.*, 16, 10, 1081–1094, 2019.
106. Bozuyuk, U., Yasa, O., Yasa, I.C., Ceylan, H., Kizilel, S., Sitti, M., Light-triggered drug release from 3D-printed magnetic chitosan microswimmers. *ACS nano*, 12, 9, 9617–9625, 2018.
107. Economidou, S.N., Lamprou, D.A., Douroumis, D., 3D printing applications for transdermal drug delivery. *Int. J. Pharm.*, 544, 2, 415–424, 2018.
108. Larush, L., Kaner, I., Fluksman, A., Tamsut, A., Pawar, A.A., Lesnovski, P., Magdassi, S., 3D printing of responsive hydrogels for drug-delivery systems. *3D Print. Med.*, 1, 4, 219–229, 2017.
109. Liaw, C.Y. and Guvendiren, M., Current and emerging applications of 3D printing in medicine. *Biofabrication*, 9, 2, 024102, 2017.
110. Mathew, E., Pitzanti, G., Larrañeta, E., Lamprou, D.A., 3D printing of pharmaceuticals and drug delivery devices. *Pharmaceutics*, 12, 3, 266, 2020.
111. Katseli, V., Economou, A., Kokkinos, C., Single-step fabrication of an integrated 3D-printed device for electrochemical sensing applications. *Electrochem. Commun.*, 103, 100–103, 2019.

112. Cardoso, R.M., Silva, P.R., Lima, A.P., Rocha, D.P., Oliveira, T.C., do Prado, T.M., Munoz, R.A., 3D-Printed graphene/polylactic acid electrode for bioanalysis: Biosensing of glucose and simultaneous determination of uric acid and nitrite in biological fluids. *Sens. Actuators B Chem.*, 307, 127621, 2020.
113. Liu, Y., Li, T., Ma, H., Zhai, D., Deng, C., Wang, J., Wu, C., 3D-printed scaffolds with bioactive elements-induced photothermal effect for bone tumor therapy. *Acta Biomaterialia*, 73, 531–546, 2018.
114. Vijayavenkataraman, S., Fuh, J.Y., Lu, W.F., 3D printing and 3D bioprinting in pediatrics. *Bioengineering*, 4, 3, 63, 2017.
115. Friedrich, L. and Begley, M., *In situ* characterization of low-viscosity direct ink writing: Stability, wetting, and rotational flows. *J. Colloid Interface Sci.*, 529, 599–609, 2018.
116. Naghieh, S. and Chen, X., Printability–A key issue in extrusion-based bioprinting. *J. Pharm. Anal.*, 11, 5, 564–579, 2021.
117. Ghilan, A., Chiriac, A.P., Nita, L.E., Rusu, A.G., Neamtu, I., Chiriac, V.M., Trends in 3D printing processes for biomedical field: Opportunities and challenges. *J. Polym Environ.*, 28, 5, 1345–1367, 2020.

12

Self-Sustained Nanobiomaterials: Innovative Materials for Biomedical Applications

Sudipto Datta[1]*, Samir Das[2] and Ranjit Barua[3]

[1]*Department of Materials Engineering, Indian Institute of Science, Bangalore, Karnataka, India*
[2]*Indian Institute of Technology, Kharagpur, India*
[3]*Indian Institute of Engineering Science and Technology, Howrah, India*

Abstract
The definition of a biomaterial is a material that interacts with the biological environment and which should be biodegradable and biocompatible. Among various types of biomaterials, nanobiomaterials are the more interesting ones for researchers. Because of their high surface area, and nano-size ranging from 1 to 100 nm, these biomaterials find lots of applications in various biomedical applications like medical imaging, gene delivery, cancer therapy, and tissue engineering. These biomaterials can be easily tuned as per the requirement of the specific applications which makes these biomaterials a supreme candidate for most biomedical applications. Furthermore, because of the multi-functional dimensions of these biomaterials, it improves the patient's life quality by reducing the harmful effects when compared to the traditional biomaterials. In the first part of this chapter, the types, morphology, and dimensions of nanobiomaterials are discussed in brief. The second part of the chapter discussed the use of nano-biomaterials in various biomedical applications. And lastly, the problems and the prospects of the nanobiomaterials are discussed briefly.

***Keywords*:** Nanobiomaterials, nanoparticles, polymer, metals and alloys, tissue engineering, biomedical applications

**Corresponding author*: sudiptodatta1990@gmail.com

Arbind Prasad, Ashwani Kumar and Manoj Gupta (eds.) Advanced Materials and Manufacturing Techniques for Biomedical Applications, (303–324) © 2024 Scrivener Publishing LLC

12.1 Introduction

The term "nanobiomaterials" can be defined by two words: "materials in nanoscale" and "biomaterials" that can consist of particles of dimensions below 100 nm [1]. Biomaterials may be natural or synthetic, substances that interact with biological systems and treat the organs or restore injured tissue to accurate bodily functionality [2]. Nanobiomaterials are the recent trend for realistic applications in chemistry, physics, biology, medicine, and engineering [3, 4]. Most particularly nanotechnology is the science that deals with nanobiomaterials to care for the condition inside the body. Nanobiomaterials like nanoparticles, nanowires, nanosurfaces, nanofibers, and nanocomposites. Those are developed and recently used for a wide variety of nanomedical tissue regeneration and restoration and regeneration, cancer therapy, drug delivery, and medical imaging applications [5]. The building unit of all nanobiomaterials is the nanoparticle. They are made up of polymers (synthetic or natural) and inorganic materials (ceramics or metals). The shape of the nanobiomaterials can be varied from needles, platelets, rods, and polygons, in which spherical nanobiomaterials are the most common ones. Unlike other conventional bulk materials, techniques of preparation of nanobiomaterials are in diverse forms such as dispersions, colloids, deposited layers, agglomerates, or suspensions, [6] which also control physical, chemical, mechanical, electrical, and optical properties for enhanced performance and modulation.

The size and shape of the nanobiomaterials depend on the fabrication process and the purpose of application. That will decide how it will interact with the tissue. Shape and size distribution is significant during the synthesis of nanobiomaterials [7]. The era of biometrical lies in the application of inert to bioactive nanomaterials for tissue engineering applications within which it can interact with the area of biological molecules and smaller organelles that are in the micron to submicron size. Nanobiomaterials can offer a smart method to prove the nanoarchitecture of biological tissues for enhancing cellular interaction. They can effectively treat to regenerate the defective tissue for biomimicking the biological structures and physiological functions at the molecular level. Using multiple post-processing, nanobiomaterials can be used for dental implant and load-bearing applications like the knee joint, shoulder joint, and hip joint as well. They can be utilized for culturing cells and analyzing blood proteins and taking the biomolecules for imaging applications [8]. Combining the abilities and functionalization of nanobiomaterials attach with the particles of biomolecules such as cell membranes, antibodies, drug molecules, and additional biological compositions is essential for exact applications. The foreign particle

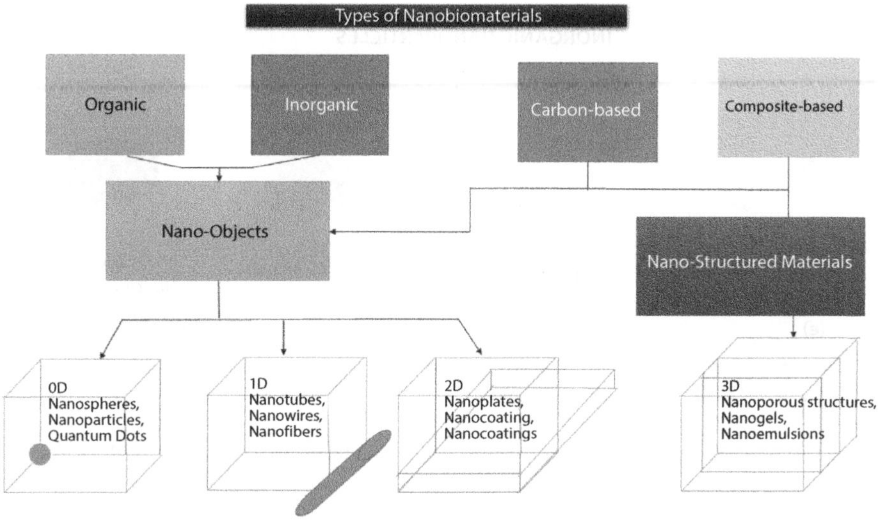

Figure 12.1 Types of nanobiomaterials [70].

reaction of nanobiomaterials depends on the design and size of the surface chemistry and the degradation profile of the implanted materials [9]. The most challenging is the nature of the nanomaterials and the safety concern during interaction with the body tissues. Further advancing the design of the nanobiomaterials for biosensing applications can be achieved at the nanoscale levels, which requires revising the design parameters and criteria of the nanobiomaterials [10]. This article will contribute design criteria and measure utility challenges toward the clinical transformation of nanobiomaterials in nanomedicine for intensifying applications. In Figure 12.1 types of nanobiomaterials are shown.

12.1.1 Classification of Nanobiomaterials

Nanobiomaterials can classify concerning their physical and chemical properties such as dimensional morphology as well as composition and morphology. Nan materials can be classified according to the electrical optical magnetic properties as well. It can be also classified according to its biomedical applications in which biocompatibility and cytotoxicity play an important role to know the *in vivo* and *in vitro* characteristics of nanobiomaterials. The modification of surface chemistry is extremely studied for immunes' system evasion reduction inside toxicity improved in cellular targeting [11]. Figures 12.2 and 12.3 represents the inorganic and organic types of nanobiomaterials. Morphological effects with exact dimensionality

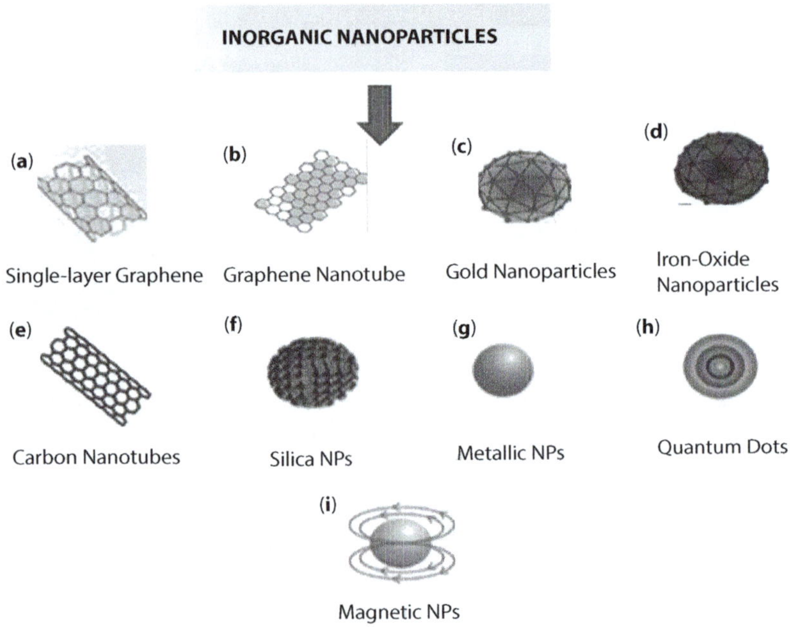

Figure 12.2 Inorganic nanoparticles [70].

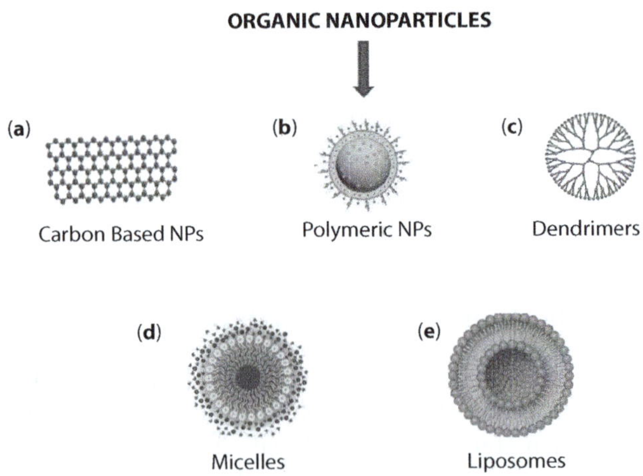

Figure 12.3 Organic nanoparticles [70].

like shape, size, and the width-to-height ratio will contribute towards tissue penetration biodistribution and cellular uptake of nanobiomaterials. Figure 12.4 represents polymeric-based and lipid-based nanobiomaterials types.

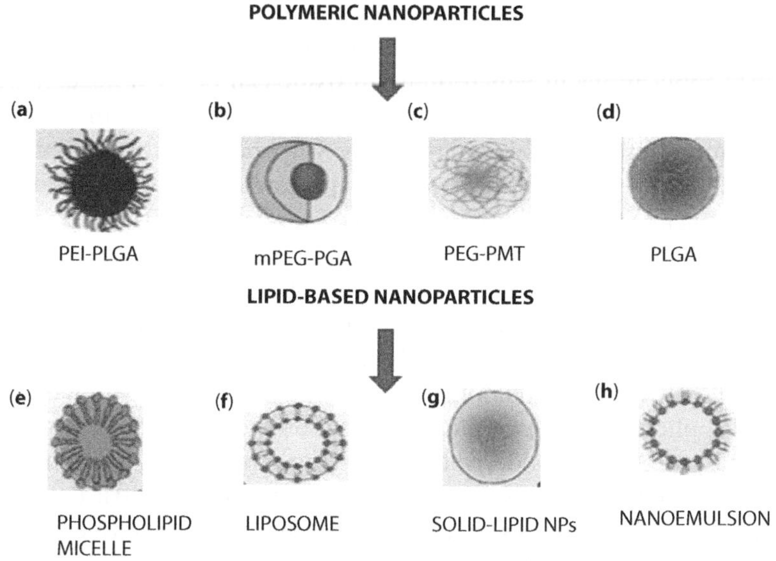

Figure 12.4 Polymeric-based and lipid-based nanoparticles [70].

12.1.2 Composition

Nanobiometricals are consisting of organic inorganic and sometimes carbon-based composite. It will vary from different morphological parameters such as shape and size. Nanobiometricals compositions consist of the core of the materials shell of the materials and the modifications of surface chemistry. Generally, the prior property that is to be customized for biocompatibility is to reduce cytotoxicity in the chemical composition, and the presence of organic materials in it. Carbon-based organic nanobiometricals that can be synthetic or natural polymers and those polymer base nanomaterials are largely investigated by pharmaceutical applications of control and sustainable drug delivery. The most favorite polymers of synthetic origin like polyglycolic acid, and polylactic acid, and polymers of natural origin like chitosan, and silk can be used for synthesizing nanobiomaterials. Those nanobiomaterials can be extensively used for drug delivery, tissue regeneration as well as diagnostic applications due to their low toxicity effect [12, 13]. The biocompatibility with tunable biodegradability can modulate the drug release kinetics, cytotoxicity, and cellular interaction of the nanobiometricals during tissue engineering applications. Nanomaterials are carbon-free and can also include metal and nonmetal as well. Silver, copper, gold, and iron-based nanoparticles are well established for diagnosis, drug delivery, and orthopedic repairs. Those have some unique properties concerning their bulk metal derivatives. Metal oxides such as titanium oxide and magnesium oxide are well established for drug delivery applications in which titanium oxide can be used for constructing synthetic processes and implants. The metal oxides are mechanically strong and show good biocompatibility with better antibacterial properties. Ceramics like HAP have a large application in natural hard tissue engineering like teeth and bone repairing. Inorganic materials like calcium and phosphate have some flexibility while applying hard tissue engineering [14]. A nanoclay-like platelet of thin layers of silicate materials of size one nanometer in thickness is recently and promisingly used as bio-compatible nanomaterials that are mechanically modified scaffolds for tissue regeneration [15]. Those Nanobiomaterials can be used as inert mineral fillers for oral applications [16]. Carbone base nanomaterials including fullerene, graphene oxides, and its derivative like nanodiamonds, Quantum dots of carbon-based have exclusive and tremendous, thermal, electrical, mechanical, and optical properties that have concerned them to tissue engineering and drug delivery applications [17]. Notwithstanding those exclusive properties, cytotoxicity of biocompatibility and biodegradations are the additional advantages of using nanobiomaterials in the medical field [18]. Nanobiomaterials of composite types can be fabricated using a combination of organic and inorganic materials to balance the weakness of a particular material and form multiphase materials indeed [19]. The permutation of different materials is capable to produce a hybrid material with customizable mechanical strength and biocompatibility. For example, artificial polymers are joined with natural polymers to increase biocompatibility, and natural polymers are combined with a synthetic polymer to modify and increase the material mechanical strength of the newly designed nanobiomaterials [19, 20].

12.1.3 Dimensionality

The dimensionality of nanobiomaterials referred to 3D external dimensions such as depth and height that lies in the nanoscale divisions such as zero-dimensional Quantum dot. Those zero-dimensional nanomaterials have a nanoscale dimension in all three external dimensions. In the case of one-dimensional nanobiomaterials two external dimensions have at nanoscales level. These consist of elongated nanofiber, nanowires, and nanotubes. In the case of nanobiomaterials with two dimensionality, one external dimension has nanoscales level-like nanofilms, nanolayers, etc. 3D nanomaterials don't contain any exterior dimension at the nanoscale level but they are considered multiple or bulk arrangement materials and encompass interior structure at the nanoscale level [21]. Current developments for cancer therapy and diagnosis in nanomedicine have proposed a fourth dimension such as "Time". The availability of nanomaterials is to vary over time, particularly for degradation and then accumulation inside the body [22]. 1D and 2D nanomaterials are rapidly used in biosensing, tissue engineering, drug delivery, and diagnosis applications with enhanced cellular and biomolecular interaction at the material boundary [23].

12.1.4 Morphology

Morphological classification can be subcategorized into shape, size, and the width-to-height ratio [24, 25]. Many studies have prominently showed the size of nanoparticle effect cellular penetration and distribution we can consider that smaller size particles having a greater surface-to-volume ratio can generate more molecular signals at the cellular level. A report by Pan and coworkers indicated that 15-nanometer colloids of gold are non-toxic though 1.2 nm and 1.4 nm nanoparticles of gold cause apoptotic cell death [26], due to their penetrating and accumulation properties anywhere in the body. For that, over the last decade, the potential of genotoxicity and cytotoxicity have been studied at the cellular level based on the nonmaterial with the cellular interactions [27]. Higher aspect ratio nanomaterials like nanotech nanofibers, nanochain typically 1D nonmaterial can accumulate anywhere in the body [28, 29, 31]. Lower Aspect ratio nanomaterials like spherical shapes nanospheres are less accumulated in the body and more interaction with the cellular entities as well [30–32]. The greatest rising groups of drug delivery nanomaterials are carbon nanodots and gold nanorods which have the largest surface-to-volume ratio that will provide larger space for numerous types of tissue functionalization to improve cellular interactions.

12.2 Nanobiomaterials Applications

In precision medicine and nanomedicine, nanotechnology is an important part. NPs with a huge range of properties, new structural changes, more surface areas, and small-scale nano-dimension provide exclusive functionalities of biomimicry and manipulability for improving the biological targeting and interactions. This gives freedom to change the nano-biointerfacial functions, microenvironmental interactivity, tunable properties, structural architecture as well as the scene of nanorobots development for doing cellular level operations [35]. The physiochemical properties advancement of the nanobiomaterials showed that it improved the tenability, biodegradability, stability, and biocompatibility because of this it gains attention in biomedical domains like oncology, tissue engineering, and drug delivery, diagnostics as well as the fabrication of various antifouling and antimicrobial biomaterials. In this section, the applications of nanobiomaterials used in various biomedical areas are discussed.

12.2.1 Drug Deliverance

The hurdle of the human body like the blood-brain barrier, blood tissue barrier, stomach acid, tears, mucous membrane, and skin make delivery of the targeted therapeutics drugs complicated to achieve. The therapeutic and pharmacological drug effects are reduced when it is directed into the bloodstream of the human body because of the previously mentioned barriers that resist the entry of the molecules of the drug to the targeted tissue. The blood-tissue barriers formed by the epithelial cells contain a thick rigid junction, discriminatory transporters of small molecules and ions, and a lesser number of pinocytotic vesicles which results in crossing obstruction of the macromolecules through the barrier [36, 37]. Nanocarriers i.e. the nanobiomaterials take the complex responsibility to carry the therapeutic molecules of the drug through the blood-tissue barriers to the targeted tissue location for which the drug is specially made. The small nano size of the nanobiomaterials and their exclusive properties allows them to pass the blood-tissue barrier and to deliver the drug to the specific tissue target location in the body. Nanocarriers are modified nanomaterial developed for drug delivery. It can carry both hydrophobic as well as hydrophilic drugs through the physical entrapment inside the core of the nanomaterial. Also, they attached the drug to the surface of the nanocarrier by covalent or non-covalent interactions through nondegradable and degradable bonds for controlled release functions, targeted drug delivery, and optimized drug use [37, 38]. There are various types of nanocarriers they are a) nanospheres, b) dendrimers, c) liposomes, d) micelles, and e) nanocapsules [39]. The adjustable shape, surface, and size properties of the nanocarriers allow the development of nanopharmaceuticals having good biocompatibility, solubility, and stability considering the biological fluids [38]. The older technologies for opening the barrier tight junction and disturbing the barrier function for targeted drug delivery for improving the permeability are now replaced with nanotechnological tactics. The administration of the nanocarrier into the human body is via the following methods which are parenteral delivery, intraperitoneal delivery, skin, and oral delivery. The entry of nanocarriers into the targeted location can be done by two processes active and passive targeting [40]. For active targeting, physical methods are commonly used for enhancing the nanocarrier permeability via the membrane of the cells for reaching the targeted location [40]. In passive targeting, the accumulation of the nanodrug or the nanobiomaterial into the affected area is done, which is hugely related to how much time the drug can sustain in the bloodstream [40]. The problem blood-tissue barriers which are solved by the nanocarriers are mainly the blood-retinal barrier for the treatment of diabetic retinopathy, macular degeneration for old aged people, glaucoma, and the blood-brain barrier for neurodegenerative infections and disorders treatment [41, 42]. This results in optimization of the drug's therapeutic efficacy, reduction of toxicity and unwanted side-effects, and resistance of the drug [39].

12.2.2 Oncology

Nanobiomaterials are also used in oncology for cancer treatment and drug delivery. It improves the chemotherapy drug's performance with respect to safety, specificity, and bioavailability. With the help of nanobiomaterials, the synthesis and design of multifunctional and extremely capable drug delivery nanosystems are possible which mark the location of the tumor with a diversity of moieties, like imaging probes, cytotoxic agents, antibodies, tumor-specific ligands, etc. [43]. For successful cancer treatment, rational drug therapy and early detection of the tumor are necessary. The main problem is that the biological barriers resist the diagnostic and therapeutic element's entry into the targeted cancer tissue. The orthodox cancer detection techniques like computed tomography, endoscopy, ultrasound, MRI, and X-ray do not detect cancerous changes until and unless there is a macroscopic change in the tissue [44]. As long as the diffusion of the diagnostic and therapeutic elements inside the mingling tumor cells is achieved the treatment and the primary recognition of the cancer site is not possible. Nanobiomaterials i.e. the nanocarriers are very efficient to cross the cancer biological barriers like tumor cell membrane penetration, extracellular matrix (ECM) intratumoral pressure, and the mononuclear phagocyte system [33–35]. The main function of the nanobiomaterials is to execute two functions a) delivery of the genes and therapeutic drugs to the site of cancer and b) detect the changes at the molecular level because of the metastasis of the tumor, this involves the evolving domain of theranostics. Other problems of efficient therapies for cancer involve lysosomal escape, systemic toxicity, and low cellular uptake linked with the present chemotherapy treatments. Nanocarriers which are mainly modified according to many shapes and sizes and with particular surface functionalities can overcome these problems by giving improved times of blood circulation, improving the surface area for maximum capacities of drugs, high cellular uptake, and efficient tumor targeting [35]. For cancer drug delivery, nanocarriers that are designed may include inorganic and organic nanoparticles, peptide nanoparticles, peptide, and polymeric nanoparticles, viral nanocarriers, nanoemulsions, nanodiamonds, carbon nanotubes, and dendrimers, nanoparticles, solid and lipid nanoparticles [38]. The conventional efficacy of nanocarriers has directed the constitutional support and medical use of various formulations of nanobiomaterials (employing PEGylation, albumin, and liposomes for better solubility, delivery, and circulation half-lives of the currently obtainable chemotherapeutic agents like irinotecan, mifamurtide, vincristine, paclitaxel, daunorubicin, and doxorubicin [38, 45, 46]. Though nanocarriers for chemotherapeutic drug delivery have gotten much attention for their good efficiency, researchers and scientists are still struggling to bring this technology for clinical uses for cancer early detection further detailed research is still needed. Nanocarriers give many advantages like distended aspect ratio and nanoscale size which allows the area of the surface to get thickly populated by various moieties, like aptamers, small molecules, peptides, and antibodies for binding and identifying particular molecules of cancer. The design of the nanocarriers is done in such a way that they can arrest circulating tumor cells, circulating tumor DNA, exosomes, and cancer biomarkers. For diagnostics cancer, the nanocarriers commonly used are magnetic nanoparticles, gold nanoparticles, polymer dots, and quantum dots [47]. Though still there are no clinical uses of the mentioned nanocarriers presently for the diagnostics of cancer, the potential is well apparent seeing the progress of further nanodiagnostic gears.

12.2.3 Diagnostics

Nanobiomaterials are applied in identifies consist of different modern clinical diagnostic like detection, medical imaging, sensing, etc. While any outer material placed into the living body should be non-toxic, biocompatible, biodegradable, and non-immunogenic. Nanostructures exist as 0, 1, and 2 dimensions materials, which show the size variety in which biomolecules, for instance, proteins, microbes, and nucleic acids present at the nanoscale. *In vitro* diagnostics use materials to identify biochemical transforms, performance, and concentrations of exacting ingredients in biological specimen samples collected from the living body for identifying a disease or health circumstance [40]. Present *in vivo* diagnostics contain more persistent measures to check biomolecules (for example, antigen test for skin or biopsies test, etc.), and for imaging, intentions to follow the incidence or development of a syndrome (for example, MRI, X-ray, and CT-scan for the imagines infections of pulmonary tuberculosis or any cancers in different body parts, correspondingly). Significant therapeutic imaging relies on complicated investigations to distinguish biological procedures and syndrome sequences on a molecular and cellular level. Nanoscale probes demonstrate advanced arrangements, judged to particular molecular complementary agents, for better comparison and biomarker revelation, amplified time of circulation, and high loading surface area ability and associate by fluorophores, or else theranostic functionality of the molecules of drug [7, 48]. Similarly, the purpose of nanomaterials for biosensing functions is to present proficient investigative methods for disease or infection detection by depending on the improved ratio of surface and volume [49]. Nanoparticle probes, for example, quantum dots, and fluorescent nanobeads develop indication brightness (signal), and photostability, and propose multiplexing competencies compared to traditional fluorescent molecules and organic dyes [51]. Biosensors are distinct as diagnostic devices applied for the recognition of a substance analysis or material by exchanging a biological or biochemical effect into a scientific and quantifiable signal physicochemical. These are applied in drugs to distinguish molecules of the biological samples, (for instance, blood glucose), pathogens (for example viruses or bacteria), or erstwhile pollutant agents [52]. A biosensor device includes two elements, a transducer component, and a bioreceptor component. The bioreceptor includes moieties like antibodies, microorganisms, enzymes, aptamers, DNAs, or cells familiar with the element target. A transducer includes nanomaterials of semiconductors, along with a system of electronics, and is composed of a signal processor, and an amplifier, which exhibits that exchanges the reaction to a quantifiable signal [53]. Nanobiomaterials, for example, magnetic nanoparticles, carbon-based nanostructures, and metal nanoparticles are examined as transporters for improved signal amplification. Metal and carbon-based nanomaterials are efficient in advancing direct electron transfer involving the surface of electrodes and biomolecules. The main requirements for biosensing nanobiomaterials comprise chemical constancy, strong mechanical strength, biocompatibility, and good electrical conductivity [52]. The biosensing applications of nanobiomaterials will add to the progress of highly receptive and precise technology. This will permit the outset of a high-performance strategy having lower system difficulty, improved sensitivity, along with condensed time of examination for medical utilization outside the laboratory [53] or limited resource surroundings.

12.2.4 Application in Tissue Engineering

Tissue engineering is a quickly budding area to facilitate intends to renovate living soft biological organs and tissues by applying artificial or typical materials as constructs to substitute or redevelop imperfect living organs and living tissues. The tissue construct must be steady substrates that allow cell replication and differentiation [5]. The ideal construct should contain suitable opens of adequate dimensions for blood vessels and cells to go through and connect with bones. Various polymeric materials like polyglycolic acid, polylactic acid, and elastomers of polyglycerol have been applied in the modern tissue engineering domain. Artificial hydrogels are likewise being considered for the advancement of tissue constructs due to their outstanding biocompatibility, as well as tissue mimic features [5, 48]. For summarization, of the bioelectrical features of cardiac, ECM, bone, skeletal, and neuronal muscle tissues, electro-conductive nanomaterials are considered the present invention of biomaterials for tissue engineering applications [54]. The nanobiomaterials have a dimension of <100 nm, and because of their similar size to biomolecules as well as former biological nano and microstructures, which can interrelate with and be affected by cells for encouraging a cellular reaction [48]. Because of their minuscule size and bio-mimetic quality, nanomaterials can encourage receptors of cells and perform as supply signals directions to begin exact cellular characteristics [55]. The objective of tissue engineering is to expand a biomimetic biological alternative able of reinstating, upholding, or recovering the native purpose of the defective tissue. Conventional scaffold production methods frequently do not succeed to copycat the 3D ECM microstructure to give the most favorable surroundings for proliferation, adhesion, and discrimination of cells. To rise above this drawback, nanomaterials endorse attention in design activities for higher tissue engineering scaffolds (nanostructured) for the most favorable biomimetic, bioactivity, and biomechanical characteristics. Dissimilar arrangements of usual and artificial polymers and a variety of nanofabrication methods are applied to make nanofibrous scaffolds that look like normal ECM. Nanoporous are applied to rebuild the spongy and fibrillar nanoscale ECM structure for applying fibers of suitable diameters precise in an extremely organized, permeable architecture to permit waste removal and mass transfer through the initial tissues [56]. Influencing biomaterials to make nanoscale structures and planes (nanomaterials) imitate the inhabitant nano and micro-environment (chemical and structural) of tissues and cells to persuade cell growth, proliferation, adhesion, and separation for most select tissue remedial [54]. While the biological element is entrenched inside the body, it creates a foreign body response. The amount of this response relays on the size, shape, surface chemistry, design, electric charge, roughness, composition, porosity, hydrophilic or hydrophobic nature, contact degradation and duration, and characteristics of the entrenched device [9, 57, 58]. Considering such nanomaterial characteristics in addition to the aptitude to manage them, directs the connections which arise between the nanomaterial boundary and the cells and tissues. As well, material reactions to the congregation, like degradation of materials, swelling, as well as being free of the monomers, also take place at the cell–biomaterial boundary. All of these interfacial measures are significant for the accomplishment of nanomaterials as they associate with substance cytocompatibility as well as host-resistant reactions that decide its effectiveness and security *in-vivo* [5, 48].

12.2.5 Antifouling and Antimicrobial Nanobiomaterials

Remedy or drug, in the precedent and these days, is reliant on the accessibility of effectual antibiotics to administer contaminations, mainly in offensive traditional surgeries. The unreasonable application of antibiotics has behaved in the appearance of antibiotic-defiant microbes, which, therefore, requires the inventive proposal of antifouling and antimicrobial biomaterials to avoid biofilms from an appearance on the medical supplies, surfaces and grafts. The rising utilization of medical grafts (tissue engineering assembles, and artificial joints), pacemakers like medical instruments, and various materials of medical applications (dental or catheter materials) offer external surfaces *in-vivo* for microbial accessories, thus raising the threat of infections and require for antibiotic management [59]. With the progress in nanotechnology and material science is the invention of new nanomaterials with inherent antifouling and antimicrobial characteristics to easiness microbial executive in biomedical applications. The area of dentistry is mainly achieving impetus in the application of antifouling and antimicrobial nanomaterials for the avoidance of biofilm creation and repeated caries in dental endodontic, tissue regenerative, and restorative uses of the oral cavity. Apart from the antibiotic-free materials, the chemical formation of nanomaterials may also have inherent antifouling and antimicrobial activity, for example, polymeric biocides, metallic nanoparticles, as well as biocidal polymers, that use their antifouling and antimicrobial act upon straight microorganism contact. Metallic nanoparticles (for example, Ag, iron, Zn, and Ti metal oxide) interrupt microbial membranes and produce ROS (reactive oxygen species) that create protein denaturation, mitochondrial damage, cell membrane damage, and mitochondrial damage [61, 62]. It is recommended that composite constructs including Ag nanoparticles can normalize bacterial contamination throughout the reconstructive bone operation and perform like covering for safety in contradiction of succeeding sepsis, infection, or disturbing of grafts [63]. Other action mechanisms comprise reticence of metabolic procedures, interruption of the oxidation of macromolecules, electron transport system, and inhibition of replication of DNA [64] dissimilar antibiotics, nanomaterials offer a prospect to boundary microbial enlargement by the inception of disease, and consequently, decrease the possibility of initial microbial confrontation [60].

12.3 Challenge in the Clinical Rendition of Nanobiomaterials

12.3.1 Nanotoxicity

The applications of nanotechnology have made significant progress. However, nanomaterial exposure may have possible threats. Due to its outcomes, the study of health and toxicology risk estimation has become relevant [10]. It is the uniqueness of nanoscale materials that has made it advantageous as well as probably risky for a few cases. "Nanotoxicity" can be defined as the possible negative impacts on human health caused by exposure to new nanomaterials. Thus, the safety and reliability of engineered nanomaterials are attracting particular concern. Also, the nanophase surface attributes must correlate with other characteristics like stability, cytotoxicity, and biodistribution. The formation of free radicals and consequent oxidative stress may be influenced by manipulating nanomaterial properties like shape, size, surface chemistry, and degree of aggregation. The oxidative stress may become inflammatory, further leading to genotoxicity. There may be an unexpected reorganization of nanoparticles from their target area, which can bypass natural phagocytic defenses and alter the protein structures. Also, existing literature has reported that nanosized particles are generally poisonous to the bone marrow, lungs, heart, liver, kidneys, and spleen [10]. Hence, discovering non-toxic and safe nanoparticle amounts needs comparatively more realistic and long-term explorations. Additionally, new values are required to study the interface properties of nanofabrication and nanoparticles techniques [65]. In the field of orthopedics, wear of artificial joints causes the aseptic loosening of orthopedic implants. It is well-known that nano-sized wear particles are greatly inflammatory, which may cause osteolysis. Also, it may move to another body part and cause harmful reactions. Mostly, the literature has reported short-term studies of nanotoxicity so far. Thus, further research has attracted attention to anticipate expected long-term impacts. Recent proposals stress the significance of analytical approaches and a detailed understanding of the nanotoxicity mechanisms, which are validated completely. Thus, soon it will become possible to apply nanobiomaterials for diagnostic and therapeutic equipment safely [10]. This prediction has raised the need for nanobiomaterials which are comparatively more biocompatible and non-toxic to natural tissues [66]. Hence, long-term progress has become popular toward green nanobiotechnology to fabricate such nanobiomaterials.

12.3.2 Regulatory Considerations

The regulation of nanobiomaterials design is highly challenging due to the lack of proper guidelines. Safety protocols generally depend on the bulk-material data instead of the nano-state nanobiomaterials, as issued by regulatory bodies like the FDA and the European Science Foundation [67]. This becomes challenging as their operations may differ extensively due to the variation in pharmacodynamics and pharmacokinetics [67]. Thus, the existing safety guidelines may fail to replicate accurate data about the operation and other efficiency parameters of nano-biomaterials in a clinical setup and attract attention to the formation of specific regulations [67]. Also, the testing of bioactivity in the traditional animal trial is considered unethical therefore *in vitro* testing is currently the essential approach to taste biomaterials at the preliminary level as yes it is less costly with higher control of the environment. Regulatory bodies with other institutions and medical organizations have played a major part to determine the guideline for nanomedicine. Clear leadership and guidance will provide the necessary documentation to finalize the nanobiomaterial-based new products to enter the market.

12.3.3 Commercialization

Nanobiomaterials can be standard and exist for clinical use after clearing all the regulatory laws and compliance. Nanomedicines including antifungal and anticancer means have been verified to have an improved ratio of risk-to-benefit and concern about the accumulations and permission of the nanomedicines. The gap in clinical information will lead to difficulty to regulate the design of customized nanobiomaterial-based devices for testing. The toxicity and stability are the important aspects to develop nanobiomaterials in the field of nanomedicine. More lack of information is the difficulty faced in future studies and clinical trials. This global results lack formal regulation and regulatory authorities during the variation from country to country in the field of nanomedical devices. Another constraint of nanobiomaterials advancement is the economical lag. The Control way of using nanomaterials with the 3D printing process can create a unique opportunity to produce nanomedicine at a lower cost [55] for regenerative therapy.

12.4 Conclusion and Future Directions

Biocompatibility and size structure are the two important factors in determining the success of nanobiomaterials. To avoid the non-toxicity of nanomaterial structural modification of the composition and functionality experience is important. Other than the structural composition elemental size and shape is a very important factor during the readjustment and reformulation of nanomaterials for circulation of half-life, biodegradation, and elimination from the biological body. The escalating concern of nontoxicity required standardized protocol for *in-vivo* analysis of clinical trials. To ensure the safety and reliability of the nanomedicine based on nanobiomaterials regulatory guidelines should be strictly followed. Another challenge of the commercialization of Patient-specific nanomedicine is to determine the pharmacokinetics and pharmacodynamics inconsistency across, individuals and populations. Nanobiomaterials have the potential that proves their significance and success in imaging, diagnosis application, drug delivery, and artificial blood formulation which leads to the next level of tissue engineering application based on nanobiomaterials. Advance-level biosensors based on antimicrobial properties are based on functionalized Nanobiomaterials based on biosensing devices. The emerging trend of hybrid bionic cell systems is the nano biomechanics or nanobots to perform integrated cellular functions, diagnosis, and therapeutic agents in the blood systems. Investigating the electrical and optical properties of the nanobiomaterials combined with the nano hydrogel technology advances the biosensing and bio-electrical device functionality in the felid biomedical appliance. Nanobiomaterials have shown a great asset in the field of biomedical from the first decades as they are very much compatible with biological tissue. Nanobiomaterials can attain a level of accuracy due to their smaller dimension compared to other conventional medicine. Their abilities in drug targeting drastically reduce the undesirable effect caused by certain treatment processes such as chemotherapy [40]. The individuality of nonobiomaterial can benefit to resolve many problems connected to human well-being and can incorporate into tissue engineering applications. It can also be effective to treat novel viruses such as COVID-19 [50]. Nanobiomaterials can be extremely appreciated in the case of biomedical applications though more research and trials in clinics are desired to create the proper concern of nanobiomaterials. The regulation certificate estimated the wave of medicines that can make sure about the reliability and the safe use of this product [68, 69].

References

1. Hochella, M.F., There's plenty of room at the bottom: Nanoscience in geochemistry. *Geochim. Cosmochim. Acta*, 66, 5, 735–43, 2002.
2. Hulbert, S.F., Ceramics in clinical applications: Past, present and future, in: *High Tech Ceramics*, P. Vincenzini (Ed.), pp. 189–213, Elsevier, Amsterdam, 1987.
3. Thakur, S. and Agrawal, R., Application of nanotechnology in pharmaceutical formulation design and development. *Curr. Drug Ther.*, 10, 1, 20–34, 2015.
4. Bayda, S., Adeel, M., Tuccinardi, T., Cordani, M., Rizzolio, F., The history of nanoscience and nanotechnology: From chemical– physical applications to nanomedicine. *Molecules*, 25, 1, 112, 2020.
5. Singh, N. and Sharma, U.C., Introduction to nanobiomaterials, in: *Nanotechnology Biomaterials*, vol. 11, N.K. Navani and S. Sinha (Eds.), pp. 1–12, Studium Press LLC, Texas, 2014.
6. Saleh, T.A., Nanomaterials: Classification, properties, and environmental toxicities. *Environ. Technol. Innov.*, 20, 101067, 2020.
7. Salata, O.V., Applications of nanoparticles in biology and medicine. *J. Nanobiotechnol.*, 2, 1, 3, 2004.
8. Ratner, B.D., Hoffman, A.S., Schoen, F.J., Lemons, J.E., *Biomaterials science: An introduction to materials in medicine*, 2nd edition, Elsevier, USA, 2004.
9. Balasundaram, G. and Webster, T.J., Nanotechnology and biomaterials for orthopedic medical applications. *Nanomedicine*, 1, 2, 169–76, 2006.
10. Saji, V.S., Choe, H.C., Yeung, K.W.K., Nanotechnology in biomedical applications: A review. *Int. J. Nano Biomater.*, 3, 2, 119–39, 2010.
11. Schaeublin, N.M., Braydich-Stolle, L.K., Maurer, E.I., Park, K., MacCuspie, R.I., Afrooz, A.R.M.N. *et al.*, Does shape matter? Bioeffects of gold nanomaterials in a human skin cell model. *Langmuir*, 28, 6, 3248–58, 2012.
12. Virlan, M.J.R., Miricescu, D., Radulescu, R., Sabliov, C.M., Totan, A., Calenic, B. *et al.*, Organic nanomaterials and their applications in the treatment of oral diseases. *Molecules*, 21, 2, 207, 2016.
13. Operti, M.C., Bernhardt, A., Grimm, S., Engel, A., Figdor, C.G., Tagit, O., PLGA-based nanomedicines manufacturing: Technologies overview and challenges in industrial scale-up. *Int. J. Pharm.*, 605, 120807, 2021.

14. Kumar, P., Saini, M., Dehiya, B.S., Sindhu, A., Kumar, V., Kumar, R. *et al.*, Comprehensive survey on nanobiomaterials for bone tissue engineering applications. *Nanomaterials*, 10, 10, 2019, 2020.
15. Murali, A., Lokhande, G., Deo, K.A., Brokesh, A., Gaharwar, A.K., Emerging 2D nanomaterials for biomedical applications. *Mater Today*, 50, 276–302, 2021.
16. Massaro, M., Cavallaro, G., Colletti, C.G., Lazzara, G., Milioto, S., Noto, R. *et al.*, Chemical modification of halloysite nanotubes for controlled loading and release. *J. Mater Chem. B*, 6, 21, 3415–33, 2018.
17. Patel, K.D., Singh, R.K., Kim, H.-W., Carbon-based nanomaterials as an emerging platform for theranostics. *Mater Horiz.*, 6, 3, 434–69, 2019.
18. Maiti, D., Tong, X., Mou, X., Yang, K., Carbon-based nanomaterials for biomedical applications: A recent study. *Front. Pharmacol.*, 9, 1401, 2019.
19. Sen, M., Nanocomposite materials, in: *Nanotechnology and the Environment*, M. Sen (Ed.), pp. 107–12, IntechOpen, London, 2020.
20. Turlybekuly, A., Pogrebnjak, A.D., Sukhodub, L.F., Sukhodub, L.B., Kistaubayeva, A.S., Savitskaya, I.S. *et al.*, Synthesis, characterization, *in vitro* biocompatibility and antibacterial properties study of nanocomposite materials based on hydroxyapatitebiphasic ZnO micro- and nanoparticles embedded in alginate matrix. *Mater Sci. Eng. C.*, 104, 109965, 2019.
21. Liu, Y., Yu, Q., Chang, J., Wu, C., Nanobiomaterials: From 0D to 3D for tumor therapy and tissue regeneration. *Nanoscale*, 11, 29, 13678–708, 2019.
22. Torresan, V., Forrer, D., Guadagnini, A., Badocco, D., Pastore, P., Casarin, M. *et al.*, 4D multimodal nanomedicines made of nonequilibrium Au–Fe alloy nanoparticles. *ACS Nano*, 14, 10, 12840–53, 2020.
23. Liu, M., Fakhrullin, R., Novikov, A., Panchal, A., Lvov, Y., Tubule nanoclay-organic heterostructures for biomedical applications. *Macromol Biosci.*, 19, 4, 1800419, 2019.
24. Yokota, J. and Kyotani, S., Influence of nanoparticle size on the skin penetration, skin retention and anti-inflammatory activity of non-steroidal anti-inflammatory drugs. *J. Chin. Med. Assoc.*, 81, 6, 511–9, 2018.
25. Meng, Q., Meng, H., Pan, Y., Liu, J., Li, J., Qi, Y. *et al.*, Influence of nanoparticle size on blood–brain barrier penetration and the accumulation of anti-seizure medicines in the brain. *J. Mater. Chem. B*, 10, 2, 271–81, 2022.
26. Jahnen-Dechent, W. and Simon, U., Function follows form: Shape complementarity and nanoparticle toxicity. *Nanomedicine*, 3, 5, 601–3, 2008.
27. Pan, Y., Neuss, S., Leifert, A., Fischler, M., Wen, F., Simon, U. *et al.*, Size-dependent cytotoxicity of gold nanoparticles. *Small*, 3, 11, 1941–9, 2007.
28. Kim, J.S., Lee, K., Lee, Y.H., Cho, H.S., Kim, K.H., Choi, K.H. *et al.*, Aspect ratio has no effect on genotoxicity of multi-wall carbon nanotubes. *Arch. Toxicol.*, 85, 7, 775–86, 2011.
29. Bauer, L.A., Birenbaum, N.S., Meyer, G.J., Biological applications of high aspect ratio nanoparticles. *J. Mater. Chem.*, 14, 4, 517–26, 2004.

30. Fratila, R.M., Rivera-Fernández, S., Jesús, M., Shape matters: Synthesis and biomedical applications of high aspect ratio magnetic nanomaterials. *Nanoscale*, 7, 18, 8233–60, 2015.
31. Luanpitpong, S., Wang, L., Davidson, D.C., Riedel, H., Rojanasakul, Y., Carcinogenic potential of high aspect ratio carbon nanomaterials. *Environ. Sci. Nano*, 3, 3, 483–93, 2016.
32. Yi, X., Shi, X., Gao, H., A universal law for cell uptake of onedimensional nanomaterials. *Nano Lett.*, 14, 2, 1049–55, 2014.
33. Sanchez, V.C., Weston, P., Yan, A., Hurt, R.H., Kane, A.B., A 3-dimensional *in vitro* model of epithelioid granulomas induced by high aspect ratio nanomaterials. *Part Fibre Toxicol.*, 8, 1, 17, 2011.
34. Wan, J., Wang, J.-H., Liu, T., Xie, Z., Yu, X.-F., Li, W., Surface chemistry but not aspect ratio mediates the biological toxicity of gold nanorods *in vitro* and *in vivo*. *Sci. Rep.*, 5, 1, 11398, 2015.
35. Wang, Y., Sun, S., Zhang, Z., Shi, D., Nanomaterials for cancer precision medicine. *Adv. Mater.*, 30, 17, 1705660, 2018.
36. Auría-Soro, C., Nesma, T., Juanes-Velasco, P., Landeira- Viñuela, A., Fidalgo-Gomez, H., Acebes-Fernandez, V. *et al.*, Interactions of nanoparticles and biosystems: Microenvironment of nanoparticles and biomolecules in nanomedicine. *Nanomaterials*, 9, 10, 1365, 2019.
37. Fröhlich, E., Aufbau und funktion von blut-gewebe-schranken. *Dtsch Med. Wochenschr*, 127, 49, 2629–34, 2002. (German).
38. Bregoli, L., Movia, D., Gavigan-Imedio, J.D., Lysaght, J., Reynolds, J., Prina-Mello, A., Nanomedicine applied to translational oncology: A future perspective on cancer treatment. *Nanomed. Nanotechnol. Biol. Med.*, 12, 1, 81–103, 2016.
39. Ugwu, M.C., Amadi, N.C., Okolo, C.R., Dada, M.O., Egejuru, C.C.J., Ogbodo, E.C. *et al.*, Nanomedicine: Application of nanotechnology. *Eur. J. BioMed. Pharm. Sci.*, 4, 12, 136–43, 2017.
40. Melchor-Martínez, E.M., Torres Castillo, N.E., Macias-Garbett, R., Lucero-Saucedo, S.L., Parra-Saldívar, R., Sosa-Hernández, J.E., Modern world applications for nano-bio materials: Tissue engineering and COVID-19. *Front. Bioeng. Biotechnol.*, 9, 393, 2021.
41. Khiev, D., Mohamed, Z.A., Vichare, R., Paulson, R., Bhatia, S., Mohapatra, S. *et al.*, Emerging nano-formulations and nanomedicines applications for ocular drug delivery. *Nanomaterials*, 11, 1, 173, 2021.
42. Zhang, W., Mehta, A., Tong, Z., Esser, L., Voelcker, N.H., Development of polymeric nanoparticles for blood–brain barrier transfer – strategies and challenges. *Adv. Sci.*, 8, 10, 2003937, 2021.
43. Riggio, C., Pagni, E., Raffa, V., Cuschieri, A., Nano-oncology: Clinical application for cancer therapy and future perspectives. *J. Nanomater.*, 2011, 17, 2011.
44. Zhang, Y., Li, M., Gao, X., Chen, Y., Liu, T., Nanotechnology in cancer diagnosis: Progress, challenges and opportunities. *J. Hematol. Oncol.*, 12, 1, 137, 2019.

45. Anselmo, A.C. and Mitragotri, S., Nanoparticles in the clinic: An update. *Bioeng. Transl. Med.*, 4, 3, e10143, 2019.
46. Lôbo, G.C.N.B., Paiva, K.L.R., Silva, A.L.G., Simões, M.M., Radicchi, M.A., Báo, S.N., Nanocarriers used in drug delivery to enhance immune system in cancer therapy. *Pharmaceutics*, 13, 8, 1167, 2021.
47. Zhang, Q., Yu, H., Barbiero, M., Wang, B., Gu, M., Artificial neural networks enabled by nanophotonics. *Light: Sci. Appl.*, 8, 1, 42, 2019.
48. Yang, L., Zhang, L., Webster, T.J., Nanobiomaterials: State of the art and future trends. *Adv. Eng. Mater.*, 13, 6, B197–B217, 2011.
49. Shafiei-Irannejad, V., Soleymani, J., Azizi, S., KhoubnasabJafari, M., Jouyban, A., Hasanzadeh, M., Advanced nanomaterials towards biosensing of insulin: Analytical approaches. *TrAC - Trends Anal. Chem.*, 116, 1–12, 2019.
50. Lei, J. and Ju, H., Signal amplification using functional nanomaterials for biosensing. *Chem. Soc. Rev.*, 41, 6, 2122–34, 2012.
51. Holzinger, M., Le Goff, A., Cosnier, S., Synergetic effects of combined nanomaterials for biosensing applications. *Sensors*, 17, 5, 1010, 2017.
52. Hwang, H.S., Jeong, J.W., Kim, Y.A., Chang, M., Carbon nanomaterials as versatile platforms for biosensing applications. *Micromachines*, 11, 9, 814, 2020.
53. Carneiro, P., Morais, S., Pereira, M.C., Nanomaterials towards biosensing of Alzheimer's disease biomarkers. *Nanomaterials*, 9, 12, 1663, 2019.
54. Mostafavi, E., Medina-Cruz, D., Kalantari, K., Taymoori, A., Soltantabar, P., Webster, T.J., Electroconductive nanobiomaterials for tissue engineering and regenerative medicine. *Bioelectricity*, 2, 2, 120–49, 2020.
55. Di Marzio, N., Eglin, D., Serra, T., Moroni, L., Bio-fabrication: Convergence of 3D bioprinting and nanobiomaterials in tissue engineering and regenerative medicine. *Front. Bioeng. Biotechnol.*, 8, 326, 2020.
56. Dvir, T., Timko, B.P., Kohane, D.S., Langer, R., Nanotechnological strategies for engineering complex tissues. *Nat. Nanotechnol.*, 6, 1, 13–22, 2011.
57. Hudita, A., Galateanu, B., Costache, M., Chapter 1 – Nanobiomaterials for tissue engineering, in: *Materials for Biomedical Engineering*, A.-M. Holban and A.M. Grumezescu (Eds.), pp. 1–21, Elsevier, Amsterdam, 2019.
58. Nikolova, M.P. and Chavali, M.S., Recent advances in biomaterials for 3D scaffolds: A review. *Bioact. Mater.*, 4, 271–92, 2019.
59. Wang, J. and Vermerris, W., Antimicrobial nanomaterials derived from natural products – A review. *Materials*, 9, 4, 255, 2016.
60. Ogunsona, E.O., Muthuraj, R., Ojogbo, E., Valerio, O., Mekonnen, T.H., Engineered nanomaterials for antimicrobial applications: A review. *Appl. Mater. Today*, 18, 100473, 2020.
61. Dickson, M.N., Liang, E.I., Rodriguez, L.A., Vollereaux, N., Yee, A.F., Nanopatterned polymer surfaces with bactericidal properties. *Biointerphases*, 10, 2, 021010, 2015.
62. Sánchez-López, E., Gomes, D., Esteruelas, G., Bonilla, L., Lopez- Machado, A.L., Galindo, R. *et al.*, Metal-based nanoparticles as antimicrobial agents: An overview. *Nanomaterials*, 10, 2, 292, 2020.

63. Hasan, A., Morshed, M., Memic, A., Hassan, S., Webster, T.J., Marei, H.E.-S., Nanoparticles in tissue engineering: Applications, challenges and prospects. *Int. J. Nanomed.*, 13, 5637–55, 2018.
64. Hemeg, H.A., Nanomaterials for alternative antibacterial therapy. *Int. J. Nanomed*, 12, 8211–25, 2017.
65. Grainger, D.W. and Castner, D.G., Nanobiomaterials and nanoanalysis: Opportunities for improving the science to benefit biomedical technologies. *Adv. Mater.*, 20, 5, 867–77, 2008.
66. Patra, J.K. and Baek, K.-H., Green nanobiotechnology: Factors affecting synthesis and characterization techniques. *J. Nanomater.*, 2014, 219, 2015.
67. Foulkes, R., Man, E., Thind, J., Yeung, S., Joy, A., Hoskins, C., The regulation of nanomaterials and nanomedicines for clinical application: Current and future perspectives. *Biomater. Sci.*, 8, 17, 4653–64, 2020.
68. Giubilato, E., Cazzagon, V., Amorim, M.J.B., Blosi, M., Bouillard, J., Bouwmeester, H. *et al.*, Risk management framework for nanobiomaterials used in medical devices and advanced therapy medicinal products. *Materials*, 13, 20, 4532, 2020.
69. Talebian, S., Rodrigues, T., das Neves, J., Sarmento, B., Langer, R., Conde, J., Facts and figures on materials science and nanotechnology progress and investment. *ACS Nano.*, 15, 10, 15940–52, 2021.
70. Ramburrun, P., Khan, R.A., Choonara, Y.E., Design, preparation, and functionalization of nanobiomaterials for enhanced efficacy in current and future biomedical applications. *Nano. Rev.*, 11, 1, 1802–1826, 2022, https://doi.org/10.1515/ntrev-2022-0106.

13

Residual Stress Analysis in Titanium Alloys Used for Biomedical Applications

Gulshan Kumar[1]*, Rohit Kumar[2] and Arshpreet Singh[3]

[1]*Birla Institute of Technology & Science, Pilani-Dubai Campus, Dubai, United Arab Emirates*
[2]*Dr. B R Ambedkar National Institute of Technology, Jalandhar, Punjab, India*
[3]*Thapar Institute of Engineering & Technology, Patiala, Punjab, India*

Abstract

Biomedical applications like cranial plate, knee arthroplasty, ankle replacement, elbow and knee support braces, other orthopedic implants, and braces are required to be customized depending on the respective human body. These items can be made up of sheet metal. The slightest modification to the overall shape of sheet metal components leads to a reduction in production rate, raises costs, and impairs dimensional accuracy. Commercially pure titanium due to its excellent corrosion resistance, high formability properties, and high strength-to-weight ratio is considered to be an excellent material for a medical prosthesis. Single point incremental forming (SPIF), is an emerging technology that consists of a basic hemispherical tool, numerically monitored by a CNC machine that performs a progressing highly localized deformation. The poor dimensional accuracy would attribute to the development of residual stresses which may cause severe conditions for cranioplasty patients if the cranial plate is deformed. Even a small deformation due to residual stress may result in surgical replacement or even severe pain. Therefore, a better knowledge of the residual stresses is necessary to improve the geometrical precision of the process. The primary field of study is to thoroughly examine the surface residual stresses affected by the different process parameters of single-point incremental forming. In addition, microstructural behavior and forces induced during SPIF were also observed in relation to the strength of sheet metal.

***Keywords*:** Titanium grade-2, SPIF, microstructure, residual stresses, EBSD, X-ray diffraction, biomedical applications

**Corresponding author*: gulshankumar@dubai.bits-pilani.ac.in

Arbind Prasad, Ashwani Kumar and Manoj Gupta (eds.) Advanced Materials and Manufacturing Techniques for Biomedical Applications, (325–344) © 2024 Scrivener Publishing LLC

13.1 Introduction

With the advancement in the medical sector and its collaboration with engineering sciences, different types of biomedical implants made through advanced manufacturing techniques are used to cure patients' illnesses [1]. The key attributes of any bio-medical implant replacing the original bone or tissue are that they should be non-toxic, have almost the same mechanical and bio-medical properties; should be customized to the exact specifications according to the need; should be bio-compatible; should be economical and easy to manufacture for ready mass acceptance [2].

CNC machining technologies are used to fabricate high-quality orthopedic implants, but a longer period for manufacturing, material waste, high cost, and poor mechanical properties cannot be neglected [3]. Conventional press forming techniques were employed to overcome the issues related to time and material properties, but the complex design of the die and the manufacturing operation limits the application of such technology [4]. In recent times, the advent of additive manufacturing and 3D printing has revolutionized the manufacturing industry. It is found to be a suitable technique to build complex and large-shaped implants [5]. However, limited by high cost and a more time-consuming process. Considering the certain shortcomings and incompatibilities such as poor customizability, low efficiency, high cost, and inadequate process compatibility which results in poor mechanical properties of the product, a more flexible manufacturing technique is needed.

Incremental sheet forming (ISF), a flexible forming technique is attempted to manufacture customized macro and micro-sized orthopedic implants [6]. Incremental sheet forming (ISF) enables low cost, high customization, and low volume manufacturing which is done through accumulating the sheet's localized plastic deformation [7]. The sheet metal component is formed using a hemispherical tool operated by a computer numerical controlled machine. Three-dimensional shape split-up into two dimensional shapes in such a way that the hemispherical tool travels to contour one by one. The distance between the two neighboring contours is known as step size [8]. Single point incremental forming (SPIF) is the basic version of the flexible, customized part-oriented forming technique [1]. Despite SPIF, two points incremental forming can be used for more intricate shapes [9]. Figure 13.1 depicts the schematic of SPIF set-up showing the dimensional attributes and major process parameters such as Initial sheet thickness (t_i), Final formed sheet thickness (t_f), Stretch forming angle (ϕ), and Incremental step depth in each pass (Δz).

Single point incremental forming has considerable advantages over conventional forming, which consist of dedicated tools and die such as increased

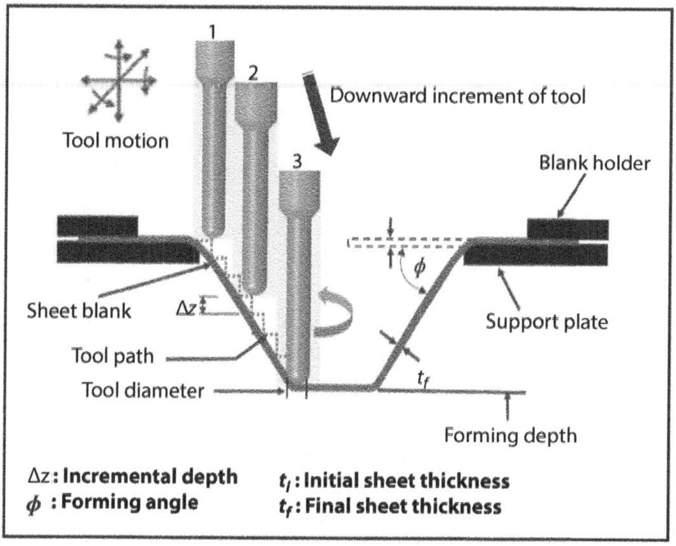

Figure 13.1 Schematic of single point incremental forming operation.

process formability, and considerable cost reduction compared to other forming processes [10]; especially in tooling and equipment as expensive dies, molds, and punches are avoided in this process; the process allows design flexibility; changes in design can be incorporated easily; deforming forces are considerably lower [11, 12]; process accuracy, repeatability, and dimensions of the fabricated components are only restricted by the machine tool specifications process outcomes such as surface quality, residual stresses in the components can be easily altered or tailored according to the application [13, 14].

The process cycle of manufacturing bio-medical implants employing incremental forming mainly involves an X-ray, a CT, or MRI scan of the deceased part; creating a virtual CAD model from the 3D scan; generation of the NC part program from the CAD model; transferring the NC program onto the CNC forming setup and generating the desired profile; optimizing the forming parameters for the desired outcome in the form of surface texture, accuracy, grain structure and residual stress (Figure 13.2). As depicted in Figure 13.2, the initial stage in creating a customized cranial implant is to get medical images using computer tomography (CT). Using the DICOM (digital imaging communication in medicine) protocol, the CT scans were exported. After creating a CAD model from the skull bone, a numerical control coding file was created. It was subsequently installed in CNC for ISF's tool path generation. The cranial implant must be trimmed to remove the extra material after the sheet has been formed [15].

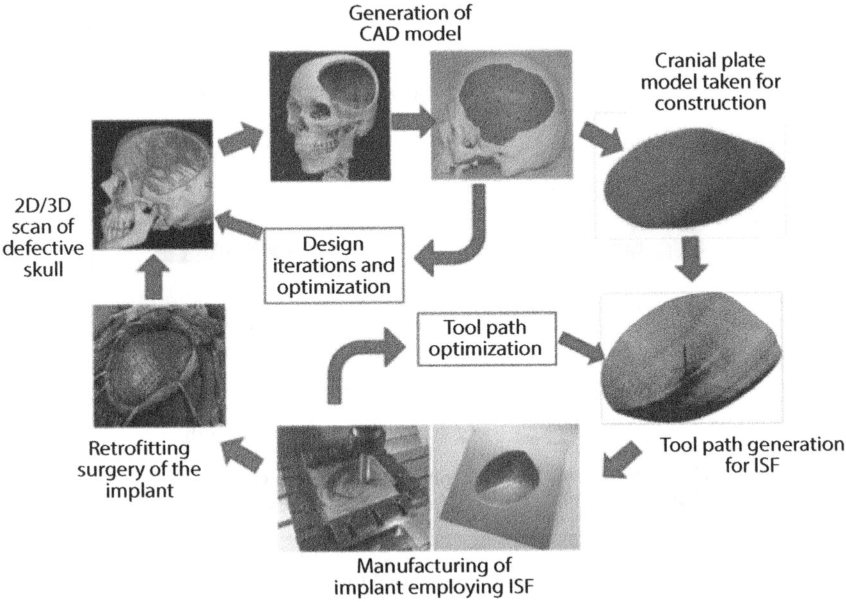

Figure 13.2 Sequence of manufacturing of a cranial implant through incremental forming [16, 17].

In previous studies, researchers have explored and proved the practicality of incremental forming for manufacturing implants in a variety of areas such as making these implants with diverse biomedical materials [6], shapes and sizes [18], input cost, and process quality [13]. Metals and polymers are the two main types of materials utilized in biomedical implants. The stainless steel used to be the most popular implant material due to its low cost, good mechanical performance, and reasonable biocompatibility [19]. However, the poor anti-corrosion properties of stainless steel in comparison to the advanced materials increase the risk of infection and are subsequently employed for a short period. To overcome the issue with stainless steel, new polymer materials such as PEEK (Polyether ether ketone), and PMMA (Poly methyl methacrylate) have been developed, which are light in weight, have excellent biocompatibility, and have comparable properties to a human bone or tissue. The issue with such materials is the high and sometimes unaffordable cost associated [2].

Due to excellent corrosion resistance, good mechanical properties, high biocompatibility, biological performance, and availability at a reasonable cost [20], titanium has been adapted world widely by various industries such as structural, aerospace, manufacturing, and biomedical industries. It has emerged as an excellent alternative to steel and synthetic materials.

Biomedical implants require a good biocompatible, corrosion-resistant, high formability, and high strength-to-weight ratio, and titanium enriched with all these characteristics specifically commercially pure titanium (Cp-Ti) due to its high formability which is needed while making the biomedical prosthesis using SPIF process. In this study, Cp-Ti grade 2 has been preferred over other commercially available titanium grades due to its good formability and Osseo integration capability with bio-tissue in the human body. In an effort to improve the mechanical characteristics of Cp-Ti grade 2, Rack *et al.* [21] developed a strengthening approach utilizing equal channel angular pressing (ECAP). The mechanical characteristics of CP-Ti grade 2 in its various microstructural states are shown in Table 13.1.

SPIF induces stress shielding and undue residual stresses in high-strength titanium alloys [22]. This leads to inhomogeneity and geometrical discrepancies as and when these stresses are relieved from the structure with time. Once the implant is in the body of the patient, a slight geometrical variation can have severe consequences in the form of unbearable pain and sometimes could be fatal also. Residual stresses play a vital role in the non-homogeneous deformation during SPIF. The residual stresses are in direct correlation with the material grain structure [23]. Therefore, internal grain structural formation and its development during the manufacturing of implants are very critical. The evaluation of induced residual stresses during manufacturing also plays a vital role. The stresses that remained in the component after unloading the body are known as residual stresses [24, 25]. These residual stresses develop due to the non-uniform and inhomogeneous plastic deformation involved in the incremental forming. The type and magnitude of residual stress distribution in any deformation process have direct implications on the elastic recovery and overall strengthening of the thin structures and sheet metal components. Elastic recovery and stress distribution in components undergoing repetitive operations are mostly influenced by the residual stresses remaining in the component from the previous work cycle. Since incremental forming is a repetitive process, it is critical to investigate the magnitude and nature of residual stresses for the process involved in the fabrication. The investigations have proven that the residual stress distribution across the thickness of the components formed by incremental forming is comparable to the conventional forming techniques [26]. Although, there is a need to critically correlate the deforming forces required in the incremental forming of the component with the microstructure, grain evolution, and induced residual stresses.

This work is an attempt to determine the correlation of residual stresses with the microstructure and grain evolution of the commercially pure

Table 13.1 Mechanical properties of CP Ti grade 2 at different microstructural states [21].

Microstructure type	Hv (MPa)	UTS (MPa)	YS (MPa)	EL (%)	RA (%)	Fatigue limit
Coarse-grained	1800	460	380	26	60	238
UFG (Equiaxed, submicron-grained)	2700	460	625	14	60	403
UFG (Fibrous w. high dislocation density)	2821	960	725	10	45	434
UFG (subgrained w. internal cells)	2850	1100	915	9	40	500
Ti-6Al-4V ELI (annealed)	-	965	875	10–15	25–47	515

titanium mainly used for biomedical applications, manufactured through incremental forming. Since, this work is focused on microstructural and core residual stress evolution aspects in incremental forming, only the effect of parameters influencing the most were considered. Other parameters that seem to have minimal bearing on the core structural properties were kept constant. Parameters such as tool rotation, tool diameter (*d*), feed rate (*fr*), and lubrication, which only have a superficial impact were also kept constant. The change in the above-mentioned process parameters for a specific application can be controlled by favorable stresses which can improve the mechanical properties.

13.2 Methodology

13.2.1 Material Selection

In the present study, commercially pure titanium grade 2 has been chosen due to its greater applicability in the biomedical sector [22]. Titanium is also well recognized for its good biocompatibility and sensitivity to corrosion caused by oxidation of the outer layer when exposed to atmospheric conditions [27]. Table 13.2 listed the weight percentage of the element constituents present in the Cp-Ti grade 2 used for this study.

13.2.2 Experimental Details

This study's experimental setup (see Figure 13.3) was used to investigate the effect of different process parameters of incremental forming (i.e. forming angle and step size) on residual stresses and force produced in SPIF-manufactured components. Other forming parameters such as tool rotation, tool diameter (*d*), feed rate (*fr*), and lubrication, were kept constant during the experimentation. A total of 16 experiments were conducted for four different forming angles and incremental step depths. The process parameters and their corresponding values are presented in Table 13.3.

Table 13.2 Weight percentage of the commercially pure titanium grade-2 used in the present study.

Element	%C	%Cr	%Fe	%N	%Ni	%Sn	%W	%Ti
Wt.%	0.006	0.004	0.072	0.014	0.018	0.006	0.007	Balance

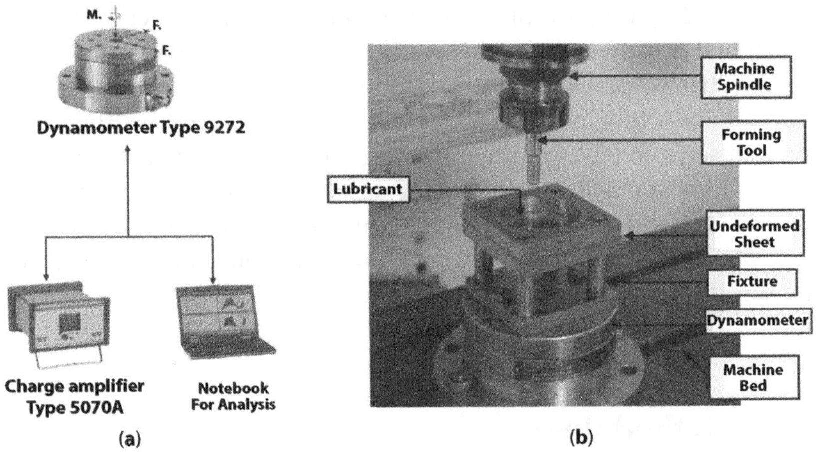

Figure 13.3 Experimental setup used for the incremental forming process (a) force measurement dynamometer (b) SPIF experimental set up.

Table 13.3 Process parameters and levels employed in experimentation.

Parameter	Levels
Forming angle (ϕ)	30°, 45°, 60°, 75°
Incremental step depth (Δz)	0.25, 0.5. 0.75, 1.0 mm
Feed rate (*fr*)	100 mm/min
Spindle rpm	300
Sheet thickness (*t*)	0.97 mm
Tool diameter (*d*)	8 mm

13.2.2.1 *Experimental Setup and Force Calculation*

To study the relationship of residual stresses and formation of related forces with process parameters on commercially pure titanium, the experiments were carried out on a 3-axis CNC vertical milling center (Model: BFW, Chandra +) with an 8 mm hemispherical tool (Figure 13.3). Forces were measured using a force dynamometer (Model: Kistler 9272) equipped with a multichannel amplifier of 5070 amperes to record the forces in three coordinate directions (i.e., x-axis, y-axis, and z-axis) with a measuring range of -5 to 20 KN (see Figure 13.3a). An assembled view of the set-up used for performing the experiments is shown in Figure 13.3b. To execute

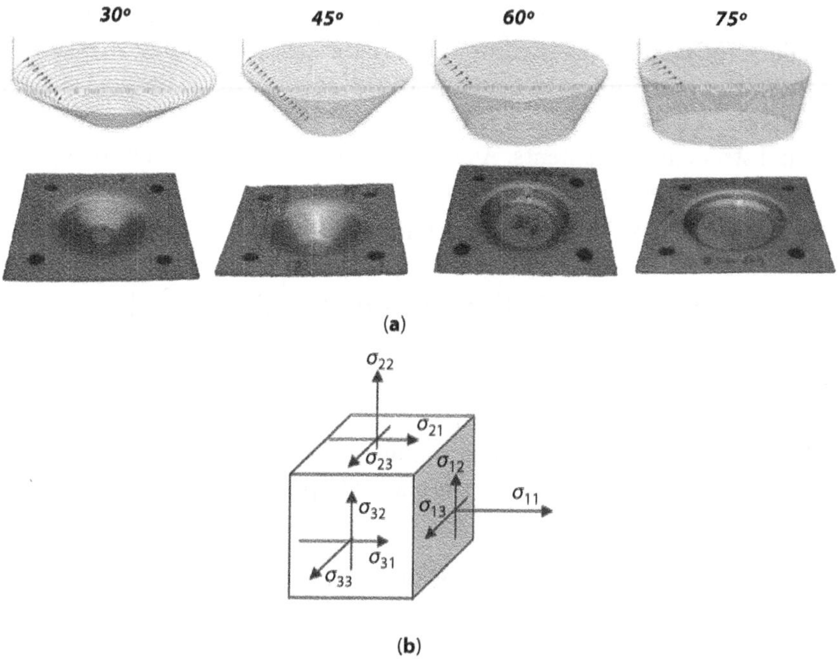

Figure 13.4 (a) Contour tool path for different forming angles along with fabricated test samples (b) tensorial representation of normal and shear stresses.

the operation on a CNC machine, a fixture made of mild steel plates and supporting pillars was used (see Figure 13.3b).

13.2.2.2 Sample Preparation

The initially received metal sheet with a thickness of 0.97 mm, and the test samples with dimensions of 75 mm on each side were used for the experiments. A contour tool path with increasing step depth was chosen to form inverted cone shape samples. Figure 13.4 (a) shows the contour tool path along with different forming angles and the corresponding fabricated test samples by SPIF. To measure the achieved depth of the test sample, a coordinating measuring machine (Model: CMM Accurate spectra) was used. Inverted cone was formed at different forming angles in the working area of 50 mm diameter.

13.2.2.3 Residual Stress and Microstructure Evaluations

The residual stress measurements were done to investigate the change in the surface residual stresses at different process parameters of the SPIF process. A micro-focus X-ray with laser tracker (Brukers™) and a Vantec™ area detector system was used to measure the surface residual stresses.

The measurements were done at 100 ± 25 mm from the top surface. The standard d-sin²ψ method (by X-ray diffraction) [27–33] was used for residual stress measurements. A tensorial representation of the stress components is shown in Figure 13.4(b).

Interplanar spacing d_{hkil} (ψ, φ) is represented as:

$$d_{hkil}(\psi,\varphi) = d_0[1+S_1(\sigma_{11}+\sigma_{12})+\frac{1}{2}S_2\sigma_\varphi \text{Sin}^2\psi] \quad \text{(i)}$$

where, ψ = the angle which defines the orientation of the sample surface, the angle between the normal to the sample surface, and the bisector of the incident and diffraction beam. φ = angle of rotation about the specimen surface normal. S_1 and $1/2S_2$ are the elasticity parameters, given as:

$$S_1 = \left(-\frac{\nu}{E}\right) \quad \text{and} \quad \frac{1}{2}S_2 = \left(\frac{1+\nu}{E}\right) \quad \text{(ii)}$$

A value of 0.37 and 110 GPa were used for ν and E respectively for the stress analysis. The d-Sin²ψ plots will give the normal (σ_φ) residual stresses and the shear (τ_φ) stresses from the slope of the curve and through the so-called ψ-splitting [27–33] respectively.

Electron backscatter diffraction (EBSD) was used to characterize the microstructure of the samples. The electro-polished samples in the weight ratio (80:20) of methyl alcohol and perchloric acid were used for the characterization.

Struers™ Lectropol-5 system at 22V DC and -35°C was used for electro-polishing. The EBSD measurements were done on the FEI™ Quanta-3D FEG (field emission gun) SEM (Scanning electron microscopy) system and TSL-OIM™ analysis package was used to analyze for estimating grain size and grain misorientation in the EBSD scans keeping the parameters settings constant among all the measurements. The continuous presence of an area of the microstructure bounded by >5° misorientation was defined as a grain. An average point-to-point misorientation method was used to determine the grain average misorientation in this study. At least an area of 400 × 400 mm was covered for each sample through multiple scans at a step size of 0.1mm. The IPF (Inverse pole figure) data points having confidence index (CI) above 0.1 were considered for analyzing the scans. CI used to measure an automated indexing [34–36] and data points >0.1 CI showed accuracy more than 95%. The data points <0.1 are defined as non-indexed data points.

13.3 Results and Discussion

13.3.1 Deforming Forces Analysis in Incremental Forming

Analysis of the forming forces is especially critical in choosing the correct materials and suitable process parameters to ensure the accuracy and stability of the incremental forming process. It was observed that the axial force (F_z) is larger in magnitude intensity compared to the other two lateral forces (F_x and F_y) in SPIF. It is a more prominent force for modeling or selecting a suitable setup for incremental forming (Figure 13.5). Originally F_z begins at zero, after traveling some contour it achieves its peak value and then further decreases up to the stabilized state. The tool doesn't have a completely developed contact area unless a range of contours has been formed. Interestingly, the effect of bending occurs initially that transforms a steady forming after a certain forming depth. A sinusoidal nature of F_x and F_y have been observed as they replicate their maximum and minimum force values owing to the location of the forming tool traveled in the direction of the contour tool path. Figure 13.6 shows the average resultant force for different samples for a range of step depth varied from 0.25 mm to 1.0 mm at different forming angles. The graph predicts the relationship between the process parameters of SPIF which is further discussed in the subsequent section.

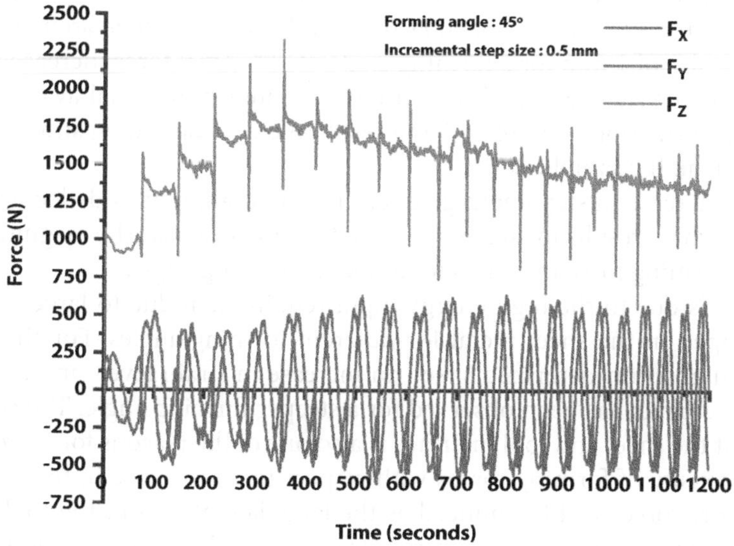

Figure 13.5 Experimental trend of three components of force in SPIF.

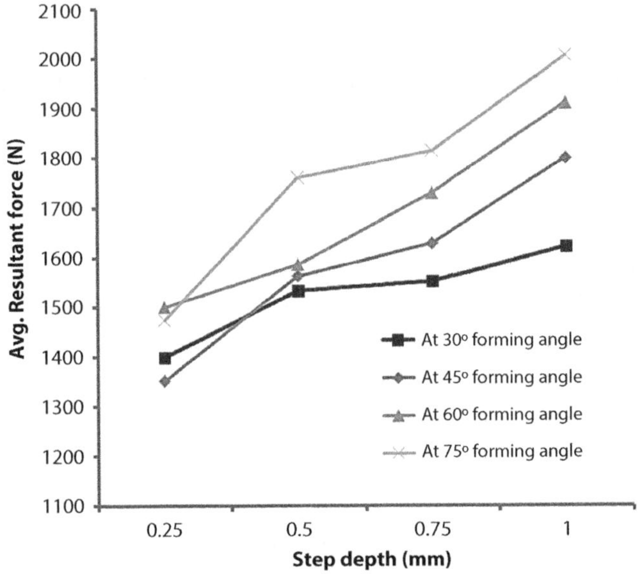

Figure 13.6 Average resultant force for the different forming angles at different incremental step depths.

The average resulting force for all the samples manufactured at different forming angles of 30°, 45°, 60°, and 75° are shown in Figure 13.7 (a-d). A maximum resultant force of 2009.76 N has been recorded at an incremental step size of 1 mm for a forming angle of 75°. Axial force increased with a rise in vertical step depth because the amount of material available for local deformation is more and as a result, the corresponding strain is also more in all the samples.

An increase in step depth gives an increased amount of deflection and strain, which results in large stresses and the load values. The magnitude of the deforming force increases for a steeper forming angle.

This is due to the high energy input requirement due to large stresses developed in the sheet metal for different forming angles. For the same margin of an un-deformed sheet, an increase in strain levels and material displacements has been observed for a steeper forming angle. The contact zone at the sheet and tool interface is also one of the reasons for increasing the intensity of forming forces with respect to an increase in tool diameter. More forces will be induced as the large lateral area of the tool point touches the sheet metal which results in greater local deformation when the higher forming angles are taken in the process.

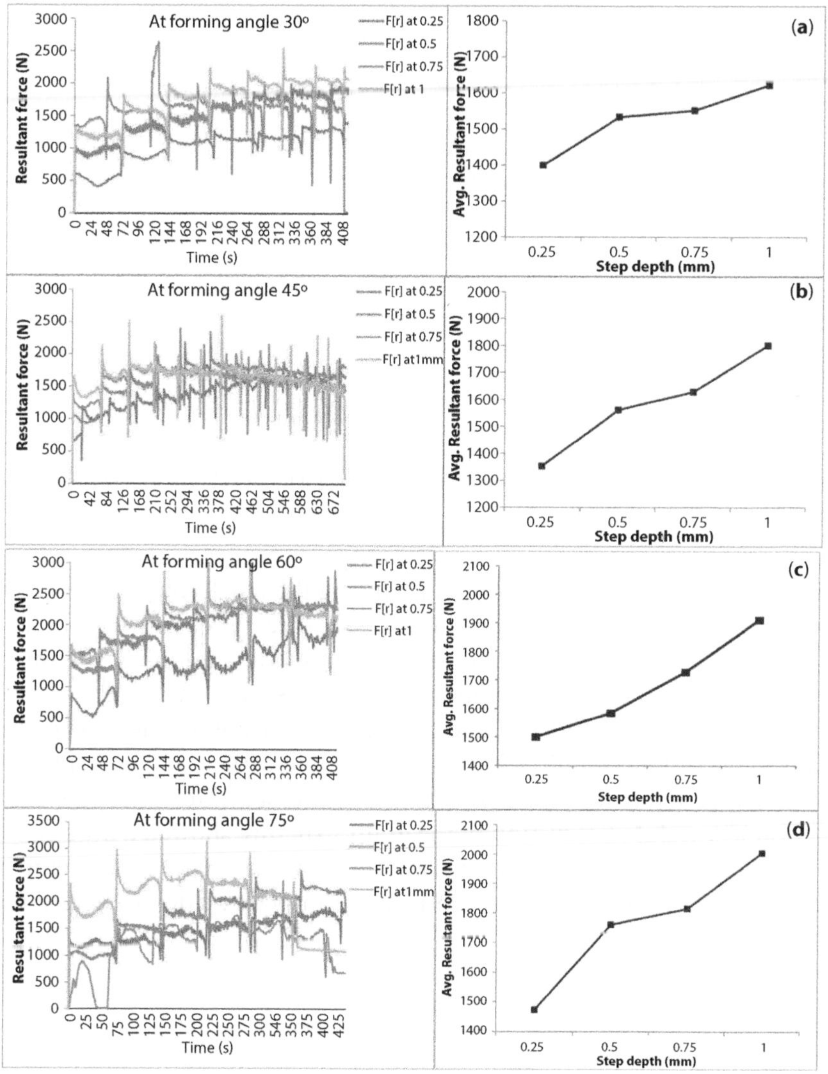

Figure 13.7 Resultant deforming force and average resultant force at different forming angles of (a) 30° (b) 45° (c) 60° and (d) 75° at various incremental step depths.

13.3.2 Microstructural Analysis

Figure 13.8 shows the initial microstructure of the as received metal sheet. It showed a fully recrystallized, bimodal structure consisting of a few smaller grains with a diameter of 0.94 mm and coarser grains with a diameter of 17.92 μm. These are generated in the microstructure due to

Figure 13.8 EBSD (electron backscattered diffraction) data in IPF (inverse pole figure) map of the as received metal sheet.

the non-uniform plastic deformation during the single-point incremental forming operation It also created in-grain misorientation and non-indexed data points [37]. The color scheme of the individual grains of the microstructure shows the misorientation (Figure 13.9) and the grain fragmentation due to the creation of large dislocation densities in the deformation leads to non-indexed points [35]. As mentioned earlier, forming angle and incremental depth are the main influencing factors which affect the deformation. A careful observation of the grain structure showed a little change with a range of forming angle varied from 30° to 75° at an incremental depth of 0.25 mm (see Figure 13.9). A rise in plastic flow of the grains is

Figure 13.9 IPF and IQ maps of SPIF fabricated test samples.

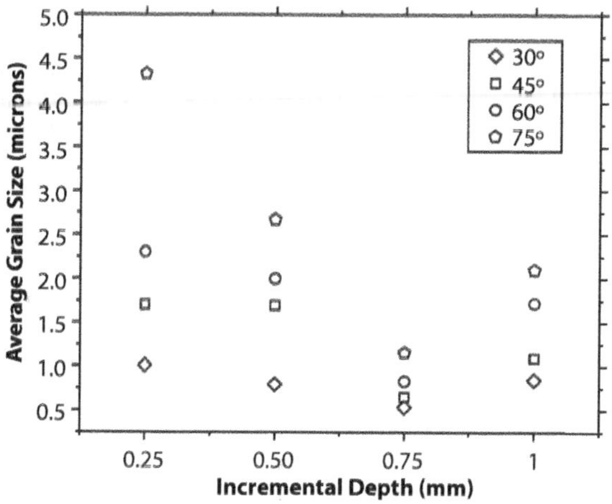

Figure 13.10 Average grain size at different forming angles and various Incremental step depths.

clearly visible at 75° which leads to the development of plastic strain gradient in the material. This non-homogeneous deformation caused in-grain misorientation and the non-indexed points.

The EBSD data was further analyzed to bring out the gradient induced in grain size during the SPIF operation (Figure 13.10). A noticeable decrease in the average grain size is observed across the range of forming angles and step depth. When measured from the top surface of the metal sheet, grain size is highest (4.3 mm) at a depth of 0.25 mm and lowest (0.54 mm) at a depth of 0.75 mm. A similar trend has been observed in the gradient of grain size also, highest at a depth of 0.25 mm and lowest at 0.75 mm. A slight increment in the gradient has been seen for a step depth of 1 mm compared to 0.75 mm.

13.3.3 Residual Stresses

As discussed earlier, d-sin$^2\psi$ technique has been used for the residual stress measurements. The behavior of surface residual stresses for all the 16 samples is collated in Figure 13.11. A significant compressive stress has been observed for a step depth of 0.25 mm at a forming angle of 30°. As soon as forming angle increase from 30° to 60°, a transition in the nature of stresses from compressive to tensile has been observed. This transition was more effective until a step depth of 0.75 mm. Further, a slight decrease in the stress occurs at 75° which leads to end up with less compressive stresses in

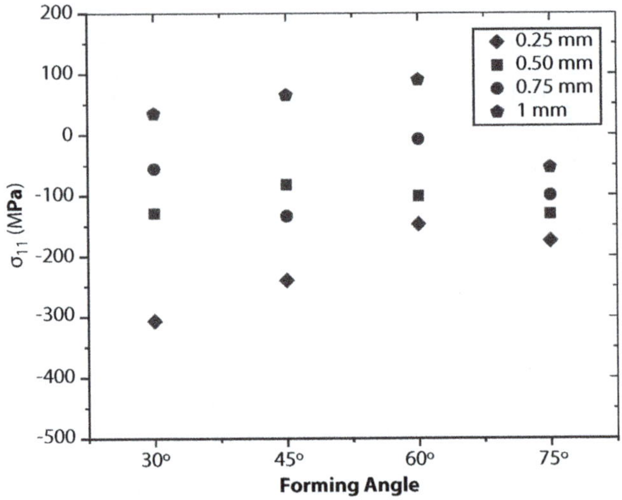

Figure 13.11 Gradient in residual stresses with forming angle at a particular incremental step depth in SPIF operation.

the body. On the other hand, it is to be noted that a distinct stress gradient has been observed at each different forming angle and step depth. This also strengthens the literature fact that reduced incremental depth and forming angle have a favorable impact on dimensional accuracy, which described in terms of surface residual stress. This is because the amplitude of the residual stress changes as a result of the increasing amount of strain and material displacement that occurs at steeper forming angles and at higher step sizes.

Singh and Agrawal [38] investigated the surface residual stresses in Al 6063-T6 manufactured by deformation machining (DM) technique (a combination of incremental forming and thin structure machining). They observed the growth of the surface tensile residual stresses with forming depth, feed rate and forming angle. Moreover, Radu *et al.* [39] conducted research on residual stresses induced in SPIF-formed aluminum sheets. They used hole drilling method to measure the residual stresses in two different aluminum metal sheets. They found tensile stresses with increase in tool size and incremental depth and leads to distortion and geometrical inaccuracy in the products made by SPIF. Mahnoush Alinaghian *et al.* [40] found a trend in Al/Cu bimetal that a smaller increment step size leads to small residual stress forces. In the present study, an attempt has been made to determine the controlling factors that would result in a maximum residual stress reduction to achieve dimensional stability. A change in the process parameters provides the customization in the residual stresses for a particular application. These findings may help to advance the development of more comprehensive knowledge of the interaction of process factors, which will, in turn, lead to the development of novel process and experimental designs in the future.

13.4 Conclusions

The influences of surface residual stress, microstructure, and forces on SPIF were experimentally investigated. The findings of the present study are summarized below:

- The local deformation is more pronounced for a higher forming angle which results in a large deforming force required to deform the sheet metal. A similar trend has been observed for step depth: an increase in step depth leads to more local deformation.
- A clear gradient in grain structure has been found and decreases with an increase in step depth. However, a reduction in grain size has been observed with incremental forming. It is highest at a step depth of 0.25 mm, lowest at a depth of 0.75 mm and finally, a slight increase has been noticed at a step depth of 1 mm. As a consequence of this, the sheets that are created will have increased strength in comparison to the original raw material.
- A large compressive residual stress has been observed near the surface and with the increase in forming angle from 30° to 60° degrees, the stresses undergo a transition from a compressive to a tensile value.
- A noticeable effect of different process parameters of the incremental forming operation has been established in the study. As a result, residual stresses can be managed or modified to meet the requirements of a particular application by adjusting the process parameters as needed.

References

1. Jeswiet, J., Micari, F., Hirt, G., Bramley, A., Duflou, J., Allwood, J., Asymmetric single point incremental forming of sheet metal. *CIRP Ann.*, 54, 2, 88–114, 2005.

2. Lu, B., Ou, H., Shi, S.Q., Long, H., Chen, J., Titanium based cranial reconstruction using incremental sheet forming. *Int. J. Mater. Form.*, 9, 3, 361–370, 2016.
3. Ambrogio, G., Conte, R., De Napoli, L., Fragomeni, G., Gagliardi, F., Forming approaches comparison for high customised skull manufacturing. *Key Eng. Mater.*, 651, 925–931, 2015.
4. Joffe, J., Harris, M., Kahugu, F., Nicoll, S., Linney, A., Richards, R., A prospective study of computer-aided design and manufacture of titanium plate for cranioplasty and its clinical outcome. *Br. J. Neurosurg.*, 13, 6, 576–580, 1999.
5. Cheng, C.-H., Chuang, H.-Y., Lin, H.-L., Liu, C.-L., Yao, C.-H., Surgical results of cranioplasty using three-dimensional printing technology. *Clin. Neurol. Neurosurg.*, 168, 118–123, 2018.
6. Bagudanch, I., García-Romeu, M.L., Ferrer, I., Ciurana, J., Customized cranial implant manufactured by incremental sheet forming using a biocompatible polymer. *Rapid Prototyp. J.*, 24, 120–129, 2018.
7. Echrif, S.B.M. and Hrairi, M., Research and progress in incremental sheet forming processes. *Mater. Manuf. Process.*, 26, 1404–1414, 2011.
8. Emmens, W.C., Sebastiani, G., van den Boogaard, A.H., The technology of incremental sheet forming—A brief review of the history. *J. Mater. Process. Technol.*, 210, 981–997, 2010.
9. Reddy, N.V., Lingam, R., Cao, J., Incremental metal forming processes in manufacturing, in: *Handbook of Manufacturing Engineering and Technology*, pp. 411–452, Springer, London, 2015.
10. Filice, L., Fratini, L., Micari, F., Analysis of material formability in incremental forming. *CIRP Ann.*, 51, 199–202, 2002.
11. Bansal, A., Lingam, R., Yadav, S.K., Venkata Reddy, N., Prediction of forming forces in single point incremental forming. *J. Manuf. Process.*, 28, 486–493, 2017.
12. Silva, M.B., Skjoedt, M., Martins, P.A.F., Bay, N., Revisiting the fundamentals of single point incremental forming by means of membrane analysis. *Int. J. Mach. Tools Manuf.*, 48, 73–83, 2008.
13. Hussain, G., Al-Ghamdi, K.A., Khalatbari, H., Iqbal, A., Hashemipour, M., Forming parameters and forming defects in incremental forming process. *Part B. Mater. Manuf. Process.*, 29, 454–460, 2014.
14. Kumar, R., Kumar, G., Singh, A., An assessment of residual stresses and micro-structure during single point incremental forming of commercially pure titanium used in biomedical applications. *Mater. Today Proc.*, 28, 1261–1266, 2019.
15. Echrif, S.B.M. and Hrairi, M., Significant parameters for the surface roughness in incremental forming process. *Mater. Manuf. Process.*, 29, 697–703, 2014.
16. Bonda, D.J., Manjila, S., Selman, W.R., Dean, D., The recent revolution in the design and manufacture of cranial implants: Modern advancements and future directions. *Neurosurgery*, 77, 824, 2015.

17. Elshanawany, A.M. and Essa, A.A., Traumatic acute subdural hematoma: Treatment by evacuation with decompressive craniotomy and cranioplasty, case series and surgical outcome analysis. *Open J. Mod. Neurosurg.*, 08, 331–341, 2018.
18. Han, F., Mo, J.H., Wang, P., Deng, Y.Z., A digital manufacture technology for skull prosthesis using incremental sheet forming method. *Adv. Mater. Res.*, 102–104, 348–352, 2010.
19. Dewidar, M.M., Khalil, K.A., Lim, J.K., Processing and mechanical properties of porous 316L stainless steel for biomedical applications. *Trans. Nonferrous Met. Soc. China.*, 17, 468–473, 2007.
20. Chiba, D., Yamada, N., Mori, Y., Oyama, M., Ohtsu, S., Kuwahara, Y., Itoi, E., Mid-term results of a new femoral prosthesis using Ti-Nb-Sn alloy with low Young's modulus. *BMC Musculoskelet. Disord.*, 22, 987, 2021.
21. Rack, H.J. and Qazi, J., I, Titanium alloys for biomedical applications. *Mater. Sci. Eng. C.*, 26, 1269–1277, 2006.
22. Che-Haron, C.H. and Jawaid, A., The effect of machining on surface integrity of titanium alloy Ti–6% Al–4% V. *J. Mater. Process. Technol.*, 166, 188–192, 2005.
23. Kumar, G., Kanjarla, A.K., Lodh, A., Singh, J., Singh, R., Srivastava, D., Samajdar, I., Burst ductility of zirconium clads: The defining role of residual stress. *Metall. Mater. Trans. A.*, 47, 3882–3896, 2016.
24. Withers, P.J., Residual stress and its role in failure. *Rep. Prog. Phys.*, 70, 2211–2264, 2007.
25. Rossini, N.S., Dassisti, M., Benyounis, K.Y., Olabi, A.G., Methods of measuring residual stresses in components. *Mater. Des.*, 35, 572–588, 2012.
26. Singh, A. and Agrawal, A., Comparison of deforming forces, residual stresses and geometrical accuracy of deformation machining with conventional bending and forming. *J. Mater. Process. Technol.*, 234, 259–271, 2016.
27. Verlinden, B., Driver, J., Samajdar, I., Doherty, R.D., *Thermo-mechanical processing of metallic materials*, Jordon Hill, United Kingdom, Elsevier, 2007.
28. Bacon, D.J. and Dieter, G.E., *Mechanical metallurgy*, McGraw-Hill, New York, 1988.
29. Noyan, I.C. and Cohen, J.B., *Measurement by diffraction and interpretation*, Springer, New York, 1987.
30. Stock, S.R. and Cullity, B.D., *Elements of x-ray diffraction*, New Jersey, Prentice Hall, 2001.
31. Van Houtte, P. and De Buyser, L., The influence of crystallographic texture on diffraction measurements of residual stress. *Acta Metall. Mater.*, 41, 323–336, 1993.
32. James, M.R. and Cohen, J.B., Geometrical problems with a position-sensitive detector employed on a diffractometer, including its use in the measurement of stress. *J. Appl. Cryst.*, 12, 4, 339–345, 1979.
33. Balasingh, C. and Singh, A.K., Residual stresses and their measurements by X-ray diffraction methods. *Met. Mater. Process.*, 12, 269–280, 2000.

34. Wright, S., I and Adams, B.L., Automatic analysis of electron backscatter diffraction patterns. *Metall. Trans. A.*, 23, 759–767, 1992.
35. Nowell, M.M. and Wright, S., I, Orientation effects on indexing of electron backscatter diffraction patterns. *Ultramicroscopy*, 103, 41–58, 2005.
36. Field, D.P., Recent advances in the application of orientation imaging. *Ultramicroscopy*, 67, 1–9, 1997.
37. Kumar, G., Lodh, A., Singh, J., Singh, R., Srivastava, D., Dey, G.K., Samajdar, I., Experimental characterization and finite element modeling of through thickness deformation gradient in a cold rolled zirconium sheet. *CIRP J. Manuf. Sci. Technol.*, 19, 176–190, 2017.
38. Singh, A. and Agrawal, A., Investigation of surface residual stress distribution in deformation machining process for aluminum alloy. *J. Mater. Process. Technol.*, 225, 195–202, 2015.
39. Radu, C., Tampu, C., Cristea, I., Chirita, B., The effect of residual stresses on the accuracy of parts processed by SPIF. *Mater. Manuf. Process.*, 28, 572–576, 2013.
40. Alinaghian, M., Alinaghian, I., Honarpisheh, M., Residual stress measurement of single point incremental formed Al/Cu bimetal using incremental hole-drilling method. *Int. J. Light. Mater. Manuf.*, 2, 131–139, 2019.

14

Challenges and Perspective of Manufacturing Techniques in Biomedical Applications

Francis Luther King M.[1*], G. Robert Singh[1], A. Gopichand[1] and Srinivasan V.[2]

[1]Department of Mechanical Engineering, Swarnandhra College of Engineering and Technology, Narsapur, Andhra Pradesh, India
[2]Department of Manufacturing Engineering, Annamalai University, Faculty of Engineering and Technology, Chidambaram, Tamil Nadu, India

Abstract
The healthcare, biomedical, aerospace, manufacturing, and construction sectors are among of the fastest-growing users of additive manufacturing technology. The development of 3D printers is one example of the technological progress that the industrial age ushered forth. It's been around for a long time and has become routine. As time has progressed, this amazing manufacturing method has become more popular for use in the healthcare industry. 3D printing, the new technology has been contrasted have been discussed with more established approaches to areas such as surgery, pharmaceuticals, disease modeling, development of customized implants and prostheses, organ printing, veterinary medicine, and tissue engineering. Concerns with binder type and choice, minimal mechanical characteristics, low dimensional consistency, agglomeration, injector dimensions, dissemination size, restricted choice of materials, personalization of fit and design, layer height, and build failure are just some of the key challenges identified in this chapter when applying additive manufacturing to biomedical products. Few more administrative challenges exist in the production of biological goods using 3D printing technology, including retraining of workforce, product price, limited regulations, information security issues, infrastructure, and trademarks and copyright. This chapter contains potential future prospects that are predicted to become more prevalent in the future.

Keywords: AM challenges, biomaterial, bio printing, biomedical applications

**Corresponding author*: francis.mech@gmail.com

Arbind Prasad, Ashwani Kumar and Manoj Gupta (eds.) Advanced Materials and Manufacturing Techniques for Biomedical Applications, (345–382) © 2024 Scrivener Publishing LLC

14.1 Introduction

Modern additive manufacturing [1] uses material layers and 3D model data to create complex shapes and structures. The American Dental Association predicts a $9.7 billion increase in dental additive manufacturing by 2027 (increasing 35% annually). Each day, the medical field sees more and further use of 3D printing. The report estimates that around 11% of medical sector income comes from the use of 3D-printed components such as medical devices and implants. Additive manufacturing is expanding [2] due to rising demand for personalized medicinal components. Additive manufacturing goes by many names; rapid prototyping, on-demand manufacturing, digital fabrication, desktop manufacturing, solid freeform manufacturing, layer manufacturing, direct manufacturing technology, or 3D printing are just a few examples [1, 3]. The term "additive manufacturing" refers to a wide range of processes, including stereo lithography (SLA), digital light processing (DLP), selective laser sintering (SLS), Electron beam melting (EBM), Fusion deposition modeling (FDM), multi-jet/polyjet 3D printing, and selective laser melting (SLM). A wide variety of materials are employed in these processes [4]. These technologies use data from 3D models to create complex 3D components of any shape, which are otherwise difficult to construct using conventional methods. Data from 3D models may be easily printed with additive manufacturing if they are converted to a standard triangulate language format. All of these methods have the same underlying principle, which is that the final desired part is constructed by depositing materials in successive layers. However, CT scans, MRI, and X-rays may be translated to 3D digital print files, opening the door to the creation of individualized anatomical, intricate, and therapeutic architecture [5]. The medical area is one that has lately benefited greatly from additive manufacturing, and it is anticipated to revolutionize healthcare. Among the most popular biomedical applications of additive manufacturing are the fabrication of tissue and organs; anatomical models; the formation of customized prosthetics and implants; and advancements in the pharmaceutical field, in particular the development of drug dosage forms, drug discovery, and drug delivery.

Medical product personalization and customization, biocompatibility, cost effectiveness, increased productivity, accessibility, short production time, simple assembly, collaboration, and democratization [6–11] are just a few of the reasons why additive manufacturing has become so popular in the medical field. Additive manufacturing is still not often used for mass production [12–17] because of its poor speed. But this limitation is considered as an advantage in the medical area, where high precision and individualized goods are required in limited numbers. Because each individual patient has unique medical requirements and deserves individualized care, this is the case. Therefore, clinical and biological applications [18–20] are ideal for additive manufacturing and see substantial use. Numerous scientists and researchers have attempted to make 3D models of human organs and are currently on the hunt for bio-comparable materials with which to construct vascularized human beings. However, there are certain scientific and regulatory roadblocks. Surgeons may benefit from medical models in several ways [21, 22], including better understanding of the disease, reduced costs, streamlined procedures, more efficient use of surgical equipment, and individualized device design for implant. Additive manufacturing in medicine has several applications [23]. Some examples are custom-fit masks, innovative organ creation, and practice surgeries.

14.2 3D Printing Applications in the Biomedical Field

3D printing in biomedical applications has been the subject of a great deal of study, and many organs and tissues printed with this technique continue to present significant hurdles [24]. From cancer therapy to patient-specific prostheses, 3D printing is being used to enhance the quality of life or save the lives of patients in several medical fields.

14.2.1 Surgical Applications

A good better understanding of the inpatient's anatomy is possible thanks to 3D printing since it enables a more precise imaging system for preoperative planning. A surgical template is a tool used in surgery to help in the identification of anatomical landmarks, the estimation of appropriate angles, and the determination of the bone's orientation and size. Under normal situations, it is tedious to pinpoint the source of blood flow and predict the shape of bones. As was said, this methodology is often put to use in the medical field as a surgical guide and for 3D modeling in the surgical field. Therefore, this technology is preferred in vascular surgery, tumor resections, orthopedic surgery, and neurosurgery, among other surgical applications such as cardiovascular surgery, neurosurgery, general surgery, and plastic and cosmetic surgery. When technology is used successfully in these settings, it has the potential to raise surgical outcomes and minimizes medical errors, leading to greater patient safety. The use of this strategy during and before surgery is also expected to help the practice of future doctors and surgical technicians. These benefits of technology are being taken into account; for example, 3D printers are becoming more popular for use in surgical applications, and studies pertaining to this topic are being done [25–33].

14.2.2 Disease Modeling

Using 3D printing technology, a duplicate of the patient's unique anatomy is created prior to difficult surgeries or to better comprehend the ailment. Thus, the copy of the organ may be evaluated more precisely than using imaging techniques such as tomography can be practiced before to surgery, and the margin of error can be decreased. A research involves the printing of renal disease models. High-fidelity kidney models allow for the planning and simulation of complicated surgical procedures. This strategy facilitates medical students' understanding of the illness via modeling and affords them the chance to investigate the complicated anatomy and pathology of the kidneys. Modeling thus plays a significant part in the identification of kidney tumors and the evaluation of the patient's status. Numerous studies indicate that modeling will be of considerable value not just for renal illness, but also for cardiovascular disease and liver disease. Thus, it may shorten the length of the surgery, minimize the risk of mistake, and ensure patient safety [34–41].

14.2.3 Medical Devices

In the creation of several medical equipment, the use of 3D printing technology is a prevalent preference. It is used mostly in the creation of medical equipment that are difficult to manufacture using traditional techniques. In addition, economical fabrication of devices suited to the anatomical structure of the patient is supplied. Due to this, the majority of hearing aids developed specifically for the patient, such as hearing aids that must be compatible with the patient's ear anatomy, are manufactured using this approach [42, 43]. The use of printing in the production of several forms of medical equipment, such as hearing aids, is advantageous in many ways, including cost, patient safety, and device effectiveness. In the near future, it is projected that 3D printers will be used for an increasing number of human demands.

14.2.4 Implants

The production of patient-specific prosthesis is crucial to the medical use of 3D printing technology. In dental implant applications, the use of a personal implant application offers ease for both the healthcare staff and the patient. This technology removed conventional measuring techniques and lowered waiting times, which is one of its benefits. Knee joints, tibia bone, femur bone, fibula bone implants, and dental implants favor the printing technique. This technique enables for biocompatible manufacturing that is precise and patient-specific. Surgical imaging enables the collection of patient anatomical data, the generation of a digital model, and the prevention of difficulties and mistakes during implant placement. These implants may be manufactured via a printed mold, or they can be printed directly [44].

14.2.5 Patient-Specific Implants

It's claimed that prosthetics have been used since ancient Egypt. Prostheses are often made using tried-and-true methods, like as casting. But now, prosthetics are being built using 3D printing technology, and the results are encouraging. The broad commercialization of 3D printing in prosthetic design has allowed for rapid prototyping of many desired designs, which can be examined rapidly and affordably. Ear, nose, tooth, bone, hand, and foot prostheses are all things that can be made. The multi-material printing technique also makes it possible to adjust the skin tone of the person who will be wearing the prosthesis so that it blends in with their natural skin color. The scope of this method has its bounds. The artificial skin feels rougher than real skin [45]. In addition, the prostheses aren't modifiable, so they won't continue to fit the patient once he or she reaches a different height.

14.2.6 Vet Medicine

As stated in earlier sections, 3D printing is increasingly used in human medicine, particularly in vascular surgery, neurosurgery, disease modeling, and the production of patient-specific implants and prostheses. Not only are the method's advantages promising for human treatment, but also for animal medicine. This imaging technology is chosen in order to better manage and comprehend the anatomy of the animal and the veterinarian's practice prior to surgery [46–48]. The dog's tumor removal procedure used a 3D printer so that the defect region could be detected, the mass could be removed with the precise surgical planning, and the patient could be treated [49, 50]. Not only are prostheses created for animals with severed limbs, but also for those born without limbs. Moreover, in many instances, it will be impossible or economically infeasible to manufacture implants with complex forms using technologies other than additive manufacturing. AM is primarily used in veterinary orthopedics for the creation of bone models, sophisticated patient-specific implants, and surgical guidance. The usage of this technology in veterinary medicine is anticipated to rise in the near future [51, 52].

14.2.7 Bioengineering and Stem Cell Research

Tissue engineering (TE) is the discipline that applies engineering principles and biological research to the restoration of damaged tissues, the regulation of organ functions, and the maintenance of normal organ functioning. The goal is to grow living cells in a lab on biomaterial supports, then implant them in the injured area. The fundamental goal of this research is to use cell and tissue scaffolds to create artificial tissues and organs that can perform their intended roles in the body [53–55]. Bioprinting is an emerging technology with a wide range of potential uses, including the creation of functional tissue structures for the repair or replacement of sick or damaged tissues. It's premised on the idea of stacking layers of components and living cells to create a functional biological structure. It is common practice for bioprinters to use a hydrogel solution of biomaterials or a mix of biomaterials to encapsulate the necessary cell types for tissue designs. Hydrogels containing cells are printed in precise shapes. Several different methods, including photolithography, magnetic 3D bioprinting, stereolithography, and direct cell extrusion, are used for cell 3D bioprinting [56–62]. Scaffold is a material that is often used in TE. By directing cells to a precise position, the scaffold may act as a scaffold for tissue regeneration. Maintaining chemical and biological properties and cell viability is crucial for the success of any tissue regeneration project, making tissue scaffolds invaluable. Scaffolds are necessary for cells because of their biocompatibility, porosity, and mechanical support. By controlling the pore size of printed scaffolds, it may be possible to achieve the necessary mechanical characteristics. Although a nonporous shape would be preferable, the same porous architecture weakens it, making it impossible to attain the necessary strength. In addition to printing cell-free tissue scaffolds, cartilage, bone, and skin have all been successfully tested using this technique [63–69]. Spinal cord injuries, for which there is no effective therapy, may similarly benefit from the scaffolds formed by AM [70].

14.2.8 Organ Printing

Recent developments have allowed the 3D printing of biocompatible materials and cells on living tissues. With the use of bio printing, scientists are working to create transplantable organs and tissues. Material choice, growth and differentiation factors, cell types, and technological issues with living cells are only some of the additional complexities introduced by bio printing. The production of multi-layered heart tissue, cartilage, and bone structures has been achieved using this technique. Organ pressure needs the resolution of a variety of challenges, notwithstanding the medical benefits of 3D printers and breakthroughs in tissue engineering. It will be difficult for a 3D-printed duplicate of the organ to perform the same function as the real thing. The most significant challenge confronted by tissue engineering techniques is the development of functional vascular systems. Without a proper circulatory system, 3D-printed organs and other thick tissues cannot get the nourishment, waste removal, and gas exchange they need to develop normally. This may lead to problems with cell viability and functional organs [71–76].

14.3 Multi-Functional Materials in 3D Printing

The use of drop-casting, spin-coating, or screen-printing techniques for the creation of particle-based composite multiferroics is limited to the generation of films with predetermined geometries [77–79]. Additionally, there are persistent difficulties in making composite multiferroics that have a magnetic filler within a polymer matrix [80]. These difficulties include settling, homogeneous mixing, percolation, and buoyancy. Iron oxide and photopolymer have specific gravities of 4.8 and 1.03, respectively, hence magnetic particles are expected to settle rapidly. When the polymer matrix reaches its gelling point, the viscosity of the polymer prevents particle mobility, halting (or greatly delaying) settling. This settling issue requires constant agitation throughout the manufacturing process. In a similar vein, the homogeneity of a mixture is affected by the distribution of magnetic particles inside a solvent, leading to particle agglomeration and, ultimately, electrical breakdown due to percolation. Because of the problems with both the manufacturing process and the availability of raw materials, contemporary production techniques have only been used to a limited extent up to this time.

14.3.1 Metal Alloys

Making complex components from expensive metals like titanium and titanium based alloys that are essential in the aviation and healthcare sectors, is simplified by metal additive manufacturing. New techniques, alloys, and applications for metal AM are being unveiled on a regular basis, resulting in notable quality improvements and lower production times. Combining AM with more conventional production techniques allows for lump production of quality products. Metal additive manufacturing has several benefits, including lower tooling costs, the capacity to create intricate and lightweight constructions, and the amalgamation of many components, which may eliminate the need for part assembly. The melting and solidification processes produced in microstructures that are inappropriate, such as columnar grains and periodic fissures [81], which means that the vast majority of metal alloys in use today (over 5,500) cannot survive AM procedures. Recent research by Martin *et al.* [82] proposes that employing AM-generated nanoparticles to manage alloy solidification may solve these problems. Utilizing this method, they were able to successfully 3D print strong Al alloys (Al 7075 and 6061) using AM. Crystallographic information was used to choose the nucleants. The manufactured Al alloys had crack-free, equiaxed, fine-grained microstructures with mechanical properties comparable to those of the wrought materials. The technique was implemented using selective laser melting (SLM), but it might just as easily be applied to electron beam melting (EBM) or digitally enhanced design (DED) processes, or even to more conventional manufacturing techniques. This will allow AM to be utilized with materials such as nickel superalloys, which cannot be welded.

In general, the yield and ultimate strengths of AM metal components are greater than those of traditionally manufactured ones because to their finer microstructures. However, its microstructure is strongly influenced by the direction of formation, therefore it is anisotropic. This results in the usual anisotropy [83] of material strength properties, with greater tensile strength and strain along the direction of printing. It is important to note that surface roughness and material flaws also affect the fracture strength behavior and fatigue strength of AM parts. Raised surface roughness leads to stress accumulations and early collapse under fatigue loading [84]. The fatigue strength resistance of AM parts is lowered by faults in the material itself and by insufficient layer bonding. In order to reduce any remaining porosity, alter the microstructure, and smooth out the surface, post-processing treatments are used. Hot isostatic pressing (HIP) is the process of applying high isostatic pressure on a component while also subjecting it to high temperatures and heat treatments, which may decrease residual porosities, raise ductility at the expense of strength, and boost fatigue resistance [85, 86]. Polishing or chemical etching the surface may also increase fatigue resistance.

14.3.2 Polymer

Future 3D printing progress may include the use of fiber reinforcement to improve the mechanical strength properties of the polymers [87–89]. There has been a recent difficulty in producing continuous reinforced fibers for enhancing the mechanical features of 3D-printed polymer parts and components [90]. Tekinalp *et al.* [91] looked at composite products made from carbon fiber and ABS resin feedstock and highlighted the challenges of 3D printing fiber reinforced composites by analyzing their load-bearing capacity. Made using FDM, the samples' strength, and rigidity were greatly improved by compression-molding (CM). Fibre orientation's effect on tensile properties was found to be less than that of porosity, as shown by the fact that the tensile strength of the CM samples was higher. In addition to increasing the strength and stiffness of the final 3D printed parts and less distortion [92], mixing CF with polymer feedstock is recommended.

14.3.3 Nanomaterials

Nanomaterial's, which may decrease sintering temperatures and improve mechanical and electrical characteristics, are another new material for 3D printing [93–96]. Nanomaterial's may be added to a 3D printed item either manually, with the possibility of interruptions, or automatically, with the possibility of pre-mixing with the host matrix. There are several advantages to employing nanomaterials with AM, but one of the best is the improvement in product consistency. Nano composites' attractive qualities, which include excellent thermal conductivity, better fire resistance, high strength, and low weight [97, 98], have sparked the attention of a number of different industries. In order to print conductive 3D microstructures of varying shapes and sizes from PLA-based polymer nano-composites at a cheap cost, Postiglione *et al.* [99] build up a 3D printing technique called Liquid deposition modeling (LDM). Using this method, a woven construction for a simple electrical circuit was built. Nano SiO_2, montmorillonite, and attapulgite were all investigated by Weng *et al.* [100] as potential nano-composites to be created by incorporating into stereolithographic resin. They determined that nano SiO_2 provided the most reinforcement. This method also provided the green light for the usage of nano-composites in desktop SLA 3D printing. It was looked at whether or not a nano-composite based on Nylon-6 could be used as an FDM filament. Using SEM, we can see that the nanoparticles are dispersed uniformly. Nylon-6 is thus often utilized in 3D printing since it may serve as a suitable alternative to ABS [101]. While much progress has been made in developing nano-fillers for 3D printing composites, the fundamental obstacle to the widespread use of this novel material is still the aggregation of nanoparticles and the severe light scattering induced by the high nano-filler content [95].

14.4 Merits of AM in Medical Field

While the advantages of AM are many and varied, they all have a common thread of enabling significant reductions in waste and/or savings in resources. Just a few of additive manufacturing's numerous advantages are listed here.

14.4.1 Personalization and Customization

The term "personalization" is used to describe the practice of tailoring a service to each individual consumer in order to increase their level of satisfaction. Instead, customization refers to making adjustments so that an item works better for a certain user or role. Implants, surgical tools, and fixtures that are built specifically for a patient might potentially reduce the duration of surgery and the time needed for patients to recover [102].

14.4.2 Budget-Friendly

Additive manufacturing allows for the cost-effective production of products and components. For small-scale implants or prosthesis, such as those used to treat dental and craniofacial abnormalities, 3D printing is less expensive. Additionally, additive manufacturing reduces fabrication costs by reducing the use of superfluous materials. Some drugs may also be printed in a dose form that is simpler to administer to patients and more cost efficient [102].

14.4.3 Productivity Gains

Traditional methods of making things like prostheses and implants, which required forging, milling, and a lengthy delivery time, are quicker with 3D printing technology. In addition, the additive manufacturing method's speed, reliability, and resolution are all increasing [102].

14.4.4 Democratization and Collaboration

More individuals, including those in medical areas, can utilize nothing more than a 3D printing machine and their ideas to develop and build unique things for commercial and personal usage [103] as materials become more readily available and costs decrease.

14.4.5 Precision, Adaptability, and Usability

The medical model created using 3D printing technology is more accurate and has a better surface polish [104, 105]. One material may be swapped with another in additive manufacturing to improve the properties of a medical implant [106]. The 3D model made possible by 3D printing is quick and simple to use. Since there is no need for tooling or fixtures [107].

14.5 Major Challenges of AM in the Medical Field

In particular, additive manufacturing technology is useful in medical applications. However, several other uses, such as organ printing, need more investigation. Although, AM offers a variety of benefits to businesses, including cost savings and product design optimization, a shorter supply chain, customization, and the capacity to create complicated geometry, among others. Additive manufacturing has grown quickly in the medical field over the past decade, but there are still certain barriers to adoption. Drug release may be affected by the shape of a 3D printed drug delivery device [108]. While 3D-printed medication delivery systems have numerous advantages, they nevertheless face a number of challenges that must be addressed before their full potential can be realized. Fused deposition modeling (FDM) is arguably the most researched and developed 3D printing method because of its adaptability in the manufacturing of drug-eluting devices/drug delivery devices with various geometries. However, 3D printing technology may not be suited for pharmaceuticals that are thermolabile. The thermal analysis of FDM formulations revealed that high processing temperatures may significantly contribute to the degradation of thermally fragile drugs [89]. In order to remove the remaining solvent from digital drug delivery systems (DDS) made using printing-based inkjet or PAM systems, an extreme (high temperature) drying step may be required. The Wang group [109] dried its cube-shaped device in an oven set to 50°C for 12 hours. Lam *et al.* [110] had to dry their scaffold at 100°C for an hour, while Kastra and Rowe *et al.* [111] dried their dosage forms in a nitrogen-filled vacuum chamber for almost a day. All these heated after-processing steps ultimately slowed manufacturing and hastened the degradation of the integrated drugs. After each usage, regardless of the kind of 3D printing technique used, the device must be cleaned to eliminate any traces of support material and remaining monomers. In rare cases, further steps like as tumbling or sandblasting may be necessary for a flawless outcome. It was important to take into account technical factors including laser beam energy density, scanning speed, deposition velocity, and humidity to maintain the product's physical stability [112, 113]. One further problem is that not all 3D printers, not only binder jetting, can be scaled up easily, so there's a limit to how many people can use one device at once.

14.5.1 Cost-Effective Only in Case of Customized Part

Davis coined the term "mass customization" in the 1980s. The capacity to generate functioning components that satisfy individual customer satisfaction and demands at a fair cost is referred to as mass customization. The key phases of mass customization include embedded customization, modular manufacturing, and customized product services. Mass customization has become a popular method of product development. Because of the immense rivalry, globalization, and fast technical developments, most product manufacturing organizations have been obligated to make rapid adjustments. AM is only cost efficient if the product is made to the customer's exact specifications; otherwise, it is very costly and wasteful. Customization aids in the production of small batches and necessitates regular and dynamic reconstruction of the manufacturing units [114–119].

14.5.2 Material Changing Option is Limited

Titanium and its alloys, stainless steels, cobalt alloys, cobalt chrome, titanium, and other materials are employed in biomedical applications (Wohlers, 2008). Bio functioning, biocompatibility, corrosion resistance, and other factors influence material selection for medical implant manufacture. Metallic biomaterials, which are often utilized to replace and maintain skeletal elements, have excellent fracture toughness, tensile characteristics, and fatigue strength when compared to ceramic and polymeric materials. The cobalt alloy has a number of appealing qualities, including wear resistance, corrosion resistance, and heat resistance. However, since cobalt alloys are difficult to make, their uses have been restricted [120–124].

14.5.3 Limitation in Neurosurgery

Creating 3D neuroanatomical models is a time-consuming process, which is a major drawback of employing such models. The price of AM printing

CAD software and hardware, as well as the speed and precision of materials, must improve significantly in order to make AM printing equipment appealing to neurosurgeons [125]. Additionally, this method is less exact than other machining processes [126, 127], calls for highly skilled people, and is not suited to batch production [128, 129]. Therefore, further study is needed to address the challenges presented by 3D printing technology in healthcare.

14.6 Major Challenges of AM

There are several drawbacks to Additive manufacturing (AM) that require additional research and technological improvement, despite its many benefits, such as design freedom, customization, and the ability to produce complex structures. Among the many drawbacks are high costs, limited applications outside of massive construction and bulk production, low and anisotropic mechanical strength properties, material limits, and defects. For instance, AM of a component often takes extra time than more conventional methods like casting, extrusion, fabrication, or injection-molding. Powder bed processing and stereolithography are more time-consuming than inkjet printing and fused deposition modeling. The high resolution of powder-bed (SLS or SLM) and other 3D printing methods results in higher material and energy costs. When compared to more conventional methods, 3D printing's long processing time and greater cost are significant obstacles to the mass manufacturing of any repetitive components. When it comes to personalized products with complex structures, like a 3D printed scaffold for bone tissue engineering, AM may be more cost-effective than traditional manufacturing methods [130]. Four significant challenges attributed to the nature of AM are investigated which should be assessed in each particular application.

14.6.1 Formation of a Void

Formation of gaps between materials as they are layered is a major drawbacks of 3D printing. The increased porosity caused by AM may have a substantial influence on strength performance because of the lower interaction adhesion between printed layers [131]. The 3D printing method and printed material will both have a role in the occurrence and severity of voids. One of the key flaws that leads to subpar and anisotropic mechanical properties [131, 132] is the creation of voids, which is more common in techniques that use filaments of materials, such as FDM or contour carving.

Delamination between layers might occur after printing if these voids aren't filled [133]. The tensile strength of a 3D printed composite utilizing the FDM technique decreased and its ability to absorb water increased when the filament thickness was increased [134]. Better interlayer bonding and less void growth may be achieved by the additive production of concrete by raising layer thickness and extending the duration between consecutive layers. In contrast, decreasing the height of each layer during AM powder-bed printing of alumina/glass composite may dramatically reduce the material's high porosity [135]. Because of the reduced height, laser light can more easily reach the top layer, and ceramic powders may be more evenly distributed between the layers, reducing the likelihood that voids will form in the structure. As an added bonus, Paul *et al.* [136] showed that, because of the full contact between the layers, rectangular nozzles result in less gaps than cylindrical nozzles. On the other side, 3D printing intricate items, especially at joints, is more difficult when using a rectangular nozzle.

While the increased porosity of 3D printed things isn't ideal, it may be useful in applications where limited porosity is taken as an advantage of AM, such as the porous scaffold design utilized in tissue engineering. The voids produced by 3D printing were used to the advantage by Minas *et al.* [137]. Bigger pores were placed into a lattice structure on top of the microscopic holes formed by the air bubbles in the foam filament. 3D printed bio-composites, due to their increased porosity, may exhibit hygroscopic properties by enhancing their ability to store water [138].

14.6.2 Layer by Layer Aesthetics

Due to the nature of manufacturing processes, layer-by-layer aesthetics is also an issue. This flaw is seen in a 3D-printed concrete building. If the 3D printed object is buried in the end use, such as scaffolds for tissue engineering, the look may not be significant. A flat surface, as opposed to a layer-by-layer look, is desirable in various applications such as structures, toys, and aircraft. Sintering, for example, is a chemical or physical post-processing procedure that may eliminate this fault [139], but it increases processing time and expense. To minimize the layer-by-layer look, Khoshnevis [140] used a trowel-like tool coupled to a contour-crafting print-head. The layer surface thickness and the height of the part together establish the stack of layers. Because of this, the harshness of the layer-by-layer appearance may be mitigated by minimizing the count of layers. Filament-based 3D printing techniques, such as fused deposition modeling, inkjet printing, and contour-based manufacturing, are more likely to provide a layered appearance than powder-bed and stereo-lithography procedures.

14.6.3 Conception to Completion

Computer aided Design software is the primary instrument for creating a 3D printed item. It's possible that the printed product will have certain defects that weren't taken into account during design due to the constraints of additive manufacturing. The CAD system integrates solid geometry with boundary information. Tessellation ideas are often employed as approximations to the model. However, faults and imperfections are common when moving from CAD to a 3D printed object, especially when dealing with curved surfaces. Although a fine tessellation could help with this issue, it will be tough to process and print on a computer. So, occasionally it's thought of post-processing (heating, laser, chemicals, or sanding) to fix these flaws. Planning and determining the ideal orientation for the component, dividing it into adequate layers, and providing supporting materials that can be easily removed after printing are all important steps in reducing the gap between design and implementation. Using unbound powder as the support in the powder-bed approach has the added benefit of being readily removed using air pressure after the print job is complete. External support must be created for FDM, counter-crafting, and inject printing, and it is not always simple to remove. Furthermore, the aesthetics and mechanical strength properties of the 3D printed structures can be significantly impacted by process parameters like pressure, orientation, laser, layer thickness, direction, temperature, and printing speed, as well as material strength properties (e.g. rheology, thermo-plasticity, powder packing, etc.) [141, 142].

14.6.4 Mechanical Characteristics and Anisotropic Microstructure

Managing anisotropic behavior is a major challenge in AM. Because of this layer-by-layer printing technique, each layer microstructure is different from that at the boundaries between layers. Anisotropic microstructure behavior causes a 3D printed object to behave differently in tension or compression when seen from above or below than when viewed from the side. Metals and alloys 3D printed using selective laser melting (SLM) or selective laser sintering (SLS) have a unique grain microstructure and anisotropic behavior due to temperature gradients [143] since the borders of the layers added on top of one another are reheated. In order to control the sintering process and lessen anisotropic behavior [144], it is essential that the laser beam's heat penetrate each layer. Transverse (build) direction morphological and textural changes improve tensile strength and ductility in SLM 3D printed titanium alloy over longitudinal (lengthwise) modifications [144, 145]. Materials with anisotropic behavior include polymers, ceramics, and alloys [146, 147]. Ceramic particles, for instance, take on an anisotropic behavior when their shape deviates significantly from a spherical ideal, affecting their orientation along the printing direction. It was discovered by Hambach and Volkmer [148] that 3D printing carbon fiber-reinforced concrete in a direction perpendicular to the linear direction of the beam greatly enhances flexural strength in comparison to a cross-hatch pattern. It proves that the compressive strength of 3D printed specimens is significantly affected by whether the force is applied parallel to or perpendicular to the printing direction. Alternatively, the anisotropic behavior of 3D-printed materials may be useful in certain contexts. It is possible, for instance, to impart a one-of-a-kind anisotropic wettability to a surface by manipulating the characteristics of 3D printed filaments (such as speed and spacing) on the surface [149]. For uses such breathable water resistant surfaces, a 3D-printed polydimethylsiloxane layer with superhydrophobic and anisotropic properties and exceptional thermal durability has been developed.

14.7 Problems Encountered When Processing

As long as you're utilizing 3D printing for prototypes, this isn't a major problem. One of the major obstacles to developing AM production lines is the need to scale and automate post-processing as the technology evolves into an end-part manufacturing process.

14.7.1 Materials

Following the literatures, a number of obstacles were discovered, with the selection of acceptable binders, which varies depending on the kind of product, serving as the key emphasis.

14.7.2 Selection of Binders

"Selecting an appropriate binder is one of the most difficult challenges for engineers and designers since each prototype made have its own type of binder which are biocompatible. Binders must be durable, chemically stable, and provide just the proper amount of residue for the task at hand. The powdered substance needs a strong binder to maintain it in the right form. It is true that "strong enough" is shape-dependent since smaller parts and greater features suffer lower stresses than larger portions and finer features. The process of depowdering, in which the part is extracted from the powder bed, is the most taxing on it. This is because the component has to be lifted from a bed of possibly dense powder, forcefully brushed, and carefully scrutinized. If the binder is to resist burning during pyrolysis, it must be chemically stable up to several hundred degrees, or whatever temperature the powder it is binding together may start to sinter. In addition, the metal binder shouldn't react with or change the printed substance by leaving behind a lot of residue after pyrolysis. The selection of a binder is determined by the intended use and specific need of a printing method what we adopt. Binders for biodegradable components should be non-toxic, ease of use, and easily accessible. Regarding renewable, the binder has to be derived from natural resources. Because of these limitations, ordinary binders cannot be used [150]. Metal, polymer, hydrogels, resin, glass, ceramic, or polymer may be used as building materials for 3D-printed items. The head of the 3D printing machine applies the binder layer by layer to the powder.

14.7.3 Mechanical Properties

3D printing technology also faces challenges with regards to mechanical strength. extensive study into the strength properties of 3D printed materials, including mechanical tensile strength, flexural, elongation, compression , shear, hardness, wear, and friction [151–154]. The implications of this anisotropy on the mechanical characteristics of 3D printed objects were examined by only a small number of researchers at first [155, 156]. Mechanical anisotropy in FDM-printed items may be caused by many variables, including air gaps/voids, build orientation, and raster angle. Changes in construction orientation, however, are the single most important element in drawing attention to mechanical anisotropy. Mechanical anisotropy was also seen with other AM technologies such as PolyJet [157] and stereolithography [158], however FDM remained the most sensitive to build orientation changes. Failure of a 3D printed component, including layer delamination, begins at the item's free surface. The direction and severity of the crack formation rely on the raster angle and intensity of the pores, respectively [159]. Altering the raster angle is a potential method for reducing crack growth and structural damage in FDM-printed components. Cracks that propagate along the boundaries may be halted in their tracks by misalignments in the pore alignments caused by filament misalignment [160]. This modification affects measurements in the same direction as print deposition, however the layer delamination problem persists when looking at the object from the side.

14.7.4 Distribution of Size

The pore diameters of porous titanium implants printed through SLM-type 3D printing technology are known to be quite precise ranging from 400 to 1000 μm [161] that display exceptional osteointegration performance *in vivo*. Good powder density, including flowability, has a direct impact on the printing process's ability to generate quality layers. Particles' optical and thermal characteristics may be influenced by the particle size distribution and form. The powder's layer characteristics are mostly determined by the powder's flowability and density.

Mostafaie *et al.* [162] stated that micro powder particles produce a multitude of different pores that are equally spaced all across the full section, whereas large powder particles produce a small number of large pores that are unevenly distributed throughout the product. It is simpler to move and deposit particles that are consistently sized and spherical. The packing density of the powder decreases and voids and inhomogeneities appear in the final component if the size range is too narrow. Larger than desired particles may compromise the integrity of the powder's thin layer and the overall structure of the end product [163]. Therefore, it is vital for all firms to pick a powders with the appropriate size and shape.

14.7.5 Choice of Materials

Each material has its unique qualities, which may differ in their suitability for 3D-printed biomedical product production. All existing additive manufacturing techniques include the use of some kind of energy transformation, whether it heat, light, or something else. Photopolymers, powdered thermoplastics, filament thermoplastics, and metals are the four primary material classes that map to distinct technologies. Technology and materials are intertwined in such a way that the best method for your project will usually be decided by what you need to make.

Applications, aesthetics, functionality, and certifications are just a few of the things to consider while selecting the best material. It is not always easy to keep track of all the material properties that must be met by a given component, much alone figure out which ones do. Our Material Wizard is ready to help you out with that. This all-inclusive tool may be used to sift through a wide range of qualities and attributes in order to zero in on the best possible material for your project.

14.7.6 Texture and Color

Furthermore, commodities' textures and color play a key function in differentiating things in 3D printing. Customers, it seems, wish to simulate real-life emergency procedures by purchasing artificial organs. The ancient machines provide printing with a variety of materials, but just a few color options. Therefore, they built new 3D printing equipment to enable the production of colored objects.

The longevity of materials was a newly identified obstacle while implementing 3D printing technology in this investigation. Each material utilized has a finite lifespan, according to respondent R3, and this is a crucial consideration. Theoretically, if a substance is utilized beyond its designated expiration date, its qualities may be altered, which might result in patient-harming products. Severe deterioration, breakage, or discoloration might occur in the clinic as a consequence.

14.7.7 Printers

By using 3D printing technology to generate 3D printed healthcare and medical products is hampered by insufficient dimensional precision. The design plays a crucial role in the production of extremely precise 3D-printed items.

14.7.8 Dimension Accuracy

The design of 3D-printed biomedical items determines their precision. Variations in curing and chilling, for instance, might result in shrinkage or warping. Materials also have an effect on accuracy. For example, regular SLA resin produces more precise items in terms of dimensions than ordinary resin. Therefore, it can be inferred that the development of precise form, size, shape, and minute surface roughness on synthetic implants is crucial in their optimal functioning [164]. Using arbitrarily chosen machinery and materials makes it challenging to develop biomedical goods with precise dimensions and structures. Consequently, if the dimensions precision is poor, the product will not fit inside the body and its clinical success rate will be affected. To attain the desired degree of precision, attention must be taken in selecting the suitable machines and materials. As per Bertol *et al.* [165], the approximate geometrical precision of the 3D-printed implants assessed by laser scanning was 200 µm, allowing for their use in craniofacial structures. In the meanwhile, Osman *et al.* [166] report that digital light processing (DLP) has proven effective for manufacturing bespoke zirconia dental implants with appropriate dimensional precision. Consequently, producing 3D-printed biological goods is hampered by poor dimensional precision.

14.7.9 Nozzle Size

In order to prevent powder agglomeration, the particle distribution must be established with care [167]. The size of the printer nozzle is an additional obstacle while using 3D printing technology for biological items. The diameter of the needle has a direct effect on the extrusion width of each line produced by a 3D printer. Theoretically, decreasing nozzle diameters do enable (us) to attain effective accuracy. If you use a 3D printer to produce big amounts of biomedical items, you will need to ensure that the extruder is putting down the correct quantity of material. Depending on the 3D printer, it is quite simple to swap out various nozzles. The size of the nozzle is crucial for ensuring the smooth manufacture of 3D-printed biomedical items. The development of biocompatible and biomimetic complex tissues is facilitated by the use of smaller nozzle diameters.

14.7.10 Layer Height

When the default resolution of certain printing techniques (such SLS, Material Jetting, or SLM/DMLS) is sufficient for most applications, or since the layer height is already pre-set by the machine manufacturer, choosing a layer height is not particularly necessary. For various 3D printing techniques (including FDM and SLA), layer height is a crucial design parameter that affects the printing time, cost, appearance, and physical qualities of a manufactured item. Parts produced at 100 mm and 200 mm typically have a negligible visible difference. However, printing the item at 100 mm will double the time required (since the 3D printer will have to trace twice as many cross sections), which will affect the price. Because 3D printing is done in layers, the effect of layer height is more pronounced on curved or angled surfaces than on vertical walls. The smoothness of a portion depends on its vertical resolution, which in turn is affected by the layer height.

Selecting a lower layer height is optimal if visual appeal is the primary priority as this will provide a more refined end product. When 3D printing a functional item, however, a greater layer height is recommended since it reduces printing time and costs while also enhancing the part's mechanical performance. Layer height decisions should also take post-processing of the pieces into account. If you plan on sanding, acetone smoothing, or painting the item, a greater layer height may be preferable. The layer height is average. The higher the printing speed, the worse the product quality. If we want to build a delicate model that is 30 mm in height, then we must plan for sluggish manufacturing. Because the greater the layer, the lesser the quality, it is necessary to select the appropriate surface height and material options. The reason for this is that when printing speeds increase, the quality of the final output decreases. It's important to take your time throughout production if you're working with a fragile model. The bigger the number of layers, the worse the quality of the final result. Engineers face this challenge and set out to find a solution by conducting several experiments to discover the optimum layer height. It is also the responsibility of the experts to verify that the nozzle size is appropriate for the layer height.

14.7.11 Build Failure

"Build failure" is now raising difficulty in the AM industry. If a piece doesn't adhere properly to the construction plate, it might cause a number of problems. The construction plate has no plastic bonded to it, not even a thin coating. For metal AM, there is a significant lack of automated build preparation tools. Having an experienced operator set up a build operation correctly is crucial. Building time, surface quality, and support material requirements are just a few of the variables that make determining the "optimal approach" difficult. Important construction criteria including component orientation, support structure type and position, and the most effective laser scan procedures all fall within the purview of the operator. These criteria are now determined as much by expert opinion as by hard data.

14.8 Challenges in Management

Extremely vital for the production of 3D printed products is a high degree of software knowledge and proficiency. When adopting 3D printing technology, retraining employees might be considered a hurdle. The company's employers are committed to enhancing their workers' competency. In making of healthcare goods, staff must have specialized knowledge, such as compatibility of resources, the 3D printing technique, and the design of biomedical products. Since these firm must strictly adhere to regulations and standards for their manufactured products.

14.8.1 Training

For the effective implementation of a broad variety of 3D printing technologies, although the availability of workers in India and Asia is not an issue, it might be difficult to find qualified designers, engineers, and technicians to run the equipment. Educational institutions may respond to this need by introducing courses in 3D printing-related fields such as design, programming, machine operation/optimization, etc. The skills necessary to succeed in 3D printing may be developed via the construction of educational centers and the creation of, or acceleration of, a 3D curriculum in bachelor engineering programs.

As part of its Atal Innovation Mission, the Indian government has mandated that 3D printers be installed at "Atal Tinkering Labs" in a number of public schools around the country. Under the Atal Incubation Mission, the government is also providing funding for fast prototyping equipment valued at INR 10.00 Cr. Such measures would greatly facilitate the eventual widespread use of 3D printing in India.

14.8.2 High Priced Products

The resources used for 3D printing are quite expensive. The amount for purchasing a 3D printer is important costs associated with using 3D printing technology in the industrial sector. The pricing range for a 3D printer is between 116,000 and 232,000 US dollars. The price of the machine relies on its capacity to manufacture a product with certain characteristics. The lower the price, the lesser the print quality, materials, size, and usefulness. Stratasys Direct conducted a survey of industrial customers of 3D printers and found that high equipment prices were their primary concern. Whereas printing with plastic is simple and inexpensive, printing with metal often necessitates pricey machinery that is out of reach for most businesses. Producing something still takes a long time, even when firms have access to these tools. In comparison to conventional manufacturing methods, the cost per unit of output for 3D-printed goods is quite high when mass-producing them. Depending on the complexity of the product and the printing speed, 3D printing can take anywhere from a few hours to several days to complete. Metal printers of today can take days to print due to the slow rate of construction (only 5–40 centimeters per hour is typical, even for modern machines using metal powders).

14.8.3 Lack of Guidelines

The management of 3D printing technology enterprises must also contend with the difficulty posed by the necessary processes and standards for employing the technology. Currently, the needed processes and standards for employing 3D printing technology are complex. Each firm has its own rules for the distribution of medical items. For instance, customer-supplied pharmaceuticals must be safe for human ingestion. For industrial items, organizations must guarantee that their product meets the specifications. Distinct items have varying materials and production processes, as well as different standards.

14.8.4 Cyber Security Issues

Moreover, the administration of 3D printing technology for biological items is also hampered by cyber security concerns. The proliferation of cloud-based processing has increased the feasibility of distant, discrete manufacturing settings. In a sense, this is good news, but it also suggests that stronger cybersecurity measures are much more urgently required than before. Additive manufacturing technologies allow businesses to make better use of raw resources, but their heavy reliance on digital data streams has led to a few complications. Many AM machines maintain an active network connection to facilitate remote job queuing, troubleshooting, and tracking. This interdependence makes systems more susceptible to assault and expands the attack surface. Commercial 3D printing services may be entrusted with large-scale projects, but their reliability is questionable. Examining the entire production cycle is necessary because of the privacy risks that arise due to the inherent challenge in validating credibility in the new era of AM, where designs may be developed by multiple parties but manufactured at any suitable printer.

14.8.5 Infrastructure

Manufacturers need the proper digital way to support their 3D printing project successfully before they can use it in manufacturing. Several businesses now use commercially available information technology to lay the groundwork for this kind of network. Such technologies are typically not optimal for 3D printing processes since they were created with the needs of conventional production in mind. The industry has been developing 3D printing-specific production software solutions to address this problem. Production scheduling, post-processing management and supplier communication are just some of the tasks that are facilitated by this kind of software. Therefore, an enterprise may have a centralized production planning and monitoring system, providing for improved traceability by keeping track of components and viewing projects. Among the many options necessary to build the digital infrastructure for AM manufacturing, workflow management software is rising to the top. It improves the effectiveness of routine company operations by facilitating the division of labor. Integrating AM into a digital manufacturing environment is greatly facilitated by the synchronization of data across systems.

14.8.6 Patent, Trademark, and Copyright Issues

To keep up with the ever-changing nature of technology, laws, particularly those pertaining to intellectual property, are perpetually playing catch-up.

As a result, efforts are made to find creative ways to interpret intellectual property laws in order to accommodate technological progress. The rapidly expanding field of 3D printing technology has given rise to some intriguing intellectual property concerns. In contrast to Europe and the United States, the commercialization of 3D printing technology for producing biomedical items in Asia is still in its infancy. Therefore, every firm in Asia must develop strong marketing tactics (promotions and commercials) so that all are aware that 3D printing technology exists in the region. A strong marketing plan is the only way for the 3D printing sector to achieve its objective. This development has set up a system in which the production of any product may be highly localized, raising the stakes for intellectual property owners in terms of the quality violation on a wide range of IP rights.

14.8.6.1 Patent Protection

Due to the fact that even simple items may indeed be developed on computers and since 3D printed products may have the specific pace of innovation that is inherently intellectual registered, they potentially infringe on patent databases. A violation of the Patents Act of 1970 occurs when such documents are made, sold, offered for sale, or used. Nevertheless, the exclusions excluding violations, such as study or experimenting, or, imparting instructions, can apply if the printed objects were created in a private setup. Finally but not least, the transcribed data suggests that issues over patents and copyrights provide a fresh barrier to the widespread use of 3D printing technology for the manufacture of biological products. Inventive designs, methods, tools, and materials made possible by 3D printing in the medical field are all eligible for patent protection [168]. The core concept is that patents protect ideas, not merely their representations.

14.8.6.2 Trademark Protection

A 3D printer user who illegally copies a registered model may potentially face trademark infringement charges. Forms and texturing are beginning to get trademark protection in India and elsewhere, albeit this is still in its early stages. These unconventional markings may be registered and enforced as trademarks so long as the shape/texture communicates meaning and differentiates the products or services of one person from those of another. The first apparent threat is that 3D printing might be used to produce counterfeit goods by infringing on registered form and texture marks. Furthermore, if a product is copied using a 3D printer, the trademark of the original product's manufacturer might be stolen (affixed on the product). An accusation of counterfeiting could result from this.

14.8.7 Copyright Protection

Safeguarding from unauthorized duplication and distribution is provided by intellectual property laws. Protected copyright status for work is established in India the moment it is created. Copyright protection may be obtained without formally registering the copyright. It's fascinating to think about the implications of the "tension" or overlap between copyright and design rights, particularly in the context of 3D creations. Ownership would be violated if a non-functional two-dimensional creative work were converted into a three-dimensional piece. On the other hand, according to section 15(2) of the Copyright Act, 1957, if an artistic work is registrable as a design, then copyright in the work would be extinguished if more than 50 copies were made.

14.9 Conclusion

Current biomaterials R&D has benefited greatly from AM for prototyping intricate and individualized structures to meet the needs of individual patients. However, it faces challenges including a scarcity of supply and regulation concerns. While additive manufacturing has many potential applications, there are a few drawbacks that need to be addressed before it can be widely adopted. Mechanical performance may suffer as a consequence of increased porosity caused by the loss of interfacial connection between printed layers. Anisotropic behavior is another common AM challenge that causes different mechanical behavior in vertical tension or compression compared to horizontal tension or compression. The approximation of the design in CAD's tessellation concept means that the translated CAD model into a 3D-printed object frequently has errors and defects, especially on curved surfaces. In addition, the AM's layer-by-layer appearance is unattractive in applications like as buildings, toys, and aeroplanes. Research and development into new materials and technologies have helped alleviate some of these problems, but there is always room for advancement. Although 3D printing is an innovative technology for one-of-a-kind things and niche applications, it still has a ways to go before it can compete with traditional manufacturing techniques for mass production of commonplace items. In conclusion, the findings demonstrate that, in terms of processing and materials', adopting 3D printing technology to manufacture biomedical items has many challenges. The size of a corporation or organization has little effect on printing production. However, marketing, patents, and copyright were identified as additional obstacles. This research is significant for the biomedical manufacturing industry because it provides information on the application of 3D printing technology in developing nations for the production of biomedical goods. This research might serve as a reference for the manufacturing, human resources, and management industries. For new firms planning to embrace this technology, this study's qualitative sharing experience will give early insight into the challenges they may face. It is hoped that the outcomes of this research will provide emerging countries with succinct information on the use of 3D printing technology in the industrial sector.

It is crucial to address both the novice's preparedness to create and deploy this technology and the consumers' willingness to acquire the items. This article addressed the fact that 3D printing technology is safe and effective for manufacturing medical products. Consequently, the article believes that the mentioned obstacles would inspire and equip newcomers, policymakers, and government sectors to embrace this technology cautiously and react effectively to customer trust and demand.

References

1. Gross, B.C., Erkal, J.L., Lockwood, S.Y., Evaluation of 3D printing and its potential impact on biotechnology and the chemical sciences. *Anal. Chem.*, 86, 3240–53, 2014.
2. Javaid, M. and Haleem, A., Additive manufacturing applications in medical cases: A literature based review. *Alexandria J. Med.*, 54, 411–22, 2018.
3. Chua, C.K., Leong, K.F., An, J., Introduction to rapid prototyping of biomaterials. *Rapid Prototyp. Biomater.*, Woodhead Publishing, 1–15, 2014.
4. Zengguang, L., Yanqing, W., Chunzhi, C., A critical review of fused deposition modeling 3D printing technology in manufacturing polylactic acid parts. *Int. J. Adv. Manuf. Technol.*, 2, 1–13, 2019.
5. Cui, X., Boland, T., Dlima, D.D., Lotz, M.K., Thermal inkjet printing in tissue engineering and regenerative medicine. *Recent Pat. Drug Deliv. Formul.*, 6, 149–55, 2012.
6. Banks, J., Adding value in additive manufacturing: Researchers in the United Kingdom and Europe look to 3D printing for customization. *IEEE Pulse*, 6, 22–6, 2013.
7. Mertz, L., Dream it, design it, print it in 3-D: What can 3-D printing do for you? *IEEE Pulse*, 4, 15–21, 2013.

8. Ibrahim, A.M., Jose, R.R., Rabie, A.N., Gerstle, T.L., Lee, B.T., Lin, S.J., Three-dimensional printing in developing countries *Plast. Reconstr. Surg. Glob. Open J.*, 3, 222–33, 2015.
9. Rankin, T.M., Glovinco, N.A., Cucher, D.J., Watts, G. *et al.*, Three-dimensional printing surgical instruments: Are we there yet. *J. Surg. Res.*, 189, 193–7, 2014.
10. Ursan, I., Chiu, L., Pierce, A., Three-dimensional drug printing: A structured review. *J. Am. Pharm. Assoc.*, 53, 2, 136–44, 2013.
11. Cui, X., Boland, T., Lotz, M.K., Thermal inkjet printing in tissue engineering and regenerative medicine. *Recent Pat. Drug Deliv. Formul.*, 6, 2, 149–55, 2012.
12. Lee, A.Y., An, J., Chua, C.K., Two-way 4D printing: A review on the reversibility of 3Dprinted shape memory materials. *Engineering*, 3, 663–74, 2017.
13. Perez, B., Nykvist, H., Brogger, A.F., Larsen, M.B., Falkeborg, M.F., Impact of macronutrients printability and 3D-printer parameters on 3D-food printing: A review. *Food Chem.*, 287, 249–57, 2019.
14. Chen, Z., Li, Z., Li, J. *et al.*, 3D printing of ceramics: A review. *J. Eur. Ceram. Soc.*, 39, 661–87, 2019.
15. Blok, L.G., Longana, M.L., Yu, H., Woods, B.K.S., An investigation into 3D printing of fibre reinforced thermoplastic composites. *Addit. Manuf.*, 22, 176–86, 2018.
16. Zarek, M., Mansour, N., Shapira, S., Cohn, D., 4D printing of shape memory-basedm personalized endoluminal medical devices. *Rapid Commun.*, 38, 233–44, 2017.
17. Pop, M.A., Croitoru, C., Bedo, T., Geaman, V. *et al.*, Structural changes during 3D printing of bioderived and synthetic thermoplastic materials. *J. Appl. Polym ScimSymp*, 136, 473–82, 2019.
18. Kumar, R., Kumar, S. *et al.*, Trending applications of 3D printing: A study. *Asian J. Eng. Appl. Technol.*, 9, 1, 1–12, 2020.
19. Milewski, J.O., Trends in AM, government, industry, research, business, additive manufacturing of metals, in: *Springer Series in Materials Science*, vol. 258, pp. 255–92, 2017.
20. Wu, Y.H., Chiu, Y.C., Lin, Y.H. *et al.*, 3D-printed bioactive calcium silicate/poly- "-caprolactone bioscaffolds modified with biomimetic extracellular matrices for bone regeneration. *Int. J. Mol. Sci.*, 20, 942, 2019.
21. Ackland, D.C., Robinson, D., Redhead, M. *et al.*, A personalized 3D-printed prosthetic joint replacement for the human temporomandibular joint: From implant design to implantation. *J. Mech. Behav. BioMed. Mater.*, 69, 404–11, 2017.
22. Diment, L.E., Ompson, M.S., Bergmann, J.H.M., Clinical efficacy and effectiveness of 3D printing: A systematic review. *BMJ Open*, 7, 122–32, 2017.
23. Kumar, L., Haleem, A., Tanveer, Q., Javaid, M., Shuaib, M., Kumar, V., Rapid manufacturing: Classification and recent development. *Int. J. Adv. Eng. Res. Sci. (IJAERS)*, 4, 29–40, 2017.

24. Ventola, C.L., Medical applications for 3D printing: Current and projected uses. *P&T*, 39, 704–711, 2014.
25. Qiu, K., Haghiashtiani, G., McAlpine, M.C., 3D printed organ models for surgical applications. *Annu. Rev.*, 11, 287–306, 2018.
26. Hoang, D., Perrault, D., Stevanovic, M., Ghiassi, A., Surgical applications of three-dimensional printing: A review of the current literature & how to get started. *Ann. Transl. Med.*, 4, 456, 2016.
27. Aimar, A., Palermo, A., Innocenti, B., The role of 3D printing in medical applications: A state of the art. *J. Healthc. Eng*, 10, 1–10, 2019.
28. Li, C., Cheung, T.F., Fan, V.C., Sin, K.M., Wong, C.W.Y., Leung, G., Applications of three-dimensional printing in surgery. *Surg. Innov.*, 24, 1–7, 2016.
29. Tack, P., Victor, J., Gemmel, P., Annemans, L., 3D-printing techniques in a medical setting: A systematic literature review. *BioMed. Eng. Online*, 115, 1–21, 2016.
30. Chae, M.P., Rozen, W.M., McMenamin, P.G., Findlay, M.W., Spycal, R.T., Smith, D.J.H., Emerging applications of bedside 3D printing in plastic surgery. *Front. Surg.*, 2, 25, 2015.
31. *3D printed models used to plan kidney cancer surgery*, Global Data, New York, 2019, https://www.medicaldevice-network.com/news/3d-printed-kidney-tumuor-removal/.
32. Zheng, B., Wang, X., Zheng, Y., Feng, J., 3D printout models vs. 3D-rendered images: Which is better for preoperative planning? *J. Surg. Educ.*, 1, 73, 3, 518–523, 2018.
33. Sun, Z. and Liu, D., A systematic review of clinical value of three-dimensional printing in renal disease. *Quant. Imaging Med. Surg.*, 8, 311–325, 2018.
34. Giannopoulos, A.A., Mitsouras, D., Yoo, S.J., Liu, P.P., Chatzizisis, Y.S., Rybicki, F.J., Applications of 3D printing in cardiovascular diseases. *Nat. Rev. Cardiol.*, 13, 701–718, 2016.
35. Cantinotti, M., Valverde, I., Kutty, S., Three-dimensional printed models in congenital heart disease. *Int. J. Cardiovasc. Imaging*, 33, 137–144, 2017.
36. Yap, Y.L., Tan, Y.S.E., Tan, H.K.J., Peh, Z.K., Low, X.Y., Yeong, X.Y. *et al.*, 3D printed bio-models for medical applications. *Rapid Prototyp. J.*, 23, 227–235, 2017.
37. Ma, X., Liu, J., Zhu, W., Tang, M., Lawrence, N., Yu, C. *et al.*, 3D bioprinting of functional tissue models for personalized drug screening and *in vitro* disease modelling. *Adv. Drug Deliv. Rev.*, 132, 235–251, 2018.
38. Perica, E.R. and Sun, Z., A systematic review of three-dimensional printing in liver disease. *J. Digit. Imaging*, 31, 692–701, 1, 2018.
39. Kim, G.B., Lee, S., Kim, H., Yang, D.H., Kim, Y.H., Kyung, Y.S. *et al.*, Three-dimensional printing: Basic principles and applications in medicine and radiology. *Korean J. Radiol*, 17, 182–197, 2016.
40. Banks, J., Adding value in additive manufacturing: Researchers in the United Kingdom and Europe look to 3D printing for customization. *IEEE Pulse*, 6, 22–26, 1, 2013.

41. Lipson, H., New world of 3-D printing offers "completely new ways of thinking": Q&A with author, engineer, and 3-D printing expert Hod Lipson. *IEE Pulse*, 6, 12–14, 2013.
42. Dodziuk, H., Applications of 3D printing in healthcare. *Kardiochirurgia i Torakochirurgia Polska/Polish. J. Thoracic Cardiovasc. Surg.*, 13, 3, 283–293, 2016.
43. Auricchio, F. and Marconi, S., 3D printing: Clinical applications in orthopaedics and traumatology. *EOR*, 1, 123–127, 2016.
44. Thomas, D.J., 3D printing durable patient specific knee implants. *J. Orthop.*, 14, 182–183, 2017.
45. Honigmann, P., Sharma, N., Okolo, B., Popp, U., Msallem, B., Thieringer, F.M., Patient-specific surgical implants made of 3D printed PEEK: Material, technology, and scope of surgical application. *BioMed. Res. Int.*, 2018, 1–8, 2018.
46. Haleem, A., Javaid, M., Suman, R., Additive manufacturing applications in orthopaedics: A review. *J. Clin. Orthop. Taruma*, 58, 171–173, 2019.
47. Reilly, M.O., Reese, S., Herlihy, T., Geoghegan, T., Cantwell, C.P., Feeney, R.N.M. *et al.*, Fabrication and assessment of 3D printed anatomical models of the lower limb for anatomical teaching and femoral vessel access training in medicine. *Anat Sci. Educ.*, 9, 71–79, 2016.
48. Dismukes, D.I., Fox, D.B., Tomlinson, J.L., Essman, S.C., Use of radiographic measures and three-dimensional computed tomographic imaging in surgical correction of an antebrachial deformity in a dog. *J. Am. Vet. Assoc.*, 232, 68–73, 2008.
49. Kim, S.E., Shim, K.M., Jang, K., Shim, J.H., Kang, S.S., Three-dimensional printing-based reconstruction of a maxillary bone defect in a dog following tumor removal. *In Vivo*, 32, 63–70, 2018.
50. Xie, S., Cai, B., Rasidi, E., Yen, C.C., Hsu, C., Chow, W.T. *et al.*, The use of a 3D-printed prosthesis in a Great Hornbill (Buceros bicornis) with squamous cell carcinoma of the casque. *PLoS One*, 14, 8, p.e0220922, 2019.
51. Popov, V.V., Kamskii, G.M., Demyanetz, A.K., Kovalevsky, A., Usov, S., Trofimcow, D. *et al.*, Additive manufacturing to veterinary practice: Recovery of bony defects after the osteosarcoma resection in canines. *BioMed. Eng. Lett.*, 9, 97–108, 2019.
52. Sendemir, A., Seçkin, U.D., Uyanıkgil, Y., Gördün, C., 3D bioprinting and cultivation of keratinocytes for skin tissue engineering. *Dicle. Med. J.*, 45, 9–18, 2018.
53. O'Brien, F.J., Biomaterials & scaffolds for tissue engineering. *Mater. Today*, 14, 88–95, 2011.
54. Kim, L.Y., Seo, S.J., Moon, H.S., Yoo, M.K., Park, L.Y., Kim, B.C. *et al.*, Chitosan and its derivatives for tissue engineering applications. *Biotechnol. Adv.*, 26, 1–21, 2008.
55. Mano, J.F., Silva, G.A., Azevedo, H.S., Malafaya, P.B., Sousa, R.S., Silva, S.S. *et al.*, Natural origin biodegradable systems in tissue engineering and

regenerative medicine: Present status and some moving trends. *J. R. Soc. Interface*, 4, 999–1030, 2007.

56. Chua, C.K. and Yeong, W.Y., *Bioprinting principles and applications*, World Scientific Publishing Co, Singapore, 2015.
57. Deo, K., Singh, D.A., Peak, C.W., Gaharwar, A.K., Alge, D.L., Bioprinting 101: Design, fabrication, and evaluation of cell-laden 3D bioprinted scaffolds. *Tissue Eng. Part A*, 5–6, 318–338, 2020.
58. Mahfouzi, S.H., Tali, S.H.S., Amoabediny, G., 3D bioprinting for lung and tracheal tissue engineering: Criteria, advances, challenges and future directions. *Bioprinting*, 21, p.e00124, 2021.
59. Ozbolat, I.T., Peng, W., Ozbolat, V., Application areas of 3D bioprinting. *Drug Discov. Today*, 21, 8, 1257–1271, 2016.
60. Seol, Y.J., Kang, H.W., Lee, S.J., Atala, A., Yoo, J.J., Bioprinting technology and its applications. *Eur. J. Cardiothorac Surg.*, 46, 3, 342–348, 2014.
61. Parak, A., Pradeep, P., Toit, L.C., Kumar, P., Choonara, Y.E., Pillay, V., Functionalizing bioinks for 3D bioprinting applications. *Drug Discov. Today*, 24, 1, 198–205, 2019.
62. Hong, N., Yang, G., Lee, J., Kim, G., 3D bioprinting and its *in vivo* applications. *J. BioMed. Mater. Res.*, 106, 1, 444–459, 2018.
63. An, J., Teoh, J.E.M., Suntornnord, R., Chua, C.K., Design and 3D printing of scaffolds and tissues. *Engineering*, 2, 261–268, 2015.
64. Bose, S., Vahabzadeh, S., Bandyopadhyay, A., Bone tissue engineering using 3D printing. *Mater Today*, 16, 496–503, 2013.
65. Melches, F.P.W., Domingos, M.A.N., Klein, T.J., Malda, J., Bartolo, P.J., Hutmacher, D.W., Additive manufacturing of tissues and organs. *Prog. Polym. Sci.*, 37, 1079–1104, 2012.
66. Giannitelli, S.M., Mozetic, P., Trombetta, M., Rainer, A., Combined additive manufacturing approaches in tissue engineering. *Acta Biomater.*, 24, 1–11, 2015.
67. Madrid, A.P.M., Vrech, S.M., Sanchez, M.A., Radriguez, A.P., Advances in additive manufacturing for bone tissue engineering scaffolds. *Mater. Sci. Eng. C*, 100, 631–644, 2019.
68. Tytgat, L., Damme, L.V., Hoorick, J.V., Declercq, H., Thienpont, H., Ottevaere, H. *et al.*, Additive manufacturing of photo-crosslinked gelatin scaffolds for adipose tissue engineering. *Acta Biomater.*, 2019, 94, 340–350, 2019.
69. Altun, E., Ekren, N., Kuruca, S.E., Gunduz, O., Cell studies on Electrohydrodynamic (EHD)-3D-bioprinted bacterial cellulose/polycaprolactone scaffolds for tissue engineering. *Mater. Lett.*, 234, 163–167, 2019.
70. Bedir, T., Ulag, S., Ustundag, C.B., Gunduz, O., 3D bioprinting applications in neural tissue engineering for spinal cord injury repair. *Mater. Sci. Eng. C*, 110, 1–10, 2020.
71. Kumar, P., Rajak, D.K., Abubakar, M., Syed, G.M.A., Hussain, M., 3D printing technology for biomedical practice: A review. *J. Mater. Eng. Perform.*, 192, 30, 7, 5342–5355, 2021.

72. Miranov, V., Boland, T., Trusk, T., Forgacs, G., Markwald, R.R., Organ printing: Computer-aided jet-based 3D tissue engineering. *Trends Biotechnol.*, 21, 157–161, 2003.
73. Miranov, V., Kasyanov, V., Drake, C., Markwald, R.R., Organ printing: Promises and challenges. *Regener. Med.*, 3, 93–103, 2007.
74. Murphy, S.V., 3D bioprinting of tissues and organs. *J. Nat. Biotechnol.*, 32, 773–785, 2014.
75. Fedorovich, N.E., Alblas, J., Hennink, W.E., Öner, F.C., Dhert, W.J.A., Organ printing: The future of bone regeneration? *Trends Biotechnol.*, 29, 601–606, 2011.
76. Miranov, V., Kasyanov, V., Markwald, R.R., Organ printing: From bioprinter to organ biofabrication line. *Curr. Opin. Biotechnol.*, 22, 667–673, 2011.
77. Martins, P., Lopes, A.C., Lanceros-Mendez, S., Electroactive phases of poly(vinylidene fluoride): Determination, processing and applications. *Prog. Polym. Sci.*, 39, 4, 683–706, 2014.
78. Singh, D., Deepak, Garg, A., An efficient route to fabricate fatigue-free P(VDF-TrFE) capacitors with enhanced piezoelectric and ferroelectric properties and excellent thermal stability for sensing and memory applications. *Phys. Chem. Chem. Phys.*, 19, 11, 7743–7750, 2017.
79. Rashmi, P.N., Jain, A., Prashanth, K.J., Sharma, A.Kr, Jain, A., Dielectric and piezoelectric properties of PVDF/PZT composites: A review. *Polym. Eng. Sci.*, 55, 7, 1589–1616, 2015.
80. Newacheck, S. and Youssef, G., Synthesis and characterization of polarized novel 0–3 Terfenol-D/PVDF-TrFE composites. *Compos. Part B Eng.*, 172, 97–102, 2019.
81. Zhang, H., Zhu, H., Qi, T., Hu, Z., Zeng, X., Selective laser melting of high strength Al–Cu–Mg alloys: Processing, microstructure and mechanical properties. *Mater. Sci. Eng.: A.*, 656, 47–54, 2016.
82. Martin, J.H., Yahata, B.D., Hundley, J.M., Mayer, J.A., Schaedler, T.A., Pollock, T.M., 3D printing of high-strength aluminium alloys. *Nature*, 549, 7672, 365, 2017.
83. Körner, C., Helmer, H., Bauereiß, A., Singer, R.F., Tailoring the grain structure of IN718 during selective electron beam melting. *MATEC Web of Conferences*, EDP Sciences, p. 08001, 2014.
84. Wycisk, E., Solbach, A., Siddique, S., Herzog, D., Walther, F., Emmelmann, C., Effects of defects in laser additive manufactured Ti-6Al-4V on fatigue properties. *Phys. Procedia*, 56, 371–8, 2014.
85. Wycisk, E., Siddique, S., Herzog, D., Walther, F., Emmelmann, C., Fatigue performance of laser additive manufactured Ti–6Al–4V in very high cycle fatigue regime up to 109 cycles. *Front. Mater.*, 2, 72, 2015.
86. Leuders, S., Thöne, M., Riemer, A., Niendorf, T., Tröster, T., Richard, H., Maier, H., On the mechanical behaviour of titanium alloy TiAl6V4 manufactured by selective laser melting: Fatigue resistance and crack growth performance. *Int. J. Fatigue*, 48, 300–7, 2013.

87. Ferreira, R.T.L., Amatte, I.C., Dutra, T.A., Bürger, D., Experimental characterization and micrography of 3D printed PLA and PLA reinforced with short carbon fibers. *Compos. Part B: Eng.*, 124, 88–100, 2017.
88. Tian, X., Liu, T., Yang, C., Wang, Q., Li, D., Interface and performance of 3D printed continuous carbon fiber reinforced PLA composites. *Compos. Part A: Appl. Sci. Manuf.*, 88, 198–205, 2016.
89. Hou, Z., Tian, X., Zhang, J., Li, D., 3D printed continuous fibre reinforced composite corrugated structure. *Compos. Struct.*, 184, 1005–10, 2018.
90. Parandoush, P. and Lin, D., A review on additive manufacturing of polymer-fiber composites. *Compos. Struct.*, 182, 36–53, 2017.
91. Tekinalp, H.L., Kunc, V., Velez-Garcia, G.M., Duty, C.E., Love, L.J., Naskar, A.K., Blue, C.A., Ozcan, S., Highly oriented carbon fiber–polymer composites via additive manufacturing. *Compos. Sci. Technol.*, 105, 144–50, 2014.
92. Love, L.J., Kunc, V., Rios, O., Duty, C.E., Elliott, A.M., Post, B.K., Smith, R.J., Blue, C.A., The importance of carbon fiber to polymer additive manufacturing. *J. Mater. Res.*, 29, 17, 1893–8, 2014.
93. Chiappone, A., Roppolo, I., Naretto, E., Fantino, E., Calignano, F., Sangermano, M., Pirri, F., Study of graphene oxide-based 3D printable composites: Effect of the *in situ* reduction. *Compos. Part B: Eng.*, 124, 9–15, 2017.
94. Dul, S., Fambri, L., Pegoretti, A., Fused deposition modelling with ABS–graphene nanocomposites. *Compos. Part A: Appl. Sci. Manuf.*, 85, 181–91, 2016.
95. Han, Y., Wang, F., Wang, H., Jiao, X., Chen, D., High-strength boehmite-acrylate composites for 3D printing: Reinforced filler-matrix interactions. *Compos. Sci. Technol.*, 154, 104–9, 2018.
96. Chen, Y., Mao, J., Wu, J., Microwave transparent crosslinked polystyrene nanocomposites with enhanced high voltage resistance via 3D printing bulk polymerization method. *Compos. Sci. Technol.*, 157, 160–7, 2018.
97. Nguyen, Q.T., Ngo, T., Tran, P., Mendis, P., Zobec, M., Aye, L., Fire performance of prefabricated modular units using organoclay/glass fibre reinforced polymer composite. *Constr. Build. Mater.*, 129, 204–15, 2016.
98. Song, J., Chen, C., Zhang, Y., High thermal conductivity and stretchability of layer-bylayer assembled silicone rubber/graphene nanosheets multilayered films. *Compos Part A: Appl. Sci. Manuf.*, 105, 1–8, 2018.
99. Postiglione, G., Natale, G., Griffini, G., Levi, M., Turri, S., Conductive 3D microstructures by direct 3D printing of polymer/carbon nanotube nanocomposites via liquid deposition modeling. *Compos Part A: Appl. Sci. Manuf.*, 76, 110–4, 2015.
100. Weng, Z., Zhou, Y., Lin, W., Senthil, T., Wu, L., Structure-property relationship of nano enhanced stereolithography resin for desktop SLA 3D printer. *Compos Part A: Appl. Sci. Manuf.*, 88, 234–42, 2016.
101. Boparai, K.S., Singh, R., Fabbrocino, F., Fraternali, F., Thermal characterization of recycled polymer for additive manufacturing applications. *Compos. Part B: Eng.*, 106, 42–7, 2016.

102. Ventola, C.L., Medical applications for 3D printing: Current and projected uses. *Med. Appl. 3D Print*, 39, 705–12, 2014.
103. Forbes, H. and Schaefe, D., Social product development: The democratization of design, manufacturing and innovation. *Procedia Cirp*, 60, 404–9, 2017.
104. Mallepree, T. and Bergers, D., Accuracy of medical RP models. *Rapid Prototyp. J.*, 15, 325–32, 2009.
105. Salmi, M., Paloheimo, K.S., Tuomi, J., Wolff, J., Mäkitie, A., Accuracy of medical models made by additive manufacturing. *J. Craniomaxillofac Surg.*, 41, 603–9, 2013.
106. Lei, S.C., Frank, M.D., Anderson, D., Brown, T., A method to represent heterogeneous materials for rapid prototyping: The matryoshka approach. *Rapid Prototyp. J.*, 20, 390–402, 2014.
107. Gibson, I., Cheung, L.K., Chow, S.P., The use of rapid prototyping to assist medical applications. *Rapid Prototyp. J.*, 12, 53–8, 2006.
108. Goyanes, A., Robles Martinez, P., Buanz, A., Basit, A.W., Gaisford, S., Effect of geometry on drug release from 3D printed tablets. *Int. J. Pharm.*, 494, 2, 657–63, 2015.
109. Wang, C.-C., Tejwani, M.R., Roach, W.J. *et al.*, Development of near zero-order release dosage forms using three-dimensional printing (3-DPTM) technology zero-order release dosage forms using 3-DPTM. *Drug Dev. Ind. Pharm.*, 32, 3, 367–76, 2006.
110. Lam, C.X.F., Mo, X.M., Teoh, S.H., Hutmacher, D.W., Scaffold development using 3D printing with a starch-based polymer. *Mater. Sci. Eng. C*, 20, 1-2, 49–56, 2002.
111. Rowe, C.W., Katstra, W.E., Palazzolo, R.D., Giritlioglu, B., Teung, P., Cima, M.J., Multimechanism oral dosage forms fabricated by three dimensional printing. *J. Control Release*, 66, 1, 11–7, 2000.
112. https://www.fda.gov/downloads/MedicalDevices/Device,Regulationand Guidance/GuidanceDocuments/ucm070642.pdf.
113. Ventola, C.L., Medical applications for 3D printing: Current and projected uses. *P&T*, 39, 10, 704–11, 2014.
114. Ostuzzi, F., Rognoli, V., Saldien, J., Levi, M., +TUO project: Low-cost 3D printers as helpful tool for small communities with rheumatic diseases. *Rapid Prototyp. J.*, 21, 491–505, 2015.
115. Davis, S., From future perfect: Mass customizing. *Plan. Rev.*, 17, 16–21, 1989.
116. Pine, J., Victor, B., Boyton, A., Making mass customization work. *Harv. Bus. Rev.*, 71, 108–11, 1993.
117. Eastwood, M., Implementing mass customization. *Comput. Ind.*, 30, 171–4, 1996.
118. Hart, C., Mass customization: Conceptual underpinnings, opportunities and limits. *Int. J. Serv. Ind. Manag.*, 6, 36–45, 1995.
119. Pine, J., Mass customizing products and services. *Plan. Rev.*, 21, 6–13, 1993.

120. Salmi, M., Tuomi, J., PaloheimoKaija, S. *et al.*, Patient specific reconstruction with 3D modeling and DMLS additive manufacturing. *Rapid Prototyp. J.*, 18, 209–14, 2012.
121. Lei, S.C., Frank, M.D., Anderson, D., Brown, T., A method to represent heterogeneous materials for rapid prototyping: The matryoshka approach. *Rapid Prototyp. J.*, 20, 390–402, 2014.
122. Bombač, D., Brojan, M., Fajfar, P., Kosel, F., Turk, R., Review of materials in medical applications. *RMZ – Mater. Geoenviron.*, 54, 471–99, 2007.
123. Salmi, M., Tuomi, J., Paloheimo, K. *et al.*, Patient-specific reconstruction with 3D modeling and DMLS additive manufacturing. *Rapid Prototyp. J.*, 18, 3, 209–14, 2012.
124. Ahangar, P., Cooke, M.E. *et al.*, Current biomedical applications of 3D printing and additive manufacturing. *Appl. Sci. (Basel)*, 9, 1–23, 2019.
125. Pucci, J.U., Christophe, B.R., Sisti, J.A., Connolly, E.S., Three-dimensional printing: Technologies, applications, and limitations in neurosurgery. *Biotechnol. Adv.*, 35, 5, 521–9, 2017.
126. Mallepree, T. and Bergers, D., Accuracy of medical RP models. *Rapid Prototyp. J.*, 15, 325–32, 2009.
127. Salmi, M., Paloheimo, K.S., Tuomi, J., Wolff, J., Mäkitie, A., Accuracy of medical models made by additive manufacturing (rapid manufacturing.) *J. Craniomaxillofac Surg.*, 41, 603–9, 2013.
128. Kochan, A., Rapid prototyping gains speed, volume and precision. *Assem. Autom.*, 20, 295–9, 2000.
129. Gibson, I., Cheung, L.K. *et al.*, The use of rapid prototyping to assist medical applications. *Rapid Prototyp. J.*, 12, 45–53, 2006.
130. Hollister, S.J., Porous scaffold design for tissue engineering. *Nat. Mater.*, 4, 7, 518–24, 2005.
131. Wang, X., Jiang, M., Zhou, Z., Gou, J., Hui, D., 3D printing of polymer matrix composites: A review and prospective. *Compos. Part B: Eng.*, 110, 442–58, 2017.
132. Zareiyan, B. and Khoshnevis, B., Interlayer adhesion and strength of structures in contour crafting - Effects of aggregate size, extrusion rate, and layer thickness. *Autom. Constr.*, 81, 112–21, 2017.
133. Gibbons, G.J., Williams, R., Purnell, P., Farahi, E., 3D printing of cement composites. *Adv. Appl. Ceramics*, 109, 5, 287–90, 2010.
134. Le Duigou, A., Castro, M., Bevan, R., Martin, N., 3D printing of wood fibre biocomposites: From mechanical to actuation functionality. *Mater. Des.*, 96, 106–14, 2016.
135. Zhang, W., Melcher, R., Travitzky, N., Bordia, R.K., Greil, P., Three-dimensional printing of complex-shaped alumina/glass composites. *Adv. Eng. Mater.*, 11, 12, 1039–43, 2009.
136. Paul, S.C., Tay, Y.W.D., Panda, B., Tan, M.J., Fresh and hardened properties of 3D printable cementitious materials for building and construction. *Arch. Civ. Mech. Eng.*, 18, 1, 311–9, 2018.

137. Minas, C., Carnelli, D., Tervoort, E., Studart, A.R., 3D printing of emulsions and foams into hierarchical porous ceramics. *Adv. Mater.*, 28, 45, 9993–9, 2016.
138. Le Duigou, A., Castro, M., Bevan, R., Martin, N., 3D printing of wood fibre biocomposites: From mechanical to actuation functionality. *Mater. Des.*, 96, 106–14, 2016.
139. Oropallo, W. and Piegl, L.A., Ten challenges in 3D printing. *Eng. Comput.*, 32, 1, 135–48, 2016.
140. Khoshnevis, B., Automated construction by contour crafting - Related robotics and information technologies. *Autom. Constr.*, 13, 1, 5–19, 2004.
141. Paul, S.C., Tay, Y.W.D., Panda, B., Tan, M.J., Fresh and hardened properties of 3D printable cementitious materials for building and construction. *Arch. Civ. Mech. Eng.*, 18, 1, 311–9, 2018.
142. Coniglio, N., Sivarupan, T., El Mansori, M., Investigation of process parameter effect on anisotropic properties of 3D printed sand molds. *Int. J. Adv. Manuf. Technol.*, 94, 2175–2185, 2018.
143. Carroll, B.E., Palmer, T.A., Beese, A.M., Anisotropic tensile behavior of Ti–6Al–4V components fabricated with directed energy deposition additive manufacturing. *Acta Mater.*, 87, 309–20, 2015.
144. Mühler, T., Gomes, C.M., Heinrich, J., Günster, J., Slurry-based additive manufacturing of ceramics. *Int. J. Appl. Ceram. Technol.*, 12, 1, 18–25, 2015.
145. Niendorf, T., Leuders, S., Riemer, A., Richard, H.A., Tröster, T., Schwarze, D., Highly anisotropic steel processed by selective laser melting. *Metall. Mater. Trans. B*, 44, 4, 794–6, 2013.
146. Cooke, W., Tomlinson, R.A., Burguete, R., Johns, D., Vanard, G., Anisotropy, homogeneity and ageing in an SLS polymer. *Rapid Prototyp. J.*, 17, 4, 269–79, 2011.
147. Guessasma, S., Belhabib, S., Nouri, H., Ben Hassana, O., Anisotropic damage inferred to 3D printed polymers using fused deposition modelling and subject to severe compression. *Eur. Polym. J.*, 85, 324–40, 2016.
148. Hambach, M. and Volkmer, D., Properties of 3D-printed fiber-reinforced portland cement paste. *Cem. Conc. Compos.*, 79, 62–70, 2017.
149. He, Z., Chen, Y., Yang, J., Tang, C., Lv, J., Liu, Y., Mei, J., Lau, W-M, Hui, D., Fabrication of Polydimethylsiloxane films with special surface wettability by 3D printing. *Compos. Part B: Eng.*, 129, 58–65, 2017.
150. Zeidler, H., Klemm, D., Bottger-Hiller, F., Fritsch, S., Guen, M.J.L., Singamneni, S., 3D printing of biodegradable parts using renewable biobased between used capacity. *Proc. Manuf.*, 21, 117–124, 2018.
151. Rybachuk, M., Mauger, C.A., Fiedler, T., Öchsner, A., Anisotropic mechanical properties of fused deposition modeled parts fabricated by using acrylonitrile butadiene styrene polymer. *J. Polym. Eng.*, 37, 699–706, 2017.
152. Torrado, A.R. and Roberson, D.A., Failure analysis and anisotropy evaluation of 3D-printed tensile test specimens of different geometries and print raster patterns. *J. Fail. Anal. Prev.*, 16, 154–164, 2016.

153. Guessasma, S., Belhabib, S., Nouri, H., Ben Hassana, O., Anisotropic damage inferred to 3D printed polymers using fused deposition modelling and subject to severe compression. *Eur. Polym. J.*, 85, 324–340, 2016.
154. Balderrama-Armendariz, C.O., MacDonald, E., Espalin, D., Cortes-Saenz, D., Wicker, R., Maldonado-Macias, A., Torsion analysis of the anisotropic behavior of FDM technology. *Int. J. Adv. Manuf. Technol.*, 96, 307–317, 2018.
155. Odell, D., Wright, P.K., Montero, M., Roundy, S., Ahn, S.H., Anisotropic material properties of fused deposition modeling ABS. *Rapid Prototyp. J.*, 8, 248–257, 2002.
156. Thomas, J.P., Rodríguez, J.F., Renaud, J.E., Mechanical behavior of acrylonitrile butadiene styrene (ABS) fused deposition materials. Exp. investigation. *Rapid Prototyp. J.*, 7, 148–158, 2001.
157. Das, S.C., Ranganathan, R., Murugan, N.J.R.P.J., Effect of build orientation on the strength and cost of PolyJet 3D printed parts. *Rapid Prototyp. J.*, 24, 5, 832–839, 2018.
158. Ambrosio, D., Gabrion, X., Malécot, P., Amiot, F., Thibaud, S., Influence of manufacturing parameters on the mechanical properties of projection stereolithography–manufactured specimens. *Int. J. Adv. Manuf. Technol.*, 106, 265–277, 2020.
159. Ning, F., Cong, W., Hu, Y., Wang, H., Additive manufacturing of carbon fiber-reinforced plastic composites using fused deposition modeling: Effects of process parameters on tensile properties. *J. Compos. Mater.*, 51, 451–462, 2016.
160. Guessasma, S., Belhabib, S., Nouri, H., Significance of pore percolation to drive anisotropic effects of 3D printed polymers revealed with X-ray μ-tomography and finite element computation. *Polymer*, 81, 29–36, 2015.
161. Zhang, B., Pei, X., Zhou, C., Fan, Y., Jiang, Q., Ronca, A., D'Amora, U., Chen, Y., Sun, Y., Zhang, X., The biomimetic design and 3D printing of customized mechanical properties porous Ti6Al4V scaffold for load-bearing bone reconstruction. *Mater. Des.*, 152, 30–39, 2018.
162. Mostafaei, A., De Vecchis, P.R., Nettleship, I., Chmielus, M., Effect of powder size distribution on densification and microstructural evolution of binder-jet 3Dprinted alloy 625. *Mater. Des.*, 162, 375–383, 2019.
163. Gopinathan, J. and Noh, I., Recent trends in bioinks for 3D printing. *Biomater. Res.*, 22, 1, 1–15, 2018.
164. Singh, S. and Ramakrishna, S., Biomedical applications of additive manufacturing: Present and future. *Curr. Opin. Biomed. Eng*, 2, 105–115, 2017.
165. Bertol, L.S., Schabbach, R., Dos Santos, L.A.L., Dimensional evaluation of patientspecific 3D printing using calcium phosphate cement for craniofacial bone reconstruction. *J. Biomater. Appl.*, 31, 6, 799–806, 2017.
166. Osman, R.B., Van Der Veen, A.J., Huiberts, D., Wismeijer, D., 3D-printing zirconia implants; A dream or a reality? An *in-vitro* study evaluating the dimensional accuracy, surface topography and mechanical properties of

printed zirconia implant and discs. *J. Mech. Behav. Biomed. Mater.*, 75, 521–528, 2017.
167. Zeidler, H., Klemm, D., Bottger-Hi ller, F., Fritsch, S., Guen, M.J.L., Singamneni, S., 3D printing of biodegradable parts using renewable biobased between used capacity. *Proc. Manuf.*, 21, 117–124, 2018.
168. MyIPO, *The official portal of intellectual property corporation of Malaysia*, Vantage Asia Publishing Limited, HongKong, 2017, https://law.asia/3d-printing-design-rights-and-challenges.

15

Metal 3D Printing for Emerging Healthcare Applications

Sudipto Datta[1]*, Yusuf Olatunji Waidi[1] and Arbind Prasad[2]

[1]Department of Materials Engineering, Indian Institute of Science, Bangalore, Karnataka, India
[2]Department of Mechanical Engineering, Katihar Engineering College (under DST, Govt. of India), Katihar, Bihar, India

Abstract
The technology of additive manufacturing also known as 3D printing has got a lot of attention if the last few decades in various engineering applications. It is used in both industry and academics because of advantages like less time of fabrication, complex geometry, and structure printing, less cost of manufacturing, and lastly customization in fabrication can be easily done. Metallic 3D printing technology is used in biomedical applications for the development of metallic implants for various applications. In this book chapter, we will briefly discuss the present 3D printing technology and the materials which are used for the development of medical metallic devices and implants. Lastly, the challenges and the future of this technology are also represented briefly.

Keywords: (3D) printing, metallic implants, biomaterials, additive manufacturing (AM)

15.1 Introduction

The present global population makes is around 10% of people aged 65 and above. And this figure is predicted to double by 2050 [1]. Because of their age, these older people are more prone to suffer from health concerns including tissue loss and bone fractures. Hence, fixing, reconstructing, or

**Corresponding author*: sudiptodatta1990@gmail.com

Arbind Prasad, Ashwani Kumar and Manoj Gupta (eds.) Advanced Materials and Manufacturing Techniques for Biomedical Applications, (383–410) © 2024 Scrivener Publishing LLC

replacing these issues is then necessary. Metals are typically employed to give support, permit fast movement of patients, and avoid the emergence of problems due to the hard tissue mechanical demands [2, 3]. Traditionally, stainless steels [4], titanium alloys [6, 7], cobalt-chromium alloys [5, 6], nitinol [8, 9], niobium [12, 13], and tantalum [10, 11] account for about 70–80% of clinically administered implants. Lately, biodegradable metals such as iron [15], magnesium [14], calcium [19], and zinc [16–18] have attracted lots of interest and numerous novel findings have been published over the previous decade [14, 20, 21]. Various researchers have used biodegradable polymer-based composites for scaffold, implants fabrication and other orthopedic application, in this regards various nanomaterials, which is biocompatible and bioactive are used for further addition of mechanical strength and their stability [137–146]. However, it is challenging to achieve complex exterior shapes and sophisticated interior architecture using traditional powder metallurgy and casting techniques. But, with the advancement of the 3D printing fabrication method, it has now become feasible to create a porous metallic implant having mechanical strength similar to natural bone, hence decreasing stress shielding concerns [22]. 3D bioprinting technique offers various advantages such as fast production time, low cost, and great reproducibility [28]. In addition, it is coupled with computer-aided design (CAD), allowing personalized designs with a tremendous degree of freedom to be created. Consequently, 3D printing offers a special potential to improve personalized therapy by generating tailored metallic devices which can match certain tissue flaws [23]. The fast advancement of this technique has enabled the overview of various 3D printing approaches that solve the drawbacks of electron beam and laser printing technologies that are generally used. The subsequent sections and subsections addressed the uses of biometals created using classic 3D printing technology as well as the novel possibilities drawn from evolving 3D printing technology. Currently, bio metals used to produce implants are used in surgical tools, orthopedics, dentistry, and tissue repair are shown in Figure 15.1.

15.2 Metallic 3D Printing Methods for Biomedical Applications

3D printing technologies enable the large manufacture of metallic implants with complicated forms and interior parts, in conjunction with the development of personalized medical implants that match particular individual demands, making them extremely intriguing for therapeutic

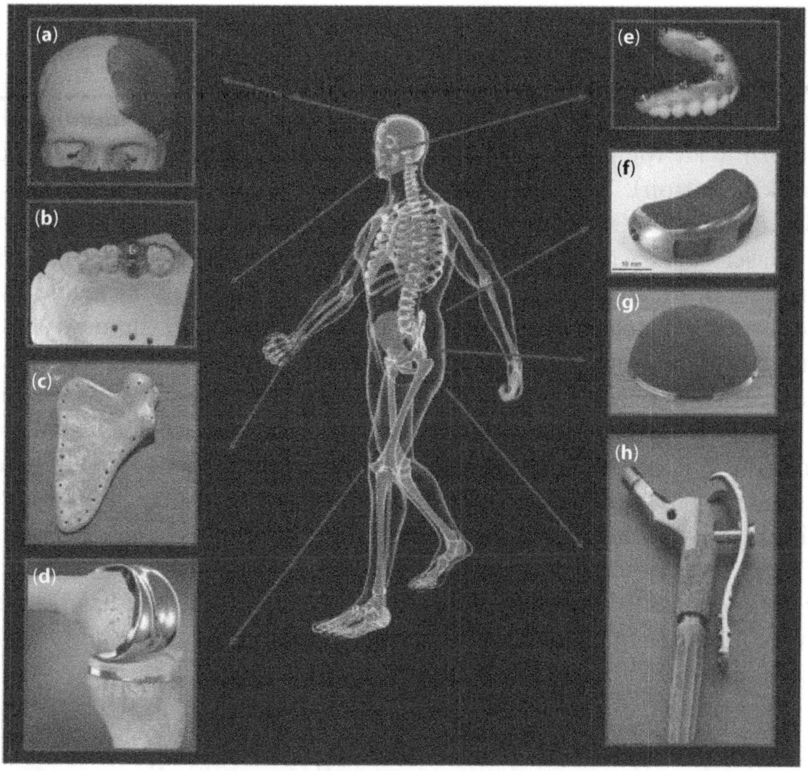

Figure 15.1 Biometals used in biomedical applications (a) prosthesis for cranial; (b) medical guide; (c) prosthesis for scapula; (d) prosthesis for knee; (e) implants for dental; (f) interbody fusion cage of interbody fusion type; (g) acetabular cup; and (h) prosthesis for the hip. Republish with authorization from Copyright Clearance Centre [155].

interventions [24]. For instance, using 3D printing technology, orthopedic surgeons may mass produce precise anatomic models for use in surgery preparedness. Additionally, customized implants may be created via a 3D printing technique using magnetic resonance imaging (MRI) or X-ray computed tomography (X-ray CT) data from specific patients to generate a CAD file that is being used.

15.2.1 (SLS) Selective Laser Sintering 3D Printing

In 1989, (SLS) was originally suggested and subsequently patented, whereby the metallurgical mechanism was based on liquid-phase sintering. However, during the forming step, the metal powder is mildly melted while keeping its solid state interior further followed by powder densification which is accomplished through successive solid-phase particle reorganization as well as bonding via liquid-phase solidification [25, 26]. The widely employed devices in the SLS technique are forming cylinder and powder cylinder. The powder cylinder's piston lifts and the powdering roller uniformly lay the powder material on the piston of the working cylinder (working piston). Depending on the sample slicing design, a laser beam which is a computer-controlled two-dimensional (2D) scanning trajectory of a laser beam was used to selectively sinter the solid powder material, resulting in the formation of a build layer. After finishing the single layer, the forming piston is dropped by a single layer's depth, the new precipitation is deposited, and then the next layer is created using laser beam scans. Such procedure is repeated, and layers are built until the required 3D object is formed. Since the SLS method uses a liquid-phase semisolid sintering system, whereby the material is not entirely molten. Even though the forming material's thermal stress generated may be decreased to a certain degree, the developed part comprises solid-phase particles, resulting in high porosity and less density. As a result, manufacturing flaws like poor surface roughness and low tensile strength may occur. Nonetheless, SLS is favorable due to the wide range of forming materials available and the ease of the fabrication procedure (required no support) [27, 61].

15.2.2 (SLM) Selective Laser Melting 3D Printing

In 1995, the concept of selective laser melting (SLM) was originally presented [26]. But since this approach was inspired by SLS, the core notions of both are identical. Although, for fast laser manufacturing of metal parts, SLM uses sophisticated fiber lasers with high energy, highly precision paving powder, as well as metallurgical procedures which completely melt particles Figure 15.2 [29]. To prevent other gases from reacting with the forming product at high temperatures, the SLM procedure is performed in a chamber protected with a vacuum or inert gas. SLM techniques are vast in producing functional metal components directly without the use of intermediary steps. Given the narrower focusing point and an increased density of laser energy, functional components having greater quality of dimension as well as lower roughness on the surface are produced due to the total melting of metal powder, the resultant metal portions possess bonded architectures metallurgically, excellent masses, and high material strength, requiring no further treatments after printing [30, 31]. The powder substance might be one or multiple component products, and the used ingredients are not required to be processed in any specific way. SLM may dramatically cut production time and costs while providing design

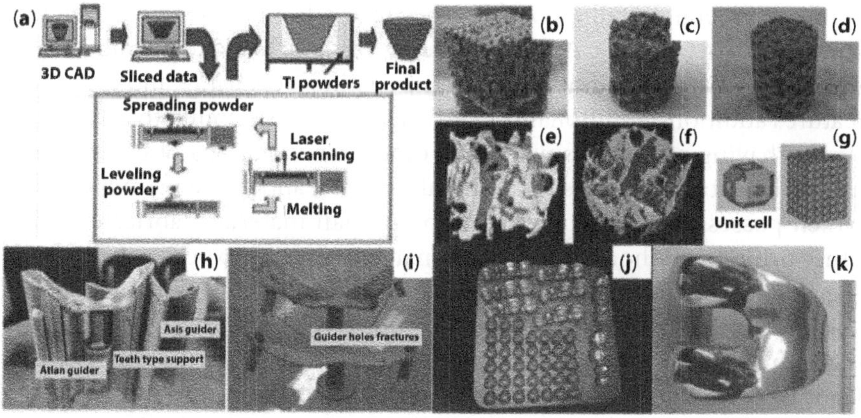

Figure 15.2 (a) SLM working flowchart; (b)–(d) SLM porous titanium 3D printed; (e)–(f) human cancellous bones imaging by Micro-CT; (g) using SLM arranged cubes of hollow cubes types; one kind of spinal surgery pattern; (h) the surgical pattern with its supporting structure; (i) after abrasive blasting the surgical pattern; (j) using SLM the dental repairs are made; (k) modified femoral constituent crosswise surface after polishing by hand. Reprint with permission from the copyright clearance center with license number 5453580670025 [155].

flexibility. Additionally, SLM may be employed to produce a broad range of materials while also minimizing material waste. Furthermore, the wide spectrum of materials such as metals [32], polymers [33], metal-ceramic combinations [26], and metal-polymer combinations [34], may all be printed using this process. Several medical metals like cobalt-based alloys, stainless steel, nickel-based superalloys, and titanium alloys have been extensively fabricated via the SLM method.

15.2.3 (LDMD) Laser Direct Metal Deposition 3D Printing

The Sandia National Laboratory originally suggested laser direct metal deposition (LDMD) since1990s. This method has been subsequently well-expanded to numerous countries around the globe [35]. Numerous institutions and universities have separately explored LDMD, which has resulted in several names, including but not limited to laser fast forming (LRF), directed metal deposition (DMD), and laser-engineered net shaping (LENS). Despite their different names, their fundamental principles are similar. At the time of the fabrication steps, the powder is collected by a nozzle on its working plane and the laser beam focuses on one position. Although SLS/SLM are powder-based processes, LDMD on the other hand uses a nozzle to dispense metal powder. This technique is very beneficial because of its capacity to create massive sections of components and combine several metallic powder types inside the nozzle to manufacture specialized metal alloys. Aside from being versatile to make full metal components, they also find applications in coating, repairing, and structural features addition. Additionally, LDMD also offers the ability to create constructs having ranges of small pores otherwise variations of diverse metals, such as stainless steels, titanium alloys [36], and shape memory alloys, which possess higher geometric mobility than materials fabricated using other printing methods [37]. Unfortunately, materials experience fast heating and cooling in the course of the fabrication process due to batching of laser or electron beam and stepwise scanning, thus resulting in a huge thermal gradient, invariably creating a large residual stress distribution that leads to defects such as cracks and metal parts deformation [38, 39]. Both mechanical characteristics and corrosion resistance will also be impacted by the highly generated residual stress [38]. The following lists are some of the current approaches for decreasing or removing residual stresses [40, 41]: (a) optimum design: to prevent residual stress during the fabrication process, it should be considered during the design phase; (b) Laser scanning mode modification: from one printing layer to another, the laser scanning vector is transferred to avoid stress being focused on a single plane; (c) minimizing huge areas and continuous sintering, and (d) metal components stress reduction annealing and powder bed preheating.

15.2.4 (LIFT) Laser-Induced Forward Transfer 3D Printing

The printing of metallic materials is usually confined to lower melting points materials. Till now, 3D printing metallic materials like copper or gold has been found either to be extremely challenging or highly costly. Therefore, laser-induced forward transfer (LIFT) became invented. The LIFT working mechanism is a straightforward technique that may deposit a variety of metallic substances following laser encounters [43]. This method is distinct from previous metal 3D printing methods earlier mentioned in that it does not use metallic powder substances. A strong pulsed laser is directed onto a light layer of a metallic substance known as the "donor film," causing an evaporation or thermal stress wave which results in the expulsion of a micron-sized liquid droplet upon a clear platform (called receiver). The droplet hardens by cooling upon impact [44]. This technique has therefore been extensively explored for printing various metallic materials like titanium [45], aluminum [46], nickel [47], chromium [48], and gold [49]. In addition, this approach can be employed to print vertical wire precisely, but difficult to produce overhang or big 3D objects [50]. These architectural constraints result from the nature of the system, thus, could necessitate the use of materials for sacrificial support. Furthermore, it can be employed to construct biosensors [45]. For instance, it was reported recently that it was employed to print copper/silver tiny objects on a sacrificial support of copper [51]. In the context of precision, published LIFT metal constructs possess lateral measurements of large micrometers [135]; the least size of approximately 3 μm [52].

15.2.5 (SEBM) Selective Electron Beam Melting 3D Printing

Alike SLM as well as SLS, SEBM, is a fast fabrication that employs rays of high-speed electrons with high energy for precise attack and melts the powder substance to produce 3D objects. SEBM, on the other hand, consists of four major steps: primarily, a single powder layer is dispersed on the flagging platform; then, the rays of an electron are preferentially lined up to the powders depending on the 3D CAD file cross-sectional outline data input through a computer; ultimately, the powder of the metal is melted in a layer-wise manner, joining to the designed section beneath with a magnetic guided electron rays (having power of 3 kW) under a vacuum environment up to the whole segment, and finally, the remaining material is isolated to achieve the required 3D construct [42]. SEBM operates via scanning by employing a high-intensity electron beam as a manufacturing heating element to operate a coil that is magnetically deflected having no indolence. Additionally, electron beams have a greater energy consumption rate, a stronger powder rate of absorption, improved constancy, as well as cheaper operating and repair costs as paralleled to lasers. In comparison to other techniques, SEBM technology offers greater efficiency, less component distortion, no molding process support, and compact microstructure development.

15.2.6 (ADAM) Atomic Diffusion Additive Manufacturing

Recently, ADAM was presented first as a novel 3D printing process for layer-by-layer printing of metallic powder materials encapsulated in plastic polymer binders [53]. Upon printing, the polymer binder is eliminated via sintering in a furnace, thus the resultant metallic materials are sintered to about 95–99% density. Sintering the whole part at once accelerates the formation of crystals of metals via its layers of adhesive, thereby eliminating the strength issue imperfection at the interlayer seen in conventional 3D printing methods. Furthermore, with a printing highest dimension of 250 mm x 220 mm x 200 mm, the ADAM printer is capable of producing geometrical proportions that conventional 3D metal printing technologies cannot. It similarly possesses cloud software that enables an online system laser detection. Thus, the program allows users to inspect and oversee every printing layer. Both the metal material management technology, as well as a cloud-based remote access approach, are incorporated into the printer. The printer can 3D-print 17-4 and 303 stainless steel, while suitability with A-2, Ti6Al4V, and D-2 steels is being researched. The key benefits of ADAM over conventional metal 3D printing processes are as follows: it is quicker 100 folds more than the conventional machining process, one-tenth the price of typical metal 3D printing, creates greater quality component surfaces with no post-processing, builds accurate, complicated structures, has outstanding isotropic behavior, and is appropriate for the batch production process.

15.2.7 (NPJ) Nanoparticle Jetting

In 2016, NPJ was originally introduced as a revolutionary 3D printing technology for metal. In contrast to the conventional 3D metal printing methods which employ powder of metal substances, this technique adopts fluid ink which encapsulates metallic granules [53]. Large metal components are broken down into nanoparticles, which are then infused through a binder to generate a homogeneous ink ready for printing [53]. Consistently, metal particles are then dispersed in a state of suspension inside the ink, or when suitably homogenized, the obtained ink is expelled via an outlet for allowing for layerwise deposition. Employing the ink of the printer in such a manner is advantageous since it improves the overall quality of the output. During the printing procedure, a standard inkjet printhead may be employed as a printing instrument. Following printing, the heat generated inside the chamber is employed to remove the surplus binder, keeping just the metal component. Such a technique is capable of 1-μm layer resolutions and has a temperature of forming of just around 300 degrees Celsius. NPJ is 5 times quicker than standard laser printing since the technique may inject ink droplets of 221 million in a second. In addition, it also lowers wastage of materials, saves money, and therefore may be used to manufacture practically any complicated design. Their components have excellent accuracy and surface quality, and NPJ's operating processes are safer and simpler. One downside of this process is that the product has a lesser tolerance for temperature than typical 3D metal printing techniques. Nonetheless, NPJ is a revolutionary technology that allows for a straightforward and efficient process that does not necessitate elaborate structures for support to be designed and removed.

15.2.8 Inkjet 3DP/Binder Jetting

A cartridge containing ink precisely prints a tiny water jet, in a layer-by-layer version, upon a metallic powder platform guided by the direction information in the provided CAD file [54]. Following that, the metal components printed are furnace heat treated to produce the mechanical strength necessary. Unlike conventional metal 3D printing technologies, this method does not require an electron beam or a laser to liquefy metallic granules. One primary benefits of 3D Inkjet printing are the inexpensive facility costs and regulated thermal sintering operation. Despite having lower precision than SLM, LENS, and SEBM 3D Inkjet printing's ability to quickly production of bio metallic devices at a cheap cost makes it a viable technology for the biomedical field [55].

15.3 Biometals 3D Printing

Because of its bulk and personalized production capabilities technique, coupled with imaging and scanning equipment (CT, ultrasonic tests, MRI, etc.) offers swifts, cost-effective, and beneficial medical care remedies, transforming the established medical sector's supply chain. The uses of 3D printing invention are becoming more diverse. Numerous 3D-printed healthcare gadgets have already received market acceptance by the US Food and Drug Administration (FDA).

15.3.1 Titanium Alloy

It has been well established that titanium and titanium alloys have been broadly utilized as implantable device components ever since the 1970s, due to their desirable biomedical features including great biocompatibility, resistance corrosion, and high relative strength [6, 7, 56]. These alloys are commonly applied in the biomedical field like dental and orthopedic devices, bone plates and screws, artificial hearts and joints, pacemakers, cornea backplates, and cardiac valve prostheses [57–61]. Titanium Ti6Al4V alloy (Ti-64) and plain commercial titanium (CP–Ti) are now the dominant metallic materials for biomedical purposes and are employed in about 90% of conventional orthopedic implants. As their stiffness, density and elastic modulus are substantially greater than that of bone tissue, therefore, metallic dense implant produces stress shielding [62, 136]. Furthermore, it has been demonstrated by various studies that implants made of porous titanium alloy have high bone ingrowth potential which encourages the development of cells of human bone and generates a powerful grip among the bones as well as the implants, considerably enhancing the implant's success rate [63]. In addition, 3D-printed biofunctional surfaces having small pores on therapeutic metallic implants may improve their long-term durability and efficacy [64, 65]. Hollander *et al.* in their study investigated the SLM fabrication technique employing powder of Ti6Al4V, and their findings revealed vertebrae made of titanium alloy and created by SLM had high biocompatibility and might be utilized to substitute biological components [66]. Furthermore, the titanium alloy implants' biocompatibility produced by SLM or EBM was already established [67–71]. A female patient of 83-year-old had the first patient-specific jaw implant clinical transplant procedure in 2011 which was created using a 3D printer [72]. The first 3D-printed titanium operation to implant a heel prosthesis was conducted on a 71-year-old male patient in 2014 [73], who was suffering from the amputation of the leg under the knee due to calcaneus cancer.

15.3.2 Tantalum

Tantalum (Ta) has already been explored extensively as radiographic bone markers, vascular clips, cranial-defect repair materials, and nerve repair materials in dentistry and orthopedic applications since the early 1940s due to its outstanding chemical stability and biocompatibility [3, 10, 11, 74]. Although, its exploitation was previously constrained by huge production costs and modular implant fabrication concerns [75]. Up to now, porous tantalum implants have not only found applications for osteogenesis use, like alternatives for bone graft, but also widely employed for knee, spinal, and hip prostheses in surgery procedures. It has been recently established in a canine model that porous tantalum promotes the ingrowth of dense bone and robust bone-implant interface through cylindrical implants and acetabular cups [76, 77]. For the first time, Balla *et al.* [75] used LENS to coat Ta on Ti, and thus the Ta coating dramatically enhanced osseointegration characteristics as compared to the bare surface Ti. They then employed LENS to create varied porosity on Ta porous structures [78]. Thijs *et al.* [147] in their work presented the fabrication of functional tantalum components using SLM. Given the tantalum powder's ability to dissolve thoroughly, SLM may be used to manufacture robust components with very minimal porosity. In another study, Wauthle *et al.* [79] create pure tantalum implants with interconnecting open holes by using SLM. In comparison to porous Ti6Al4V, pure porous tantalum manufactured using SLM has superior bone osteoconductive characteristics, good ductility, and better-normalized fatigue strength. Therefore, it is envisaged that the porous tantalum implants generated using the SLM technology would be widely employed in orthopedics due to their customizable mechanical qualities and anticipated biological features. A male patient aged 84 underwent the first world operation to fix a critically defective knee joint in 2017 via deploying a 3D tantalum-printed cushion artificially made into his body. The fundamental importance of this procedure is that it was the first-time tantalum was employed as an implant in a clinical setting [148–150].

15.3.3 Alloy of Cobalt–Chromium

These alloy types are mostly made up of chromium as well as cobalt, as these alloys are also referred to as superalloys. They offer exceptional mechanical qualities as well as corrosion resistance. In addition, Co-Cr alloy components offer great mechanical properties, excellent biocompatibility, high-temperature resistance, and good wear protection, making them suitable for load bearing in the medical field such as implants [5, 80–83]. They were initially employed for artificial joints but are currently widely applied in various oral and dental tools [84, 85]. Furthermore, Co-Cr alloy 3D-printed porcelain teeth have emerged as a topmost option product made from porcelain manufactured from non-precious metals since they do not include dangerous components like antimony and nickel [86, 87]. Recently, Espana *et al.* [88] for the first time manufactured a porous core CoCrMo alloy object and a solid wall that had a modulus of 33–43 Gpa and an average of 10–18% using LENS. Furthermore, there have been considerable reports on the microstructure [87, 89–93], corrosion properties [141, 145], biocompatibility [96], and mechanical features [87, 90, 91, 94, 95] of Co–Cr alloys fabricated using SLM for dental interventions [86, 97]. SLM was also used by Hazlehurst *et al.* [98] to create cellular structures with cobalt-chrome-molybdenum having volumetric porosities ranging between 25% and 95%. These cellular structures have strength and stiffness similar to human femur cancellous bone and cortical bone. Sahasrabudhe *et al.* [99] in their study established that the wear resistance of CoCrMo alloys might be increased by adding calcium monophosphide (CaP). By using LENS 3D printing to create CoCrMo alloys having 3% CaP, it was seen that the rate of wear of the sample was just one-third that of neat CoCrMo alloys. The Co2þ and Cr2þ discharge was also decreased to four times those of plain CoCrMo alloy lacking CaP.

15.3.4 Liquid Metals

The term "liquid metal" refers to a functional material type with different physical and chemical properties. Mercury (Hg, 39°C), gallium alloys (30°C), mercury (Hg, 39°C), gallium (Ga, 30°C), cesium (Cs, 29°C), rubidium (Rb, 40°C), francium (Fr, 27°C) are all members of this group [134]. Mercury happens to be the largest abundant, but due to its toxicity, its uses were been constrained. Gallium-based alloys exhibiting minimal toxicity, like gallium-indium-tin (Galinstan) and eutectic gallium-indium (EGaIn) are viable options for Hg [134]. Several Liquid metals have been used to tackle many key biomedical challenges, ushering conceptual breakthroughs in medical technology. For instance, liquid metal has already been employed to join and repair nerves. Besides that, it has been widely explored for other various medical therapeutics conditions like angiography, thermochemical ablation of alkali metal fluid for tumor therapy, flexible exoskeletons, embolization for vascular tumor treatment, flexible exoskeletons, injection-based solid-liquid transformation for metal cement with the low melting point, printed bendable radiation protection, direct-printed electronic circuits on human skin, and injectable implantable medical electronics. Liu *et al.* [135] recently improved signal transmission in mice via silicone tube suturing enclosing liquid gallium through the two terminals of a sciatic injured nerve. It was observed that the muscle atrophy of mice was 2 months slowed when compared to the control group revealed by the electrophysiological testing following liquid metal surgery. In a different study, Li *et al.* [136] used liquid metal conduction on the skin based on low electrical voltage impact to successfully cure malignant melanoma. They also examined the packaging and printing characteristics, washing qualities, and liquid metal restrictions on various textiles like clothes. Regarding liquid metals 3D printing, Yu *et al.* [147] were able to use hydrogel as a clear support platform to accomplish suspension liquid metal 3D printing. These hydrogels have mechanical qualities intermediate between those of liquid and solid, as well as the ability to self-heal. The printing process addressed technical difficulties posed by the liquid metal ink's low viscosity, high surface tension, heavyweight, and facile flow.

15.3.5 Iron

Iron is an important trace element in the human body that partakes in several physiological responses. Presently, owing to their superior mechanical qualities and degradability, iron-based materials are now regarded as a viable choice for the development of biodegradable implants. Preliminary in vivo studies revealed that pure iron stents placed into pig aorta caused neither local nor systemic damage [15, 110–113]. This can be related to the presence of iron ions produced during breakdown and may be consumed in the human body and metabolized with no accumulation. Unfortunately, the breakdown rate of pristine iron metal in physiological conditions is smaller compared to the body's required rate [15, 110, 111, 114]. Many alloys of Fe-Mn have already been devised recently to increase the biodegradable process of iron-based parts [110, 114]. Metallic degradable made of iron was one of the initial metals exploited for 3D printed constructs in the biomedical field [115]. Furthermore, Chou *et al.* [116] used CALPHAD (calculation of phase diagram), a theoretical model, to investigate the influence of Fe-Mn alloy composition on deterioration rate. It was observed that when Ca and Mg elements were added, Fe-Mn, Fe-Mn-Mg, and Fe-Mn-Ca alloys all displayed better cytocompatibility. In another study, Zhang *et al.* [117] printed composite constructs comprising bioactive glass/polycaprolactone and nanoparticles of Fe3O4 (Fe3O4/MBG/PCL) using a 3D Bioplotter. These composite products exhibited 13–16 MPa compressive strength and an evenly spread 60 % porosity of 400 μm. In a similar version, Liu *et al.* [118] employed a 3D extrusion printer to print artificial magnetic cilium from polydimethylsiloxane doped with iron particles. In conclusion, 3D-printed porous biodegradable constructs built of iron alloys based can support bone regeneration without problems and hence make them suitable candidates for orthopedic implants [151–154].

15.3.6 Magnesium

Titanium alloys and stainless steels are generally internal fixation implants utilized globally for clinical orthopedic procedures. Implants manufactured from either material are non-biodegradable inside the human body and most often necessitate a revisited surgery to modify or take them away. Additionally, the usage of these materials (inert biometals) may result in protracted challenges. As a result of these issues, implants constituted of biodegradable biometals are necessary [14]. Therefore, people are increasingly interested in magnesium alloys. In comparison to titanium alloys or stainless steel, magnesium alloys possess the highest human-bone compatibility biomechanically (because of their comparable density) and produce the least amount of discomfort [14, 100, 101]. Moreover, they can be taken in vivo by humans and then discharged as magnesium ions to increase the osteoblast's differentiation and proliferation, hence increasing bone development and repair [102–104]. Aside from being used for medical devices like bone plates and screws, they may also be employed for the progress of cardiovascular stents which are biodegradable [105, 106]. Additionally, Jauer *et al.* [107] used SLM to successfully create WE43 having scaffold-like structures possessing linked porosity. Furthermore, Li *et al.* [108] on the other hand developed an organized topographically and linked porous WE43 biodegradable scaffold created by SLM and after around 4 weeks lost 20% of its volume while maintaining mechanical qualities essential for trabecular bone support. Additionally, Li *et al.* [109] established degradable magnesium-containing polymer for bone healing using low-temperature 3D printing techniques. Composite of newberyite (MgHPO4.3H2O) and struvite (MgNH4PO4.6H2O) was created using an inkjet 3D printing process via a hydraulic reaction setting around binder liquid Mg3(PO4)2 powder containing 2M K2HPO4, 20% H3PO4, or 0.5M (NH4)2HPO4. The 3D printed post-hardening struvite compressive strength improved to 10 MPa, whereas the compressive strength of the post-hardening newberyite enhanced to 36 MPa, indicating a growth in the region of 1.3–2.8 MPa. This matrix of magnesium phosphate developed through 3D printing is projected to act as a resorbable bone replacement due to its excellent properties.

15.3.7 Zinc

Zn is one of the important components of the human body. It has a direct role in the body's enzyme production, gene expression, metabolism of nucleic acid, signal transmission, apoptosis control, tissue regeneration, and growth stimulation [119]. A few years ago, Hennig *et al.* [120] revealed that Zn had substantial antiatherogenic effects. Even though there are still worries regarding using Zn metal in humans might have negative implications, researchers have proven that zinc toxicity is low. In addition, Zn has also been shown to be antimicrobial [121]. Several pieces of research have been performed on the design [122, 123], production [14, 123], and degradation processes of zinc alloys *in vitro* [124–126] and *in vivo* [127–131] conditions. Demir *et al.* [132] reported the construction of natural Zn biodegradable implants via SLM as well as presented the production procedure configurations. 3D-printed zinc components have the same oxidation level as pristine zinc powder (approximately 2 wt %). Similarly, Wen *et al.* [133] further reported the effects of SLM printing constraints on pure Zn products like surface quality, densification, and mechanical characteristics.

15.4 Future Direction and Challenges

In the present day, due to its capacity to adapt biomedical devices to every person's demand, 3D printing is now becoming a vital tool in the medical profession. Currently, as the world's older population grows, so does the need for individualized medication and personalized 3D-printed health items, particularly orthopedic tools and surgical implants. As a result, metallic implants enabled by 3D printing are already playing an important part in the medical sector. Towards the construction of metallic medical equipment, 3D printing technology has numerous unmatched benefits, such as the possibility to have personalized models, the production of complicated geometrical structures, and the decrease of production duration and healthcare expenses. At this moment, with even the most widely used EBM approach, there is still much opportunity for advancement. Till now, the precision and effectiveness of metal 3D printing manufacturing are insufficient. Furthermore, Material qualities, manufacturing and equipment capacity, and other factors end up making it challenging to effectively create components having great accuracy and outstanding finished surface while also attaining large production performance, since there will ever be a trade-off involving speed printing and accuracy. Additionally, to prevent potential biological dangers, these materials must fulfill biological standards for medical device applications. However, just a well-established few materials for 3D metal printing are now accessible, like stainless steel, titanium alloys, and aluminum alloys. Another pressing problem that must be addressed is the absence of biosafety standards for 3D-printed metallic medical products. Any materials aimed at clinic applications must meet highly strict specifications. Therefore, to fulfill the criteria for manufacturing and clinical usage, biocompatibility, materials' safety, biological activity, and degradation performance must be considered before and after printing. Furthermore, careful sterility of implants is among the most important factors in ensuring implant success. An orthopedic joint prosthesis and implant failure could occur as a result of wrong sterilization handling because such a step could give a negative influence on the implant surface features. Presently, the soft and hard biomaterials combination interface is critical for improving therapeutic outcomes in several domains, including tissue engineering. Despite all these limitations of metallic 3D printed products in medical applications, 3D biological printing may present an unparalleled chance due to the continual growth and enhancement of the materials in the biomedical sector and 3D printing innovation at large.

It must also acknowledge that metallic implants 3D printed is still in the infant stage. However, it is realistic to expect 3D printing invention to take a larger function in the domain of metallic devices for biomedical applications, ultimately becoming a key component in terms of digital production and the modern medical device market which is being incorporated with the internet, artificial intelligence, and big data.

References

1. Ashammakhi, N., Ahadian, S., Darabi, M.A., El Tahchi, M., Lee, J., Suthiwanich, K., Sheikhi, A., Dokmeci, M.R., Oklu, R., Khademhosseini, A.J., Minimally invasive and regenerative therapeutics. *Adv. Mater.*, 31, 1, 1804041, 2019.
2. Niinomi, M., Recent metallic materials for biomedical applications. *Metall. Mater. Trans.*, 33, 3, 477, 2002.
3. Niinomi, M., Nakai, M., Hieda, J.J., Development of new metallic alloys for biomedical applications. *Acta Biomater.*, 8, 11, 3888–3903, 2012.
4. Shih, C.-C., Shih, C.-M., Su, Y.-Y., Su, L.H.J., Chang, M.-S., Lin, S.-J.J., Effect of surface oxide properties on corrosion resistance of 316L stainless steel for biomedical applications. *Corros. Sci.*, 46, 2, 427–441, 2004.
5. Disegi, J.A., Kennedy, R.L., Pilliar, R., *Cobalt-base alloys for biomedical applications*, ASTM, Danvers, 1999.
6. Niinomi, M.J., Mechanical biocompatibilities of titanium alloys for biomedical applications. *J. Mech. Behav. Biomed. Mater.*, 1, 1, 30–42, 2008.
7. Elias, C., Lima, J., Valiev, R., Meyers, M., Biomedical applications of titanium and its alloys. *Jom*, 60, 3, 46–49, 2008.
8. Yoneyama, T. and Miyazaki, S., *Shape memory alloys for biomedical applications*, Elsevier, UK, 2008.
9. Barras, C. and Myers, K.J., Nitinol–its use in vascular surgery and other applications. *Eur. J. Vasc. Endovasc. Surg.*, 19, 6, 564–569, 2000.
10. Black, J.J., Biologic performance of tantalum. *Clin. Mater.*, 16, 3, 167–173, 1994.
11. Levine, B.R., Sporer, S., Poggie, R.A., Della Valle, C.J., Jacobs, J.J., Experimental and clinical performance of porous tantalum in orthopedic surgery. *J. Biochem. (JB)*, 27, 27, 4671–4681, 2006.
12. O'Brien, B., Stinson, J., Carroll, W.J., Development of a new niobium-based alloy for vascular stent applications. *J. Mech. Behav. Biomed. Mater.*, 1, 4, 303–312, 2008.
13. O'Brien, B.J., Stinson, J.S., Boismier, D.A., Carroll, W.M., Characterization of an NbTaWZr alloy designed for magnetic resonance angiography compatible stents. *J. Biochem. (JB)*, 29, 34, 4540–4545, 2008.

14. Zheng, Y., Gu, X., Witte, F.J.M.S., Reports, E.R., Biodegradable metal. *Materials Science and Engineering R Reports*, 77, 1–34, 2014.
15. Hermawan, H., Dube, D., Mantovani, D.J., Developments in metallic biodegradable stents. *Acta Biomater.*, 6, 5, 1693–1697, 2010.
16. Bowen, P.K., Shearier, E.R., Zhao, S., Guillory, R.J., Zhao, F., Goldman, J., Drelich, J.W.J., Biodegradable metals for cardiovascular stents: From clinical concerns to recent Zn-Alloys. *Adv. Healthc. Mater.*, 5, 10, 1121–1140, 2016.
17. Hwang, I., Guan, Z., Li, X.J., Fabrication of zinc–tungsten carbide nanocomposite using cold compaction followed by melting. *J. Occup. Med. (JOM) Sci. Eng.*, 140, 8, 084503, 2018.
18. Hwang, I., Guan, Z., Li, X.J., Scalable manufacturing of zinc-tungsten carbide nanocomposites. *Philos. Med. (P & M)*, 26, 140–145, 2018.
19. Xin, Y., Hu, T., Chu, P.J., *In vitro* studies of biomedical magnesium alloys in a simulated physiological environment: A review. *Acta Biomater.*, 7, 4, 1452–1459, 2011.
20. Gu, X.-N. and Zheng, Y.-F.J., A review on magnesium alloys as biodegradable materials. *Front. Mater. Sci. China*, 4, 2, 111–115, 2010.
21. Guan, Z., Hwang, I., Pan, S., Li, X.J., Nano-manufacturing, scalable manufacturing of AgCu40 (wt%)–WC nanocomposite microwires. *J. Occup. Med. (JOM)*, 6, 3, 031008, 2018.
22. Ryan, G., Pandit, A., Apatsidis, D.P., Fabrication methods of porous metals for use in orthopaedic applications. *J. Biochem. (JB)*, 27, 13, 2651–2670, 2006.
23. Bose, S., Roy, M., Bandyopadhyay, A.J., Recent advances in bone tissue engineering scaffolds. *Trends Biotechnol.*, 30, 10, 546–554, 2012.
24. Tofail, S.A., Koumoulos, E.P., Bandyopadhyay, A., Bose, S., O'Donoghue, L., Charitidis, C.J., Additive manufacturing: Scientific and technological challenges, market uptake and opportunities. *Mater. Today*, 21, 1, 22–37, 2018.
25. Kruth, J.-P., Mercelis, P., Van Vaerenbergh, J., Froyen, L., Rombouts, M.J., Binding mechanisms in selective laser sintering and selective laser melting. *Rapid Prototyp. J.*, 11, 1, 26–36, 2005.
26. Kruth, J.-P., Wang, X., Laoui, T., Froyen, L.J., Lasers and materials in selective laser sintering. *Assemb. Autom.*, 23, 4, 357–371, 2003.
27. Frazier, W.E., Metal additive manufacturing: A review. *J. Mater. Eng. Perform.*, 23, 6, 1917–1928, 2014.
28. Datta, S., Jana, S., Das, A., Chakraborty, A., RoyChowdhury, A., Datta, P., Bioprinting of radiopaque constructs for tissue engineering and understanding degradation behaviour by use of Micro-CT. *Bioactive Mater.*, 5, 3, 569–576, 2020.
29. Pattanayak, D.K., Fukuda, A., Matsushita, T., Takemoto, M., Fujibayashi, S., Sasaki, K., Nishida, N., Nakamura, T., Kokubo, T.J., Bioactive Ti metal analogous to human cancellous bone: Fabrication by selective laser melting and chemical treatments. *Acta Biomater.*, 7, 3, 1398–1406, 2011.

30. Yap, C.Y., Chua, C.K., Dong, Z.L., Liu, Z.H., Zhang, D.Q., Loh, L.E., Sing, S.L.J., Review of selective laser melting: Materials and applications. *Appl. Phys. Rev.*, 2, 4, 041101, 2015.
31. Murr, L.E., Gaytan, S.M., Ramirez, D.A., Martinez, E., Hernandez, J., Amato, K.N., Shindo, P.W., Medina, F.R., Wicker, R.B.J., Metal fabrication by additive manufacturing using laser and electron beam melting technologies. *J. Mater. Sci. Technol.*, 28, 1, 1–14, 2012.
32. Vandenbroucke, B. and Kruth, J.-P.J., Selective laser melting of biocompatible metals for rapid manufacturing of medical parts. *Read. Psychol. J.*, 13, 4, 196–203, 2007.
33. Tan, K., Chua, C., Leong, K., Cheah, C., Gui, W., Tan, W., Wiria, F.J., Engineering, selective laser sintering of biocompatible polymers for applications in tissue engineering. *Biomed. Mater. Eng.*, 15, 1, 2, 113–124, 2005.
34. Ko, S.H., Pan, H., Grigoropoulos, C.P., Luscombe, C.K., Frechet, J.M., Poulikakos, D., All-inkjet-printed flexible electronics fabrication on a polymer substrate by lowtemperature high-resolution selective laser sintering of metal nanoparticles. *Philos. Soc. Sci. Ed.*, 18, 34, 345202, 2007.
35. Selcuk, C.J., Laser metal deposition for powder metallurgy parts. *Philos. Med. (P & M)*, 54, 2, 94–99, 2011.
36. Dinda, G., Song, L., Mazumder, J.J.M., Fabrication of Ti-6Al-4V scaffolds by direct metal deposition. *Metall. Mater. Trans.*, 39, 12, 2914–2922, 2008.
37. Ahsan, M.N., Paul, C.P., Kukreja, L., Pinkerton, A.J.J., Porous structures fabrication by continuous and pulsed laser metal deposition for biomedical applications: Modelling and experimental investigation. *J. Mater. Process. Technol.*, 211, 4, 602–609, 2011.
38. Gu, D., Meiners, W., Wissenbach, K., Poprawe, R.J., Laser additive manufacturing of metallic components: Materials, processes and mechanisms. *Int. Mater. Rev.*, 57, 3, 133–164, 2012.
39. Sames, W.J., List, F., Pannala, S., Dehoff, R.R., Babu, S.S.J., The metallurgy and processing science of metal additive manufacturing. *Int. Mater. Rev.*, 61, 5, 315–360, 2016.
40. Ali, H., Ghadbeigi, H., Mumtaz, K.J.M.S., Effect of scanning strategies on residual stress and mechanical properties of selective laser melted Ti6Al4V. *Agric. Eng.*, 712, 175–187, 2018.
41. DebRoy, T., Wei, H., Zuback, J., Mukherjee, T., Elmer, J., Milewski, J., Beese, A.M., Wilson-Heid, A., De, A., Zhang, W.J., Additive manufacturing of metallic components–process, structure and properties. *Prog. Mater. Sci.*, 92, 112–224, 2018.
42. Gao, C., Wang, C., Jin, H., Wang, Z., Li, Z., Shi, C., Leng, Y., Yang, F., Liu, H., Wang, J.J., Additive manufacturing technique-designed metallic porous implants for clinical application in orthopedics. *RSC Adv.*, 8, 44, 25210–25227, 2018.

43. Zergioti, I., Mailis, S., Vainos, N., Fotakis, C., Chen, S., Grigoropoulos, C.J., Microdeposition of metals by femtosecond excimer laser. *Appl. Surf. Sci.*, 127, 601–605, 1998.
44. Delaporte, P. and Alloncle, A.-P.J., Laser-induced forward transfer: A high resolution additive manufacturing technology. *Opt. Laser. Technol.*, 78, 33–41, 2016.
45. Papazoglou, S. and Zergioti, I.J., Laser Induced Forward Transfer (LIFT) of nano-micro patterns for sensor applications. *Microelectron. Eng.*, 182, 25–34, 2017.
46. Bullock, A.B. and Bolton, P.R.J., Laser-induced back ablation of aluminum thin films using picosecond laser pulses. *J. Appl. Phys.*, 85, 1, 460–465, 1999.
47. Sano, T., Yamada, H., Nakayama, T., Miyamoto, I.J., Experimental investigation of laser induced forward transfer process of metal thin films. *Appl. Surf. Sci.*, 186, 1–4, 221–226, 2002.
48. Toth, Z., Hopp, B., Szoerenyi, T., Bor, Z., Shakhno, E.A., Veiko, V.P., *Pulsed laser ablation mechanisms of thin metal films, computer-controlled microshaping*, pp. 18–27, International Society for Optics and Photonics, Germany, 1999.
49. Unger, C., Koch, J., Overmeyer, L., Chichkov, B.N.J., Time-resolved studies of femtosecond-laser induced melt dynamics. *Opt. Express*, 20, 22, 24864–24872, 2012.
50. Hirt, L., Reiser, A., Spolenak, R., Zambelli, T.J., Additive manufacturing of metal structures at the micrometer scale. *Adv. Mater.*, 29, 17, 1604211, 2017.
51. Zenou, M., Sa'ar, A., Kotler, Z., Laser transfer of metals and metal alloys for digital microfabrication of 3D objects. *Small*, 11, 33, 4082–4089, 2015.
52. Visser, C.W., Pohl, R., Sun, C., R€omer, G.W., Huis in 't Veld, B., Lohse, D.J., Toward 3D printing of pure metals by laser-induced forward transfer. *Adv. Mater.*, 27, 27, 4087–4092, 2015.
53. Dilberoglu, U.M., Gharehpapagh, B., Yaman, U., Dolen, M.J., The role of additive manufacturing in the era of industry 4.0. *Philos. Med. (P & M)*, 11, 545–554, 2017.
54. Xia, M. and Sanjayan, J., Design, method of formulating geopolymer for 3D printing for construction applications. *Mater. Des.*, 110, 382–390, 2016.
55. El-Hajje, A., Kolos, E.C., Wang, J.K., Maleksaeedi, S., He, Z., Wiria, F.E., Choong, C., Ruys, A.J.J., Physical and mechanical characterisation of 3D-printed porous titanium for biomedical applications. *J. Mater. Sci. Mater. Med.*, 25, 11, 2471–2480, 2014.
56. Banerjee, D. and Williams, J.J., Perspectives on titanium science and technology. *Acta Mater.*, 61, 3, 844–879, 2013.
57. Khorasani, A.M., Goldberg, M., Doeven, E.H., Littlefair, G.J., Titanium in biomedical applications—Properties and fabrication: A review. *J. Biomimetics, Biomater. Tissue Eng.*, 5, 8, 593–619, 2015.

58. Gepreel, M.A.-H. and Niinomi, M.J., Biocompatibility of Ti-alloys for long-term implantation. *J. Mech. Behav. Biomed. Mater.*, 20, 407–415, 2013.
59. Mohammed, M.T., Khan, Z.A., Siddiquee, A.N.J., Beta titanium alloys: The lowest elastic modulus for biomedical applications: A review. *Int. J. Chem. Mol. Nucl. Mater. Metall. Eng.*, 8, 8, 32340, 2014.
60. Niinomi, M.J.M.S., Mechanical properties of biomedical titanium alloys. *Agric. Eng.*, 243, 1–2, 231–236, 1998.
61. Agarwala, M., Bourell, D., Beaman, J., Marcus, H., Barlow, J.J., Direct selective laser sintering of metals. *Read. Psychol. J.*, 1, 1, 26–36, 1995.
62. Zhang, L.-C., Attar, H., Calin, M., Eckert, J.J., Review on manufacture by selective laser melting and properties of titanium based materials for biomedical applications. *Mater. Technol.*, 31, 2, 66–76, 2016.
63. Tan, X., Tan, Y., Chow, C., Tor, S., Yeong, W.J.M.S., Metallic powder-bed based 3D printing of cellular scaffolds for orthopaedic implants: A state-of-the-art review on manufacturing, topological design, mechanical properties and biocompatibility. *Mater. Sci. Eng.*, 76, 1328–1343, 2017.
64. Yavari, S.A., van der Stok, J., Chai, Y.C., Wauthle, R., Birgani, Z.T., Habibovic, P., Mulier, M., Schrooten, J., Weinans, H., Zadpoor, A., Bone regeneration performance of surface-treated porous titanium. *Biomaterials*, 35, 24, 6172–6181, 2014.
65. Gorgin Karaji, Z., Speirs, M., Dadbakhsh, S., Kruth, J.-P., Weinans, H., Zadpoor, A., Amin Yavari, S.J., Additively manufactured and surface biofunctionalized porous nitinol. *ACS Appl. Mater. Interfaces*, 9, 2, 1293–1304, 2017.
66. Hollander, D.A., Von Walter, M., Wirtz, T., Sellei, R., Schmidt-Rohlfing, B., Paar, O., Erli, H.-J., Structural, mechanical and *in vitro* characterization of individually structured Ti–6Al–4V produced by direct laser forming. *J. Biochem.*, 27, 7, 955–963, 2006.
67. Murr, L.J., Metallurgy of additive manufacturing: Examples from electron beam melting. *Acta Mater.*, 5, 40–53, 2015.
68. Taniguchi, N., Fujibayashi, S., Takemoto, M., Sasaki, K., Otsuki, B., Nakamura, T., Matsushita, T., Kokubo, T., Matsuda, S.J.M.S., Effect of pore size on bone ingrowth into porous titanium implants fabricated by additive manufacturing: An *in vivo* experiment. *Mater. Sci. Eng*, 59, 690–701, 2016.
69. Wysocki, B., Idaszek, J., Zdunek, J., Ro_zniatowski, K., Pisarek, M., Yamamoto, A., Swięszkowski, W.J., The influence of selective laser melting (SLM) process parameters on *in-vitro* cell response. *Int. J. Mol. Sci.*, 19, 6, 1619, 2018.
70. Ponader, S., Vairaktaris, E., Heinl, P., Wilmowsky, C.v., Rottmair, A., Körner, C., Singer, R.F., Holst, S., Schlegel, K.A., Neukam, F.W.J., Effects of topographical surface modifications of electron beam melted Ti-6Al-4V titanium on human fetal osteoblasts. *J. Biomed. Mater. Res. Part A Off. J. Soc. Biomater. Jn. Soc. Biomater. Aust. Soc. Biomater. Korean Soc. Biomater.*, 84, 4, 1111–1119, 2008.

71. Ponader, S., Von Wilmowsky, C., Widenmayer, M., Lutz, R., Heinl, P., Körner, C., Singer, R.F., Nkenke, E., Neukam, F.W., Schlegel, K.A., *In vivo* performance of selective electron beam-melted Ti-6Al-4V structures. *J. Biomed. Mater. Res. Part A Off. J. Soc. Biomater. Jn. Soc. Biomater. Aust. Soc. Biomater. Korean Soc. Biomater*, 92, 1, 56–62, 2010.
72. Gadia, A., Shah, K., Nene, A.J., Emergence of three-dimensional printing technology and its utility in spine surgery. *Asian. J. Spine*, 12, 2, 365–371, 2018.
73. Imanishi, J. and Choong, P.F.J.I., Three-dimensional printed calcaneal prosthesis following total calcanectomy. *Int. J. Surg. Case Rep.*, 10, 83–87, 2015.
74. Maccauro, G., Iommetti, P.R., Muratori, F., Raffaelli, L., Manicone, P.F., Fabbriciani, C.J., An overview about biomedical applications of micron and nano size tantalum. *Recent Pat. Biotechnol.*, 3, 3, 157–165, 2009.
75. Balla, V.K., Banerjee, S., Bose, S., Bandyopadhyay, A.J., Direct laser processing of a tantalum coating on titanium for bone replacement structures. *Acta Biomater.*, 6, 6, 2329–2334, 2010.
76. Bobyn, J.D., Toh, K.-K., Hacking, S.A., Tanzer, M., Krygier, J.J.J., Tissue response to porous tantalum acetabular cups: A canine model. *J. Arthroplast.*, 14, 3, 347–354, 1999.
77. Bobyn, J., Stackpool, G., Hacking, S., Tanzer, M., Krygier, J.J., Characteristics of bone ingrowth and interface mechanics of a new porous tantalum biomaterial. *J. Bone Jt. Surg.*, 81, 5, 907–914, 1999.
78. Balla, V.K., Bodhak, S., Bose, S., Bandyopadhyay, A., Porous tantalum structures for bone implants: Fabrication, mechanical and *in vitro* biological properties. *Acta Biomater.*, 6, 8, 3349–3359, 2010.
79. Wauthle, R., Van Der Stok, J., Yavari, S.A., Van Humbeeck, J., Kruth, J.-P., Zadpoor, A.A., Weinans, H., Mulier, M., Schrooten, J., Additively manufactured porous tantalum implants. *Acta Biomater.*, 14, 217–225, 2015.
80. Niinomi, M., Narushima, T., Nakai, M., *Advances in metallic biomaterials*, Springer, Heidelberg, DE, 2015.
81. Hsu, R.W.-W., Yang, C.-C., Huang, C.-A., Chen, Y.-S., Electrochemical corrosion studies on Co–Cr–Mo implant alloy in biological solutions. *Mater. Chem. Phys.*, 93, 2–3, 531–538, 2005.
82. Zangeneh, S., Lashgari, H., Roshani, A., Microstructure and tribological characteristics of aged Co–28Cr–5Mo–0.3 C alloy. *Mater. Des.*, 37, 292–303, 2012.
83. Metikoš-Hukovič, M. and Babic, R., Passivation and corrosion behaviours of cobalt and cobalt–chromium–molybdenum alloy. *Corros. Sci.*, 49, 9, 3570–3579, 2007.
84. Sinnett-Jones, P., Wharton, J., Wood, R.J., Micro-abrasion–corrosion of a CoCrMo alloy in simulated artificial hip joint environments. *Wear*, 259, 7–12, 898–909, 2005.
85. Sadiq, K., Black, R.A., Stack, M.J., Bio-tribocorrosion mechanisms in orthopaedic devices: Mapping the micro-abrasion–corrosion behaviour of a

simulated CoCrMo hip replacement in calf serum solution. *Wear*, 316, 1–2, 58–69, 2014.
86. Koutsoukis, T., Zinelis, S., Eliades, G., Al-Wazzan, K., Rifaiy, M.A., Al Jabbari, Y.S., Selective laser melting technique of Co-Cr dental alloys: A review of structure and properties and comparative analysis with other available techniques. *J. Prosthodont.*, 24, 4, 303–312, 2015.
87. Lu, Y., Wu, S., Gan, Y., Li, J., Zhao, C., Zhuo, D., Lin, J.J.M.S., Investigation on the microstructure, mechanical property and corrosion behavior of the selective laser melted CoCrW alloy for dental application. *Mater. Sci. Eng.*, 49, 517–525, 2015.
88. España, F.A., Balla, V.K., Bose, S., Bandyopadhyay, A.J.M.S., Design and fabrication of CoCrMo alloy based novel structures for load bearing implants using laser engineered net shaping. *Mater. Sci. Eng.*, 30, 1, 50–57, 2010.
89. Xin, X., Chen, J., Xiang, N., Gong, Y., Wei, B.J., Surface characteristics and corrosion properties of selective laser melted Co–Cr dental alloy after porcelain firing. *Dent. Mater.*, 30, 3, 263–270, 2014.
90. Kajima, Y., Takaichi, A., Kittikundecha, N., Nakamoto, T., Kimura, T., Nomura, N., Kawasaki, A., Hanawa, T., Takahashi, H., Wakabayashi, N., Effect of heat-treatment temperature on microstructures and mechanical properties of Co–Cr–Mo alloys fabricated by selective laser melting. *Mater. Sci. Eng.*, 726, 21–31, 2018.
91. Lu, Y., Wu, S., Gan, Y., Zhang, S., Guo, S., Lin, J., Lin, J., Microstructure, mechanical property and metal release of As-SLM CoCrW alloy under different solution treatment conditions. *J. Mech. Behav. Biomed. Mater.*, 55, 179–190, 2016.
92. Liverani, E., Fortunato, A., Leardini, A., Belvedere, C., Siegler, S., Ceschini, L., Ascari, A., Fabrication of Co–Cr–Mo endoprosthetic ankle devices by means of Selective Laser Melting (SLM). *Mater. Des.*, 106, 60–68, 2016.
93. Darvish, K., Chen, Z., Phan, M., Pasang, T., Selective laser melting of Co-29Cr-6Mo alloy with laser power 180–360 W: Cellular growth, intercellular spacing and the related thermal condition. *Mater.Char.*, 135, 183–191, 2018.
94. Kajima, Y., Takaichi, A., Nakamoto, T., Kimura, T., Yogo, Y., Ashida, M., Doi, H., Nomura, N., Takahashi, H., Hanawa, T., Fatigue strength of Co–Cr–Mo alloy clasps prepared by selective laser melting. *J. Mech. Behav. Biomed. Mater.*, 59, 446–458, 2016.
95. Song, C., Zhang, M., Yang, Y., Wang, D., Jia-kuo, Y., Morphology and properties of CoCrMo parts fabricated by selective laser melting. *Mater. Sci. Eng.*, 713, 206–213, 2018.
96. Hedberg, Y.S., Qian, B., Shen, Z., Virtanen, S., Wallinder, I.O., *In vitro* biocompatibility of CoCrMo dental alloys fabricated by selective laser melting. *Dent. Mater.*, 30, 5, 525–534, 2014.

97. Qian, B., Saeidi, K., Kvetkova, L., Lofaj, F., Xiao, C., Shen, Z., Defects-tolerant Co-Cr-Mo dental alloys prepared by selective laser melting. *Dent. Mater.*, 31, 12, 1435–1444, 2015.
98. Hazlehurst, K., Wang, C.J., Stanford, M., Evaluation of the stiffness characteristics of square pore CoCrMo cellular structures manufactured using laser melting technology for potential orthopaedic applications. *Mater. Des.*, 51, 949–955, 2013.
99. Sahasrabudhe, H., Bose, S., Bandyopadhyay, A., Laser processed calcium phosphate reinforced CoCrMo for load-bearing applications: Processing and wear induced damage evaluation. *Acta Biomater*, 66, 118–128, 2018.
100. Li, N. and Zheng, Y., Novel magnesium alloys developed for biomedical application: A review. *J. Mater. Sci. Technol.*, 29, 6, 489–502, 2013.
101. Chen, Y., Xu, Z., Smith, C., Sankar, J., Recent advances on the development of magnesium alloys for biodegradable implants. *Acta Biomater.*, 10, 11, 4561–4573, 2014.
102. Witte, F., Ulrich, H., Palm, C., Willbold, E., Biodegradable magnesium scaffolds: Part II: Peri-implant bone remodeling. *J. Biomed. Mater. Res. A*, 81, 3, 757–765, 2007.
103. Han, P., Cheng, P., Zhang, S., Zhao, C., Ni, J., Zhang, Y., Zhong, W., Hou, P., Zhang, X., Zheng, Y., *In vitro* and *in vivo* studies on the degradation of high-purity Mg (99.99wt.%) screw with femoral intracondylar fractured rabbit model. *Biomaterials*, 64, 57–69, 2015.
104. Cheng, P., Han, P., Zhao, C., Zhang, S., Wu, H., Ni, J., Hou, P., Zhang, Y., Liu, J., Xu, H., High-purity magnesium interference screws promote fibrocartilaginous entheses regeneration in the anterior cruciate ligament reconstruction rabbit model via accumulation of BMP-2 and VEGF. *Biomaterials*, 81, 14–26, 2016.
105. Erbel, R., Di Mario, C., Bartunek, J., Bonnier, J., de Bruyne, B., Eberli, F.R., Erne, P., Haude, M., Heublein, B., Horrigan, M., Temporary scaffolding of coronary arteries with bioabsorbable magnesium stents: A prospective, non-randomised multicentre trial. *Lancet*, 369, 9576, 1869–1875, 2007.
106. Haude, M., Ince, H., Abizaid, A., Toelg, R., Lemos, P.A., von Birgelen, C., Christiansen, E.H., Wijns, W., Neumann, F.-J., Kaiser, C., Safety and performance of the second-generation drug-eluting absorbable metal scaffold in patients with denovo coronary artery lesions (BIOSOLVE-II): 6 month results of a prospective, multicentre, non-randomised, first-in-man trial. *Lancet*, 387, 10013, 31–39, 2016.
107. Jauer, L., Meiners, W., Vervoort, S., Gayer, C., Zumdick, N.A., Zander, D., Selective laser melting of magnesium alloys, european congress and exhibition on powder metallurgy. *European PM Conference Proceedings, The European Powder Metallurgy Association*, pp. 1–6, 2016.

108. Li, Y., Zhou, J., Pavanram, P., Leeflang, M., Fockaert, L., Pouran, B., Tümer, N., Schr€oder, K.-U., Mol, J., Weinans, H., Additively manufactured biodegradable porous magnesium. *Acta Biomater.*, 67, 378–392, 2018.
109. Li, C., Lai, Y., Li, L., Cao, H., Long, J., Wang, X., Qin, L.J., The *in vitro* biocompatibility and osteoinductive activity study of magnesium composed PLGA/TCP porous scaffold for bone regeneration. *J. Orthop. Trans.*, 7, 78, 2016.
110. Schinhammer, M., Hänzi, A.C., Löffler, J.F., Uggowitzer, P., Design strategy for biodegradable Fe-based alloys for medical applications. *Acta Biomater.*, 6, 5, 1705–1713, 2010.
111. Kraus, T., Moszner, F., Fischerauer, S., Fiedler, M., Martinelli, E., Eichler, J., Witte, F., Willbold, E., Schinhammer, M., Meischel, M., Biodegradable Fe-based alloys for use in osteosynthesis: Outcome of an *in vivo* study after 52 weeks. *Acta Biomater.*, 10, 7, 3346–3353, 2014.
112. Hermawan, H., Purnama, A., Dube, D., Couet, J., Mantovani, D., Fe–Mn alloys for metallic biodegradable stents: Degradation and cell viability studies. *Acta Biomater.*, 6, 5, 1852–1860, 2010.
113. Peuster, M., Hesse, C., Schloo, T., Fink, C., Beerbaum, P., von Schnakenburg, C., Longterm biocompatibility of a corrodible peripheral iron stent in the porcine descending aorta. *Biomaterials*, 27, 28, 4955–4962, 2006.
114. Liu, B. and Zheng, Y., Effects of alloying elements (Mn, Co, Al, W, Sn, B, C and S) on biodegradability and *in vitro* biocompatibility of pure iron. *Acta Biomater.*, 7, 3, 1407–1420, 2011.
115. Do, A.V., Khorsand, B., Geary, S.M., Salem, A.K., 3D printing of scaffolds for tissue regeneration applications. *Adv. Healthc. Mater.*, 4, 12, 1742–1762, 2015.
116. Chou, D.-T., Wells, D., Hong, D., Lee, B., Kuhn, H., Kumta, P.N., Novel processing of iron–manganese alloy-based biomaterials by inkjet 3-D printing. *Acta Biomater.*, 9, 10, 8593–8603, 2013.
117. Zhang, J., Zhao, S., Zhu, M., Zhu, Y., Zhang, Y., Liu, Z., Zhang, C.J., 3D-printed magnetic Fe 3 O 4/MBG/PCL composite scaffolds with multifunctionality of bone regeneration, local anticancer drug delivery and hyperthermia. *J. Mater. Chem.*, 2, 43, 7583–7595, 2014.
118. Liu, F., Alici, G., Zhang, B., Beirne, S., Li, W., Structures, Fabrication and characterization of a magnetic micro-actuator based on deformable Fe-doped PDMS artificial cilium using 3D printing. *Smart Mater. Struct.*, 24, 3, 035015, 2015.
119. Hambidge, K.M. and Krebs, N.F.J.T., Zinc deficiency: A special challenge. *J. Nutr.*, 137, 4, 1101–1105, 2007.
120. Hennig, B., Toborek, M., McClain, C.J., Antiatherogenic properties of zinc: Implications in endothelial cell metabolism. *Philos. Soc. Sci. Ed.*, 12, 10, 711–717, 1996.
121. Xiang, R., Ding, D., Fan, L., Huang, X., Xia, K.J.C., Antibacterial mechanism and safety of zinc oxide. *Chin. J. Tissue Eng. Res.*, 3), 470–475, 2014.

122. Zhao, S., McNamara, C.T., Bowen, P.K., Verhun, N., Braykovich, J.P., Goldman, J., Drelich, J.W., Structural characteristics and *in vitro* biodegradation of a novel Zn-Li alloy prepared by induction melting and hot rolling. *Metall. Mater. Trans.*, 48, 3, 1204–1215, 2017.
123. Sikora-Jasinska, M., Mostaed, E., Mostaed, A., Beanland, R., Mantovani, D., Vedani, M., Fabrication, mechanical properties and *in vitro* degradation behavior of newly developed ZnAg alloys for degradable implant applications. *Mater. Sci. Eng.*, 77, 1170–1181, 2017.
124. Vojtech, D., Kubasek, J., Serak, J., Novak, P., Mechanical and corrosion properties of newly developed biodegradable Zn-based alloys for bone fixation. *Acta Biomater.*, 7, 9, 3515–3522, 2011.
125. Törne, K., Örnberg, A., Weissenrieder, J., Influence of strain on the corrosion of magnesium alloys and zinc in physiological environments. *Acta Biomater.*, 48, 541–550, 2017.
126. Törne, K., Larsson, M., Norlin, A., Weissenrieder, J.J., Degradation of zinc in saline solutions, plasma, and whole blood. *J. Biomed. Mater. Res. Part B: Appl. Biomater.*, 104, 6, 1141–1151, 2016.
127. Mostaed, E., Sikora-Jasinska, M., Drelich, J.W., Vedani, M., Zinc-based alloys for degradable vascular stent applications. *Acta Biomater.*, 71, 1–23, 2018.
128. Bowen, P.K., Drelich, J., Goldman, J., Zinc exhibits ideal physiological corrosion behavior for bioabsorbable stents. *Adc. Mater.*, 25, 18, 2577–2582, 2013.
129. Bowen, P.K., Guillory II, R.J., Shearier, E.R., Seitz, J.-M., Drelich, J., Bocks, M., Zhao, F., Goldman, J., Metallic zinc exhibits optimal biocompatibility for bioabsorbable endovascular stents. *Mater. Sci. Eng.*, 56, 467–472, 2015.
130. Drelich, A.J., Zhao, S., Guillory II, R.J., Drelich, J.W., Goldman, J., Long-term surveillance of zinc implant in murine artery: Surprisingly steady biocorrosion rate. *Acta Biomater.*, 58, 539–549, 2017.
131. Yang, H., Wang, C., Liu, C., Chen, H., Wu, Y., Han, J., Jia, Z., Lin, W., Zhang, D., Li, W., Evolution of the degradation mechanism of pure zinc stent in the one-year study of rabbit abdominal aorta model. *Biomaterials*, 145, 92–105, 2017.
132. Demir, A.G., Monguzzi, L., Previtali, B., Selective laser melting of pure Zn with high density for biodegradable implant manufacturing. *Addit. Manuf.*, 15, 20–28, 2017.
133. Wen, P., Voshage, M., Jauer, L., Chen, Y., Qin, Y., Poprawe, R., Schleifenbaum, J.H., Laser additive manufacturing of Zn metal parts for biodegradable applications: Processing, formation quality and mechanical properties. *Mater. Des.*, 155, 36–45, 2018.
134. Joshipura, I.D., Ayers, H.R., Majidi, C., Dickey, M.D.J., Methods to pattern liquid metals. *J. Mater. Chem.*, 3, 16, 3834–3841, 2015.
135. Liu, F., Yu, Y., Yi, L., Liu, J., Liquid metal as reconnection agent for peripheral nerve injury. *Springer*, 61, 12, 939–947, 2016.

136. Li, J., Guo, C., Wang, Z., Gao, K., Shi, X., Liu, J., Electrical stimulation towards melanoma therapy via liquid metal printed electronics on skin. *Clin. Transl. Med.*, 5, 1, 21, 2016.
137. Bhoi, S., Prasad, A., Kumar, A., Sarkar, R.B., Mahto, B., Meena, C.S., Pandey, C., Experimental study to evaluate the wear performance of UHMWPE and XLPE material for orthopedics application. *Bioengineering*, 9, 676, 2022, https://doi.org/10.3390/bioengineering9110676.
138. Bhoi, S., Kumar, A., Prasad, A., Meena, C.S., Sarkar, R.B., Mahto, B., Ghosh, A., Performance evaluation of different coating materials in delamination for micro-milling applications on high-speed steel substrate. *Micromachines*, 13, 1277, 2022, https://doi.org/10.3390/mi13081277.
139. Gupta, A. *et al.*, Multifunctional nanohydroxyapatite-promoted toughened high-molecular-weight stereocomplex Poly(Lactic acid)-based bionanocomposite for both 3D-printed orthopedic implants and high-temperature engineering applications. *ACS Omega*, 2, 7, 4039–52, 2017.
140. Prasad, A., Bioabsorbable polymeric materials for biofilms and other biomedical applications: Recent and future trends. *Mater. Today: Proc.*, 44, 2447–53, 2021a, https://doi.org/10.1016/j.matpr.2020.12.489.
141. Prasad, A., State of art review on bioabsorbable polymeric scaffolds for bone tissue engineering. *Mater. Today: Proc.*, 44, 1391–1400, 2021b, https://doi.org/10.1016/j.matpr.2020.11.622.
142. Prasad, A., Bhasney, S., Katiyar, V., Ravi Sankar, M., Biowastes processed hydroxyapatite filled Poly (Lactic acid) bio-composite for open reduction internal fixation of small bones. *Mater. Today: Proc.*, 4, 9, 10153–57, 2017. http://linkinghub.elsevier.com/retrieve/pii/S2214785317311410.
143. Prasad, A., Devendar, B., Ravi Sankar, M., Robi, P.S., Micro-scratch based tribological characterization of Hydroxyapatite (HAp) fabricated through fish scales. *Mater. Today: Proc.*, 2, 4–5, 1216–24, 2015. http://linkinghub.elsevier.com/retrieve/pii/S2214785315002795.
144. Prasad, A., Ravi Sankar, M., Katiyar, V., State of art on solvent casting particulate leaching method for orthopedic scaffolds fabrication. *Mater. Today: Proc.*, 4, 2, 898–907, 2017. http://dx.doi.org/10.1016/j.matpr.2017.01.101.
145. Prasad, A., Chakraborty, G., Kumar, A., Bio-based environmentally benign polymeric resorbable materials for orthopedic fixation applications, in: *Advanced Materials for Biomedical Applications*, vol. Chapter 15, A. Kumar, Y. Gori, A. Kumar, C.S. Meena, N. Dutt (Eds.), pp. 251–266, CRC Press, Boca Raton, FL, USA, 2022, https://doi.org/10.1201/9781003344810-15.
146. Prasad, A., Biomaterial-based nanofibers scaffolds in tissue engineering application, in: *Functional Biomaterials*, pp. 245–264, Springer, Singapore, 2022.
147. Yu, Y., Liu, F., Zhang, R., Liu, J., Suspension 3D printing of liquid metal into selfhealing hydrogel. *Adv. Mater. Technol.*, 2, 11, 1700173, 2017.
148. Sidambe, A.T., Biocompatibility of advanced manufactured titanium implants—A review. *Materials*, 7, 12, 8168–8188, 2014.

149. Thijs, L., Sistiaga, M.L.M., Wauthle, R., Xie, Q., Kruth, J.-P., Van Humbeeck, J.J., Strong morphological and crystallographic texture and resulting yield strength anisotropy in selective laser melted tantalum. *Acta Mater.*, 61, 12, 4657–4668, 2013.
150. Kumar, A., Gangwar, A.K.S., Kumar, A., Meena, C.S., Singh, V.P., Dutt, N., Prasad, A., Gori, Y., Biomedical study of femur bone fracture and healing, in: *Advanced Materials for Biomedical Applications*, vol. Chapter 14, A. Kumar, Y. Gori, A. Kumar, C.S. Meena, N. Dutt (Eds.), pp. 235–250, CRC Press, Boca Raton, FL, USA, 2022, https://doi.org/ 10.1201/9781003344810-14.
151. Kumar, R., Kumar, A., Kant, L., Prasad, A., Bhoi, S., Meena, C.S., Singh, V.P., Ghosh, A., Experimental and RSM-based process-parameters optimisation for turning operation of EN36B Steel. *Materials*, 16, 339, 2023, https://doi.org/10.3390/ma16010339.
152. Prasad, A., Chakraborty, G., Kumar, A., Gajrani, K.K., Introduction to biodegradable polymers, in: *Biodegradable Composites for Packaging Applications*, vol. Chapter 1, A. Prasad, A. Kumar, K.K. Gajrani (Eds.), pp. 1–12, CRC Press, Boca Raton, FL, USA, 2022, https://doi.org/10.1201/9781003227908-1.
153. Chakraborty, G., Prasad, A., Kumar, A., Processing of biodegradable composites, in: *Biodegradable Composites for Packaging Applications*, vol. Chapter 3, A. Prasad, A. Kumar, K.K. Gajrani (Eds.), pp. 33–48, CRC Press, Boca Raton, FL, USA, 2022, https://doi.org/ 10.1201/9781003227908-3.
154. Paul, A.K., Prasad, A., Kumar, A., Review on artificial neural network and its application in the field of engineering. *J. Mechanical Engineering: PRAKASH*, 1, 1, 53–61, 2022, https://doi.org/10.56697/JMEP.2022.1107.
155. Ni. J, Ling. H, Zhang. S, Wang. Z, Peng. Z, Benyshek. C, Zan. R, Miri. A.K., Ni, J., Ling, H., Zhang, S., Wang, Z., Peng, Z., Benyshek, C., Zan, R., Miri, A.K., Li, Z., Zhang, X., Lee, J., Lee, K.J., Kim, H.J., Tebon, P., Hoffman, T., Dokmeci, M.R., Ashammakhi, N., Li, X., Khademhosseini, A., Three-dimensional printing of metals for biomedical applications. *Mater. Today Bio*, 3, 100024, 2019, https://doi.org/10.1016/j.mtbio.2019.100024

16

Additive Manufacturing for the Development of Artificial Organs

Sudipto Datta[1]*, Ranjit Barua[2] and Arbind Prasad[3]

[1]Department of Materials Engineering, Indian Institute of Science, Bangalore, Karnataka, India
[2]Centre for Healthcare Science and Technology, Indian Institute of Engineering Science and Technology, Howrah, West Bengal, India
[3]Department of Mechanical Engineering, Katihar Engineering College (under DST, Govt. of Bihar), Katihar, Bihar, India

Abstract

Dimensional Printing or (3D) printing also referred to as additive manufacturing (AM), is a manufacturing technology that has lots of applications in various engineering fields. Among them, bioprinting, which is a 3D printing technology for biomedical applications, is being used to develop artificial organs and tissues for various tissue engineering applications. Furthermore, the current development in AM has enabled more expansion of patient-specific solutions in healthcare. Modifications of many products of healthcare like *in vitro* models, prosthetics, medical instruments, drug delivery devices, and implants are possible because of AM technology. In this book chapter, we will discuss briefly the various bio-inks, various bio-printing techniques, and the application of bioprinting to develop organs and tissues for various tissue engineering applications. We will also highlight the challenges and prospects of 3D printing technology in healthcare.

***Keywords*:** Additive manufacturing, 3D bio-printing, drug delivery, bio-inks, implant, biomaterials

**Corresponding author*: sudiptodatta1990@gmail.com

Arbind Prasad, Ashwani Kumar and Manoj Gupta (eds.) Advanced Materials and Manufacturing Techniques for Biomedical Applications, (411–428) © 2024 Scrivener Publishing LLC

16.1 Introduction

A huge figure of patients expires because of the unavailability of needed organs like the liver, heart, kidney, etc., however, these organs are extremely reformed, and at this time, modern 3D Printing has wide convenience [1–5]. In health and clinical area, additive manufacturing technology is an advanced method applied for the fabrication of essential biological organs and tissues [6, 7]. It is appropriate for decidedly difficult innate organs that can be implanted into the patient body in the surgical room. Novel biocompatible materials are applied to construct tissue scaffolds and organs and exhibit healthier biocompatibility [8–10, 84]. Metallic fixation has lot of limitations so to avoid the limitations as an alternative polymeric bioabsorbable materials are widely used for fabrication of scaffolds and implants. The combination of biocompatible polymer and nanomaterials from various sources mimics the bone like structure shows usefulness towards application in biomedical applications and artificial organ developments [70–80]. In the tissue engineering and biomedical engineering area, additive manufacturing is hastily accepted because of its complex tissue scaffolds, organs, and implants. This technique can fabricate materials having outstanding eco-friendly and biomechanical characteristics. Additive manufacturing develops inexpensively proficiently end when applied for exact uses and is the greatest technique for generating a detailed model. Additive manufacturing generates an invention from a CAD model that is typically in (standard triangulate language) .stl format. This design is introduced into the 3D printer device to make a construct or object layer-by-layer layer material deposition [11–13, 85, 86]. At this time, the fabrication process depends on the layer thickness as well as the geometrical shape of the object. Additive manufacturing creates dedicated implements, comprehensive quantities, and associated personalized medical maneuvers.

Additive manufacturing employs dissimilar categories of 3D printing expertise to construct organs and tissues by means of a method resembling an inkjet process or extrusion process. FDM or fused deposits modeling process is used generally which are lesser in price and has reasonable suppleness. It usages polymer filaments delivered through a hot extruder nozzle, and the molten material is then extruded and dropped on the bed layerwise manner; these amounts of an essential layer together fabricate an object [14]. There are dissimilar categories of materials applied in the fused deposits modeling process like metal, thermoplastic, hydrogel, etc.

Figure 16.1 shows the advancement of additive manufacturing technology at different times. The foremost use of sustenance material is to

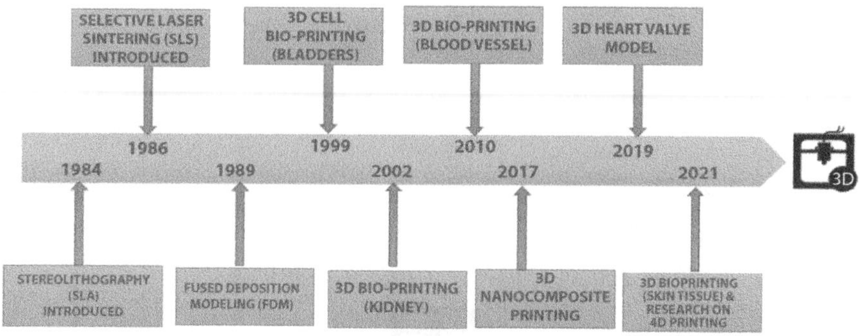

Figure 16.1 The expansion of additive manufacturing technology at different times.

construct complex design objects deprived of any imperfection exactly [15, 16]. Another additive manufacturing expertise used for organ and tissue printing is 3D bioprinting. It usages a head part that has dissimilar types of living cells which allows the creation of multicellular tissue/organs and it is applied for the construction of complicated tissue scaffolds and organs like aortic valves, liver tissue, bone, and cartilage [17, 18].

16.2 3D Printing of Biomaterials

The procedure independence obtainable by additive manufacturing techniques delivers numerous chances for the construction of biomaterial creates with multifaceted and exactly measured outer and inner profiles. Additive manufacturing suggests the further advancement and prospect to correspondingly manage the inner profile or micro-design of the produced constructions, which may affect the regeneration of the tissue. Furthermore, the inner microarchitecture will influence the biomechanical, biological, and physical characteristics of permeable biomaterials; [19, 20] E.g., fatigue character, static mechanical belongings [21, 22, 88], and permeability characteristics [23] have been revealed to be purposes of their geometrical limitations. Two main challenges are essential to be reported to employ the maximum possible additive manufacturing performances for successful the enactment of biomaterials. Mainly, the ideal micro-design for the presentation of all biomaterial is habitually indistinct. Investigative and simulation analysis methods are essential to be applied to define the greatest micro-design for any exact use [24]. Preferably, applicable, biological, physical, and mechanical characteristics of the biomaterial should be careful at the same time when planning the microarchitecture.

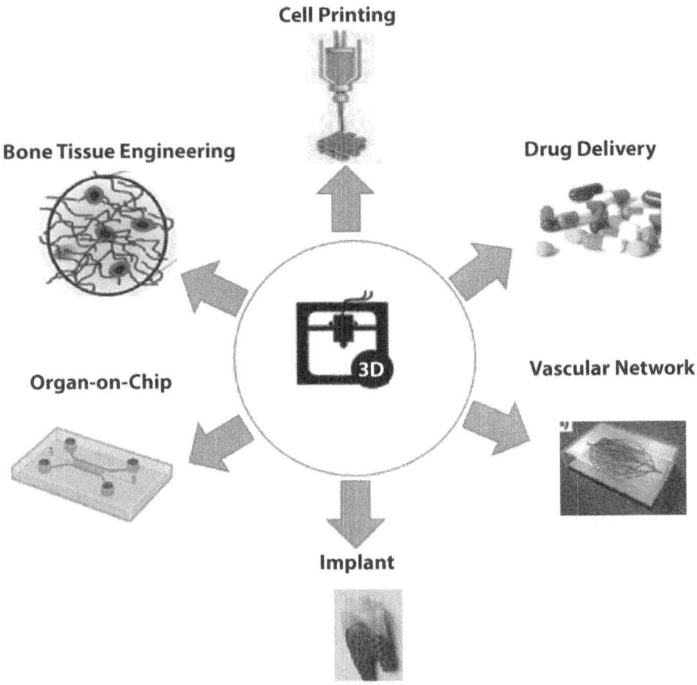

Figure 16.2 Application of 3D printing in the different biomedical sectors.

Furthermore, there is still restricted access to materials that are well-suited to additive manufacturing procedures [26]. Figure 16.2 shows the application of additive manufacturing in modern biomedical engineering. Usually applied biomaterials can frequently not be processed with additive manufacturing methods, at the same time as the best-performing constituents in additive manufacturing machinery, in terms of precision and utility, are not bio-comfortable and don't show the essential biodegradation performance. Thus, essential not only to increase the resource of existing biomaterials but also to familiarize existing additive manufacturing technologies to improve procedures for the best accessible biomaterials [27]. So, advances in both additive manufacturing materials and methods are essential to developing the possibility of additive manufacturing to its complete extent.

16.3 Main Mechanisms of 3D Printing for Organ and Tissue Printing

This technology of 3D printing can handle fast scaling and also has good material flexibility. This technology uses most types of materials subject to the arability and requirements of the technologies [28–30]. Also, it can scale up or down designs as per the expenses to the size ratio. The procedure of organ and tissue printing starts using biomaterials that solve various needs like porosity, size, shape, biodegradability, and bioactivity, etc. [31–33]. This technology has transformed healthcare for developing patient-suitable complex prosthetics and implants [34, 35]. The use of biomaterials is to give support to the cells for the specific function inside the body of the patient. The cell adhesion capabilities of these biomaterials are based on the three approaches which are surface coatings, chemical modification, and physical modification. In Table 16.1, all the lists necessary components of this technology are mentioned which are used for organ and tissue printing. Researchers and scientists are commercializing and developing this technology to develop a genuine structure. This technology seems a useful means for tissue engineering to develop structures having needed volume, shape, and structure which will mimic regular tissue. This technology prints various materials and at the time of printing it keeps cells with the bioinks alive [36, 76]. This can also keep the medical part's mechanical properties through new materials. This technology is also used for infarcted cardiac tissue replacement and progenitor cell regenerator. This technology can accurately develop tissues, implants, and organs to achieve the desired structure depending upon the tissue engineering application [37, 38].

Table 16.1 Various components of 3D bio printing technology.

Necessary parts of bio printing	Example	References
Hardware	3D (Extrusion, Inkjet, Laser-Based), 4D, 5D printer, 3D Scanner,	[4–6], [15], [21], [34], [84–86, 88]
Software	Solidworks, Ansys, Abaqus, Cura, Mimic, 3D Builder, Sketchup, Autocad.	[6, 7], [21], [84–86, 88, 93]
Bioink	GelMA, Gelatin, Alginates, Chitosan, Cellulose Silk, Hyaluronic acids, Cell-laden Bioinks	[22], [33], [36] [84, 87]
Biomaterials	Biomaterials should be Biocompatible, Biodegderable, Antibacterial, Anti-inflammatory	[43], [46], [57] [89–92]

16.4 Techniques to Fabricate Tissues and Organs Using 3D Printing

Computed Tomography (CT) and Magnetic Resonance Imaging (MRI) are the two helpful techniques used to develop and design patient-specific constructs. This technology can develop needed functional tissues and organs with the help of various cells and biomaterials [39, 40, 76]. These biomaterials have good mechanical, chemical, and biological properties. By 3D printing, bone-like properties materials are fabricated [41, 75]. A few examples of these materials are bio glass, zirconium oxide, graphene, and silica.

16.5 Application of 3D Printing in Implants and Drug Delivery

Many technologies for the administration of drugs and devices for delivery including topical drug delivery systems2 [42], implantable drug delivery vehicles [43], and solid dosage forms [44] are being benefitted from this 3D printing technology. The current acceptance of drug products using this technology was approved by FDA in August 2015 which marked the beginning of a new era in these more 3D-printed drugs can be expected to be used in clinics [45]. In the old-fashioned drug delivery area of investigation, the core aim is on managing the control summary by many techniques, concerning these techniques the most important criterion is to create a biomaterial with predictable, controllable, and distinct release profiles.

But the technology of 3D printing offers techniques for developing new systems of drug delivery with release profiles that are tailorable by changing the microarchitecture and 3D structure of the system of drug delivery and also by changing the active agents' spatial distribution [46]. This technology has added a new measurement to manufacture and design implants in broad as well as patient-related implants in specific. Implants of patient-specific types [47] are designed in such a way that they can perfectly acceptable the structure or additional criteria of a particular patient, which is one of the main sections for daily scientific applications [4, 7]. Current approaches close the pipeline coil which goes from implant manufacturing, implant design, image processing, and image acquisition by way of the total procedure can be monitored over computer-aided design structures which assemble all or some of the needed phases. The flexible manner of 3D printing technology allows implants having difficult structures and geometries to be generated cost-effectively, reliably, and quickly [47]. Companies that participate with many features that needed patient-specific 3D printing technology are already present in the marketplace also their fabricated implants are in the clinic already. Metallic materials are commonly used in this technology where load-bearing and functional implants are required. For this purpose, selective laser melting (SLM) and electron beam melting (EBM) are commonly used [49, 50].

16.6 Application 3D Printing in Orthotics and Prosthetics

The technology of 3D printing is used to fabricate external deformity-correcting devices (orthotics) and artificial organ replacements (prosthetics) [51]. This technology to develop prosthetics offers not only reasonable and easy customization choices but also allows the material prosthetics to perfectly fit the patient's required anatomical structure needed like the dimensions of the prosthetics can be easily adjusted by this technology [52, 53]. Subject to whether high or low-end markers are aimed, this technology can play various roles. In the area of the high-end market, better aesthetics, comfort, and functionality can be attained by using sophisticated 3D printing technology which involves multi-material and multicolor jetting printers [54]. Similarly, for low-end market, general 3D printing methods like FDM along with general prosthetics design can be used for making accessible prosthetics to the maximum population part, specifically given the point that major individual fractions require the devices of prosthetics living lower income underdeveloped countries having lesser access to clinical services and products. In the literature, various low-cost 3D printing prosthetics are reported [55, 56].

16.7 3D Printing Application in Tissue Engineering

Bioprinting means generation of automated tissue constructs using bio-assembly along with subsequent maturation. It gives the chance to create constructs that closely mimic the structure and composition of the natural tissue [57, 58, 84–86]. The foremost variance between additive manufacturing and 3D bioprinting is that in 3D bioprinting cells incorporation is done with the bioinks that are used in various applications of tissue engineering applications [59]. The most universally used 3D bioprinting

printing technologies using bioinks are robotic dispensing, inkjet printing, and laser-induced forward transfer (LiFT) [60]. The pros, and cons of this technology in tissue engineering are discussed in detail in this special issue [61]. The printed construct combined with cells, biomolecules, and biomaterials are expected to mature and grow into the required tissue Figure 16.3. Mixing biomaterials generally gives the necessary primary structural and mechanical support for gas and mass transfer also the activation of suitable mechano transductory pathways. Simultaneously, the incorporation of biomolecules in bioinks gives the necessary biological cues to support the process of tissue regeneration [62]. Many types of bioinks along with various cells can be spread into the same construct of tissue for getting the good process of tissue engineering and also to allow complicated regeneration of tissue constructions. More advanced techniques of imaging can easily do the shape quantification of defects in the organs or tissues and also specifically their exact composition. The developed image can be processed further to get the computer-aided design (CAD) format file which defines the accurate geometrical shape for the needed patient-specific organs and tissue constructs [63]. This patient-specific feature is highlighted by the possibility to use the source of cells of analogue types, reducing the rejection chances of the fabricated organs and tissues. Bioprinting technique is presently being studied to develop many types of constructs of tissues like periodontal tissues [64], skin [65], bone [66], cartilage [67], cardiovascular tissues, and vascularized tissues [68, 69]. Furthermore, this technology is also used *in vitro* to develop disease models to study toxicity, and drug screening, which reduces the studies on animals for doing *in-vivo* experiments.

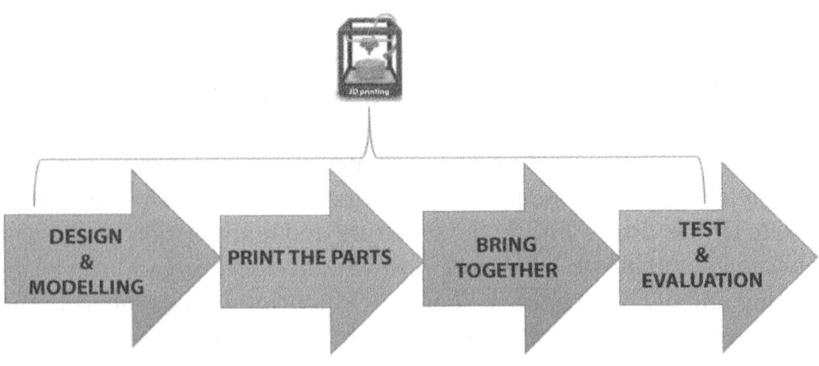

Figure 16.3 The procedure of 3D bioprinting.

16.8 Future Scope

This technology will improve the chances of custom-made tissue engineering. This will allow momentous development in therapeutics and diagnostics. As per the criteria of the patient, this technology can develop desired vascular, organs, and tissue structures [70]. This also showed huge interest in the artificial heart development design using biomaterials. In forthcoming years, this technology can be used for printing microchannels that can be used to increase the number of nutrients and oxygen from nearby tissues [81, 82]. This technology of 3D bioprinting can modify the manufacturing process of various biomedical devices, organs, and tissue if it is used as demanded by the patient party. With the appropriate development and research, it will give assurance to the medical practitioner, researcher, surgeon, and doctor in this area of healthcare to save mankind [83]. Also in the future, this technology will be used for seeding neural cells that can be useful for repairing brain injury and the spinal cord.

16.9 Conclusion

Printing biomaterials is possible using 3D bioprinting which provides accurate technology that mix cell and rectifies their growth factor. Tissue-like structures can be bioprinted layer by layer by this technology. Bioprinting can simply print organs, tissue, and cell by using regenerated ligaments and joints. Materials of liquefied types are mixed with cells, which allows nutrients and oxygen for the cells that are commonly called bioink. The bioink is loaded into the bioprinter head which prints the layer-by-layer structure as needed for the patient which is captured using imaging techniques like MRI or CT. This printed structure is a biological construct loaded with cells, this construct can be the liver, other working blood vessels, and the kidney. The post-printing process is also required which involves the ability to survive transplants and to generate the tissues growth. Moreover, bioprinting is also used as a pioneering method of biomimicry, fabrication of blocks of mini-tissue building, and independent self-assembly. Also using this technology, scientists are working to develop living tissues and organs having outstanding mechanical and biological properties. The final goal of this technology is to generate structures like normal that should survive inside the human body. This will create quickly the cellular part of the tissue or organ which are containing approaches of self-assembly.

References

1. Zadpoor, A.A. and Malda, J., Additive manufacturing of biomaterials, tissues, and organs. *Ann. BioMed. Eng.*, 45, 1–11, 2017, https://doi.org/10.1007/s10439-016-1719-y.
2. Yavari, S.A., Ahmadi, S.M., van der Stok, J., Wauthlé, R., Riemslag, A.C., Janssen, M., Zadpoor, A.A., Effects of bio-functionalizing surface treatments on the mechanical behavior of open porous titanium biomaterials. *J. Mech. Behav. Biomed. Mater.*, 36, 109–119, 2014.
3. Javaid, M. and Haleem, A., Additive manufacturing applications in medical cases: A literature based review. *Alexandria J. Med.*, 54, 4, 411–422, 2018.
4. Barua, R., Datta, S., Datta, P., Roychowdhury, A., Study and application of machine learning methods in modern additive manufacturing processes, in: *Applications of Artificial Intelligence in Additive Manufacturing*, pp. 75–95, IGI Global, USA, 2022.
5. Barua, R., Roychowdhury, A., Datta, P., Study of different additive manufacturing processes and emergent applications in modern healthcare, in: *Advanced Manufacturing Techniques for Engineering and Engineered Materials*, pp. 239–259, IGI Global, USA, 2022.
6. Datta, S., Barua, R., Sarkar, R., Barui, A., Chowdhury, A.R., Datta, P., Design and development of alginate: Poly-l-lysine scaffolds by 3D bio printing and studying their mechanical, structural and cell viability properties, in: *IOP Conference Series: Materials Science and Engineering*, vol. 402, IOP Publishing, p. 012113, 2018, August.
7. Barua, R., Datta, S., Datta, P., Chowdhury, A.R., Scaffolds and tissue engineering applications by 3D bio-printing process: A new approach, in: *Design, Development, and Optimization of Bio-Mechatronic Engineering Products*, pp. 78–99, IGI Global, USA, 2019.
8. Martinez-Marquez, D., Mirnajafizadeh, A., Carty, C.P., Stewart, R.A., Application of quality by design for 3D printed bone prostheses and scaffolds. *PLoS One*, 13, 4, e0195291, 2018, https://doi.org/10.1371/journal.pone.0195291.
9. Barua, R., Datta, P., Chowdhury, A.R., The impact of 3D printing technology on the COVID-19 pandemic, in: *Quality of Healthcare in the Aftermath of the COVID-19 Pandemic*, pp. 135–154, IGI Global, USA, 2022.
10. Wang, Z. and Yang, Y., Application of 3D printing in implantable medical devices. *BioMed. Res. Int.*, 2021, 6653967, 2021, https://doi.org/10.1155/2021/6653967.
11. Wang, D., Han, L., Xu, G., Zhang, W., Li, H., Xu, C., Li, H., Li, J., Zhang, H., Li, J., 3D printing and computer-aided design for precision osteotomy-aided modules in bone biomechanical study. *Int. J. Bioprint.*, 8, 4, 607, 2022, https://doi.org/10.18063/ijb.v8i4.607.
12. Li, B., Zhang, M., Lu, Q., Zhang, B., Miao, Z., Li, L., Zheng, T., Liu, P., Application and development of modern 3D printing technology in the

field of orthopedics. *BioMed. Res. Int.*, 2022, 8759060, 2022, https://doi.org/10.1155/2022/8759060.

13. Lin, G.X., Chen, C.M., Rui, G., Hu, B.S., Research relating to three-dimensional (3D) printing in spine surgery: A bibliometric analysis. *Eur. Spine J.*, 32, 2, 395–407, 2022. 10.1007/s00586-022-07376-8. Advance online publication, https://doi.org/10.1007/s00586-022-07376-8.
14. Wong, K.C., 3D-printed patient-specific applications in orthopedics. *Orthop. Res. Rev.*, 8, 57–66, 2016, https://doi.org/10.2147/ORR.S99614.
15. Tack, P., Victor, J., Gemmel, P., Annemans, L., 3D-printing techniques in a medical setting: A systematic literature review. *Biomed. Eng. Online*, 15, 1, 115, 2016, https://doi.org/10.1186/s12938-016-0236-4.
16. Lador, R., Regev, G., Salame, K., Khashan, M., Lidar, Z., Use of 3-dimensional printing technology in complex spine surgeries. *World Neurosurg.*, 133, e327–e341, 2020, https://doi.org/10.1016/j.wneu.2019.09.002.
17. Martelli, N., Serrano, C., van den Brink, H., Pineau, J., Prognon, P., Borget, I., El Batti, S., Advantages and disadvantages of 3-dimensional printing in surgery: A systematic review. *Surgery*, 159, 6, 1485–1500, 2016, https://doi.org/10.1016/j.surg.2015.12.017.
18. Ballard, D.H., Mills, P., Duszak J.R, R., Weisman, J.A., Rybicki, F.J., Woodard, P.K., Medical 3D printing cost-savings in orthopedic and maxillofacial surgery: Cost analysis of operating room time saved with 3D printed anatomic models and surgical guides. *Acad. Radiol.*, 27, 8, 1103–1113, 2020, https://doi.org/10.1016/j.acra.2019.08.011.
19. Zhu, W., Ma, X., Gou, M., Mei, D., Zhang, K., Chen, S., 3D printing of functional biomaterials for tissue engineering. *Curr. Opin. Biotechnol.*, 40, 103–112, 2016, https://doi.org/10.1016/j.copbio.2016.03.014.
20. Feng, Y., Zhu, S., Mei, D., Li, J., Zhang, J., Yang, S., Guan, S., Application of 3D printing technology in bone tissue engineering: A review. *Curr. Drug Deliv.*, 18, 7, 847–861, 2021, https://doi.org/10.2174/156720181799920111310 0322.
21. Jakus, A.E., Rutz, A.L., Shah, R.N., Advancing the field of 3D biomaterial printing. *Biomed. Mater. (Bristol, England)*, 11, 1, 014102, 2016, https://doi.org/10.1088/1748-6041/11/1/014102.
22. Li, W., Mille, L.S., Robledo, J.A., Uribe, T., Huerta, V., Zhang, Y.S., Recent advances in formulating and processing biomaterial inks for Vat polymerization-based 3D Printing. *Adv. Healthc. Mater.*, 9, 15, e2000156, 2020, https://doi.org/10.1002/adhm.202000156.
23. Almetwally, A.G. and Jabbari, H., 3D-printing replication of porous media for lab-scale characterization research. *ACS omega*, 6, 4, 2655–2664, 2021, https://doi.org/10.1021/acsomega.0c04825.
24. Shi, W., Sun, M., Hu, X., Ren, B., Cheng, J., Li, C., Duan, X., Fu, X., Zhang, J., Chen, H., Ao, Y., Structurally and functionally optimized silk-fibroin-gelatin scaffold using 3D printing to repair cartilage injury *in vitro* and *in*

vivo. Adv. Mater. (Deerfield Beach, Fla.), 29, 29, 1701089, 2017. 10.1002/adma.201701089https://doi.org/10.1002/adma.201701089.

25. Mishra, A. and Srivastava, V., Biomaterials and 3D printing techniques used in the medical field. *J. Med. Eng. Technol.*, 45, 4, 290–302, 2021, https://doi.org/10.1080/03091902.2021.1893845.
26. Olmos-Juste, R., Guaresti, O., Calvo-Correas, T., Gabilondo, N., Eceiza, A., Design of drug-loaded 3D printing biomaterial inks and tailor-made pharmaceutical forms for controlled release. *Int. J. Pharm.*, 609, 121124, 2021, https://doi.org/10.1016/j.ijpharm.2021.121124.
27. Jin, Z., Li, Y., Yu, K., Liu, L., Fu, J., Yao, X., Zhang, A., He, Y., 3D printing of physical organ models: Recent developments and challenges. *Adv. Sci. (Weinheim, Baden-Wurttemberg, Germany)*, 8, 17, e2101394, 2021, https://doi.org/10.1002/advs.202101394.
28. Weinhart, M., Hocke, A., Hippenstiel, S., Kurreck, J., Hedtrich, S., 3D organ models-revolution in pharmacological research? *Pharmacol. Res.*, 139, 446–451, 2019, https://doi.org/10.1016/j.phrs.2018.11.002.
29. Carvalho, V., Gonçalves, I., Lage, T., Rodrigues, R.O., Minas, G., Teixeira, S.F.C.F., Moita, A.S., Hori, T., Kaji, H., Lima, R.A., 3D printing techniques and their applications to organ-on-a-chip platforms: A systematic review. *Sensors (Basel, Switzerland)*, 21, 9, 3304, 2021, https://doi.org/10.3390/s21093304.
30. Kim, J., Kong, J.S., Han, W., Kim, B.S., Cho, D.W., 3D cell printing of tissue/organ-mimicking constructs for therapeutic and drug testing applications. *Int. J. Mol. Sci.*, 21, 20, 7757, 2020, https://doi.org/10.3390/ijms21207757.
31. Zhang, Y.S., Yue, K., Aleman, J., Moghaddam, K.M., Bakht, S.M., Yang, J., Jia, W., Dell'Erba, V., Assawes, P., Shin, S.R., Dokmeci, M.R., Oklu, R., Khademhosseini, A., 3D Bioprinting for Tissue and Organ Fabrication. *Ann. Biomed. Eng.*, 45, 1, 148–163, 2017, https://doi.org/10.1007/s10439-016-1612-8.
32. Dey, M. and Ozbolat, I.T., 3D bioprinting of cells, tissues and organs. *Sci. Rep.*, 10, 1, 14023, 2020, https://doi.org/10.1038/s41598-020-70086-y.
33. Kassem, T., Sarkar, T., Nguyen, T., Saha, D., Ahsan, F., 3D printing in solid dosage forms and organ-on-chip applications. *Biosensors*, 12, 4, 186, 2022, https://doi.org/10.3390/bios12040186.
34. Khorsandi, D., Fahimipour, A., Abasian, P., Saber, S.S., Seyedi, M., Ghanavati, S., Ahmad, A., De Stephanis, A.A., Taghavinezhaddilami, F., Leonova, A., Mohammadinejad, R., Shabani, M., Mazzolai, B., Mattoli, V., Tay, F.R., Makvandi, P., 3D and 4D printing in dentistry and maxillofacial surgery: Printing techniques, materials, and applications. *Acta biomaterialia*, 122, 26–49, 2021, https://doi.org/10.1016/j.actbio.2020.12.044.
35. Dawood, A., Marti Marti, B., Sauret-Jackson, V., Darwood, A., 3D printing in dentistry. *Br. Dental J.*, 219, 11, 521–529, 2015, https://doi.org/10.1038/sj.bdj.2015.914.

36. Petta, D., D'Amora, U., Ambrosio, L., Grijpma, D.W., Eglin, D., D'Este, M., Hyaluronic acid as a bioink for extrusion-based 3D printing. *Biofabrication*, 12, 3, 032001, 2020, https://doi.org/10.1088/1758-5090/ab8752.
37. Matai, I., Kaur, G., Seyedsalehi, A., McClinton, A., Laurencin, C.T., Progress in 3D bioprinting technology for tissue/organ regenerative engineering. *Biomaterials*, 226, 119536, 2020, https://doi.org/10.1016/j.biomaterials.2019.119536.
38. Chae, S., Lee, S.S., Choi, Y.J., Hong, D.H., Gao, G., Wang, J.H., Cho, D.W., 3D cell-printing of biocompatible and functional meniscus constructs using meniscus-derived bioink. *Biomaterials*, 267, 120466, 2021, https://doi.org/10.1016/j.biomaterials.2020.120466.
39. Murphy, S.V. and Atala, A., 3D bioprinting of tissues and organs. *Nat. Biotechnol.*, 32, 8, 773–785, 2014, https://doi.org/10.1038/nbt.2958.
40. Sun, W. and Xu, T., Special issue on 3D cell printing. *Biofabrication*, 7, 4, 043001, 2016, https://doi.org/10.1088/1758-5090/7/4/043001.
41. Feng, Y., Zhu, S., Mei, D., Li, J., Zhang, J., Yang, S., Guan, S., Application of 3D printing technology in bone tissue engineering: A review. *Curr. Drug Deliv.*, 18, 7, 847–861, 2021, https://doi.org/10.2174/156720181799920111131 00322.
42. de Oliveira, R.S., Fantaus, S.S., Guillot, A.J., Melero, A., Beck, R.C.R., 3D-printed products for topical skin applications: From personalized dressings to drug delivery. *Pharmaceutics*, 13, 11, 1946, 2021, https://doi.org/10.3390/pharmaceutics13111946.
43. Jamróz, W., Szafraniec, J., Kurek, M., Jachowicz, R., 3D printing in pharmaceutical and medical applications - recent achievements and challenges. *Pharm. Res.*, 35, 9, 176, 2018, https://doi.org/10.1007/s11095-018-2454-x.
44. Jacob, S., Nair, A.B., Patel, V., Shah, J., 3D printing technologies: Recent development and emerging applications in various drug delivery systems. *AAPS PharmSciTech*, 21, 6, 220, 2020, https://doi.org/10.1208/s12249-020-01771-4.
45. Prasad, L.K. and Smyth, H., 3D Printing technologies for drug delivery: A review. *Drug Dev. Ind. Pharm.*, 42, 7, 1019–1031, 2016, https://doi.org/10.3109/03639045.2015.1120743.
46. Rahman, Z., Barakh Ali, S.F., Ozkan, T., Charoo, N.A., Reddy, I.K., Khan, M.A., Additive manufacturing with 3D printing: Progress from bench to bedside. *AAPS J.*, 20, 6, 101, 2018, https://doi.org/10.1208/s12248-018-0225-6.
47. Thomas, D.J. and Singh, D., 3D printing for developing patient specific cosmetic prosthetics at the point of care. *Int. J. Surg. (London, England)*, 80, 241–242, 2020, https://doi.org/10.1016/j.ijsu.2020.04.023.
48. Tahayeri, A., Morgan, M., Fugolin, A.P., Bompolaki, D., Athirasala, A., Pfeifer, C.S., Ferracane, J.L., Bertassoni, L.E., 3D printed versus conventionally cured provisional crown and bridge dental materials. *Dent. Mater. :Off. Publ. Acad. Dent. Mater.*, 34, 2, 192–200, 2018, https://doi.org/10.1016/j.dental.2017.10.003.

49. Murray, S.P., Pusch, K.M., Polonsky, A.T., Torbet, C.J., Seward, G.G.E., Zhou, N., Forsik, S.A.J., Nandwana, P., Kirka, M.M., Dehoff, R.R., Slye, W.E., Pollock, T.M., A defect-resistant Co-Ni superalloy for 3D printing. *Nat. Commun.*, 11, 1, 4975, 2020, https://doi.org/10.1038/s41467-020-18775-0.
50. Czyżewski, W., Jachimczyk, J., Hoffman, Z., Szymoniuk, M., Litak, J., Maciejewski, M., Kura, K., Rola, R., Torres, K., Low-cost cranioplasty-a systematic review of 3D printing in medicine. *Materials (Basel, Switzerland)*, 15, 14, 4731, 2022, https://doi.org/10.3390/ma15144731.
51. Shahar, F.S., Hameed Sultan, M.T., Lee, S.H., Jawaid, M., Md Shah, A.U., Safri, S.N.A., Sivasankaran, P.N., A review on the orthotics and prosthetics and the potential of kenaf composites as alternative materials for ankle-foot orthosis. *J. Mech. Behav. Biomed. Mater.*, 99, 169–185, 2019, https://doi.org/10.1016/j.jmbbm.2019.07.020.
52. Moya, D., Gobbato, B., Valente, S., Roca, R., Uso de planificación preoperatoria e impresión 3D en ortopedia y traumatología: Ingresando en una nueva era [Use of preoperative planning and 3D printing in orthopedics and traumatology: Entering a new era]. *Acta Ortopedica Mexicana*, 36, 1, 39–47, 2022.
53. Papagelopoulos, P.J., Savvidou, O.D., Koutsouradis, P., Chloros, G.D., Bolia, I.K., Sakellariou, V., I, Kontogeorgakos, V.A., Mavrodontis, I.I., Mavrogenis, A.F., Diamantopoulos, P., Three-dimensional technologies in orthopedics. *Orthopedics*, 41, 1, 12–20, 2018, https://doi.org/10.3928/01477447-20180109-04.
54. Wedekind, L., Güth, J.F., Schweiger, J., Kollmuss, M., Reichl, F.X., Edelhoff, D., Högg, C., Elution behavior of a 3D-printed, milled and conventional resin-based occlusal splint material. *Dent. Mater. :Off. Publ. Acad. Dent. Mater.*, 37, 4, 701–710, 2021, https://doi.org/10.1016/j.dental.2021.01.024.
55. O'Brien, L., Cho, E., Khara, A., Lavranos, J., Lommerse, L., Chen, C., 3D-printed custom-designed prostheses for partial hand amputation: Mechanical challenges still exist. *J. Hand. Ther. :Off. J. Am. Soc. Hand. Ther.*, 34, 4, 539–542, 2021, https://doi.org/10.1016/j.jht.2020.04.005.
56. Anderson, P.A., Hsu, W., Golish, S.R., Jakus, A.E., Mihalko, W.M., Applications of Three-dimensional printing in orthopaedic surgery. *Instr. Course Lect.*, 67, 587–594, 2018.
57. Hong, H., Seo, Y.B., Kim, D.Y., Lee, J.S., Lee, Y.J., Lee, H., Ajiteru, O., Sultan, M.T., Lee, O.J., Kim, S.H., Park, C.H., Digital light processing 3D printed silk fibroin hydrogel for cartilage tissue engineering. *Biomaterials*, 232, 119679, 2020, https://doi.org/10.1016/j.biomaterials.2019.119679.
58. Distler, T. and Boccaccini, A.R., 3D printing of electrically conductive hydrogels for tissue engineering and biosensors - A review. *Acta Biomaterialia*, 101, 1–13, 2020, https://doi.org/10.1016/j.actbio.2019.08.044.
59. Pavan Kalyan, B.G. and Kumar, L., 3D printing: Applications in tissue engineering, medical devices, and drug delivery. *AAPS PharmSciTech*, 23, 4, 92, 2022, https://doi.org/10.1208/s12249-022-02242-8.

60. Visser, C.W., Pohl, R., Sun, C., Römer, G.W., Huis in 't Veld, B., Lohse, D., Toward 3D printing of pure metals by laser-induced forward transfer. *Adv. Mater. (Deerfield Beach, Fla.)*, 27, 27, 4087–4092, 2015, https://doi.org/10.1002/adma.201501058.
61. Zenou, M. and Kotler, Z., Printing of metallic 3D micro-objects by laser induced forward transfer. *Opt. Express*, 24, 2, 1431–1446, 2016, https://doi.org/10.1364/OE.24.001431.
62. Trombetta, R., Inzana, J.A., Schwarz, E.M., Kates, S.L., Awad, H.A., 3D printing of calcium phosphate ceramics for bone tissue engineering and drug delivery. *Ann. Biomed. Eng.*, 45, 1, 23–44, 2017, https://doi.org/10.1007/s10439-016-1678-3.
63. Schipani, R., Nolan, D.R., Lally, C., Kelly, D.J., Integrating finite element modelling and 3D printing to engineer biomimetic polymeric scaffolds for tissue engineering. *Connect. Tissue Res.*, 61, 2, 174–189, 2020, https://doi.org/10.1080/03008207.2019.1656720.
64. Nesic, D., Schaefer, B.M., Sun, Y., Saulacic, N., Sailer, I., 3D printing approach in dentistry: The future for personalized oral soft tissue regeneration. *J. Clin. Med.*, 9, 7, 2238, 2020, https://doi.org/10.3390/jcm9072238.
65. Aavani, F., Biazar, E., Kheilnezhad, B., Amjad, F., 3D bio-printing for skin tissue regeneration: Hopes and hurdles. *Curr. Stem Cell Res. Ther.*, 17, 5, 415–439, 2022, https://doi.org/10.2174/1574888X17666220204144544.
66. McCulloch, R.A., Frisoni, T., Kurunskal, V., Maria Donati, D., Jeys, L., Computer navigation and 3D printing in the surgical management of bone sarcoma. *Cells*, 10, 2, 195, 2021, https://doi.org/10.3390/cells10020195.
67. Chen, P., Zheng, L., Wang, Y., Tao, M., Xie, Z., Xia, C., Gu, C., Chen, J., Qiu, P., Mei, S., Ning, L., Shi, Y., Fang, C., Fan, S., Lin, X., Desktop-stereolithography 3D printing of a radially oriented extracellular matrix/mesenchymal stem cell exosome bioink for osteochondral defect regeneration. *Theranostics*, 9, 9, 2439–2459, 2019, https://doi.org/10.7150/thno.31017.
68. Hann, S.Y., Cui, H., Esworthy, T., Miao, S., Zhou, X., Lee, S.J., Fisher, J.P., Zhang, L.G., Recent advances in 3D printing: Vascular network for tissue and organ regeneration. *Trans. Res.:J. Lab. Clin. Med.*, 211, 46–63, 2019, https://doi.org/10.1016/j.trsl.2019.04.002.
69. Puluca, N., Lee, S., Doppler, S., Münsterer, A., Dreßen, M., Krane, M., Wu, S.M., Bioprinting approaches to engineering vascularized 3D cardiac tissues. *Curr. Cardiol. Rep.*, 21, 9, 90, 2019, https://doi.org/10.1007/s11886-019-1179-8.
70. Ferrari, E., Gallo, M., Wang, C., Zhang, L., Taramasso, M., Maisano, F., Pirelli, L., Berdajs, D., von Segesser, L.K., Three-dimensional printing in adult cardiovascular medicine for surgical and transcatheter procedural planning, teaching and technological innovation. *Interact. Cardiovasc. Thorac. Surg.*, 30, 2, 203–214, 2020, https://doi.org/10.1093/icvts/ivz250.
71. Bhoi, S., Prasad, A., Kumar, A., Sarkar, R.B., Mahto, B., Meena, C.S., Pandey, C., Experimental study to evaluate the wear performance of UHMWPE and

XLPE material for orthopedics application. *Bioengineering*, 9, 676, 2022, https://doi.org/10.3390/bioengineering9110676.
72. Bhoi, S., Kumar, A., Prasad, A., Meena, C.S., Sarkar, R.B., Mahto, B., Ghosh, A., Performance evaluation of different coating materials in delamination for micro-milling applications on high-speed steel substrate. *Micromachines*, 13, 1277, 2022, https://doi.org/10.3390/mi13081277.
73. Gupta, A. *et al.*, Multifunctional nanohydroxyapatite-promoted toughened high-molecular-weight stereocomplex Poly (Lactic acid)-based bionanocomposite for both 3D-printed orthopedic implants and high-temperature engineering applications. *ACS Omega*, 2, 7, 4039–52, 2017.
74. Prasad, A., Bioabsorbable polymeric materials for biofilms and other biomedical applications: Recent and future trends. *Mater. Today: Proc.*, 44, 2447–53, 2021a, https://doi.org/10.1016/j.matpr.2020.12.489.
75. Prasad, A., State of art review on bioabsorbable polymeric scaffolds for bone tissue engineering. *Mater. Today: Proc.*, 44, 1391–1400, 2021b, https://doi.org/10.1016/j.matpr.2020.11.622.
76. Prasad, A., Bhasney, S., Katiyar, V., Ravi Sankar, M., Biowastes processed hydroxyapatite filled Poly (Lactic acid) bio-composite for open reduction internal fixation of small bones. *Mater. Today: Proc.*, 4, 9, 10153–57, 2017. http://linkinghub.elsevier.com/retrieve/pii/S2214785317311410.
77. Prasad, A., Devendar, B., Ravi Sankar, M., Robi, P.S., Micro-scratch based tribological characterization of hydroxyapatite (HAp) fabricated through fish scales. *Mater. Today: Proc.*, 2, 4–5, 1216–24, 2015. http://linkinghub.elsevier.com/retrieve/pii/S2214785315002795.
78. Prasad, A., Ravi Sankar, M., Katiyar, V., State of art on solvent casting particulate leaching method for orthopedic scaffoldsfabrication. *Mater. Today: Proc.*, 4, 2, 898–907, 2017. http://dx.doi.org/10.1016/j.matpr.2017.01.101.
79. Prasad, A., Chakraborty, G., Kumar, A., Bio-based environmentally benign polymeric resorbable materials for orthopedic fixation applications, in: *Advanced Materials for Biomedical Applications*, vol. Chapter 15, A. Kumar, Y. Gori, A. Kumar, C.S. Meena, N. Dutt (Eds.), pp. 251–266, CRC Press, Boca Raton, FL, USA, 2022, https://doi.org/10.1201/9781003344810-15.
80. Prasad, A., Biomaterial-based nanofibers scaffolds in tissue engineering application, in: *Functional Biomaterials*, pp. 245–264, Springer, Singapore, 2022.
81. Domínguez-Robles, J., Shen, T., Cornelius, V.A., Corduas, F., Mancuso, E., Donnelly, R.F., Margariti, A., Lamprou, D.A., Larrañeta, E., Development of drug loaded cardiovascular prosthesis for thrombosis prevention using 3D printing. *Mater. Sci. Eng. C, Mater. Biol. Appl.*, 129, 112375, 2021, https://doi.org/10.1016/j.msec.2021.112375.
82. Gold, K., Gaharwar, A.K., Jain, A., Emerging trends in multiscale modeling of vascular pathophysiology: Organ-on-a-chip and 3D printing. *Biomaterials*, 196, 2–17, 2019, https://doi.org/10.1016/j.biomaterials.2018.07.029.

83. Larrañeta, E., Domínguez-Robles, J., Margariti, A., Basit, A.W., Goyanes, Á., 3D printing for the development of implantable devices for cardiovascular disease treatment. *Ther. Deliv.*, 13, 7, 359–362, 2022, https://doi.org/10.4155/tde-2022-0017.
84. Datta, S., Sarkar, R., Vyas, V., Bhutoria, S., Barui, A., RoyChowdhury, A., Datta, P., Alginate-honey bioinks with improved cell responses for applications as bioprinted tissue engineered constructs. *J. Mater. Res.*, 33, 14, 2029–2039, 2018. doi: https://doi.org/10.1557/jmr.2018.202.
85. Datta, S., Das, A., Sasmal, P., Bhutoria, S., RoyCHowdhyry, A., Datta, P., Alginate-poly (amino acid) extrusion printed scaffolds for tissue engineering applications. *Int. J. Polym. Mater. Polym. Biomater.*, 69, 2, 65–72, 2018, https://doi.org/10.1080/00914037.2018.1539988.
86. Datta, S., Das, A., RoyChowdhury, A., Datta, P., Bioink formulations to ameliorate bioprinting-induced loss of cellular viability. *Biointerphases*, 14, 5, 051006, 2019, https://doi.org/10.1116/1.5111392.
87. Datta, S., Jana, S., Das, A., Chakraborty, A., RoyChowdhury, A., Datta, P., Bioprinting of radiopaque constructs for tissue engineering and understanding degradation behavior by use of micro-CT. *Bioact. Mater.*, 5, 3, 569–576, 2020, https://doi.org/10.1016/j.bioactmat.2020.04.015.
88. Banerjee, A., Datta, S., Das, A., RoyChowdhury, A., Datta, P., A micro-scale non-linear finite element model to optimize the mechanical behavior of bioprinted constructs. *3D Print. Addit. Manuf.*, 9, 6, 490–502, 2022, https://doi.org/10.1089/3dp.2021.0238.
89. Kumar, A., Gangwar, A.K.S., Kumar, A., Meena, C.S., Singh, V.P., Dutt, N., Prasad, A., Gori, Y., Biomedical study of femur bone fracture and healing, in: *Advanced Materials for Biomedical Applications*, vol. Chapter 14, A. Kumar, Y. Gori, A. Kumar, C.S. Meena, N. Dutt (Eds.), pp. 235–250, CRC Press, Boca Raton, FL, USA, 2022, https://doi.org/ 10.1201/9781003344810-14.
90. Kumar, R., Kumar, A., Kant, L., Prasad, A., Bhoi, S., Meena, C.S., Singh, V.P., Ghosh, A., Experimental and RSM-based process-parameters optimisation for turning operation of EN36B steel. *Materials*, 16, 339, 2023, https://doi.org/10.3390/ma16010339.
91. Prasad, A., Chakraborty, G., Kumar, A., Gajrani, K.K., Introduction to biodegradable polymers, in: *Biodegradable Composites for Packaging Application*, vol. Chapter 1, A. Prasad, A. Kumar, K.K. Gajrani (Eds.), pp. 1–12, CRC Press, Boca Raton, FL, USA, 2022, https://doi.org/10.1201/9781003227908-1.
92. Chakraborty, G., Prasad, A., Kumar, A., Processing of biodegradable composites, in: *Biodegradable Composites for Packaging Applications*, vol. Chapter 3, A. Prasad, A. Kumar, K.K. Gajrani (Eds.), pp. 33–48, CRC Press, Boca Raton, FL, USA, 2022, https://doi.org/ 10.1201/9781003227908-3.
93. Paul, A.K., Prasad, A., Kumar, A., Review on artificial neural network and its application in the field of engineering. *J. Mech. Eng.: PRAKASH*, 1, 1, 53–61, 2022, https://doi.org/10.56697/JMEP.2022.1107.

Index

Printed and bound by CPI Group (UK) Ltd, Croydon, CR0 4YY
17/09/2026
14964276-0006